INTERNATIONAL UNION OF CRYSTALLOGRAPHY
BOOK SERIES

IUCr BOOK SERIES COMMITTEE

IUCr Monographs on Crystallography

1 *Accurate molecular structures*
A. Domenicano, I. Hargittai, editors
2 *P.P. Ewald and his dynamical theory of X-ray diffraction*
D.W.J. Cruickshank, H.J. Juretschke, N. Kato, editors
3 *Electron diffraction techniques, Vol. 1*
J.M. Cowley, editor
4 *Electron diffraction techniques, Vol. 2*
J.M. Cowley, editor
5 *The Rietveld method*
R.A. Young, editor
6 *Introduction to crystallographic statistics*
U. Shmueli, G.H. Weiss
7 *Crystallographic instrumentation*
L.A. Aslanov, G.V. Fetisov, J.A.K. Howard
8 *Direct phasing in crystallography*
C. Giacovazzo
9 *The weak hydrogen bond*
G.R. Desiraju, T. Steiner
10 *Defect and microstructure analysis by diffraction*
R.L. Snyder, J. Fiala and H.J. Bunge
11 *Dynamical theory of X-ray diffraction*
A. Authier
12 *The chemical bond in inorganic chemistry*
I.D. Brown
13 *Structure determination from powder diffraction data*
W.I.F. David, K. Shankland, L.B. McCusker, Ch. Baerlocher, editors

14 *Polymorphism in molecular crystals*
J. Bernstein
15 *Crystallography of modular materials*
G. Ferraris, E. Makovicky, S. Merlino
16 *Diffuse X-ray scattering and models of disorder*
T.R. Welberry
17 *Crystallography of the polymethylene chain: an inquiry into the structure of waxes*
D.L. Dorset
18 *Crystalline molecular complexes and compounds: structure and principles*
F. H. Herbstein
19 *Molecular aggregation: structure analysis and molecular simulation of crystals and liquids*
Gavezzotti
20 *Aperiodic crystals: from modulated phases to quasicrystals*
T. Janssen, G. Chapuis, M. de Boissieu
21 *Incommensurate crystallography*
S. van Smaalen
22 *Structural crystallography of inorganic oxysalts*
S.V. Krivovichev
23 *The nature of the hydrogen bond: outline of a comprehensive hydrogen bond theory*
G. Gilli, P. Gilli
24 *Macromolecular crystallization and crystal perfection*
N.E. Chayen, J.R. Helliwell, E.H.Snell
25 *Neutron protein crystallography: hydrogen, protons, and hydration in bio-macromolecules*
N. Niimura, A. Podjarny
26 *Intermetallics: structures, properties, and statistics*
W. Steurer, J. Dshemuchadse
27 *The chemical bond in inorganic chemistry: The bond valence model*, second edition
I.D. Brown
28 *Aperiodic crystals: From modulated phases to quasicrystals: Structure and properties*, second edition
T. Janssen, G. Chapuis, M. de Boissieu
29 *Biological small angle scattering: Theory and practice*
Eaton E. Lattman, Thomas D. Grant, Edward H. Snell
30 *Polymorphism in molecular crystals*, second edition
J. Bernstein
31 *Diffuse X-ray scattering and models of disorder*, second edition
T.R. Welberry

IUCr Texts on Crystallography

1 *The solid state*
A. Guinier, R. Julien
4 *X-ray charge densities and chemical bonding*
P. Coppens
8 *Crystal structure refinement: a crystallographer's guide to SHELXL*
P. Müller, editor

9 *Theories and techniques of crystal structure determination*
U. Shmueli

10 *Advanced structural inorganic chemistry*
Wai-Kee Li, Gong-Du Zhou, Thomas Mak

11 *Diffuse scattering and defect structure simulations: a cook book using the program DISCUS*
R. B. Neder, T. Proffen

13 *Crystal structure analysis: principles and practice, second edition*
W. Clegg, editor

14 *Crystal structure analysis: a primer, third edition*
J.P. Glusker, K.N. Trueblood

15 *Fundamentals of crystallography, third edition*
C. Giacovazzo, editor

16 *Electron crystallography: electron microscopy and electron diffraction*
X. Zou, S. Hovmöller, P. Oleynikov

17 *Symmetry in crystallography: understanding the International Tables*
P.G. Radaelli

18 *Symmetry relationships between crystal structures: applications of crystallographic group theory in crystal chemistry*
U. Müller

19 *Small angle X-ray and neutron scattering from solutions of biological macromolecules*
D.I. Svergun, M.H.J. Koch, P.A. Timmins, R.P. May

20 *Phasing in crystallography: a modern perspective*
C. Giacovazzo

21 *The basics of crystallography and diffraction, fourth edition*
C. Hammond

22 *Atomic pair distribution function analysis, a primer*
S.J.L. Billinge, K.M.O. Jensen

23 *X-ray Structure Analysis of Bio-Macromolecular Crystals*
A. Takenaka, editor

24 *Symmetry relationships between crystal structures: applications of crystallographic group theory in crystal chemistry, second edition*
U. Müller, Gemma de la Flor

Symmetry Relationships between Crystal Structures

Applications of Crystallographic Group Theory in Crystal Chemistry

Second edition

Ulrich Müller
Fachbereich Chemie, Philipps-Universität Marburg, Germany

Gemma de la Flor
Institut für Angewandte Geowissenschaften,
Karlsruher Institut für Technologie, Karlsruhe, Germany

With texts adapted from
Hans Wondratschek and Hartmut Bärnighausen

Great Clarendon Street, Oxford, OX2 6DP,
United Kingdom

Oxford University Press is a department of the University of Oxford.
It furthers the University's objective of excellence in research, scholarship,
and education by publishing worldwide. Oxford is a registered trade mark of
Oxford University Press in the UK and in certain other countries

First Edition published in 2013

Second Edition published in 2024

First published in paperback in 2017

Translated from German (except Chapter 24): Symmetriebeziehungen zwischen Kristallstrukturen. Second edition.
Springer Spektrum, Heidelberg (2023)
Spanish edition: Relaciones de simetría entre estructuras cristalinas. Editorial Síntesis, Madrid (2013)

Published in the United States of America by Oxford University Press
198 Madison Avenue, New York, NY 10016, United States of America

British Library Cataloguing in Publication Data
Data available

Library of Congress Control Number: 2024940668

ISBN 9780192858320

DOI: 10.1093/oso/9780192858320.001.0001

Printed and bound by
CPI Group (UK) Ltd, Croydon, CR0 4YY

Some of the figures were drawn with the programs
Atoms by E. Dowty and Diamond by K. Brandenburg
Ten figures are the same as in *International Tables for Crystallography*, Volume *A*1,
2nd edition (2010), Chapter 1.6; <https//it.iucr.org/>. They are reproduced with
permission of the *International Union of Crystallography*, together with some
accompanying text written by the first author of this book

Dedicated to

Hartmut Bärnighausen

and

Hans Wondratschek

Hartmut Bärnighausen was Professor of Inorganic Chemistry at the University of Karlsruhe from 1967 to 1998. His article Group–subgroup relations between space groups: a useful tool in crystal chemistry, *published in 1980, showed the way to the application of crystallographic group theory in crystal chemistry. The present book is based on this article and on his manuscripts for several courses on this topic.*

Hans Wondratschek was Professor of Crystallography at the University of Karlsruhe from 1964 to 1991. He contributed to the derivation of the space groups of four-dimensional space and to the complete listing of the subgroups of the three-dimensional space groups. These were then published in International Tables for Crystallography, *Volume* A1.

Preface

Crystal-structure analysis has become one of the most essential tools in chemistry and related disciplines. Several hundreds of thousands of crystal structures have been determined over the course of the years. The results obtained from 1931 to 1990 were published year by year in *Strukturbericht* [1], later *Structure Reports* [2]. Nowadays, crystal structures are deposited in several large databases [3–9]. However, the mere accumulation of data is only of restricted value if it lacks a systematic order and if the scientific interpretation of the data leaves much to be desired.

Shortly after the discovery of X-ray diffraction from crystals by MAX VON LAUE, WALTHER FRIEDRICH, and PAUL KNIPPING (1912) and the subsequent pioneering work by father WILLIAM HENRY BRAGG and son WILLIAM LAWRENCE BRAGG, efforts were made to order the crystal structures found. By 1926 the number of crystal structures was already large enough for VIKTOR MORITZ GOLDSCHMIDT to formulate the basic principles of packing of atoms and ions in inorganic solids [10]. In 1928 LINUS PAULING set forth a number of structural principles, essentially for ionic crystals, which he later repeated in his famous book *The Nature of the Chemical Bond*, first published in 1938 [11]. A considerable number of other approaches to show relationships between crystal structures and to bring order into the constantly increasing amount of data were presented and developed quite successfully over time. Most of these approaches, however, have one peculiarity in common: they make no or nearly no use of the symmetry of the crystal structures.

The importance of symmetry relations in phase transitions in the solid state was realized in 1937 by LEW LANDAU [12]. Around 1968 HARTMUT BÄRNIGHAUSEN developed a procedure to work out relationships between crystal structures with the aid of symmetry relations [13]. Since then, chemists have become more and more aware of the value of these symmetry relations. Symmetry relations can be formulated mathematically. This offers a secure foundation for their application and makes it possible to develop algorithms in order to make use of computers.

The symmetry of crystals is presented in *International Tables for Crystallography*, Volume *A* [14, 15], by diagrams and with the aid of analytical geometry. The methods of analytical geometry can be applied universally; they are based on the techniques of matrix calculus and make use of the results of elementary group theory. Since 2004, the supplementary volume *A*1 of *International Tables for Crystallography* has been available [16]. For the first time they contain a complete listing of the subgroups of the space groups. This book shows how to make use of these tables.

Part I of the book presents the necessary mathematical tools: the fundamentals of crystallography, especially of symmetry; the theory of crystallographic groups; and the formalisms for the necessary crystallographic calculations. As often in the natural sciences, these tools may appear difficult if one is not accustomed to their use. However, the presented calculation techniques are nothing more than applications of simple theorems of algebra and group theory.

Group theory has deep foundations. For its application, however, the depth is not needed. The mathematical foundations are contained in the presented formalisms. Calculations can be performed and consequences can be drawn with these formalisms, without the need to duplicate their mathematical background.

Those who have some familiarity with the symmetry of crystals (i.e. who have worked with space groups, are acquainted with Hermann–Mauguin symbols, know how to handle atomic coordinates, etc.) may take a first look at Part II to obtain an impression of the results that follow from the mathematical relations. However, it is not recommended to skip the chapters of Part I. Don't be mistaken: crystallographic group theory and symbolism does have pitfalls, and calculations are susceptible to errors if they are not performed strictly in accordance with the rules.

Part II of the book gives an insight into the application to problems in crystal chemistry. Numerous examples show how crystallographic group theory can be used to disclose relations between crystal structures, to maintain order among the enormous number of crystal structures, to predict possible crystal-structure types, to analyse phase transitions, to understand the phenomenon of domain formation and twinning in crystals, and to avoid errors in crystal-structure determinations.

Part III deals with peculiarities of a certain kind of subgroup of the space groups, the isomorphic subgroups, including connections to number theory, and it offers some insight to the theory of phase transitions. Finally, a series of computer programs are presented that help to avoid the extensive browsing in *International Tables for Crystallography* and that can be used to perform crystallographic calculations.

A broad range of end-of-chapter exercises offers the possibility to apply the learned material. Worked-out solutions to the exercises can be found in the Appendices.

In the Glossary the meanings of special terms used in the field can be looked up.

One topic of group theory is not addressed in this book: representation theory. Crystallographic symmetry does not deal with time. Representation theory is needed to cover the symmetry properties of time-dependent phenomena (such as vibrations). This is dealt with in numerous books and articles; we could only repeat their content (see e.g. [17–25]). However, some remarks can be found in Chapters 16, 22, and 23. Neither do we deal with magnetic space groups (black-and-white space groups, Shubnikov groups), where a colour (black or white) is assigned to every point in space and part of the symmetry operations are connected with a colour inversion (or spin inversion); for more details see [59] and the cited references.

The book has many predecessors. It is based on earlier lectures and on courses that have been taught repeatedly since 1975 in Germany, Italy, France, Czechia, Bulgaria, Russia, and South Africa. In the beginning, the lecturers for these courses were H. BÄRNIGHAUSEN (Karlsruhe), TH. HAHN (Aachen), and H. WONDRATSCHEK (Karlsruhe), later also M. AROYO (Sofia, later Bilbao), G. CHAPUIS (Lausanne), W. E. KLEE (Karlsruhe), R. PÖTTGEN (Münster) and myself. The courses were then continued by T. DOERT (Dresden), M. RUCK (Dresden), U. SCHWARZ (Dresden), H. KOHLMANN (Leipzig), O. OECKLER (Leipzig), C. RÖHR (Freiburg), and G. DE LA FLOR (Karlsruhe).

The text of Chapters 2–7 is due to H. WONDRATSCHEK, who allowed me to use his material; he also revised these chapters after I had appended figures, examples, exercises, and a few paragraphs. These chapters partly reflect lecture notes by W. E. KLEE. Chapters 1, 11, 12, 16, and 17 essentially go back to H. BÄRNIGHAUSEN and contain text by him; he also critically checked drafts of these chapters. Parts of a script by R. PÖTTGEN, R.-D. HOFFMANN, and U. RODEWALD were included in Chapter 18. I am especially grateful to all of them. Without their manuscripts and without their consent to make use of their texts, this book could not have come into being. Chapter 24 about the Bilbao Crystallographic Server is new in this edition. It was written by GEMMA DE LA FLOR and adapted to the style of this book by myself.

Indirect contributions are made through the suggestions of G. NEBE (mathematician, Aachen), J. NEUBÜSER (mathematician, Aachen), and V. JANOVEC (physicist, Liberec), and through their numerous discussions with H. WONDRATSCHEK. In addition, I am grateful to further unnamed colleagues for suggestions and discussions.

Compared to the first edition of this book, published in 2013 and as paperback in 2017, the text and references have been updated. Chapter 8 has been divided into a chapter with more theoretic elements and a chapter with practical applications. Sections 13.4, 14.3.2, 16.7, and Chapter 24 have been added. Some other changes, corrections, and additions have been made, especially to Chapters 3, 9, 14, and 18.

Ulrich Müller
Marburg, Germany, July 2023

Contents

List of symbols

Symbol	Meaning
$\{\ldots\}$, M	sets
m, m_k	elements of the set M
P, Q, R, X, X_k	points
O	origin
$\mathbf{u}, \mathbf{w}, \mathbf{x}$	vectors
$\mathbf{t}, \mathbf{t}_i$	translation vectors
$\mathbf{a}, \mathbf{b}, \mathbf{c}, \mathbf{a}_i$	basis vectors
$(\mathbf{a})^{\mathrm{T}}, (\mathbf{a}')^{\mathrm{T}}$	row of basis vectors
$a, b, c, \alpha, \beta, \gamma$	lattice parameters
$x_i, \tilde{x}_i, x'_i, w_i$	point coordinates or vector coefficients
$\boldsymbol{r}, \boldsymbol{t}, \boldsymbol{u}, \boldsymbol{w}, \boldsymbol{x}$	columns of coordinates or vector coefficients
$\boldsymbol{x}_F$	column of coordinates of a fixed point
$\boldsymbol{o}$	column of zeros
$\mathbf{T}$	vector lattice (lattice)
$\tilde{X}, \tilde{\boldsymbol{x}}, \tilde{x}_i$	image point and its coordinates
$\boldsymbol{x}', x'_i$	coordinates in the new coordinate system
$\mathsf{I}, \mathsf{R}, \mathsf{T}, \mathsf{W}$	mappings
$\boldsymbol{I}$	unit matrix
$\boldsymbol{I}, \boldsymbol{U}, \boldsymbol{W}, \boldsymbol{V}$	(3×3) matrices
$(\boldsymbol{U}, \boldsymbol{u}), (\boldsymbol{V}, \boldsymbol{v}), (\boldsymbol{W}, \boldsymbol{w})$	matrix–column pairs
W_{ik}	matrix coefficients
$\boldsymbol{W}^{\mathrm{T}}$	transposed matrix
$\mathbb{r}, \mathbb{t}, \mathbb{x}$	augmented columns
$\mathbb{U}, \mathbb{V}, \mathbb{W}$	augmented (4×4) matrices
$\boldsymbol{P}, \mathbb{P}$	transformation matrices
$\boldsymbol{p}$	column of coefficients of an origin shift
$\boldsymbol{G}, g_{ik}$	metric tensor and its coefficients
$\mathrm{tr}(\ldots)$	trace of a matrix
$\det(\ldots)$	determinant of a matrix
$[u\,v\,w]$	lattice direction
$(h\,k\,l)$	lattice plane, crystal face
$\{h\,k\,l\}$	set of symmetry-equivalent crystal faces
φ	angle of rotation

Symbol	Meaning
$e, g, g_i, m, m_i, \ldots$	group elements; e = unit element
$\{g_1, g_2, \ldots\}$	group with elements $g_1, g_2, \ldots$
$\{\boldsymbol{W}\}$	group of the matrices $\boldsymbol{W}_1, \boldsymbol{W}_2, \ldots$
$\mathcal{G}, \mathcal{H}, \mathcal{S}$	groups
$\mathcal{T}$	group of all translations
$\mathcal{H} < \mathcal{G}$	$\mathcal{H}$ is a proper subgroup of $\mathcal{G}$
$\mathcal{H} \leq \mathcal{G}$	$\mathcal{H}$ is a subgroup of $\mathcal{G}$ or equal to $\mathcal{G}$
$\mathcal{H} \lhd \mathcal{G}$	$\mathcal{H}$ is a normal subgroup of $\mathcal{G}$
$g \in \mathcal{G}$	g is an element of $\mathcal{G}$
$g_i, g_j \in \mathcal{G}$	g_i and g_j are elements of $\mathcal{G}$
$A \subset \mathcal{G}$	A is a subset of $\mathcal{G}$
$A \subseteq \mathcal{G}$	A is a subset of $\mathcal{G}$ or equal to $\mathcal{G}$
$\lvert\mathcal{G}\rvert$	order of the group $\mathcal{G}$
$\lvert\mathcal{G} : \mathcal{H}\rvert$	index of $\mathcal{H}$ in $\mathcal{G}$
$\mathcal{G}/\mathcal{T}$	factor group of $\mathcal{G}$ with respect to $\mathcal{T}$
$\mathcal{N}_{\mathcal{G}}(\mathcal{H})$	normalizer of $\mathcal{H}$ in $\mathcal{G}$
$\mathcal{N}_{\mathcal{E}}(\mathcal{G})$	Euclidean normalizer of $\mathcal{G}$
$\mathcal{N}_{\mathcal{E}^+}(\mathcal{G})$	chirality-preserving Euclidean normalizer of $\mathcal{G}$

For group–subgroup relations:

Symbol	Meaning	
t3	*translationengleiche*	subgroup of index 3
k3	*klassengleiche*	
i3	isomorphic	
i	index of a subgroup	
p	prime number	

Physical quantities:

Symbol	Meaning
G	Gibbs free energy
H	enthalpy
S	entropy
C_p	heat capacity at constant pressure
p	pressure
T	absolute temperature
T_c	critical temperature
V	volume
η	order parameter

Introduction

Crystallography is the science of crystals. The inner (atomic and electronic) structure of crystalline solids, as well as their physical properties, are of central interest. This includes the methods of structure determination and of measurement of properties. A well-founded theoretical treatment is of special importance to understanding the connections and to applications. In part, the theories are strongly influenced by mathematics. Due to its strong interrelation with mathematics, physics, chemistry, mineralogy, materials sciences, biochemistry, molecular biology, pharmaceutics, and metrology, crystallography is more multidisciplinary than most other fields of science.

The *theory of symmetry* is of special importance among the theories in crystallography. The symmetry of crystals, which also has influence on their physical properties, is specified with the aid of *space groups*.

Crystal chemistry is the branch of chemistry that deals with the structures, properties, and other chemical aspects of crystalline solids. Geometric considerations relating to the structures attract much attention in this discipline. In this context a main objective is to disclose relationships between different crystal structures and to document the corresponding results in a concise but informative way. To this end, different approaches were presented over time, which demonstrate the similarities and the differences of distinct structures from different points of view. For example, the main attention can be directed to the coordination polyhedra and the joining of these polyhedra, or to the relative size of ions, or to the kind of chemical bonding, or to similar physical or chemical properties.

Symmetry has been a focus in the description of single structures—this is familiar to anyone who has been engaged in work with crystal structures. However, concerning the comparison of structures, symmetry considerations have long been the exception. There certainly exist diverse reasons for this astonishingly unbalanced development of crystal chemistry. The main one is likely to be that related crystal structures often have different space groups, so that their relationship becomes apparent only by consideration of the group–subgroup relations between their space groups. It was relatively late that an essential part of the necessary group-theoretical material, namely a listing of the subgroups of the space groups, became available in a useful form.

Aspects of space-group theory important to crystal chemistry were solved around 1930 by C. HERMANN and H. HEESCH and were included in the 1935 edition of *International Tables for the Determination of Crystal Structures* [26]; this comprised lists of the subgroups of the space groups. However, in the following edition of 1952 [27] they were excluded. In addition, in the edition of 1935 only a certain kind of subgroup was mentioned, namely the *translationengleiche* subgroups, called *zellengleiche* subgroups at that time. A broad application was thus hardly possible. For crystal-chemical applications another kind of subgroup, the *klassengleiche* subgroups, are at least as important. A compilation of the *klassengleiche* subgroups of the space groups was presented by J. NEUBÜSER and H. WONDRATSCHEK as much as 53 years after the discovery of X-ray diffraction [28], and the *isomorphic* subgroups, which are a special category of *klassengleiche* subgroups, were then derived by E. BERTAUT and Y. BILLIET [29].

For 18 years this material existed only as a collection of copied sheets of paper and was distributed this way among interested scientists. Finally, the subgroups of the space groups were included in the 1983 edition of Volume *A* of *International Tables for Crystallography* [14]. And yet, the listing of the subgroups in the first to the fifth editions of Volume *A* (1983–2006) has been incomplete. Beginning with the sixth edition (2016) the subgroups of the space groups are no longer included in Volume *A*.

Instead, a finally complete listing of all subgroups of the space groups has existed since 2004 in the supplementary Volume *A*1 of *International Tables for Crystallography* [16]. This includes the corresponding axes and coordinate transformations as well as the relations that exist between the Wyckoff positions of a space group and the Wyckoff positions of its subgroups. This information, which is essential for group-theoretical considerations, can indeed also be derived from the data of Volume *A*; that, however, is cumbersome and prone to errors. In addition, supporting computer programs are available in form of the *Symmetry Database* and the *Bilbao Crystallographic Server*; see Chapter 24. They offer access to databases and computer programs that display the subgroups and supergroups of space groups and other tools to resolve problems addressed in this book.

International Tables for Crystallography, Volumes *A* and *A*1, will henceforth be referred to as *International Tables A* and *International Tables A*1. Given the case, for Volume *A* its year of publication is additionally specified if reference is made to a specific edition: 2006 [14] or 2016 [15]. *International Tables* are available in printed and in electronic form, `http://it.iucr.org/`.

In this book it is shown that symmetry relations between the space groups are a useful tool for the clear derivation and the concise presentation of facts in the field of crystal chemistry. The presented examples will speak for themselves. However, it should be mentioned why the abstract framework of group theory is so successful: it is due to the symmetry principle in crystal chemistry.

1.1 The symmetry principle in crystal chemistry

The symmetry principle is an old principle based on experience that has been worded during its long history in rather different ways, such that a common root is hardly discernible at first glance (see Chapter 20 for the historical development). In view of crystal chemistry, BÄRNIGHAUSEN summarized the symmetry principle in the following way, pointing out three important partial aspects [13]:

1. In crystal structures the arrangement of atoms reveals a pronounced tendency towards the highest possible symmetry.
2. Counteracting factors due to special properties of the atoms or atom aggregates may prevent the attainment of the highest possible symmetry. However, in many cases the deviations from ideal symmetry are only small (key word: *pseudosymmetry*).
3. During phase transitions and solid-state reactions which result in products of lower symmetry, the higher symmetry of the starting material is often indirectly preserved by the formation of oriented domains.

Aspect 1 is due to the tendency of atoms of the same kind to occupy equivalent positions in a crystal, as stated by BRUNNER [30]. This has physical reasons, as follows:

Depending on the given conditions, such as chemical composition, the kind of chemical bonding, electron configurations of the atoms, relative size of the atoms, pressure, temperature, etc., there exists *one* energetically most favourable surrounding for atoms of a given species that all of these atoms strive to attain. The same surrounding for atoms in a crystal is ensured only if they are symmetry equivalent.

Aspect 2 of the symmetry principle is exploited extensively in Part II of this book. Factors that counteract the attainment of the highest symmetry include:

- stereochemically active lone electron pairs;
- distortions caused by the Jahn–Teller effect;
- Peierls distortions;
- covalent bonds, hydrogen bonds, and other bonding interactions between atoms;
- electronic effects between atoms, such as spin interactions;
- ordering of atoms in a disordered structure;
- freezing (condensation) of vibrational modes (soft modes) giving rise to phase transitions;
- ordered occupancy of originally equivalent sites by different kinds of atoms (substitution derivatives);
- partial vacation of atomic positions;
- partial occupancy of voids in a packing of atoms.

Aspect 3 of the symmetry principle has its origin in an observation by J. D. BERNAL. He noted that in the solid-state reaction $Mn(OH)_2 \rightarrow MnOOH \rightarrow MnO_2$ the initial and the product crystal had the same orientation [31]. Such reactions are called *topotactic reaction*s after F. K. LOTGERING [32] (for a more exact definition see [33]). In a paper by J. D. BERNAL and A. L. MACKAY we find the following [34]:

> One of the controlling factors of topotactic reactions is, of course, symmetry. This can be treated at various levels of sophistication, ranging from Lyubarskii's to ours, where we find that the simple concept of Buridan's ass illumines most cases.

According to the metaphor ascribed to the French philosopher JEAN BURIDAN (died *circa* 1358), the ass starves to death between two equal and equidistant piles of hay because it cannot decide between them. Referred to crystals, such asinine behaviour would correspond to an absence of phase transitions or solid-state reactions if there are two or more energetically equivalent orientations of the domains of the product. Crystals, of course, do not behave like the ass; they take both.

1.2 Introductory examples

To get an impression for the kind of considerations that will be treated in more detail in later chapters, we present a few simple examples. Many crystal structures can be related to a few simple, highly symmetrical crystal-structure types. Zinc blende (sphalerite, ZnS) has the same structural principle as diamond; alternating zinc and sulfur atoms take the positions of the carbon atoms. Both structures have the same kind of cubic unit cell, the atoms in the cell occupy the same positions, and they are bonded with one another in the same way. Whereas all atoms in diamond are symmetrically equivalent, there must be two symmetrically independent atomic positions in zinc blende, one for zinc and one for sulfur. Zinc blende cannot have the same symmetry as diamond; its space group is a subgroup of the space group of diamond. The relation is depicted in Fig. 1.1 in a way that we will make use of in later chapters and which is explained more exactly in Chapter 11.

In Fig. 1.1 a small 'family tree' is shown to the left; at its top the symmetry of diamond is mentioned, marked by the symbol of its space group $F\,4_1/d\,\bar{3}\,2/m$. An arrow pointing downwards indicates the symmetry reduction to a subgroup. The subgroup has the space-group symbol $F\,\bar{4}\,3\,m$; it has a reduced number of symmetry operations. In particular, no symmetry operation of diamond may be

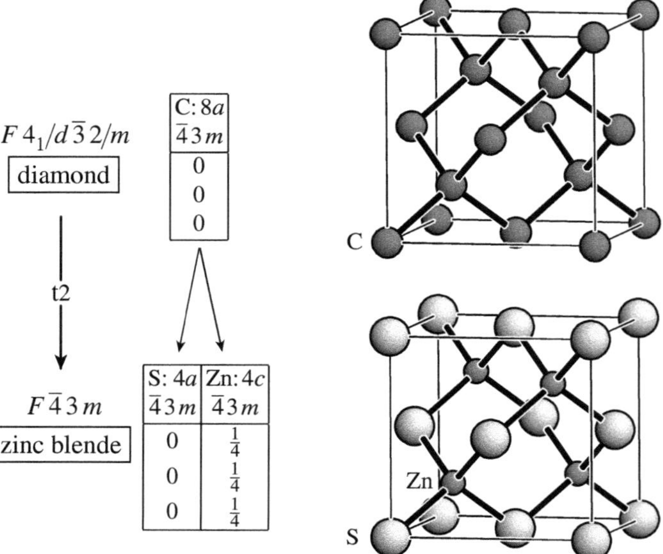

Fig. 1.1 The relation between diamond and zinc blende. The numbers in the boxes are the atomic coordinates.

retained that would convert a zinc position to a sulfur position. The *multiplicity* of the C atoms in diamond is 8, that is, the unit cell of diamond contains eight symmetrically equivalent C atoms. Their position is expressed by the *Wyckoff symbol* $8a$. The 8 marks this multiplicity, and the a is an alphabetical label according to which the positions are numbered in *International Tables A* [14]. Due to the symmetry reduction this position $8a$ splits into two independent positions $4a$ and $4c$ in the subgroup. The point symmetry of the mentioned atomic positions remains tetrahedral, symbol $\bar{4}3m$.

The 'family tree' in Fig. 1.1 is rather small; it comprises only one 'mother' and one 'daughter'. As will be shown later, larger 'family trees' can be used to depict relations among numerous crystal structures, with many 'daughters' and 'grandchildren'. This notion harmonizes with the term *family of structures* in the rather strict sense according to H. D. MEGAW [35]. For the most symmetrical structure in the family of structures MEGAW coined the term *aristotype*.[1] The derived structures are called, again after MEGAW, *hettotypes*.[2] These terms correspond to the terms *basic structure* and *derivative structure* after BUERGER [36, 37].

[1] greek *aristos* = the best, the highest.

[2] greek *hetto* = weaker, inferior.

Trees of group–subgroup relations as shown in Fig. 1.1 are called *Bärnighausen trees*.

In reality it is impossible to substitute Zn and S atoms for C atoms in a diamond crystal. The substitution takes place only in one's imagination. Nevertheless, this kind of approach is very helpful to trace back the large number of known structures to a few simple and well-known structure types and thus to obtain a general view.

On the other hand, the case that the symmetry reduction actually takes place in a sample does occur, namely in phase transitions as well as in chemical reactions in the solid state. An example is the phase transition of $CaCl_2$ that takes place at 217 °C [38–40]. It involves a mutual rotation of the coordination octahedra about **c**, which is expressed by slightly altered atomic coordinates of the Cl atoms (Fig. 1.2). Contrary to the diamond–zinc blende relation, the calcium as well as the chlorine atoms remain symmetry equivalent; no atomic position splits into several independent positions, but instead their point symmetries are reduced. Phase transitions of this kind are linked to changes of the physical properties that depend on crystal symmetry. For example, $CaCl_2$ is ferroelastic at temperatures below 217 °C.[3]

[3] Ferroelastic: The domains in a crystal differ in spontaneous strain and can be shifted by a mechanical force.

In the literature in physics the aristotype is often called the *prototype* or *parent phase*, and the hettotype the *daughter phase* or *distorted structure*. These terms are only applicable to phase transitions, that is, to processes in which one solid phase is converted to another one with the same chemical composition, with a change of symmetry.

Calcium chloride forms twinned crystals in the course of the phase transition from the high- to the low-temperature modification. The reason for this can be perceived in the images of the structures in Fig. 1.2. If the octahedron in the middle of the cell is rotated clockwise (as depicted), the tetragonal high-temperature form ($a = b$) transforms to the orthorhombic low-temperature form with decreased a and increased b axis. The same structure is obtained by anticlockwise rotation, but with an increased a and a decreased b axis (Fig. 1.3). In the initial tetragonal crystal the formation of the orthorhombic crystals sets

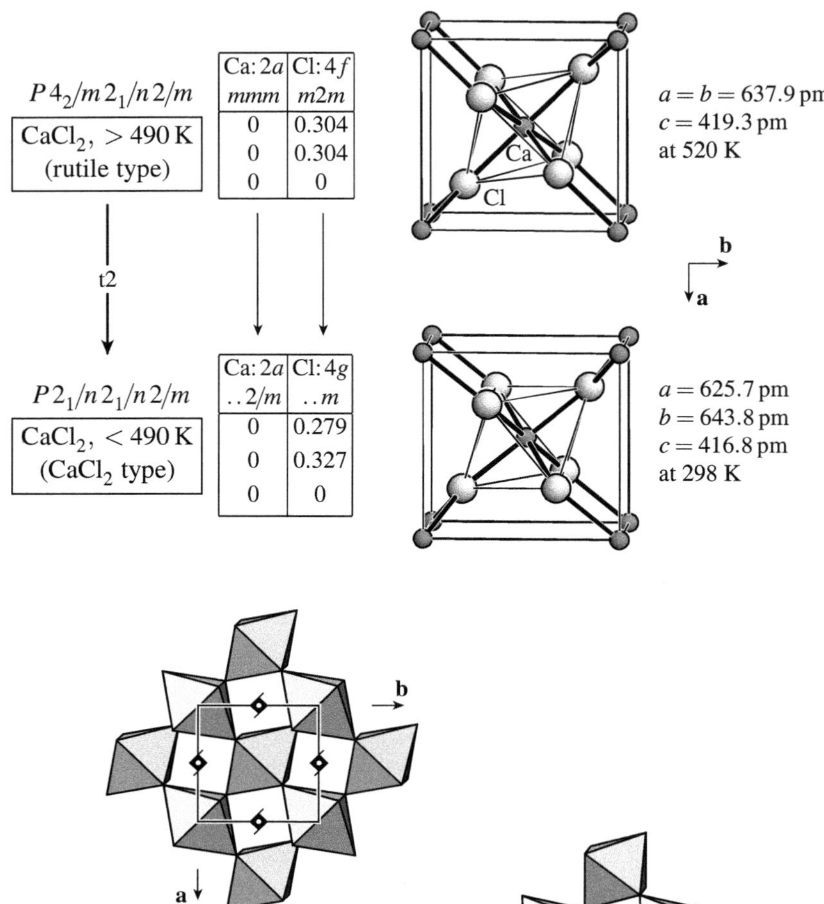

Fig. 1.2 The relation between the modifications of calcium chloride. The coordination octahedron is rotated about the direction of view (c axis), and the reflection planes running diagonally through the cell of the rutile type vanish.

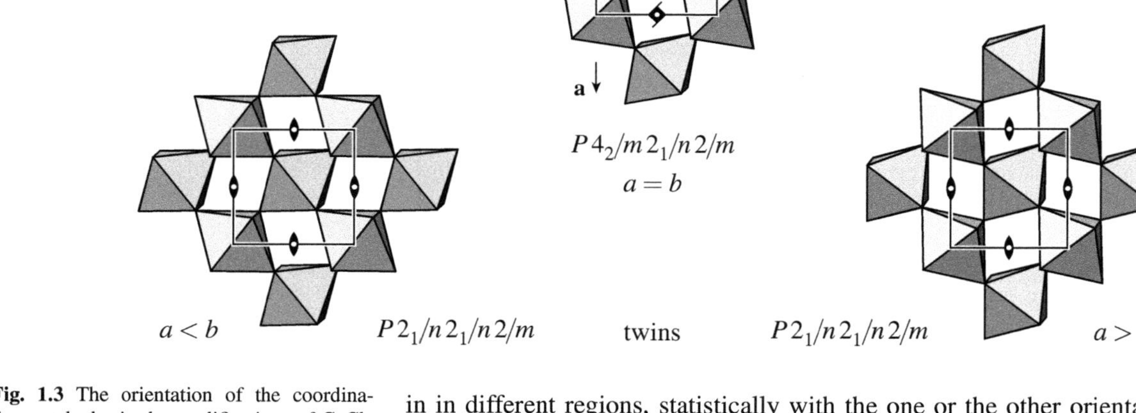

Fig. 1.3 The orientation of the coordination octahedra in the modifications of $CaCl_2$ and the relative orientation of the unit cells of the twin domains of the low-temperature modification. The marked fourfold axes of the tetragonal modification are converted to twofold axes in the orthorhombic modification.

in in different regions, statistically with the one or the other orientation. At the end the whole crystal consists of intergrown twin domains. The symmetry elements being lost during the phase transition, for example the reflection planes running diagonally through the cell of the high-temperature form, are indirectly preserved by the relative orientation of the twin domains. More details concerning this phase transition are dealt with in Chapter 16; there it is also explained that this kind of group–subgroup relation immediately shows that the formation of twinned crystals is to be expected in this case.

The occurrence of twinned crystals is a widespread phenomenon. They can severely hamper crystal-structure determination. Their existence cannot always be detected on X-ray diffraction diagrams, and systematic superposition of X-ray reflections can cause the deduction of a false space group and even a false unit cell. In spite of the false space group, often a seemingly plausible structural model can be obtained, which may even be refined. Unfortunately, faulty crystal-structure determinations are not uncommon, and undetected twins are one of the causes. The most common consequences are slight to severe errors of interatomic distances; but even wrong coordination numbers and polyhedra up to a false chemical composition may be the result. Applications that rely on certain physical properties such as the piezoelectric effect can be impeded if twinned crystals are employed. Knowledge of the group-theoretical relations can help to avoid such errors.

Another kind of phase transformation occurs when randomly distributed atoms become ordered. This is a common observation among intermetallic compounds, but it is not restricted to this class of substances. Cu_3Au offers an example. Above 390 °C the copper and gold atoms are randomly distributed among all atomic positions of a face-centred cubic packing of spheres (space group $F4/m\bar{3}2/m$; Fig. 1.4). Upon cooling an ordering process sets in; the Au atoms now take the vertices of the unit cell whereas the Cu atoms take the centres of the faces. This is a symmetry reduction because the unit cell is no longer centred. The F of the space group symbol, meaning face-centred, is replaced by a P for primitive (space group $P4/m\bar{3}2/m$).

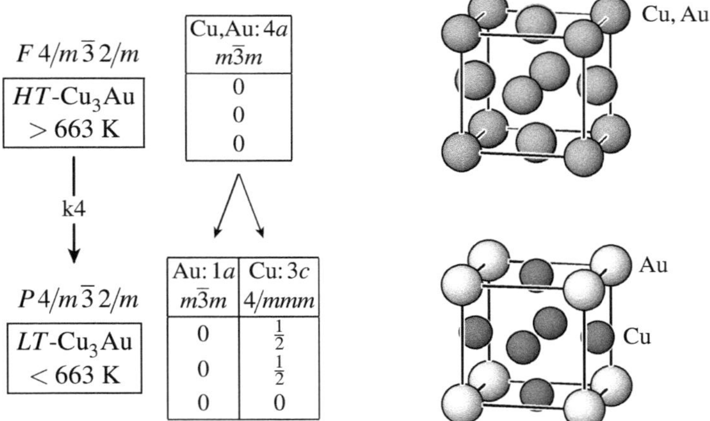

Fig. 1.4 The relation between misordered and ordered Cu_3Au. See margin note no. 2 in Section 16.1.2 (page 213) for a remark referring to the term 'misorder'.

Part I

Crystallographic Foundations

Basics of crystallography, part 1

2

2.1 Introductory remarks

Matter consists of atoms of diverse elements. These atoms occur not as isolated particles, but in organized arrays: finite arrays of interest are *molecules* (N_2, H_2O, CH_4, NH_3, C_6H_6, ...); large arrays are *crystals* that consist of equal parts that are periodically repeated in (nearly) any number.

Molecules and crystals are two kinds of appearance of matter. Molecules can assemble to form crystals. However, crystals do not necessarily consist of molecules; the crystal components may be simple ions like Na^+ and Cl^-, complex ions like CO_3^{2-} and NH_4^+, and many others. Henceforth, molecules and other such components will be called building blocks if they are components of crystals.

Other forms of appearance of matter, such as gases, liquids, glasses, partially ordered structures, modulated structures, or quasicrystals, will not be considered.

2.2 Crystals and lattices

Crystals are distinguished by the property that a shift called *translation* results in a perfect superposition of all building blocks of the crystal.

Naturally occurring crystals (quartz, rock salt, garnet, ...) and synthetically produced crystals (sugar, $SrTiO_3$, silicon, ...) can be regarded as finite blocks from infinite periodic structures. Replacement of the finite real crystal by the corresponding periodic, infinite array usually allows an excellent description of the real conditions and, therefore, is of great value, even though the infinitely extended ideal crystal does not exist. The *crystal structure* is the spatial distribution of the atoms in a crystal; usually, it is described with the model of the infinite crystal pattern. Hereafter, when we address a crystal structure, we always assume this kind of description.

Definition 2.1 The infinite, three-dimensional periodic array corresponding to a crystal is called the *crystal pattern* (or infinite ideal crystal). The lengths of the periodicities of this array may not be arbitrarily small.

The periodicity of a crystal structure implies that it comes to coincidence with itself after having been shifted in certain directions by certain distances.

The dimension $d = 3$ can be generalized to $d = 1,2,3,\ldots$. This way, planar arrangements ($d = 2$) can be included: periodic patterns of wallpaper, tilings, brick walls, tiled roofs,[1] cross-sections and projections of three-dimensional crystals, etc. Dimensions $d = 4,5,6,\ldots$ serve to formally describe incommensurate crystal structures and quasicrystals in higher-dimensional 'superspaces'. To this end, superspacegroups have been derived, including special symbols for them [41–47]; their treatment is beyond the scope of this book.

[1] Only the patterns are two dimensional; tilings, brick walls, etc. themselves are three-dimensional bodies; their symmetries are layer groups (Section 7.4).

The condition that periodicity lengths may not be arbitrarily small excludes homogeneous continua among crystal structures. Due to the finite size of the building blocks in real crystals there always exists a lower limit of the periodicity distances (> 0.1 nanometres).

The building blocks of the crystal structure may be not only points, figures, tiles, atoms, molecules, ions, etc., but also continuous functions such as electron density.

A *macroscopic (ideal) crystal*is a finite block out of a crystal pattern. Macroscopic crystals do not really exist. A *real crystal* not only has, like the macroscopic (ideal) crystal, a finite size, but is also defective. In addition, the atoms are not located at the exact positions as in the macroscopic crystal, but perform vibrational motions about these positions. The periodic pattern of atoms of the macroscopic crystal is fulfilled only by the positions of equilibrium of the vibrations.

Definition 2.2 A shift which brings a crystal structure to superposition with itself is called a *symmetry translation* (or simply *translation*) of this crystal structure. The corresponding shift vector is a *translation vector.*

Due to the periodicity, all integral multiples of a translation vector are also translation vectors. With two non-parallel translation vectors $\mathbf{t}_1$ and $\mathbf{t}_2$ all integral linear combinations are translation vectors:

$$\mathbf{t} = q\mathbf{t}_1 + r\mathbf{t}_2 \qquad q, r = \text{integers}$$

Definition 2.3 The infinite set of all translation vectors $\mathbf{t}_i$ of a crystal pattern is its *vector lattice* **T**. The translation vectors are called *lattice vectors.*

The vector lattice is often simply called the *lattice.* In chemistry (not in crystallography) the expression 'crystal lattice' is common. Frequently, the term 'lattice' has been used as a synonym for 'structure' (e.g. diamond lattice instead of diamond structure). Here we distinguish, as in *International Tables*, between 'lattice' and 'structure', and 'lattice' is something different from 'point lattice' and 'particle lattice', as defined in the next paragraph.[2]

[2] The terms 'lattice' and 'structure' should not be joined either. Do not say 'lattice structure' when you mean a framework structure consisting of atoms linked in three dimensions.

The vector lattice **T** of a crystal structure is an infinite set of vectors $\mathbf{t}_i$. With the aid of the vector lattice **T** it is possible to construct other more expressive lattices. Choose a starting point X_o with the positional vector $\mathbf{x}_o$ (vector pointing from a selected origin to X_o). The endpoints X_i of all vectors $\mathbf{x}_i = \mathbf{x}_o + \mathbf{t}_i$ make up the *point lattice* belonging to X_o and **T**. The points of the point lattice have a periodic order, they are all equal, and they all have the same surroundings. If the centres of gravity of particles are situated at the points of a point lattice, this is a *particle lattice.* All particles of the particle lattice are of the same kind.

An infinity of point lattices exists for every (vector) lattice, because any arbitrary starting point X_o can be combined with the lattice vectors $\mathbf{t}_i$. The lattice vectors may not be arbitrarily short according to Definition 2.1.

Definition 2.4 Points or particles that are transferred one to the other by a translation of the crystal structure are called *translation equivalent*.

Avoid terms like 'identical points', which can often be found in the literature, when 'translation-equivalent points' are meant. Identical means 'the very same'. Two translation-equivalent points are equal, but they are not the very same point.

2.3 Appropriate coordinate systems, crystal coordinates

To describe the geometric facts in space analytically, we introduce a coordinate system, consisting of an origin and a *basis* of three linearly independent (i.e. not coplanar) *basis vectors* $\mathbf{a}, \mathbf{b}, \mathbf{c}$ or $\mathbf{a}_1, \mathbf{a}_2, \mathbf{a}_3$. Referred to this coordinate system, each point in space can be specified by three coordinates (a coordinate triplet). The origin has the coordinates $0, 0, 0$. An arbitrary point P has coordinates x, y, z or x_1, x_2, x_3. The vector $\vec{OP}$ pointing from the origin to the point P is the position vector:

$$\vec{OP} = \mathbf{x} = x\mathbf{a} + y\mathbf{b} + z\mathbf{c} = x_1\mathbf{a}_1 + x_2\mathbf{a}_2 + x_3\mathbf{a}_3$$

In the plane, points P have coordinates x, y or x_1, x_2 referred to an origin (0, 0) and the basis $\mathbf{a}, \mathbf{b}$ or $\mathbf{a}_1, \mathbf{a}_2$.

Often a *Cartesian coordinate system* is suitable, in which the basis vectors are mutually perpendicular and have the length of 1 (*orthonormal basis*). Commonly, the angles between $\mathbf{a}, \mathbf{b}$, and $\mathbf{c}$ are denominated by α (between $\mathbf{b}$ and $\mathbf{c}$), β (between $\mathbf{c}$ and $\mathbf{a}$), and γ (between $\mathbf{a}$ and $\mathbf{b}$) or correspondingly by $\alpha_1, \alpha_2, \alpha_3$. With an orthonormal basis we then have

$$a = |\mathbf{a}| = b = |\mathbf{b}| = c = |\mathbf{c}| = 1; \quad \alpha = \beta = \gamma = 90^\circ$$

or $|\mathbf{a}_i| = 1$ and angles $(\mathbf{a}_i, \mathbf{a}_k) = 90^\circ$ for $i, k = 1, 2, 3$ and $i \neq k$.

Generally, as far as the description of crystals is concerned, Cartesian coordinate systems are not the most convenient. For crystallographic purposes, it is more convenient to use a coordinate system that is adapted to the periodic structure of a crystal. Therefore, lattice vectors are chosen as basis vectors. With any other basis the description of the lattice of a crystal structure would be more complicated.

Definition 2.5 A basis which consists of three lattice vectors of a crystal pattern is called a *crystallographic basis* or a *lattice basis* of this crystal structure.[3]

Referred to a given crystallographic basis, each lattice vector $\mathbf{t} = t_1\mathbf{a}_1 + t_2\mathbf{a}_2 + t_3\mathbf{a}_3$ is a linear combination of the basis vectors with *rational coefficients* t_i. Every vector with *integral* t_i is a lattice vector. One can even select bases such that the coefficients of all lattice vectors are integers.

[3] The term 'basis' was used erstwhile with another meaning, namely in the sense of 'cell contents'.

Among the infinity of crystallographic bases of a crystal structure, some permit a particularly simple description and thus have turned out to be the most convenient. Such bases are the foundation for the description of the space groups in *International Tables A*. These bases are selected whenever there is no special reason for another choice.

Definition 2.6 We will call crystallographic bases *conventional bases* if they correspond to those used in *International Tables A*.

Conventional settings of space group types are those that are completely tabulated in *International Tables A* (with a complete set of coordinate triplets, Wyckoff positions, etc.). That is something different from a standard setting (cf. Sections 9.1 and 10.3).

Definition 2.7 A crystallographic basis $\mathbf{a}_1, \mathbf{a}_2, \mathbf{a}_3$ of a vector lattice is called a *primitive (crystallographic) basis* if its basis vectors are lattice vectors and if *every* lattice vector $\mathbf{t}$ can be expressed as a linear combination with *integral* coefficients t_i:

$$\mathbf{t} = t_1\mathbf{a}_1 + t_2\mathbf{a}_2 + t_3\mathbf{a}_3 \qquad (2.1)$$

For any vector lattice there exist an infinite number of primitive bases. It is always possible to choose a primitive basis. However, this would not be convenient for many applications. Therefore, the chosen conventional crystallographic basis is often not primitive, but such that as many as possible of the angles between the basis vectors amount to 90°; the coefficients t_i in eqn (2.1) can then also be certain fractional numbers (mostly multiples of $\frac{1}{2}$). Frequently, the lattice is called primitive if the conventional basis of the lattice is primitive; if it is not primitive, it is called a *centred lattice*.[4] A well-known example is the face-centred cubic lattice *cF* as in the cubic-closest packing of spheres (copper type). Lattice types are treated in Section 6.2.

[4] For the sake of precise terminology, the term 'centred' should not be misused with a different meaning; do not call a cluster of atoms a 'centred cluster' if you mean a cluster of atoms with an embedded atom, nor say, 'the F_6 octahedron of the PF_6^- ion is centred on the P atom'.

After having selected a crystallographic basis and an origin, it is easy to describe a crystal structure. To this end we define as follows:

Definition 2.8 The *unit cell* of the crystal structure is the parallelepiped in which the coordinates of all points are:

$$0 \le x, y, z < 1$$

The magnitudes (lengths) a, b, c of the basis vectors and the angles γ, α, β between them are called the *lattice constants* or (better) *cell* or the *lattice parameters* of the lattice.

The selection of a basis and an origin implies the selection of a unit cell. Every point within this unit cell has three coordinates $0 \le x, y, z < 1$. For every point of the crystal structure outside of the unit cell there is a translationally equivalent point within the unit cell whose coordinates differ by integral numbers. The addition or subtraction of integral numbers to the numerical coordinate values in order to obtain values $0 \le x, y, z < 1$ is called *standardization*. We can now mentally construct a crystal structure in two different ways:

1. We start from one unit cell and add or subtract integral numbers to the coordinates of its contents. This corresponds to a shift of the unit cell by

lattice vectors. In this way the whole crystal structure is built up systematically by joining (an infinity of) blocks, all with the same contents.

2. We start at the centre of gravity of a particle in the unit cell and add equal particles in the points of the corresponding (infinite) point lattice. If there are more particles to be considered, we take the centre of gravity of one of the remaining particles together with its point lattice, etc. Due to the minimum distances between particles in the finite size of the cell, the number of particles to be considered is finite. In this way we obtain a finite number of interlaced particle lattices that make up the crystal structure.

In the first case, the crystal structure is thought to be composed of an infinity of finite cells. In the second case, the structure is thought to be composed by interlacing a finite number of particle lattices that have an infinite extension. Both kinds of composition are useful. A third kind of composition is presented in Section 6.5.

A crystal structure can now easily be described completely by specifying the metrics of the unit cell (lengths of the basis vectors and the angles between them) and the contents of the cell (kinds of particles and their coordinates within one unit cell).[5]

[5] Do not confuse the description of a crystal structure with the structure itself. Do not call a set of coordinates a 'structure'.

In order to be able to compare different or similar structures, their descriptions have to refer to equal or similar cells. The conditions for conventional cell choices are often not sufficient to warrant this. Methods to obtain a uniquely defined cell from an arbitrarily chosen cell are called reduction methods. Common methods are:

1. derivation of the *reduced cell*, see Section 9.1 and *International Tables A* 2016, Section 3.1.3 [15]; Chapter 9.2 in previous editions [14].
2. the Delaunay reduction, see *International Tables A* 2016 [15], Section 3.1.2.3; or 2006, Section 9.1.8 [14].

The cells obtained by these methods may or may not be identical. Therefore, the method of reduction should be specified.

The geometric invariants of a crystal structure, for example the distances between particles and the bond angles, are independent of the chosen coordinate system (basis and origin). How atoms are bonded with each other is manifested in these quantities. In addition, these data are useful for the direct comparison of different particles in the same crystal structure or of corresponding particles in different crystal structures.

2.4 Lattice directions, lattice planes, and reciprocal lattice

A *lattice direction* is the direction parallel to a lattice vector $\mathbf{t}$. It is designated by the symbol $[u\,v\,w]$, u, v, w being the smallest integral coefficients of the lattice vector in this direction; u, v, and w have no common divisor. [100], [010], and [001] correspond to the directions of $\mathbf{a}_1$, $\mathbf{a}_2$, and $\mathbf{a}_3$, respectively; $[\bar{1}10]$ is the direction of the vector $-\mathbf{a}_1 + \mathbf{a}_2$.

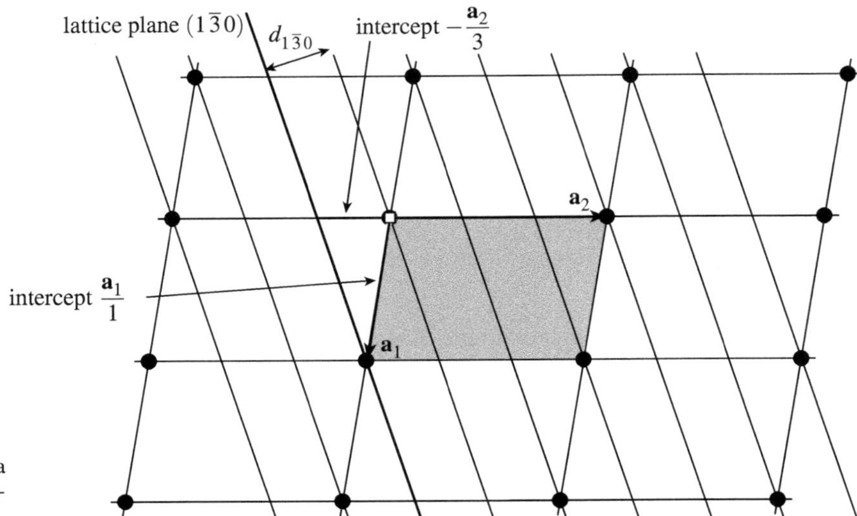

Fig. 2.1 A set of planes running through a point lattice. The third basis vector is perpendicular to the plane of the paper.

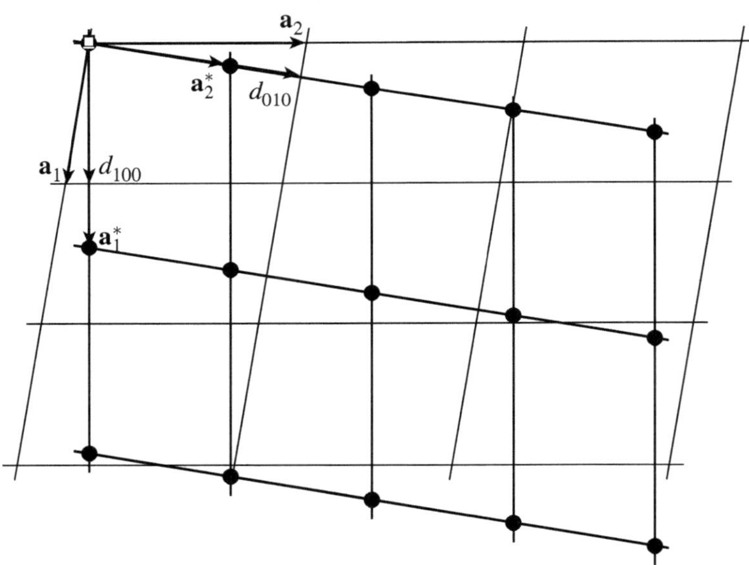

Fig. 2.2 Arrangement of the reciprocal lattice relative to a lattice when the third basis vector and the third reciprocal basis vector are perpendicular to the plane of the paper.

A lattice plane running through points of a point lattice is one out of a set of equidistant, parallel planes. The lattice plane is designated by the symbol (hkl) in parentheses; h, k, l are the integral *Miller indices*. From the set of planes, that one is selected which is closest to the origin without running itself through the origin. It intersects the coordinate axes at distances of a_1/h, a_2/k, a_3/l from the origin (Fig. 2.1). A plane running parallel to a basis vector obtains a 0 for this direction.

In order to facilitate the calculus with planes, it is convenient to represent each set of lattice planes by a vector $\mathbf{t}^*_{hkl} = h\mathbf{a}_1^* + k\mathbf{a}_2^* + l\mathbf{a}_3^*$ in the reciprocal lattice (Fig. 2.2). The *reciprocal lattice* $\mathbf{T}^*$ is a vector lattice with the reciprocal basis vectors $\mathbf{a}_1^*, \mathbf{a}_2^*, \mathbf{a}_3^*$ (or $\mathbf{a}^*, \mathbf{b}^*, \mathbf{c}^*$). $\mathbf{t}_{hkl}^*$ is perpendicular to the lattice plane (hkl) and has the length $1/d_{hkl}$, d_{hkl} being the interplanar distance between neighbouring lattice planes.

The reciprocal lattice is an essential tool for crystal-structure analysis, but less so for symmetry considerations. For more details see textbooks on crystal-structure analysis (e.g. [48–52]).

2.5 Calculation of distances and angles

Crystallographic bases are convenient for a simple description of crystals. However, the formulae for the computation of distances and angles in the crystal structure become less practical than with Cartesian coordinates.

Let Q and R be two points in a crystal structure having the coordinates x_q, y_q, z_q and x_r, y_r, z_r. Then the distance r_{qr} between Q and R is equal to the length of the vector $\mathbf{x}_r - \mathbf{x}_q = \overrightarrow{QR}$, where $\mathbf{x}_q$ and $\mathbf{x}_r$ are the position vectors (vectors from the origin) of Q and R. The length r_{qr} is the root of the scalar product of $\mathbf{x}_r - \mathbf{x}_q$ with itself:

$$\begin{aligned}
r_{qr}^2 &= (\mathbf{x}_r - \mathbf{x}_q)^2 = \left[(x_r - x_q)\mathbf{a} + (y_r - y_q)\mathbf{b} + (z_r - z_q)\mathbf{c}\right]^2 \\
&= (x_r - x_q)^2 a^2 + (y_r - y_q)^2 b^2 + (z_r - z_q)^2 c^2 + 2(x_r - x_q)(y_r - y_q) ab\cos\gamma \\
&\quad + 2(z_r - z_q)(x_r - x_q) ac\cos\beta + 2(y_r - y_q)(z_r - z_q) bc\cos\alpha
\end{aligned}$$

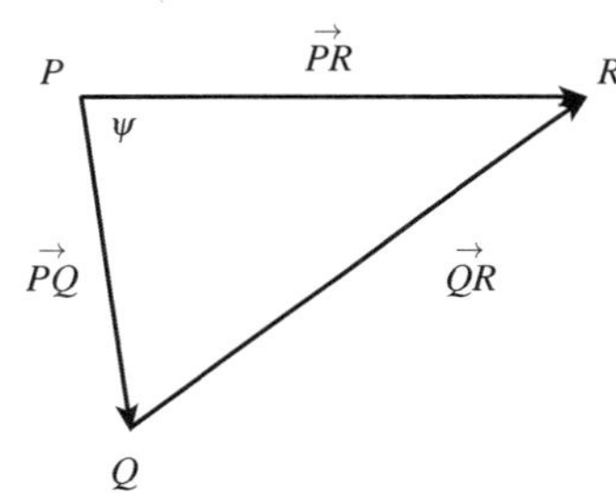

Fig. 2.3 Triangle of the points P, Q, and R, with distances PQ, PR, and QR and angle ψ.

The (bond) angle calculation ψ at the apex P between the connecting lines PQ and PR (Fig. 2.3) can be calculated with the following formula, using the scalar product of the vectors $(\mathbf{x}_q - \mathbf{x}_p)$ and $(\mathbf{x}_r - \mathbf{x}_p)$ ($\mathbf{x}_p$ is the position vector of P):

$$\begin{aligned}
(\mathbf{x}_q - \mathbf{x}_p) \cdot (\mathbf{x}_r - \mathbf{x}_p) &= r_{pq} r_{pr} \cos\psi \\
&= (x_q - x_p)(x_r - x_p)a^2 + (y_q - y_p)(y_r - y_p)b^2 + (z_q - z_p)(z_r - z_p)c^2 \\
&\quad + \left[(x_q - x_p)(y_r - y_p) + (y_q - y_p)(x_r - x_p)\right] ab\cos\gamma \\
&\quad + \left[(z_q - z_p)(x_r - x_p) + (x_q - x_p)(z_r - z_p)\right] ac\cos\beta \\
&\quad + \left[(y_q - y_p)(z_r - z_p) + (z_q - z_p)(y_r - y_p)\right] bc\cos\alpha
\end{aligned}$$

Every angle $\alpha_j = 90°$ strongly simplifies the formula. This is an advantage of an orthonormal basis; for this reason it is commonly used in crystal physics. The simplified formula then is (e = unit of length of the basis, e.g. $e = 1$ pm):

$$\begin{aligned}
r_{qr}^2 &= \left[(x_r - x_q)^2 + (y_r - y_q)^2 + (z_r - z_q)^2\right] e^2 \\
\cos\psi &= r_{pq}^{-1} r_{pr}^{-1} \left[(x_q - x_p)(x_r - x_p) + (y_q - y_p)(y_r - y_p) + (z_q - z_p)(z_r - z_p)\right] e^2
\end{aligned} \tag{2.2}$$

The volume V of the unit cell is obtained from the formula:

$$V^2 = a^2 b^2 c^2 (1 + 2\cos\alpha\cos\beta\cos\gamma - \cos^2\alpha - \cos^2\beta - \cos^2\gamma)$$

The lattice parameters $a,b,c,\alpha,\beta,\gamma$ appear in the combinations $g_{ii} = a_i^2$ or $g_{ik} = \mathbf{a}_i \cdot \mathbf{a}_k = a_i a_k \cos\alpha_j$, $i \neq j \neq k \neq i$.

For calculations, the specification of the shape of the cell by the g_{ik} values is more important than the usually quoted lattice parameters a_i and α_j, since the g_{ik} are needed for all calculations. From the a_i and α_j all g_{ik} can be calculated, conversely from the g_{ik} the a_i and α_j.

Definition 2.9 The complete set of the coefficients g_{ik} is called the *metric tensor*, formulated in the following way:

$$\boldsymbol{G} = \begin{pmatrix} g_{11} & g_{12} & g_{13} \\ g_{21} & g_{22} & g_{23} \\ g_{31} & g_{32} & g_{33} \end{pmatrix} = \begin{pmatrix} a^2 & ab\cos\gamma & ac\cos\beta \\ ab\cos\gamma & b^2 & bc\cos\alpha \\ ac\cos\beta & bc\cos\alpha & c^2 \end{pmatrix}$$

$g_{ik} = g_{ki}$ holds, since $\mathbf{a}_i \cdot \mathbf{a}_k = \mathbf{a}_k \cdot \mathbf{a}_i$.

With p_i, q_i, and r_i, $i = 1,2,3$, as the coordinates of the points P, Q, and R, we obtain the formulae:

- *Distance* $QR = r_{qr}$: $$r_{qr}^2 = \sum_{i,k} g_{ik}(r_i - q_i)(r_k - q_k) \quad (2.3)$$
- *Distance from the origin O*: $$OQ = r_q;\ r_q^2 = \sum_{i,k} g_{ik} q_i q_k$$
- *Angle QPR (apex P)*: $$\cos(QPR) = (r_{pq})^{-1}(r_{pr})^{-1} \sum_{i,k} g_{ik}(q_i - p_i)(r_k - p_k) \quad (2.4)$$
- *Volume V* of the unit cell: $$V^2 = \det(\boldsymbol{G}) \quad (2.5)$$

Application of $\boldsymbol{G}$ with the independent quantities g_{ik} instead of the six lattice parameters $a,b,c,\alpha,\beta,\gamma$ has the advantage that the g_{ik} are more homogeneous; for example, they all have the same unit Å^2 or pm^2.

The importance of the metric tensor $\boldsymbol{G}$ is not restricted to the calculation of distances and angles:

- With the aid of $\boldsymbol{G}$ it can be decided whether a given affine mapping leaves invariant all distances and angles (i.e. whether it is an isometry, see Section 3.4).
- If $\mathbf{T}^*$ is the reciprocal lattice of the lattice $\mathbf{T}$, then $\boldsymbol{G}^*(\mathbf{T}^*) = \boldsymbol{G}^{-1}(\mathbf{T})$ is the inverse matrix of $\boldsymbol{G}$: The metric tensors of the lattice and the reciprocal lattice are mutually inverse.

Mappings

3

3.1 Mappings in crystallography

3.1.1 An example

The following data can be found in *International Tables A* in the table for the space group $I4_1/amd$, No. 141, ORIGIN CHOICE 1, under the heading 'Coordinates' as a first block of entries:

$(0,0,0)+ \qquad (\frac{1}{2},\frac{1}{2},\frac{1}{2})+$

(1) x,y,z	(2) $\bar{x}+\frac{1}{2},\bar{y}+\frac{1}{2},z+\frac{1}{2}$	(3) $\bar{y},x+\frac{1}{2},z+\frac{1}{4}$	(4) $y+\frac{1}{2},\bar{x},z+\frac{3}{4}$
(5) $\bar{x}+\frac{1}{2},y,\bar{z}+\frac{3}{4}$	(6) $x,\bar{y}+\frac{1}{2},\bar{z}+\frac{1}{4}$	(7) $y+\frac{1}{2},x+\frac{1}{2},\bar{z}+\frac{1}{2}$	(8) $\bar{y},\bar{x},\bar{z}$
(9) $\bar{x},\bar{y}+\frac{1}{2},\bar{z}+\frac{1}{4}$	(10) $x+\frac{1}{2},y,\bar{z}+\frac{3}{4}$	(11) $y,\bar{x},\bar{z}$	(12) $\bar{y}+\frac{1}{2},x+\frac{1}{2},\bar{z}+\frac{1}{2}$
(13) $x+\frac{1}{2},\bar{y}+\frac{1}{2},z+\frac{1}{2}$	(14) $\bar{x},y,z$	(15) $\bar{y}+\frac{1}{2},\bar{x},z+\frac{3}{4}$	(16) $y,x+\frac{1}{2},z+\frac{1}{4}$

Following the common practice in crystallography, minus signs have been set over the symbols: $\bar{x}$ means $-x$. These coordinate triplets are usually interpreted in the following way: Starting from the point x, y, z in three-dimensional space, the following points are *symmetry equivalent*:

	$(0, 0, 0)+$	$(\frac{1}{2},\frac{1}{2},\frac{1}{2})+$
(1)	$x,\ y,\ z$	$x+\frac{1}{2},\ y+\frac{1}{2},\ z+\frac{1}{2}$
(2)	$-x+\frac{1}{2},\ -y+\frac{1}{2},\ z+\frac{1}{2}$	$-x, -y, z$
(3)	$-y,\ x+\frac{1}{2},\ z+\frac{1}{4}$	$-y+\frac{1}{2},\ x,\ z+\frac{3}{4}$
...	...	...

These entries in *International Tables A* can also be interpreted as descriptions of the symmetry operations of the space group. In the following, we will mainly adopt this kind of interpretation.

3.1.2 Symmetry operations

Definition 3.1 A *symmetry operation* is a mapping of an object such that

1. all distances remain unchanged,
2. the object is mapped onto itself or onto its mirror image.

If the object is a crystal structure, the mapping is a *crystallographic symmetry operation.*

'Mapped onto itself' does not mean that each point is mapped onto itself; rather, the object is mapped in such a way that an observer cannot distinguish between the states of the object before and after the mapping.

According to this interpretation, the entries in *International Tables A* for the space group $I4_1/amd$ mean:

$$x,y,z \rightarrow x,y,z; \qquad x,y,z \rightarrow x+\tfrac{1}{2},y+\tfrac{1}{2},z+\tfrac{1}{2};$$
$$x,y,z \rightarrow \bar{x}+\tfrac{1}{2},\bar{y}+\tfrac{1}{2},z+\tfrac{1}{2}; \qquad x,y,z \rightarrow \bar{x},\bar{y},z;$$
$$\ldots \qquad \ldots$$

or

$$\tilde{x},\tilde{y},\tilde{z} = x,y,z; \qquad \tilde{x},\tilde{y},\tilde{z} = x+\tfrac{1}{2},y+\tfrac{1}{2},z+\tfrac{1}{2};$$
$$\tilde{x},\tilde{y},\tilde{z} = \bar{x}+\tfrac{1}{2},\bar{y}+\tfrac{1}{2},z+\tfrac{1}{2}; \qquad \tilde{x},\tilde{y},\tilde{z} = \bar{x},\bar{y},z;$$
$$\ldots \qquad \ldots$$

The data of *International Tables A* thus describe how the coordinates $\tilde{x}$, $\tilde{y}$, $\tilde{z}$ of the image point result from the coordinates x, y, z of the original point.

Mappings play an important part in crystallography. A *mapping* is an instruction by which for each point of the space there is a uniquely determined point, the image point. The affine mappings and their special case, the isometric mappings or isometries, are of particular interest. Isometries are of fundamental interest for the symmetry of crystals. They are dealt with in Section 3.4.

Definition 3.2 The set of all symmetry operations of a crystal structure is called the *space group* of the crystal structure.

These symmetry operations are classified with the aid of affine mappings. Since isometries are a special case of affine mappings, we will address them first.

3.2 Affine mappings

Definition 3.3 A mapping of space which maps parallel straight lines onto parallel straight lines is called an *affine mapping*. The ratios of distances between points on a straight line are preserved.

Distances and angles are not necessarily preserved in an affine mapping. Therefore, the image can be distorted, as in the example shown in Fig. 3.1.

After having chosen a coordinate system, an affine mapping can always be represented by the following set of equations (x_1, x_2, x_3 are the coordinates of the original point; $\tilde{x}_1$, $\tilde{x}_2$, $\tilde{x}_3$ are the coordinates of the image point):

$$\text{in the plane} \quad \begin{cases} \tilde{x}_1 = W_{11}x_1 + W_{12}x_2 + w_1 \\ \tilde{x}_2 = W_{21}x_1 + W_{22}x_2 + w_2 \end{cases}$$

$$\text{in space} \quad \begin{cases} \tilde{x}_1 = W_{11}x_1 + W_{12}x_2 + W_{13}x_3 + w_1 \\ \tilde{x}_2 = W_{21}x_1 + W_{22}x_2 + W_{23}x_3 + w_2 \\ \tilde{x}_3 = W_{31}x_1 + W_{32}x_2 + W_{33}x_3 + w_3 \end{cases} \tag{3.1}$$

In matrix notation this is:

in the plane

$$\begin{pmatrix} x_1 \\ x_2 \end{pmatrix} = \boldsymbol{x}; \quad \begin{pmatrix} \tilde{x}_1 \\ \tilde{x}_2 \end{pmatrix} = \tilde{\boldsymbol{x}}; \quad \begin{pmatrix} w_1 \\ w_2 \end{pmatrix} = \boldsymbol{w}; \quad \begin{pmatrix} W_{11} & W_{12} \\ W_{21} & W_{22} \end{pmatrix} = \boldsymbol{W}$$

in space

$$\begin{pmatrix} x_1 \\ x_2 \\ x_3 \end{pmatrix} = \boldsymbol{x}; \quad \begin{pmatrix} \tilde{x}_1 \\ \tilde{x}_2 \\ \tilde{x}_3 \end{pmatrix} = \tilde{\boldsymbol{x}}; \quad \begin{pmatrix} w_1 \\ w_2 \\ w_3 \end{pmatrix} = \boldsymbol{w}; \quad \begin{pmatrix} W_{11} & W_{12} & W_{13} \\ W_{21} & W_{22} & W_{23} \\ W_{31} & W_{32} & W_{33} \end{pmatrix} = \boldsymbol{W} \quad (3.2)$$

Equation (3.1) then reads:

$$\begin{pmatrix} \tilde{x}_1 \\ \tilde{x}_2 \\ \tilde{x}_3 \end{pmatrix} = \begin{pmatrix} W_{11} & W_{12} & W_{13} \\ W_{21} & W_{22} & W_{23} \\ W_{31} & W_{32} & W_{33} \end{pmatrix} \begin{pmatrix} x_1 \\ x_2 \\ x_3 \end{pmatrix} + \begin{pmatrix} w_1 \\ w_2 \\ w_3 \end{pmatrix} \quad (3.3)$$

or for short

$$\tilde{\boldsymbol{x}} = \boldsymbol{W}\boldsymbol{x} + \boldsymbol{w} \quad \text{or} \quad \tilde{\boldsymbol{x}} = (\boldsymbol{W}, \boldsymbol{w})\boldsymbol{x} \quad \text{or} \quad \tilde{\boldsymbol{x}} = (\boldsymbol{W}|\boldsymbol{w})\boldsymbol{x} \quad (3.4)$$

The matrix $\boldsymbol{W}$ is called the *matrix part*, the column $\boldsymbol{w}$ is the *column part* of the representation of the mapping by matrices. The symbols $(\boldsymbol{W}, \boldsymbol{w})$ and $(\boldsymbol{W}|\boldsymbol{w})$ are called the *matrix–column pair* and the *Seitz symbol* of the mapping.

Calculations with matrix–column pairs $(\boldsymbol{W}, \boldsymbol{w})$ are more cumbersome than calculations with square matrices. A more transparent and more elegant way to write down general formulae is to use 4×4 matrices and four-row columns (correspondingly, in the plane, 3×3 matrices and three-row columns):

$$\boldsymbol{x} \to \mathbb{x} = \left(\begin{array}{c} x \\ y \\ z \\ \hline 1 \end{array}\right) \qquad \tilde{\boldsymbol{x}} \to \tilde{\mathbb{x}} = \left(\begin{array}{c} \tilde{x} \\ \tilde{y} \\ \tilde{z} \\ \hline 1 \end{array}\right) \qquad (\boldsymbol{W}, \boldsymbol{w}) \to \mathbb{W} = \left(\begin{array}{ccc|c} & \boldsymbol{W} & & \boldsymbol{w} \\ \hline 0 & 0 & 0 & 1 \end{array}\right)$$

$$\left(\begin{array}{c} \tilde{x} \\ \tilde{y} \\ \tilde{z} \\ \hline 1 \end{array}\right) = \left(\begin{array}{ccc|c} & \boldsymbol{W} & & \boldsymbol{w} \\ \hline 0 & 0 & 0 & 1 \end{array}\right) \left(\begin{array}{c} x \\ y \\ z \\ \hline 1 \end{array}\right) \quad \text{or} \quad \tilde{\mathbb{x}} = \mathbb{W}\mathbb{x} \quad (3.5)$$

The three-row columns have been augmented to four-row columns by appending a 1. The 3×3 matrix $\boldsymbol{W}$ has been combined with the column $\boldsymbol{w}$ to form a

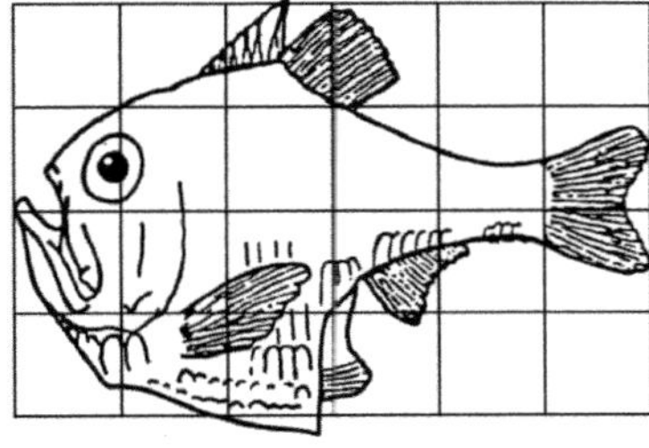

Argyropelecus olfersii

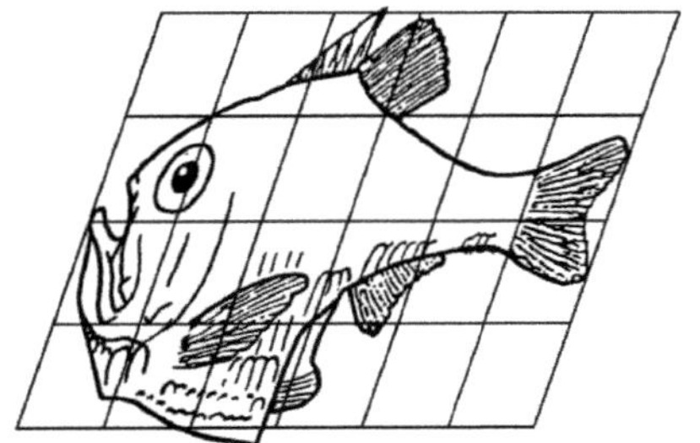

Sternoptyx diaphana.

Fig. 3.1 Example of an affine mapping. Taken from a pattern by Sir D'ARCY WENTWORTH THOMPSON (1860 – 1948), "On growth and form" (1917). The right image, compared to the original drawing (which has a circular eye), has been slightly altered in order to ensure an exact affine mapping.

3×4 matrix and then supplemented by a fourth row '0 0 0 1' to form a square 4×4 matrix. These columns are called *augmented columns*, and the matrices are called *augmented matrices* and are designated by open-face letters.

The vertical and horizontal lines in the matrices have no mathematical meaning and can be omitted. They are simply a convenience for separating the matrix part from the column part and from the row '0 0 0 1'.

If two affine mappings are performed one after the other, the result is again an affine mapping. If the first mapping is represented by $\tilde{\mathbb{x}} = \mathbb{W}\mathbb{x}$ and the second one by $\tilde{\tilde{\mathbb{x}}} = \mathbb{V}\tilde{\mathbb{x}}$, the result is:

$$\tilde{\tilde{\mathbb{x}}} = \mathbb{V}\mathbb{W}\mathbb{x} = \mathbb{U}\mathbb{x} \tag{3.6}$$

$$\text{with} \quad \mathbb{U} = \left(\begin{array}{ccc|c} & \boldsymbol{U} & & \boldsymbol{u} \\ \hline 0 & 0 & 0 & 1 \end{array}\right) = \left(\begin{array}{ccc|c} & \boldsymbol{V} & & \boldsymbol{v} \\ \hline 0 & 0 & 0 & 1 \end{array}\right)\left(\begin{array}{ccc|c} & \boldsymbol{W} & & \boldsymbol{w} \\ \hline 0 & 0 & 0 & 1 \end{array}\right) \tag{3.7}$$

Therefore, the 4×4 matrix of $(\boldsymbol{U}, \boldsymbol{u}) = (\boldsymbol{V}, \boldsymbol{v})(\boldsymbol{W}, \boldsymbol{w})$ is the product of the 4×4 matrices of $(\boldsymbol{V}, \boldsymbol{v})$ and $(\boldsymbol{W}, \boldsymbol{w})$. Let U_{ik} by the element of $\mathbb{U}$ in the ith row and kth column; i is called the row index, k is the column index. Corresponding denominations apply to $\mathbb{V}$ and $\mathbb{W}$. The rules of matrix calculus have to be observed, so in the product $\mathbb{U} = \mathbb{V}\mathbb{W}$ the element U_{ik} is said to be the product of the ith row of $\mathbb{V}$ with the kth column of $\mathbb{W}$:

$$U_{ik} = V_{i1}W_{1k} + V_{i2}W_{2k} + V_{i3}W_{3k} + V_{i4}W_{4k}$$

The sequence of the matrices may not be altered. In addition to the multiplication of two square matrices having equal numbers of rows and columns, it is also possible to multiply a square matrix with a column to its right if it has the same number of rows. The result is a column. In the product $\tilde{\tilde{\mathbb{x}}} = \mathbb{U}\mathbb{x}$ the ith element $\tilde{\tilde{x}}_i$ of the column $\tilde{\tilde{\mathbb{x}}}$ is the product of the ith row of $\mathbb{U}$ with the column $\mathbb{x}$ (of course $\mathbb{x}$, being a column, has only a row index):

$$\tilde{\tilde{x}}_i = U_{i1}x_1 + U_{i2}x_2 + U_{i3}x_3 + U_{i4}x_4$$

In the matrix formalism a column interpreted as a 3×1 or 4×1 matrix.

The calculation of the sequential execution $\tilde{\tilde{\mathbb{x}}} = \mathbb{V}\mathbb{W}\mathbb{x} = \mathbb{U}\mathbb{x}$ of two affine mappings using matrix–column pairs is more complicated. The first mapping then is $\tilde{\boldsymbol{x}} = \boldsymbol{W}\boldsymbol{x} + \boldsymbol{w}$ and the second one is $\tilde{\tilde{\boldsymbol{x}}} = \boldsymbol{V}\tilde{\boldsymbol{x}} + \boldsymbol{v}$. Then we have:

$$\begin{aligned} \tilde{\tilde{\boldsymbol{x}}} &= \boldsymbol{V}(\boldsymbol{W}\boldsymbol{x} + \boldsymbol{w}) + \boldsymbol{v} = \boldsymbol{V}\boldsymbol{W}\boldsymbol{x} + \boldsymbol{V}\boldsymbol{w} + \boldsymbol{v} \qquad (3.8) \\ &= \boldsymbol{U}\boldsymbol{x} + \boldsymbol{u} \end{aligned}$$

$$\text{with} \quad \boldsymbol{U} = \boldsymbol{V}\boldsymbol{W} \quad \text{and} \quad \boldsymbol{u} = \boldsymbol{V}\boldsymbol{w} + \boldsymbol{v} \tag{3.9}$$

Due to the row '0 0 0 1', the calculation with 4×4 matrices yields the same results. Another advantage of the $4{\times}4$ matrices is discussed in the next section.

The mapping with the 4×4 unit matrix $\mathbb{I}$ is identical to the mapping $(\boldsymbol{I}, \boldsymbol{o})$ with the unit matrix $\boldsymbol{I}$ and the zero column $\boldsymbol{o}$. It is the mapping of every point onto itself and is called the *identity mapping*:

$$\mathbb{I} = \left(\begin{array}{ccc|c} 1 & 0 & 0 & 0 \\ 0 & 1 & 0 & 0 \\ 0 & 0 & 1 & 0 \\ \hline 0 & 0 & 0 & 1 \end{array}\right)$$

$$\boldsymbol{I} = \begin{pmatrix} 1 & 0 & 0 \\ 0 & 1 & 0 \\ 0 & 0 & 1 \end{pmatrix} \qquad \boldsymbol{o} = \begin{pmatrix} 0 \\ 0 \\ 0 \end{pmatrix}$$

$$\tilde{\boldsymbol{x}} = \boldsymbol{I}\boldsymbol{x} \qquad \tilde{\mathbb{x}} = \mathbb{I}\mathbb{x}$$

A *translation* by the vector **t** is represented by the mapping $(\boldsymbol{I}, \boldsymbol{t})$ or $\mathbb{T}$ with $\boldsymbol{t} \neq \boldsymbol{o}$:

$$\boldsymbol{t} = \begin{pmatrix} t_1 \\ t_2 \\ t_3 \end{pmatrix} \qquad \mathbb{T} = \left(\begin{array}{ccc|c} 1 & 0 & 0 & t_1 \\ 0 & 1 & 0 & t_2 \\ 0 & 0 & 1 & t_3 \\ \hline 0 & 0 & 0 & 1 \end{array}\right) \qquad \tilde{\mathbb{x}} = \mathbb{T}\mathbb{x} \tag{3.10}$$

A mapping $\mathbb{V} = \mathbb{W}^{-1}$ or $(\boldsymbol{V}, \boldsymbol{v}) = (\boldsymbol{W}, \boldsymbol{w})^{-1}$ that reverses a mapping $\mathbb{W}$ is called the *inverse mapping* of $\mathbb{W}$, such that:

$$\mathbb{V}\mathbb{W} = \mathbb{W}^{-1}\mathbb{W} = \mathbb{I} \quad \text{or} \quad (\boldsymbol{V}, \boldsymbol{v})(\boldsymbol{W}, \boldsymbol{w}) = (\boldsymbol{W}, \boldsymbol{w})^{-1}(\boldsymbol{W}, \boldsymbol{w}) = (\boldsymbol{I}, \boldsymbol{o}) \tag{3.11}$$

In more detail eqn. (3.11) reads:

$$\mathbb{V}\mathbb{W} = \left(\begin{array}{ccc|c} & \boldsymbol{V} & & \boldsymbol{v} \\ \hline 0 & 0 & 0 & 1 \end{array}\right)\left(\begin{array}{ccc|c} & \boldsymbol{W} & & \boldsymbol{w} \\ \hline 0 & 0 & 0 & 1 \end{array}\right) = \left(\begin{array}{ccc|c} & \boldsymbol{I} & & \boldsymbol{o} \\ \hline 0 & 0 & 0 & 1 \end{array}\right)$$

Due to the zeros in the fourth row we obtain $\boldsymbol{V}\boldsymbol{W} = \boldsymbol{I}$ or $\boldsymbol{V} = \boldsymbol{W}^{-1}$. However, at the bottom of the fourth column of $\mathbb{W}$ there is a 1, resulting in $\boldsymbol{V}\boldsymbol{w} + \boldsymbol{v} = \boldsymbol{o}$ for the fourth column of $\mathbb{V}\mathbb{W}$. Therefore, $\boldsymbol{v} = -\boldsymbol{V}\boldsymbol{w} = -\boldsymbol{W}^{-1}\boldsymbol{w}$ and:

$$\mathbb{W}^{-1} = \left(\begin{array}{ccc|c} & \boldsymbol{W}^{-1} & & -\boldsymbol{W}^{-1}\boldsymbol{w} \\ \hline 0 & 0 & 0 & 1 \end{array}\right) \tag{3.12}$$

$\mathbb{W}$ can be inverted if $\boldsymbol{W}^{-1}$ exists, that is if $\det(\boldsymbol{W}) \neq 0$ holds. These mappings are called regular mappings or non-singular mappings. If $\det(\boldsymbol{W}) = W = 0$, the volume of the image is zero; this is called a *projection*. Projections are not considered in the following. The volume of the image of a body, for example of a unit cell, is W times as large as the original volume, with $\det(\boldsymbol{W}) = W \neq 0$.

A change of the coordinate system generally causes a change of the matrix and the column of an affine mapping. Thereby the matrix $\boldsymbol{W}$ depends only on the change of the basis, whereas the column $\boldsymbol{w}$ depends on the basis *and* the origin change of the coordinate system, see Section 3.6. The determinant $\det(\boldsymbol{W})$ and the trace $\mathrm{tr}(\boldsymbol{W}) = W_{11} + W_{22}$ (in the plane) or $\mathrm{tr}(\boldsymbol{W}) = W_{11} + W_{22} + W_{33}$ (in space) are *independent* of the choice of the coordinate system. This is also valid for the order k; that is the smallest positive integral number, for which $\boldsymbol{W}^k = \boldsymbol{I}$ holds. $\boldsymbol{W}^k$ is the product of k matrices. The determinant, trace, and order are the *invariants of a mapping*.

$$\boldsymbol{W}^k = \underbrace{\boldsymbol{W} \cdot \boldsymbol{W} \cdot \ldots \cdot \boldsymbol{W}}_{k \text{ times}} = \boldsymbol{I} \tag{3.13}$$

Definition 3.4 A point X_F that is mapped onto itself is called a *fixed point* of the mapping.

Fixed points of affine mappings can be obtained from the equation:

$$\tilde{\mathbb{x}}_F = \mathbb{x}_F = \mathbb{W}\mathbb{x}_F \tag{3.14}$$

Let $\boldsymbol{W}^k = \boldsymbol{I}$. If $(\boldsymbol{W}, \boldsymbol{w})^k = (\boldsymbol{I}, \boldsymbol{o})$ or $\mathbb{W}^k = \mathbb{I}$, the mapping has at least one fixed point. In this case, the number of points X_i with the coordinate columns

$$\mathbb{W}\mathbb{x},\ \mathbb{W}^2\mathbb{x},\ \ldots,\ \mathbb{W}^{k-1}\mathbb{x},\ \mathbb{W}^k = \mathbb{I}\mathbb{x} = \mathbb{x}$$

is finite, and a finite number of points always has its centre of gravity as a fixed point. If $(\boldsymbol{W},\boldsymbol{w})^k = (\boldsymbol{I},\boldsymbol{t})$ or $\mathbb{W}^k = \mathbb{T}$ with $\boldsymbol{t} \neq \boldsymbol{o}$, the mapping has no fixed point.

An affine mapping $(\boldsymbol{W},\boldsymbol{o})$ does not include a translation and leaves the origin unchanged, because $\tilde{\boldsymbol{o}} = \boldsymbol{W}\boldsymbol{o} + \boldsymbol{o} = \boldsymbol{o}$. Every affine mapping $(\boldsymbol{W},\boldsymbol{w})$ can be composed by the successive execution of a mapping $(\boldsymbol{W},\boldsymbol{o})$ and a translation $(\boldsymbol{I},\boldsymbol{w})$:

$$(\boldsymbol{W},\boldsymbol{w}) = (\boldsymbol{I},\boldsymbol{w})(\boldsymbol{W},\boldsymbol{o}) \tag{3.15}$$

The mapping represented by $(\boldsymbol{W},\boldsymbol{o})$ is called the *linear part*. $(\boldsymbol{I},\boldsymbol{w})$ represents the *translation part*.

3.3 Affine mappings of vectors

The mapping of a point P onto the point $\tilde{P}$ by a translation with the translation vector **t**, $\tilde{\boldsymbol{x}} = \boldsymbol{x} + \boldsymbol{t}$ causes a change of its coordinates. Of course, a translation of *two* points P and Q by the same translation leaves their mutual distance invariant. Point coordinates and vector coefficients are represented by three-row columns that cannot be distinguished. Therefore, some kind of distinctive mark is desirable. It is an advantage of the augmented columns to possess this mark.

In a given coordinate system, let $\mathbb{x}_p$ and $\mathbb{x}_q$ be the augmented columns:

$$\mathbb{x}_p = \begin{pmatrix} x_p \\ y_p \\ z_p \\ \hline 1 \end{pmatrix} \quad \text{and} \quad \mathbb{x}_q = \begin{pmatrix} x_q \\ y_q \\ z_q \\ \hline 1 \end{pmatrix}$$

Then the distance vector has the augmented column with the vector coefficients:

$$\mathbb{r} = \mathbb{x}_q - \mathbb{x}_p = \begin{pmatrix} x_q - x_p \\ y_q - y_p \\ z_q - z_p \\ \hline 0 \end{pmatrix}$$

It has a 0 in its last row since $1 - 1 = 0$. Columns of vector coefficients are thus augmented in a different way to columns of point coordinates.

A translation is represented by $(\boldsymbol{I},\boldsymbol{t})$ or $\mathbb{T}$. Let $\boldsymbol{r}$ be the column of coefficients of a distance vector and let $\mathbb{r}$ be its augmented column. Using 4×4 matrices, the distance vector becomes:

$$\tilde{\mathbb{r}} = \mathbb{T}\mathbb{r} \quad \text{or} \quad \begin{pmatrix} \tilde{r}_1 \\ \tilde{r}_2 \\ \tilde{r}_3 \\ \hline 0 \end{pmatrix} = \left(\begin{array}{ccc|c} & & & t_1 \\ & \boldsymbol{I} & & t_2 \\ & & & t_3 \\ \hline 0 & 0 & 0 & 1 \end{array}\right) \begin{pmatrix} r_1 \\ r_2 \\ r_3 \\ \hline 0 \end{pmatrix} = \begin{pmatrix} r_1 \\ r_2 \\ r_3 \\ \hline 0 \end{pmatrix} \tag{3.16}$$

By the matrix multiplication, the coefficients of $\boldsymbol{t}$ are multiplied with the 0 of the column $\mathbb{r}$ and thus become ineffective.

This is valid not only for translations, but for all affine mappings of vectors:

$$\tilde{\mathbb{r}} = \mathbb{W}\mathbb{r} \quad \text{and} \quad \boldsymbol{W}\boldsymbol{r} + 0 \cdot \boldsymbol{w} = \boldsymbol{W}\boldsymbol{r} \tag{3.17}$$

Theorem 3.5 Whereas point coordinates are transformed by $\tilde{\boldsymbol{x}} = (\boldsymbol{W}, \boldsymbol{w})\boldsymbol{x} = \boldsymbol{W}\boldsymbol{x} + \boldsymbol{w}$, the vector coefficients $\boldsymbol{r}$ are affected only by the matrix part of $\boldsymbol{W}$:

$$\tilde{\boldsymbol{r}} = (\boldsymbol{W}, \boldsymbol{w})\boldsymbol{r} = \boldsymbol{W}\boldsymbol{r}$$

The column part $\boldsymbol{w}$ is ineffective.

This is also valid for other kinds of vectors, for example for the basis vectors of the coordinate system.

Consequence: if $(\boldsymbol{W}, \boldsymbol{w})$ represents an affine mapping in point space, $\boldsymbol{W}$ represents the corresponding mapping in vector space.[1]

[1] In mathematics, 'space' has a different meaning from in everyday life. 'Point space' is the space of points; 'vector space' is the space of vectors. In point space every point has its positional coordinates. Vector space can be imagined as a collection of arrows (vectors), all of which start at a common origin, each one with a length and a direction. However, vectors are independent of their location and can be shifted arbitrarily in parallel; they do not need to start at the origin. A point as well as a vector is specified by a triplet of numbers; however, there exist mathematical operations between vectors, but not between points. For example, vectors can be added or multiplied, but not points.
The space of everyday life is the Euclidean space, where Euclidean geometry holds and where every body has its place and dimensions at a given time. In affine space, bodies can have an affine distortion, and lengths and angles are arbitrary.

3.4 Isometries

An affine mapping that leaves all distances and angles unchanged is called an *isometry*. Isometries are special affine mappings that cause no distortions of any object. In crystallography, in principle, they are more important than general affine mappings. However, we have discussed the more general class of affine mappings first, the mathematical formalism being the same.

If the image of a body is not distorted as compared to the initial body, then it also occupies the same volume. The change of volume under a mapping is expressed by the determinant $\det(\boldsymbol{W})$ of the matrix of the mapping $\boldsymbol{W}$. Therefore, for isometries we have:

$$\det(\boldsymbol{W}) = \pm 1$$

This condition, however, is not sufficient. In addition, all lattice parameters have to be retained. This means that the metric tensor $\boldsymbol{G}$ (Definiton 2.9, page 18) has to remain unchanged.

Because of $(\boldsymbol{W}, \boldsymbol{w}) = (\boldsymbol{I}, \boldsymbol{w})(\boldsymbol{W}, \boldsymbol{o})$, the matrix $\boldsymbol{W}$ alone decides whether the mapping is an isometry or not; $(\boldsymbol{I}, \boldsymbol{w})$ always represents a translation, and this is an isometry.

Let $\boldsymbol{W}$ be the matrix of an isometry $(\boldsymbol{W}, \boldsymbol{w})$ and let $\mathbf{a}_1, \mathbf{a}_2, \mathbf{a}_3$ be the basis of the coordinate system.

It is convenient to write the basis vectors in a row; the vector $\mathbf{x}$ can then be formulated as the matrix product of the row of the basis vectors with the column of the coefficients:

$$(\mathbf{a}_1, \mathbf{a}_2, \mathbf{a}_3)\begin{pmatrix} x_1 \\ x_2 \\ x_3 \end{pmatrix} = x_1\mathbf{a}_1 + x_2\mathbf{a}_2 + x_3\mathbf{a}_3 = \mathbf{x} \tag{3.18}$$

In the matrix formalism a row is interpreted as a 1×3 matrix, a column as a 3×1 matrix. In crystallography, elements of such matrices are designated by lower-case letters. A matrix $\boldsymbol{W}$ that has been reflected through its main diagonal line W_{11}, W_{22}, W_{33} is called the *transpose* of $\boldsymbol{W}$, designated $\boldsymbol{W}^{\mathrm{T}}$; if W_{ik} is an element of $\boldsymbol{W}$, then this element is W_{ki} of $\boldsymbol{W}^{\mathrm{T}}$.

$$\boldsymbol{W} = \begin{pmatrix} W_{11} & W_{12} & W_{13} \\ W_{21} & W_{22} & W_{23} \\ W_{31} & W_{32} & W_{33} \end{pmatrix}$$

$$\boldsymbol{W}^{\mathrm{T}} = \begin{pmatrix} W_{11} & W_{21} & W_{31} \\ W_{12} & W_{22} & W_{32} \\ W_{13} & W_{23} & W_{33} \end{pmatrix}$$

Consider the images $\tilde{\mathbf{a}}_i$ of the basis vectors $\mathbf{a}_i$ under the isometry represented by $(\boldsymbol{W}, \boldsymbol{w})$. According to Theorem 3.5, vectors are transformed only by $\boldsymbol{W}$:

$$\tilde{\mathbf{a}}_i = \mathbf{a}_1 W_{1i} + \mathbf{a}_2 W_{2i} + \mathbf{a}_3 W_{3i} \quad \text{or} \quad \tilde{\mathbf{a}}_i = \sum_{k=1}^{3} \mathbf{a}_k W_{ki}.$$

In matrix notation this is:

$$(\tilde{\mathbf{a}}_1, \tilde{\mathbf{a}}_2, \tilde{\mathbf{a}}_3) = (\mathbf{a}_1, \mathbf{a}_2, \mathbf{a}_3) \begin{pmatrix} W_{11} & W_{12} & W_{13} \\ W_{21} & W_{22} & W_{23} \\ W_{31} & W_{32} & W_{33} \end{pmatrix}$$

$$\text{or} \quad (\tilde{\mathbf{a}}_1, \tilde{\mathbf{a}}_2, \tilde{\mathbf{a}}_3) = (\mathbf{a}_1, \mathbf{a}_2, \mathbf{a}_3)\boldsymbol{W} \tag{3.19}$$

Note that this multiplication, contrary to the multiplication of a matrix with a column, is performed by multiplying the $\mathbf{a}_k$s by the elements of a *column* of $\boldsymbol{W}$, that is for each basis vector of the image $\tilde{\mathbf{a}}_i$ the column index i of $\boldsymbol{W}$ is constant.

The scalar product (dot product) of two image vectors is:

$$\begin{aligned} \tilde{\mathbf{a}}_i \cdot \tilde{\mathbf{a}}_k &= \tilde{g}_{ik} = \left(\sum_{m=1}^{3} \mathbf{a}_m W_{mi}\right) \cdot \left(\sum_{n=1}^{3} \mathbf{a}_n W_{nk}\right) = \\ &= \sum_{m,n=1}^{3} \mathbf{a}_m \cdot \mathbf{a}_n \, W_{mi} W_{nk} = \sum_{m,n=1}^{3} g_{mn} W_{mi} W_{nk} \end{aligned} \tag{3.20}$$

$$\tilde{\boldsymbol{G}} = \boldsymbol{W}^{\mathrm{T}} \boldsymbol{G} \boldsymbol{W} \tag{3.21}$$

Condition for an isometry:

$$\boldsymbol{G} = \boldsymbol{W}^{\mathrm{T}} \boldsymbol{G} \boldsymbol{W} \tag{3.22}$$

This can be written in matrix form if we take the transpose of $\boldsymbol{W}$, eqn (3.21). An isometry may not change the lattice parameters, therefore, $\tilde{\boldsymbol{G}} = \boldsymbol{G}$. From this follows eqn (3.22), which is the necessary and sufficient condition for $(\boldsymbol{W}, \boldsymbol{w})$ to represent an isometry. Equation (3.22) serves to find out whether a mapping represented by the matrix $\boldsymbol{W}$ in the given basis $(\mathbf{a}_1, \mathbf{a}_2, \mathbf{a}_3)$ is an isometry.

Example 3.1

Are the mappings W_1 and W_2 isometries if they refer to a hexagonal basis ($a = b \neq c, \alpha = \beta = 90°, \gamma = 120°$) and if they are represented by the matrix parts $\boldsymbol{W}_1$ and $\boldsymbol{W}_2$?

$$\boldsymbol{W}_1 = \begin{pmatrix} \bar{1} & 0 & 0 \\ 0 & 1 & 0 \\ 0 & 0 & 1 \end{pmatrix} \qquad \boldsymbol{W}_2 = \begin{pmatrix} 1 & \bar{1} & 0 \\ 1 & 0 & 0 \\ 0 & 0 & 1 \end{pmatrix}$$

The metric tensor of the hexagonal basis is:

$$\boldsymbol{G} = \begin{pmatrix} a^2 & -a^2/2 & 0 \\ -a^2/2 & a^2 & 0 \\ 0 & 0 & c^2 \end{pmatrix}$$

According to eqn (3.20), for $\boldsymbol{W}_1$ we have:

$$\begin{aligned} \tilde{g}_{11} &= g_{11}(-1)(-1) + g_{12}(-1)(0) + g_{13}(-1)(0) \\ &\quad + g_{21}(0)(-1) + g_{22}(0)(0) + g_{23}(0)(0) \\ &\quad + g_{31}(0)(-1) + g_{32}(0)(0) + g_{33}(0)(0) \\ &= g_{11}; \\ \tilde{g}_{12} &= 0 + g_{12}(-1)(+1) + 0 + 0 + 0 + 0 + 0 + 0 + 0 \\ &= -g_{12} = a^2/2 \neq g_{12}; \ldots \end{aligned}$$

Therefore, $\tilde{\boldsymbol{G}} \neq \boldsymbol{G}$; W_1 does *not* represent an isometry.

For $\boldsymbol{W}_2$ we have:

$$\tilde{\boldsymbol{G}} = \boldsymbol{W}_2^{\mathrm{T}} \boldsymbol{G} \boldsymbol{W}_2 = \begin{pmatrix} 1 & 1 & 0 \\ \bar{1} & 0 & 0 \\ 0 & 0 & 1 \end{pmatrix} \begin{pmatrix} g_{11} & g_{12} & 0 \\ g_{12} & g_{11} & 0 \\ 0 & 0 & g_{33} \end{pmatrix} \begin{pmatrix} 1 & \bar{1} & 0 \\ 1 & 0 & 0 \\ 0 & 0 & 1 \end{pmatrix}$$

This results in the six equations:

$$\tilde{g}_{11} = g_{11} + 2g_{12} + g_{22} = a^2 = g_{11}; \quad \tilde{g}_{12} = -g_{11} - g_{12} = -a^2/2 = g_{12};$$
$$\tilde{g}_{13} = g_{13} + g_{23} = 0 = g_{13}; \quad \tilde{g}_{22} = g_{11} = a^2 = g_{22}$$
$$\tilde{g}_{23} = -g_{13} = 0 = g_{23}; \quad \tilde{g}_{33} = g_{33}$$

Condition (3.22) is fulfilled; W_2 represents an isometry in this basis.

Condition (3.22) becomes rather simple for an orthonormal basis $\boldsymbol{G} = \boldsymbol{I}$ (unit matrix). In this case we obtain the condition:

$$\boldsymbol{W}^{\mathrm{T}} \boldsymbol{I} \boldsymbol{W} = \boldsymbol{I} \quad \text{or} \quad \boldsymbol{W}^{\mathrm{T}} \boldsymbol{W} = \boldsymbol{I} \tag{3.23}$$

This is exactly the condition for an inverse matrix. Theorem 3.6 follows:

Conditions of orthogonality:

$$\sum_{k=1}^{3} W_{ik} W_{mk} = \begin{cases} 1 \text{ for } i = m \\ 0 \text{ for } i \neq m \end{cases}$$

$$\sum_{k=1}^{3} W_{ki} W_{km} = \begin{cases} 1 \text{ for } i = m \\ 0 \text{ for } i \neq m \end{cases}$$

Theorem 3.6 An affine mapping, referred to an orthonormal basis, is an isometry precisely if $\boldsymbol{W}^{\mathrm{T}} = \boldsymbol{W}^{-1}$ holds.

Remark: $\boldsymbol{W}^{\mathrm{T}} = \boldsymbol{W}^{-1}$ are the known conditions of orthogonality.

Example 3.2
If the matrices $\boldsymbol{W}_1$ and $\boldsymbol{W}_2$ of Example 3.1 were to refer to an orthonormal basis, the mapping $\boldsymbol{W}_1$ would be an isometry. Then, $\boldsymbol{W}_1^{\mathrm{T}} = \boldsymbol{W}_1$ holds, and so does $\boldsymbol{W}^{\mathrm{T}} \boldsymbol{W} = \boldsymbol{I}$ due to $\boldsymbol{W}_1^2 = \boldsymbol{I}$. However, for $\boldsymbol{W}_2$ we obtain:

$$\boldsymbol{W}_2^{-1} = \begin{pmatrix} 0 & 1 & 0 \\ \bar{1} & 1 & 0 \\ 0 & 0 & 1 \end{pmatrix} \neq \boldsymbol{W}_2^{\mathrm{T}}$$

Therefore, W_2 does not represent an isometry.

Regarded geometrically, the mapping $(\boldsymbol{W}_1, \boldsymbol{o})$ of Example 3.1 is a distorting, shearing reflection in the y–z plane, while $(\boldsymbol{W}_2, \boldsymbol{o})$ is a rotation by 60°. In Example 3.2, $(\boldsymbol{W}_1, \boldsymbol{o})$ represents a reflection through the y–z plane, and $(\boldsymbol{W}_2, \boldsymbol{o})$ represents a complicated distorting mapping.

The question remains, what do mappings mean geometrically? Given a matrix–column pair $(\boldsymbol{W}, \boldsymbol{w})$, referred to a known coordinate system, what is the corresponding type of mapping (rotation, translation, reflection, ...)?

The next section deals with the types of isometries. The question of how to deduce the geometric type of mapping from $(\boldsymbol{W}, \boldsymbol{w})$ is the subject of Section 4.3. Finally, in Section 4.4 it is shown how to determine the pair $(\boldsymbol{W}, \boldsymbol{w})$ for a given isometry in a known coordinate system.

3.5 Types of isometries

In space, the following kinds of isometries are distinguished:[2]

[2] We denominate mappings with upper-case sans-serif fonts like I, T, or W; matrices with bold italic upper-case letters like $\boldsymbol{W}$ or $\boldsymbol{G}$; columns (one-column matrices) with bold italic lower-case letters like $\boldsymbol{w}$ or $\boldsymbol{x}$; and vectors with bold lower-case letters like $\mathbf{w}$ or $\mathbf{x}$. See the list of symbols after the Table of Contents.

1. The identical mapping or *identity*, 1 or I, $\tilde{\boldsymbol{x}} = \boldsymbol{x}$ for all points. For the identity we have $\boldsymbol{W} = \boldsymbol{I}$ (unit matrix) and $\boldsymbol{w} = \boldsymbol{o}$ (zero column):

$$\tilde{\boldsymbol{x}} = \boldsymbol{I}\boldsymbol{x} + \boldsymbol{o} = \boldsymbol{x}$$

Every point is a fixed point.

2. *Translations* T, $\tilde{\boldsymbol{x}} = \boldsymbol{x} + \boldsymbol{w}$ (Fig. 3.2). For translations $\boldsymbol{W} = \boldsymbol{I}$ also holds. Therefore, the identity I can be regarded as a special translation with $\boldsymbol{w} = \boldsymbol{o}$. There is no fixed point for $\boldsymbol{w} \neq \boldsymbol{o}$, because the equation $\tilde{\boldsymbol{x}} = \boldsymbol{x} = \boldsymbol{x} + \boldsymbol{w}$ has no solution.

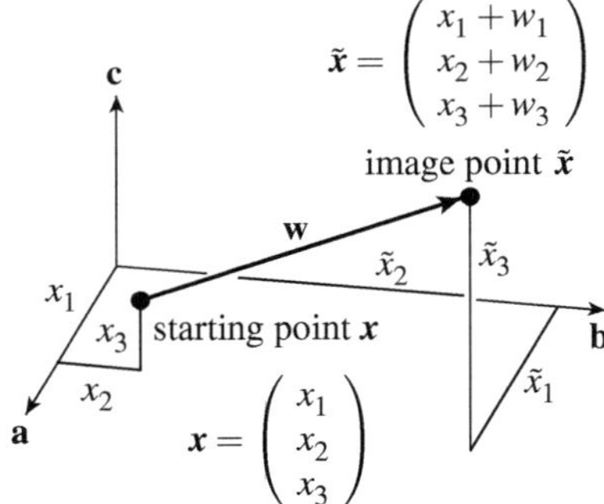

Fig. 3.2 A translation.

3. *Rotations* and *screw rotations*:

$$\tilde{\boldsymbol{x}} = \boldsymbol{W}\boldsymbol{x} + \boldsymbol{w} \quad \text{with} \quad \det(\boldsymbol{W}) = +1$$

In a Cartesian coordinate system and rotation about the $\mathbf{c}$ axis, $(\boldsymbol{W}, \boldsymbol{w})$ has the form:

$$\mathbb{W} = \left(\begin{array}{ccc|c} \cos\varphi & -\sin\varphi & 0 & 0 \\ \sin\varphi & \cos\varphi & 0 & 0 \\ 0 & 0 & +1 & w_3' \\ \hline 0 & 0 & 0 & 1 \end{array}\right)$$

The angle of rotation φ follows from the *trace* $\mathrm{tr}(\boldsymbol{W})$ of the 3×3 matrix $\boldsymbol{W}$ (i.e. from the sum of the main diagonal elements):

$$1 + 2\cos\varphi = \mathrm{tr}(\boldsymbol{W}) = W_{11} + W_{22} + W_{33} \tag{3.24}$$
$$\text{or} \qquad \cos\varphi = \frac{\mathrm{tr}(\boldsymbol{W}) - 1}{2}$$

(a) If $w_3' = 0$, the isometry is called a rotation R (Fig. 3.3).
Every rotation through an *angle of rotation* $\varphi \neq 0°$ has exactly one straight line of fixed points $\boldsymbol{u}$, the *rotation axis*. Its direction is obtained by solving the equation:

$$\boldsymbol{W}\boldsymbol{u} = \boldsymbol{u}$$

The *order* N of a rotation follows from

$$\varphi = \frac{360°}{N} j$$

with $j < N$ and j, N integral with no common divisor. The symbol of such a rotation is N^j. The order k of the mapping is $k = N$. The identity can be regarded as a rotation with $\varphi = 0°$.

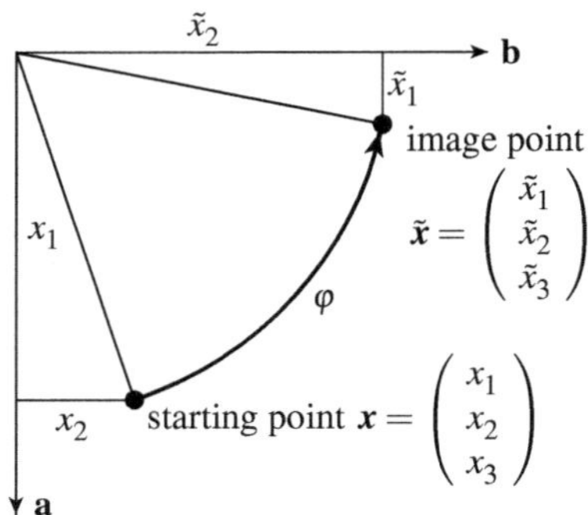

Fig. 3.3 A rotation about the $\mathbf{c}$ axis (direction of view); $\tilde{x}_3 = x_3$. For a screw rotation, the rotation is followed by a shift w_3' parallel to $\mathbf{c}$ (i.e. $\tilde{x}_3 = x_3 + w_3'$).

(b) If $w_3' \neq 0$, the isometry is called a *screw rotation* (Fig. 3.3). It can always be regarded as a rotation R by the angle φ, coupled with a translation T parallel to the rotation axis:

$$\mathbb{W} = \left(\begin{array}{ccc|c} \cos\varphi & -\sin\varphi & 0 & 0 \\ \sin\varphi & \cos\varphi & 0 & 0 \\ 0 & 0 & 1 & 0 \\ \hline 0 & 0 & 0 & 1 \end{array}\right) \left(\begin{array}{ccc|c} 1 & 0 & 0 & 0 \\ 0 & 1 & 0 & 0 \\ 0 & 0 & 1 & w_3' \\ \hline 0 & 0 & 0 & 1 \end{array}\right)$$

The angle φ is the angle of the screw rotation. The order N of R is the *order of the screw axis*. However, the order k of the symmetry operation (i.e. of the screw rotation), is always infinite because the N-fold execution does not result in the identity, but in a translation. A screw rotation has no fixed point.

4. An *inversion* $\bar{1}$ or $\bar{\mathsf{I}}$ is an isometry with $\boldsymbol{W} = -\boldsymbol{I}$: $\tilde{\boldsymbol{x}} = -\boldsymbol{x} + \boldsymbol{w}$.
Geometrically it is a 'reflection' of space through a point with the coordinates $\frac{1}{2}\boldsymbol{w}$. This one point is the only fixed point. The fixed point is called a *point of inversion, centre of symmetry*, or *inversion centre*. The relations $\boldsymbol{W}^2 = \boldsymbol{I}$ and $\bar{1} \times \bar{1} = \bar{1}^2 = 1$ hold.
In space, $\det(-\boldsymbol{I}) = (-1)^3 = -1$ holds; therefore, the inversion is a special kind of rotoinversion. However, in the plane $\det(-\boldsymbol{I}) = (-1)^2 = +1$ holds, which corresponds to a rotation with $2\cos\varphi = -2$ and $\varphi = 180°$.

5. An isometry of space with $\det(\boldsymbol{W}) = -1$ is called a *rotoinversion* $\bar{\mathsf{R}}$. It can always be interpreted as a coupling of a rotation R and an inversion $\bar{\mathsf{I}}$: $\bar{\mathsf{R}} = \bar{\mathsf{I}}\mathsf{R} = \mathsf{R}\bar{\mathsf{I}}$ (Fig. 3.4). The angle of rotation φ of R and thus of $\bar{\mathsf{R}}$ is calculated from

$$\mathrm{tr}(\boldsymbol{W}) = -(1 + 2\cos\varphi)$$

Special cases are:

- The inversion $\bar{\mathsf{I}}$ as a coupling of the inversion with a rotation of 0° (or 360°).
- The reflection m through a plane as a coupling of $\bar{\mathsf{I}}$ with a rotation through 180°; the reflection is dealt with in item 6 of this list.

With the exception of the reflection, all rotoinversions have exactly one fixed point. The corresponding axis is called the axis of rotoinversion. This axis is mapped onto itself as a whole; it runs through the fixed point, where it is reflected. The rotoinversion is called *N-fold*, if the corresponding rotation R has the order N. If N is even, then the order of the rotoinversion is $k = N$, just as in the case of rotations. If N is odd, the order is $k = 2N$: For $\bar{\mathsf{I}}$ the order is $k = 2$; for $\bar{3}$ it is $k = 6$, because $\bar{3}^3$ is not I, but $\bar{\mathsf{I}}$.

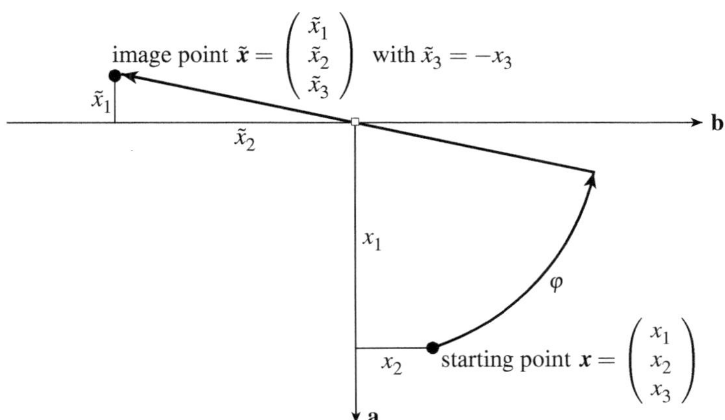

Fig. 3.4 A rotoinversion about the **c** axis (direction of view) with fixed point at the origin □. Shown is the coupling $\mathsf{R}\bar{\mathsf{I}}$ (i.e. a rotation immediately followed by an inversion); the same results from a coupling $\bar{\mathsf{I}}\mathsf{R}$.

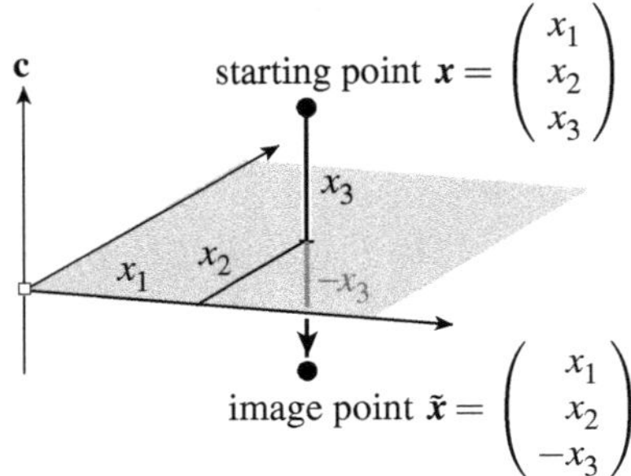

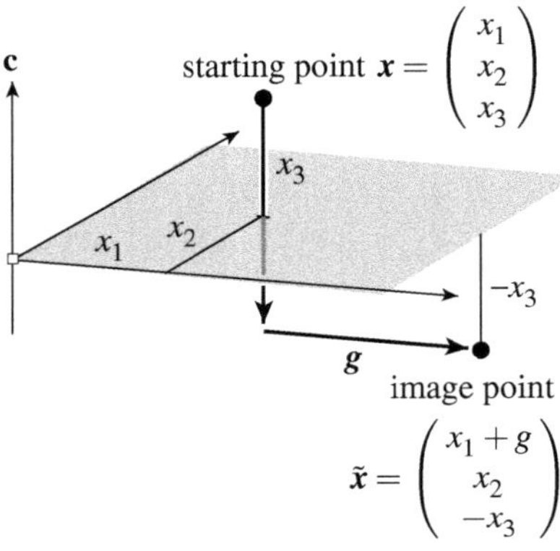

Fig. 3.5 A reflection and a glide reflection with the mirror plane and the glide plane being perpendicular to **c** and running through the origin □.

6. *Reflections* and *glide reflections*. In a Cartesian coordinate system and rotation about the **c** axis, the matrix $\boldsymbol{W}$ of a twofold rotoinversion is:

$$\boldsymbol{W} = \begin{pmatrix} 1 & 0 & 0 \\ 0 & 1 & 0 \\ 0 & 0 & \bar{1} \end{pmatrix}$$

Therefore, $\det(\boldsymbol{W}) = -1$, $\mathrm{tr}(\boldsymbol{W}) = 1$, $\boldsymbol{W}^2 = \boldsymbol{I}$, and $\boldsymbol{W} \neq \bar{\boldsymbol{I}}$. If it is performed twice, the result is a translation:

$$(\boldsymbol{W}, \boldsymbol{w})^2 = (\boldsymbol{W}^2, \boldsymbol{W}\boldsymbol{w} + \boldsymbol{w}) = (\boldsymbol{I}, \boldsymbol{t})$$

If $\boldsymbol{t} = \boldsymbol{o}$, the operation is called a *reflection*; for $\boldsymbol{t} \neq \boldsymbol{o}$ it is a *glide reflection* (Fig. 3.5).

(a) Reflection. A reflection leaves all points of a particular plane unchanged, the *mirror plane* or *plane of reflection*. It runs through the point X with $\boldsymbol{x} = \frac{1}{2}\boldsymbol{w}$.

(b) Glide reflection. A glide reflection has no fixed point. However, it has a *glide plane*, which is obtained by the *reduced mapping*:

$$(\boldsymbol{I}, -\boldsymbol{g})(\boldsymbol{W}, \boldsymbol{w}) = (\boldsymbol{W}, \boldsymbol{w} - \boldsymbol{g}) \quad \text{with} \quad \boldsymbol{g} = \tfrac{1}{2}\boldsymbol{t} = \tfrac{1}{2}(\boldsymbol{W}\boldsymbol{w} + \boldsymbol{w})$$

$\boldsymbol{g}$ is the column of coefficients of the glide vector; due to $\boldsymbol{W}\boldsymbol{g} = \boldsymbol{g}$ it is oriented parallel to the glide plane. The point with the coordinates $\boldsymbol{x} = \frac{1}{2}\boldsymbol{w}$ is situated on the glide plane.

Symmetry operations with $\det(\boldsymbol{W}) = -1$ (i.e. inversion, rotoinversion, reflection, and glide reflection) are called *symmetry operations of the second kind*.

3.6 Changes of the coordinate system

Sometimes it is necessary to change the coordinate system:

(1) If the same crystal structure has been described in different coordinate systems, the data of one of them (lattice parameters, atomic coordinates, parameters of thermal motion) have to be transformed to the other coordinate system if they are to be compared. For the comparison of similar crystal structures, the data also have to be referred to analogous coordinate systems, which may require a transformation of the coordinates.

(2) In phase transitions, the phases are often related by symmetry. Commonly, the data of both phases have been documented in conventional settings, but the conventional settings of both may differ from one another. This case requires a change of the coordinate system, if the data of the new phase are to be compared with those of the old phase.

(3) In crystal physics, it is common to use orthonormal bases (e.g. for the determination of the thermal expansion, the dielectric constant, the elasticity, the piezoelectricity). For corresponding crystal-physical calculations, the point coordinates as well as the indices of directions and planes have to be transformed from the conventional bases to orthonormal bases.

In all of these cases, either the origins or the bases or both differ and have to be converted. The corresponding formulae are derived in the following.

3.6.1 Origin shift

Let (cf. Fig. 3.6):

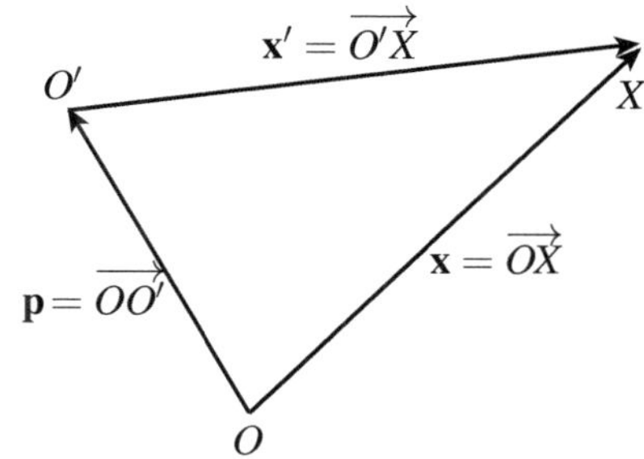

Fig. 3.6 Shift of the origin O to the new origin O'.

O	origin of the old coordinate system
$\boldsymbol{x} = \begin{pmatrix} x \\ y \\ z \end{pmatrix}$	coordinate column of the point X in the old coordinate system
O'	origin of the new coordinate system
$\boldsymbol{x}' = \begin{pmatrix} x' \\ y' \\ z' \end{pmatrix}$	coordinate column of the point X in the new coordinate system

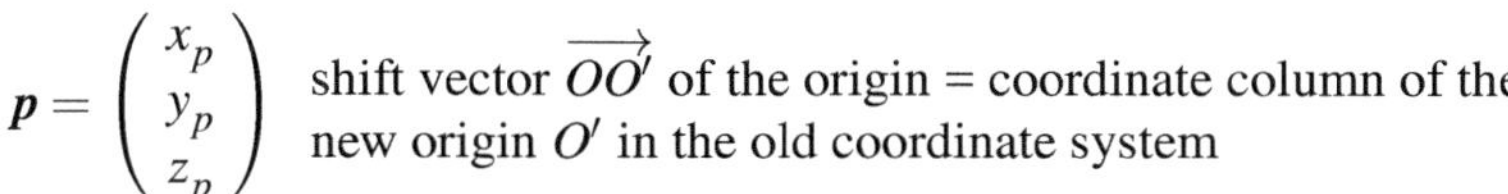

$\boldsymbol{p} = \begin{pmatrix} x_p \\ y_p \\ z_p \end{pmatrix}$ shift vector $\overrightarrow{OO'}$ of the origin = coordinate column of the new origin O' in the old coordinate system

Then we have:

$$\mathbf{x}' = \mathbf{x} - \mathbf{p} \qquad (3.25)$$

Formally, this can be written:

$$\boldsymbol{x}' = (\boldsymbol{I}, -\boldsymbol{p})\boldsymbol{x} = (\boldsymbol{I}, \boldsymbol{p})^{-1}\boldsymbol{x} \qquad (3.26)$$

Using augmented matrices, eqn (3.26) reads:

$$\mathbb{x}' = \mathbb{P}^{-1}\mathbb{x} \quad \text{with} \quad \mathbb{P} = \left(\begin{array}{ccc|c} 1 & 0 & 0 & x_p \\ 0 & 1 & 0 & y_p \\ 0 & 0 & 1 & z_p \\ \hline 0 & 0 & 0 & 1 \end{array}\right) \quad \text{and} \quad \mathbb{P}^{-1} = \left(\begin{array}{ccc|c} 1 & 0 & 0 & -x_p \\ 0 & 1 & 0 & -y_p \\ 0 & 0 & 1 & -z_p \\ \hline 0 & 0 & 0 & 1 \end{array}\right)$$

Written in full, this is:

$$\begin{pmatrix} x' \\ y' \\ z' \\ 1 \end{pmatrix} = \left(\begin{array}{ccc|c} 1 & 0 & 0 & -x_p \\ 0 & 1 & 0 & -y_p \\ 0 & 0 & 1 & -z_p \\ \hline 0 & 0 & 0 & 1 \end{array}\right) \left(\begin{array}{c} x \\ y \\ z \\ \hline 1 \end{array}\right) \qquad (3.27)$$

or $\quad x' = x - x_p, \quad y' = y - y_p, \quad z' = z - z_p.$

An origin shift of (x_p, y_p, z_p) (in the old coordinate system) causes coordinate changes by the same amounts, but with opposite signs.

The transformation $\mathbb{t}' = \mathbb{P}^{-1}\mathbb{t}$ causes no change to a distance vector because the column $\boldsymbol{p}$ remains ineffective due to the zero of $\mathbb{t}$.

$$\mathbb{t} = \left(\begin{array}{c} t_1 \\ t_2 \\ t_3 \\ \hline 0 \end{array}\right)$$

3.6.2 Basis change

A change of the basis is usually specified by a 3×3 matrix $\boldsymbol{P}$ which relates the new basis vectors to the old basis vectors by linear combinations:

$$(\mathbf{a}', \mathbf{b}', \mathbf{c}') = (\mathbf{a}, \mathbf{b}, \mathbf{c})\boldsymbol{P} \quad \text{or} \quad (\mathbf{a}')^{\mathrm{T}} = (\mathbf{a})^{\mathrm{T}}\boldsymbol{P} \qquad (3.28)$$

For a given point X let:

$$\mathbf{a}x + \mathbf{b}y + \mathbf{c}z = \mathbf{a}'x' + \mathbf{b}'y' + \mathbf{c}'z' \quad \text{or for short} \quad (\mathbf{a})^{\mathrm{T}}\boldsymbol{x} = (\mathbf{a}')^{\mathrm{T}}\boldsymbol{x}'$$

By insertion of eqn (3.28) we obtain:

$$(\mathbf{a})^{\mathrm{T}}\boldsymbol{x} = (\mathbf{a})^{\mathrm{T}}\boldsymbol{P}\boldsymbol{x}' \ \text{ or } \ \boldsymbol{x} = \boldsymbol{P}\boldsymbol{x}' \tag{3.29}$$

$$\boldsymbol{x}' = \boldsymbol{P}^{-1}\boldsymbol{x} = (\boldsymbol{P},\boldsymbol{o})^{-1}\boldsymbol{x} \tag{3.30}$$

Equations (3.28), (3.29), and (3.30) show that the matrix $\boldsymbol{P}$ transforms from the old to the new basis vectors, whereas the inverse matrix $\boldsymbol{P}^{-1}$ transforms the coordinates. For the reverse transformation it is the other way around: $\boldsymbol{P}^{-1}$ transforms from the new to the old basis vectors and $\boldsymbol{P}$ from the new to the old coordinates.

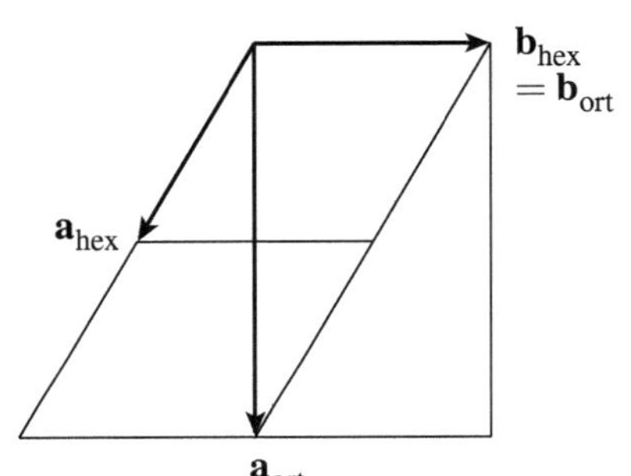

Example 3.3
Consider the transformation from a hexagonal to the corresponding (orthorhombic) *orthohexagonal* basis. According to the figure in the margin, the formulae for the conversion are:

$$\begin{aligned}(\mathbf{a}_{\text{ort}}, \mathbf{b}_{\text{ort}}, \mathbf{c}_{\text{ort}}) &= (\mathbf{a}_{\text{hex}}, \mathbf{b}_{\text{hex}}, \mathbf{c}_{\text{hex}})\boldsymbol{P} \\ &= (\mathbf{a}_{\text{hex}}, \mathbf{b}_{\text{hex}}, \mathbf{c}_{\text{hex}}) \begin{pmatrix} 2 & 0 & 0 \\ 1 & 1 & 0 \\ 0 & 0 & 1 \end{pmatrix} \\ &= (2\mathbf{a}_{\text{hex}} + \mathbf{b}_{\text{hex}}, \mathbf{b}_{\text{hex}}, \mathbf{c}_{\text{hex}})\end{aligned}$$

The coordinates are transformed according to:

$$\boldsymbol{x}_{\text{ort}} = \boldsymbol{P}^{-1}\boldsymbol{x}_{\text{hex}}$$

$$\begin{pmatrix} x_{\text{ort}} \\ y_{\text{ort}} \\ z_{\text{ort}} \end{pmatrix} = \begin{pmatrix} \frac{1}{2} & 0 & 0 \\ -\frac{1}{2} & 1 & 0 \\ 0 & 0 & 1 \end{pmatrix} \begin{pmatrix} x_{\text{hex}} \\ y_{\text{hex}} \\ z_{\text{hex}} \end{pmatrix} = \begin{pmatrix} \frac{1}{2}x_{\text{hex}} \\ -\frac{1}{2}x_{\text{hex}} + y_{\text{hex}} \\ z_{\text{hex}} \end{pmatrix}$$

Instead of inverting $\boldsymbol{P}$ by calculation, $\boldsymbol{P}^{-1}$ can be deduced from the figure by derivation of the matrix for the reverse transformation of the basis vectors from the orthohexagonal to the hexagonal cell.
Note that the components which are to be multiplied with the basis vectors $\mathbf{a}_{\text{hex}}$, $\mathbf{b}_{\text{hex}}$, and $\mathbf{c}_{\text{hex}}$ are mentioned *column by column* in the matrix $\boldsymbol{P}$, whereas the components for the coordinates are mentioned *row by row* in the inverse matrix $\boldsymbol{P}^{-1}$.

3.6.3 General transformation of the coordinate system

Generally both the basis and the origin have to be transformed. The transformation of the basis vectors is independent of origin shifts and eqn (3.28) holds as described in the preceding section. However, for the transformation of the coordinates, in addition to the inverse matrix, the origin shift $\boldsymbol{p}$ has to be taken into account. The most convenient way to do this is to apply augmented matrices. Since the origin shift $\boldsymbol{p}$ is referred to the old basis $(\mathbf{a})^{\mathrm{T}}$, it must be performed first. Therefore, the origin shift is calculated first according to eqn (3.26) and the result thereof is then transformed by multiplication from the right side to the inverse matrix of the basis transformation according to eqn (3.30). The transformation matrix for both steps thus is:

$$\mathbb{P}^{-1} = \left(\begin{array}{ccc|c} & \boldsymbol{P}^{-1} & & \boldsymbol{o} \\ \hline 0 & 0 & 0 & 1 \end{array}\right) \left(\begin{array}{ccc|c} & \boldsymbol{I} & & -\boldsymbol{p} \\ \hline 0 & 0 & 0 & 1 \end{array}\right) = \left(\begin{array}{ccc|c} & \boldsymbol{P}^{-1} & & -\boldsymbol{P}^{-1}\boldsymbol{p} \\ \hline 0 & 0 & 0 & 1 \end{array}\right) \tag{3.31}$$

Comparison with eqn (3.12), page 23, shows that $\mathbb{P}^{-1}$ actually is the inverse of $\mathbb{P}$. Therefore, we have:

$$\mathbb{x}' = \mathbb{P}^{-1}\mathbb{x} \tag{3.32}$$

The column part in eqn (3.31) is:

$$\boldsymbol{p}' = -\boldsymbol{P}^{-1}\boldsymbol{p} \tag{3.33}$$

It corresponds to the position of the old origin in the new coordinate system; this becomes evident by insertion of $\mathbb{x} = (0, 0, 0, 1)^{\mathrm{T}}$ in eqn (3.32).

Example 3.4

Consider the transformation from a cubic to a rhombohedral unit cell with hexagonal axis setting, combined with an origin shift by $\frac{1}{4}, \frac{1}{4}, \frac{1}{4}$ (in the cubic coordinate system). As can be inferred from the figure in the margin, the new basis vectors are:

$\mathbf{a}_{\text{hex}} = \mathbf{a}_{\text{cub}} - \mathbf{b}_{\text{cub}}$, $\mathbf{b}_{\text{hex}} = \mathbf{b}_{\text{cub}} - \mathbf{c}_{\text{cub}}$, $\mathbf{c}_{\text{hex}} = \mathbf{a}_{\text{cub}} + \mathbf{b}_{\text{cub}} + \mathbf{c}_{\text{cub}}$

or

$$(\mathbf{a}_{\text{hex}}, \mathbf{b}_{\text{hex}}, \mathbf{c}_{\text{hex}}) = (\mathbf{a}_{\text{cub}}, \mathbf{b}_{\text{cub}}, \mathbf{c}_{\text{cub}})\,\boldsymbol{P} = (\mathbf{a}_{\text{cub}}, \mathbf{b}_{\text{cub}}, \mathbf{c}_{\text{cub}}) \begin{pmatrix} 1 & 0 & 1 \\ -1 & 1 & 1 \\ 0 & -1 & 1 \end{pmatrix}$$

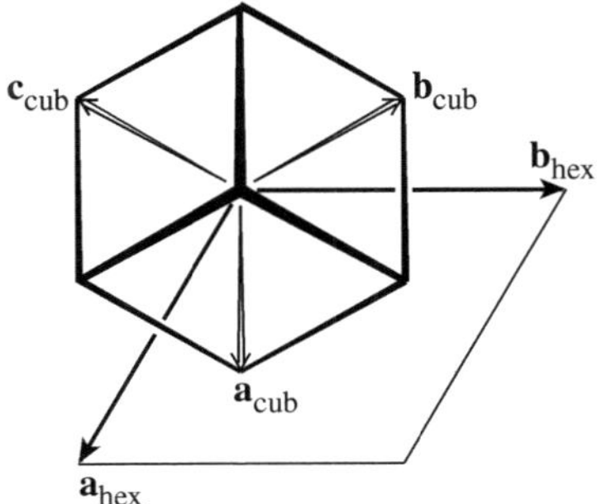

The reverse transformation hexagonal → cubic corresponds to:

$\mathbf{a}_{\text{cub}} = \frac{2}{3}\mathbf{a}_{\text{hex}} + \frac{1}{3}\mathbf{b}_{\text{hex}} + \frac{1}{3}\mathbf{c}_{\text{hex}}$, $\mathbf{b}_{\text{cub}} = -\frac{1}{3}\mathbf{a}_{\text{hex}} + \frac{1}{3}\mathbf{b}_{\text{hex}} + \frac{1}{3}\mathbf{c}_{\text{hex}}$,
$\mathbf{c}_{\text{cub}} = -\frac{1}{3}\mathbf{a}_{\text{hex}} - \frac{2}{3}\mathbf{b}_{\text{hex}} + \frac{1}{3}\mathbf{c}_{\text{hex}}$

or

$$(\mathbf{a}_{\text{cub}}, \mathbf{b}_{\text{cub}}, \mathbf{c}_{\text{cub}}) = (\mathbf{a}_{\text{hex}}, \mathbf{b}_{\text{hex}}, \mathbf{c}_{\text{hex}})\,\boldsymbol{P}^{-1} = (\mathbf{a}_{\text{hex}}, \mathbf{b}_{\text{hex}}, \mathbf{c}_{\text{hex}}) \begin{pmatrix} \frac{2}{3} & -\frac{1}{3} & -\frac{1}{3} \\ \frac{1}{3} & \frac{1}{3} & -\frac{2}{3} \\ \frac{1}{3} & \frac{1}{3} & \frac{1}{3} \end{pmatrix}$$

The column part in eqn (3.31) is:

$$-\boldsymbol{P}^{-1}\boldsymbol{p} = -\begin{pmatrix} \frac{2}{3} & -\frac{1}{3} & -\frac{1}{3} \\ \frac{1}{3} & \frac{1}{3} & -\frac{2}{3} \\ \frac{1}{3} & \frac{1}{3} & \frac{1}{3} \end{pmatrix} \begin{pmatrix} \frac{1}{4} \\ \frac{1}{4} \\ \frac{1}{4} \end{pmatrix} = \begin{pmatrix} 0 \\ 0 \\ -\frac{1}{4} \end{pmatrix}$$

Combined with the origin shift, the new coordinates result from the old ones according to:

$$\mathbb{x}_{\text{hex}} = \mathbb{P}^{-1}\mathbb{x}_{\text{cub}} = \left(\begin{array}{ccc|c} \frac{2}{3} & -\frac{1}{3} & -\frac{1}{3} & 0 \\ \frac{1}{3} & \frac{1}{3} & -\frac{2}{3} & 0 \\ \frac{1}{3} & \frac{1}{3} & \frac{1}{3} & -\frac{1}{4} \\ \hline 0 & 0 & 0 & 1 \end{array}\right) \left(\begin{array}{c} x_{\text{cub}} \\ y_{\text{cub}} \\ z_{\text{cub}} \\ \hline 1 \end{array}\right)$$

$$= \left(\begin{array}{c} \frac{2}{3}x_{\text{cub}} - \frac{1}{3}y_{\text{cub}} - \frac{1}{3}z_{\text{cub}} \\ \frac{1}{3}x_{\text{cub}} + \frac{1}{3}y_{\text{cub}} - \frac{2}{3}z_{\text{cub}} \\ \frac{1}{3}x_{\text{cub}} + \frac{1}{3}x_{\text{cub}} + \frac{1}{3}z_{\text{cub}} - \frac{1}{4} \\ \hline 1 \end{array}\right)$$

Numerical examples: $(0, 0, 0)_{\text{cub}} \to (0, 0, -\frac{1}{4})_{\text{hex}}$;
$(0.54, 0.03, 0.12)_{\text{cub}} \to$
$(\frac{2}{3}\,0.54 - \frac{1}{3}\,0.03 - \frac{1}{3}\,0.12,\ \frac{1}{3}\,0.54 + \frac{1}{3}\,0.03 - \frac{2}{3}\,0.12,$
$\frac{1}{3}\,0.54 + \frac{1}{3}\,0.03 + \frac{1}{3}\,0.12 - \frac{1}{4})_{\text{hex}} = (0.31, 0.11, -0.02)_{\text{hex}}$

3.6.4 The effect of coordinate transformations on mappings

If the coordinate system is changed, the matrix $\mathbb{W}$ of a mapping (isometry) is also changed. For a mapping, the following relations hold according to eqn (3.4), page 21:

$$\tilde{\mathbb{x}} = \mathbb{W}\mathbb{x} \quad \text{in the old coordinate system} \qquad (3.34)$$

$$\text{and} \quad \tilde{\mathbb{x}}' = \mathbb{W}'\mathbb{x}' \quad \text{in the new coordinate system} \qquad (3.35)$$

For a general transformation of the coordinate system (origin shift and change of basis), by substitution of eqn (3.32) into eqn (3.35) we obtain:

$$\tilde{\mathbb{x}}' = \mathbb{W}'\mathbb{P}^{-1}\mathbb{x}$$

The transformation matrix $\mathbb{P}^{-1}$ is valid for all points, including the image point $\tilde{\mathbb{x}}$; therefore, by analogy to eqn (3.32), $\tilde{\mathbb{x}}' = \mathbb{P}^{-1}\tilde{\mathbb{x}}$ holds. From the preceding equation we thus obtain:

$$\mathbb{P}^{-1}\tilde{\mathbb{x}} = \mathbb{W}'\mathbb{P}^{-1}\mathbb{x}$$

Multiplication on the left side with $\mathbb{P}$:

$$\tilde{\mathbb{x}} = \mathbb{P}\mathbb{W}'\mathbb{P}^{-1}\mathbb{x}$$

Comparison with eqn (3.34) yields:

$$\mathbb{W} = \mathbb{P}\mathbb{W}'\mathbb{P}^{-1}$$

Multiplication on the left side with $\mathbb{P}^{-1}$ and on the right side with $\mathbb{P}$:

$$\mathbb{W}' = \mathbb{P}^{-1}\mathbb{W}\mathbb{P} \qquad (3.36)$$

By computation we obtain the matrix and the column parts of $\mathbb{W}'$:

$$\boldsymbol{W}' = \boldsymbol{P}^{-1}\boldsymbol{W}\boldsymbol{P} \qquad (3.37)$$

$$\text{and} \quad \boldsymbol{w}' = -\boldsymbol{P}^{-1}\boldsymbol{p} + \boldsymbol{P}^{-1}\boldsymbol{w}^{-1}\boldsymbol{W}\boldsymbol{p} \qquad (3.38)$$

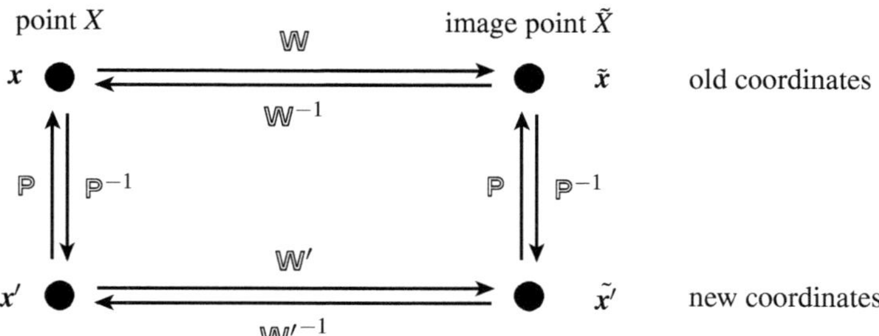

Fig. 3.7 Diagram of the 'mapping of mappings'.

The complete formalism can be depicted as shown in Fig. 3.7. The points X (left) and $\tilde{X}$ (right) are specified by the initial coordinates $\boldsymbol{x}$ and $\tilde{\boldsymbol{x}}$ (top) and the new coordinates $\boldsymbol{x}'$ and $\tilde{\boldsymbol{x}}'$ (bottom). The transformations are given next to the arrows. From left to right this corresponds to a mapping and from top to bottom to a coordinate transformation. Equation (3.36) can be derived directly from the diagram (with $\mathbb{x} = (x, y, z, 1)^{\mathrm{T}}$): On one side, $\tilde{\mathbb{x}}' = \mathbb{W}'\mathbb{x}'$ holds (lower arrow pointing to the right), on the other side we have (taking the detour through the upper part of the image):

$$\tilde{\mathbb{x}}' = \mathbb{P}^{-1}\tilde{\mathbb{x}} = \mathbb{P}^{-1}\mathbb{W}\mathbb{x} = \mathbb{P}^{-1}\mathbb{W}\mathbb{P}\mathbb{x}'$$

Equating both paths results in eqn (3.36).

Origin shift. If only the origin is shifted, the matrix part of the transformation is the unit matrix, $\boldsymbol{P} = \boldsymbol{I}$. Equation (3.36) then becomes:

$$\boldsymbol{W}' = \boldsymbol{I}^{-1}\boldsymbol{W}\boldsymbol{I} = \boldsymbol{W} \tag{3.39}$$

The column part $\boldsymbol{w}'$ of $\mathbb{W}'$ results from eqn (3.38):

$$\boldsymbol{w}' = \boldsymbol{w} + \boldsymbol{W}\boldsymbol{p} - \boldsymbol{p} \tag{3.40}$$

Consequences: An origin shift does not affect the matrix $\boldsymbol{W}$ of a mapping. The change of the column depends not only on $\boldsymbol{p}$, but also on $\boldsymbol{W}$.

Example 3.5
Space group $F\,d\,d\,d$, origin choice 1, has a twofold screw axis at $0, \frac{1}{4}, z$; it maps an atom with the coordinates x, y, z onto a symmetry-equivalent position $-x, \frac{1}{2} - y, \frac{1}{2} + z$. Switching to origin choice 2, the origin must be shifted by $(\frac{1}{8}, \frac{1}{8}, \frac{1}{8})$. What is the new mapping instruction?
The matrix and column parts for origin choice 1 read:

$$\boldsymbol{W} = \begin{pmatrix} -1 & 0 & 0 \\ 0 & -1 & 0 \\ 0 & 0 & 1 \end{pmatrix} \quad \text{and} \quad \boldsymbol{w} = \begin{pmatrix} 0 \\ \frac{1}{2} \\ \frac{1}{2} \end{pmatrix}$$

Combined with the origin shift $\boldsymbol{p}^{\mathrm{T}} = (\frac{1}{8}, \frac{1}{8}, \frac{1}{8})$, we obtain from eqn (3.40):

$$w' = w + Wp - p = \begin{pmatrix} 0 \\ \frac{1}{2} \\ \frac{1}{2} \end{pmatrix} + \begin{pmatrix} -1 & 0 & 0 \\ 0 & -1 & 0 \\ 0 & 0 & 1 \end{pmatrix} \begin{pmatrix} \frac{1}{8} \\ \frac{1}{8} \\ \frac{1}{8} \end{pmatrix} - \begin{pmatrix} \frac{1}{8} \\ \frac{1}{8} \\ \frac{1}{8} \end{pmatrix} = \begin{pmatrix} -\frac{1}{4} \\ \frac{1}{4} \\ \frac{1}{2} \end{pmatrix}$$

Together with the unchanged matrix part, the new mapping for origin choice 2 thus becomes:

$-\frac{1}{4}-x, \frac{1}{4}-y, \frac{1}{2}+z$ or (standardized to $0 \leq w_i' < 1$) $\frac{3}{4}-x, \frac{1}{4}-y, \frac{1}{2}+z$

Transformation only of the basis. If only the basis is transformed, the column part of the transformation consists only of zeros, $\boldsymbol{p} = \boldsymbol{o}$. Equations (3.36) and (3.38) then become:

A transformation of the kind $\boldsymbol{W}' = \boldsymbol{P}^{-1}\boldsymbol{W}\boldsymbol{P}$ is called a *similarity transformation.*

$$\boldsymbol{W}' = \boldsymbol{P}^{-1}\boldsymbol{W}\boldsymbol{P} \tag{3.41}$$

$$\begin{aligned} \boldsymbol{w}' &= -\boldsymbol{P}^{-1}\boldsymbol{o} + \boldsymbol{P}^{-1}\boldsymbol{w} + \boldsymbol{P}^{-1}\boldsymbol{W}\boldsymbol{o} \\ &= \boldsymbol{P}^{-1}\boldsymbol{w} \end{aligned} \tag{3.42}$$

Example 3.6

Consider the transformation of a hexagonal to an orthorhombic cell from Example 3.3. Assume the presence of a glide plane, which maps an atom from the position x, y, z to the position $x, x-y, z+\frac{1}{2}$, referred to the hexagonal coordinate system. What is the conversion formula in the orthorhombic coordinate system?

The mapping $x, x-y, z+\frac{1}{2}$ corresponds to the matrix and column:

$$\boldsymbol{W} = \begin{pmatrix} 1 & 0 & 0 \\ 1 & -1 & 0 \\ 0 & 0 & 1 \end{pmatrix} \qquad \boldsymbol{w} = \begin{pmatrix} 0 \\ 0 \\ \frac{1}{2} \end{pmatrix}$$

The transformation matrices from Example 3.3 are:

$$\boldsymbol{P} = \begin{pmatrix} 2 & 0 & 0 \\ 1 & 1 & 0 \\ 0 & 0 & 1 \end{pmatrix} \quad \text{and} \quad \boldsymbol{P}^{-1} = \begin{pmatrix} \frac{1}{2} & 0 & 0 \\ -\frac{1}{2} & 1 & 0 \\ 0 & 0 & 1 \end{pmatrix}$$

Using eqns (3.41) and (3.42), we calculate for the mapping in the orthorhombic coordinate system:

$$\begin{aligned} \boldsymbol{W}' &= \begin{pmatrix} \frac{1}{2} & 0 & 0 \\ -\frac{1}{2} & 1 & 0 \\ 0 & 0 & 1 \end{pmatrix} \begin{pmatrix} 1 & 0 & 0 \\ 1 & -1 & 0 \\ 0 & 0 & 1 \end{pmatrix} \begin{pmatrix} 2 & 0 & 0 \\ 1 & 1 & 0 \\ 0 & 0 & 1 \end{pmatrix} \\ &= \begin{pmatrix} 1 & 0 & 0 \\ 0 & -1 & 0 \\ 0 & 0 & 1 \end{pmatrix} \\ \boldsymbol{w}' &= \begin{pmatrix} \frac{1}{2} & 0 & 0 \\ -\frac{1}{2} & 1 & 0 \\ 0 & 0 & 1 \end{pmatrix} \begin{pmatrix} 0 \\ 0 \\ \frac{1}{2} \end{pmatrix} = \begin{pmatrix} 0 \\ 0 \\ \frac{1}{2} \end{pmatrix} \end{aligned}$$

This is equivalent to $x, -y, z+\frac{1}{2}$.

3.6.5 Several consecutive transformations of the coordinate system

If several consecutive coordinate transformations are to be performed, the transformation matrices for the overall transformation from the initial to the final coordinate system result by multiplication of the matrices of the single steps. If no origin shifts are involved, it is sufficient to use the 3×3 matrices; otherwise the 4×4 matrices must be taken. In the following we use the 4×4 matrices.

Let $\mathbb{P}_1, \mathbb{P}_2, \ldots$ be the 4×4 matrices for several consecutive transformations and $\mathbb{P}_1^{-1}, \mathbb{P}_2^{-1}, \ldots$ the corresponding inverse matrices. Let **a**, **b**, **c** be the original basis vectors and **a**′, **b**′, **c**′ the new basis vectors after the sequence of transformations. Let $\mathbb{x}$ and $\mathbb{x}'$ be the augmented columns of the coordinates before and after the sequence of transformations. Then the following relations hold:

$$(\mathbf{a}', \mathbf{b}', \mathbf{c}', \boldsymbol{p}) = (\mathbf{a}, \mathbf{b}, \mathbf{c}, 0)\, \mathbb{P}_1 \mathbb{P}_2, \ldots \quad \text{and} \quad \mathbb{x}' = \ldots \mathbb{P}_2^{-1} \mathbb{P}_1^{-1} \mathbb{x}$$

$\boldsymbol{p}$ is the origin shift (i.e. the coordinate column of the new origin in the old coordinate system). Note that the inverse matrices have to be multiplied in reverse order.

Example 3.7

Consider the coordinate transformation cubic → rhombohedral-hexagonal from Example 3.4, followed by another transformation to a monoclinic basis, combined with a second origin shift $\boldsymbol{p}_2^{\mathrm{T}} = (-\frac{1}{2}, -\frac{1}{2}, 0)$ (in the hexagonal coordinate system).

The matrices of the first transformation are (see Example 3.4):

$$\mathbb{P}_1 = \left(\begin{array}{c|c} \boldsymbol{P}_1 & \boldsymbol{p}_1 \\ \hline \boldsymbol{o}^{\mathrm{T}} & 1 \end{array}\right) = \left(\begin{array}{ccc|c} 1 & 0 & 1 & \frac{1}{4} \\ -1 & 1 & 1 & \frac{1}{4} \\ 0 & -1 & 1 & \frac{1}{4} \\ \hline 0 & 0 & 0 & 1 \end{array}\right)$$

and

$$\mathbb{P}_1^{-1} = \left(\begin{array}{c|c} \boldsymbol{P}_1^{-1} & -\boldsymbol{P}_1^{-1}\boldsymbol{p}_1 \\ \hline \boldsymbol{o}^{\mathrm{T}} & 1 \end{array}\right) = \left(\begin{array}{ccc|c} \frac{2}{3} & -\frac{1}{3} & -\frac{1}{3} & 0 \\ \frac{1}{3} & \frac{1}{3} & -\frac{2}{3} & 0 \\ \frac{1}{3} & \frac{1}{3} & \frac{1}{3} & -\frac{1}{4} \\ \hline 0 & 0 & 0 & 1 \end{array}\right)$$

Assume a relation between the cells for the second transformation (rhombohedral-hexagonal → monoclinic) according to the adjacent figure. From the figure we deduce:

$$(\mathbf{a}_{\mathrm{mon}}, \mathbf{b}_{\mathrm{mon}}, \mathbf{c}_{\mathrm{mon}}) = (\mathbf{a}_{\mathrm{hex}}, \mathbf{b}_{\mathrm{hex}}, \mathbf{c}_{\mathrm{hex}})\, \boldsymbol{P}_2$$
$$= (\mathbf{a}_{\mathrm{hex}}, \mathbf{b}_{\mathrm{hex}}, \mathbf{c}_{\mathrm{hex}}) \begin{pmatrix} 2 & 0 & \frac{2}{3} \\ 1 & 1 & \frac{1}{3} \\ 0 & 0 & \frac{1}{3} \end{pmatrix}$$

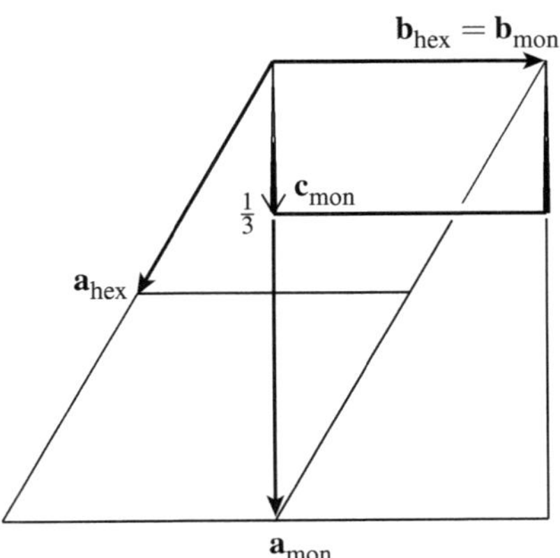

$$(\mathbf{a}_{\text{hex}}, \mathbf{b}_{\text{hex}}, \mathbf{c}_{\text{hex}}) = (\mathbf{a}_{\text{mon}}, \mathbf{b}_{\text{mon}}, \mathbf{c}_{\text{mon}})\, \boldsymbol{P}_2^{-1}$$

$$= (\mathbf{a}_{\text{mon}}, \mathbf{b}_{\text{mon}}, \mathbf{c}_{\text{mon}}) \begin{pmatrix} \frac{1}{2} & 0 & -1 \\ -\frac{1}{2} & 1 & 0 \\ 0 & 0 & 3 \end{pmatrix}$$

$$-\boldsymbol{P}_2^{-1}\boldsymbol{p}_2 = -\begin{pmatrix} \frac{1}{2} & 0 & -1 \\ -\frac{1}{2} & 1 & 0 \\ 0 & 0 & 3 \end{pmatrix} \begin{pmatrix} -\frac{1}{2} \\ -\frac{1}{2} \\ 0 \end{pmatrix} = \begin{pmatrix} \frac{1}{4} \\ \frac{1}{4} \\ 0 \end{pmatrix}$$

$$\mathbb{P}_2^{-1} = \left(\begin{array}{c|c} \boldsymbol{P}_2^{-1} & -\boldsymbol{P}_2^{-1}\boldsymbol{p}_2 \\ \hline \boldsymbol{o}^{\mathrm{T}} & 1 \end{array}\right) = \left(\begin{array}{ccc|c} \frac{1}{2} & 0 & -1 & \frac{1}{4} \\ -\frac{1}{2} & 1 & 0 & \frac{1}{4} \\ 0 & 0 & 3 & 0 \\ \hline 0 & 0 & 0 & 1 \end{array}\right)$$

The transformation of the basis vectors and the origin shift from the original to the final coordinate system resulting from the two consecutive transformations are obtained from:

$$\mathbb{P}_1\mathbb{P}_2 = \left(\begin{array}{ccc|c} 1 & 0 & 1 & \frac{1}{4} \\ -1 & 1 & 1 & \frac{1}{4} \\ 0 & -1 & 1 & \frac{1}{4} \\ \hline 0 & 0 & 0 & 1 \end{array}\right) \left(\begin{array}{ccc|c} 2 & 0 & \frac{2}{3} & -\frac{1}{2} \\ 1 & 1 & \frac{1}{3} & -\frac{1}{2} \\ 0 & 0 & \frac{1}{3} & 0 \\ \hline 0 & 0 & 0 & 1 \end{array}\right) = \left(\begin{array}{ccc|c} 2 & 0 & 1 & -\frac{1}{4} \\ -1 & 1 & 0 & \frac{1}{4} \\ -1 & -1 & 0 & \frac{3}{4} \\ \hline 0 & 0 & 0 & 1 \end{array}\right)$$

$$(\mathbf{a}_{\text{mon}}, \mathbf{b}_{\text{mon}}, \mathbf{c}_{\text{mon}}, \boldsymbol{p}) = (\mathbf{a}_{\text{cub}}, \mathbf{b}_{\text{cub}}, \mathbf{c}_{\text{cub}}, 0)\, \mathbb{P}_1\mathbb{P}_2$$

$$= (\mathbf{a}_{\text{cub}}, \mathbf{b}_{\text{cub}}, \mathbf{c}_{\text{cub}}, 0) \left(\begin{array}{ccc|c} 2 & 0 & 1 & -\frac{1}{4} \\ -1 & 1 & 0 & \frac{1}{4} \\ -1 & -1 & 0 & \frac{3}{4} \\ \hline 0 & 0 & 0 & 1 \end{array}\right)$$

This corresponds to:

$\mathbf{a}_{\text{mon}} = 2\mathbf{a}_{\text{cub}} - \mathbf{b}_{\text{cub}} - \mathbf{c}_{\text{cub}}, \quad \mathbf{b}_{\text{mon}} = \mathbf{b}_{\text{cub}} - \mathbf{c}_{\text{cub}}, \quad \mathbf{c}_{\text{mon}} = \mathbf{a}_{\text{cub}}$

combined with an origin shift of $\boldsymbol{p}^{\mathrm{T}} = (-\frac{1}{4}, \frac{1}{4}, \frac{3}{4})$ in the cubic coordinate system. The corresponding coordinate transformations (cubic $\rightarrow$ monoclinic) are calculated according to:

$$\mathbb{x}_{\text{mon}} = \mathbb{P}_2^{-1}\mathbb{P}_1^{-1}\mathbb{x}_{\text{cub}}$$

$$\begin{pmatrix} x_{\text{mon}} \\ y_{\text{mon}} \\ z_{\text{mon}} \\ \hline 1 \end{pmatrix} = \left(\begin{array}{ccc|c} \frac{1}{2} & 0 & -1 & \frac{1}{4} \\ -\frac{1}{2} & 1 & 0 & \frac{1}{4} \\ 0 & 0 & 3 & 0 \\ \hline 0 & 0 & 0 & 1 \end{array}\right) \left(\begin{array}{ccc|c} \frac{2}{3} & -\frac{1}{3} & -\frac{1}{3} & 0 \\ \frac{1}{3} & \frac{1}{3} & -\frac{2}{3} & 0 \\ \frac{1}{3} & \frac{1}{3} & \frac{1}{3} & -\frac{1}{4} \\ \hline 0 & 0 & 0 & 1 \end{array}\right) \begin{pmatrix} x_{\text{cub}} \\ y_{\text{cub}} \\ z_{\text{cub}} \\ \hline 1 \end{pmatrix}$$

$$= \left(\begin{array}{ccc|c} 0 & -\frac{1}{2} & -\frac{1}{2} & \frac{1}{2} \\ 0 & \frac{1}{2} & -\frac{1}{2} & \frac{1}{4} \\ 1 & 1 & 1 & -\frac{3}{4} \\ \hline 0 & 0 & 0 & 1 \end{array}\right) \begin{pmatrix} x_{\text{cub}} \\ y_{\text{cub}} \\ z_{\text{cub}} \\ \hline 1 \end{pmatrix}$$

This corresponds to:

$x_{\text{mon}} = -\frac{1}{2}y_{\text{cub}} - \frac{1}{2}z_{\text{cub}} + \frac{1}{2}, \quad y_{\text{mon}} = \frac{1}{2}y_{\text{cub}} - \frac{1}{2}z_{\text{cub}} + \frac{1}{4},$
$z_{\text{mon}} = x_{\text{cub}} + y_{\text{cub}} + z_{\text{cub}} - \frac{3}{4}$

3.6.6 Calculation of origin shifts from coordinate transformations

Group–subgroup relations between space groups often involve basis transformations and origin shifts. In *International Tables* A1, Part 2, the origin shifts are listed after the basis transformations in the form of numerical triplets $\boldsymbol{p}^{\mathrm{T}} = (x_p, y_p, z_p)$. These refer to the original coordinate system. In Part 3 of the same tables, the coordinate transformations are given in addition to the basis transformations; however, the origin shifts are only listed together with the coordinate transformations, namely as additive numbers to the coordinate values. These additive numbers are nothing other than the vector coefficients of the shift vector $\boldsymbol{p}'^{\mathrm{T}} = (x'_p, y'_p, z'_p)$ in the *new* coordinate system of the subgroup.

To calculate the corresponding origin shift $\boldsymbol{p}$ in the *initial* coordinate system from the coordinate transformation listed in Part 3 of *International Tables* A1, eqn (3.33), page 33, has to be applied:

$$\boldsymbol{p}' = -\boldsymbol{P}^{-1}\boldsymbol{p} \quad \text{and thus} \quad \boldsymbol{p} = -\boldsymbol{P}\boldsymbol{p}' \tag{3.43}$$

Unfortunately, often there exist several choices for the basis transformation and for the origin shift for the very same group–subgroup relation. In Part 2 of *International Tables* A1 often a different basis transformation and/or a different origin shift have been chosen as compared to in Part 3; this is not always obvious due to the different kinds of presentation. Therefore, where required, $\boldsymbol{p}$ must be calculated from $\boldsymbol{p}'$; one cannot simply look up the values listed in the corresponding space group table in Part 2.[3]

[3] The different choices for the basis transformations and origin shifts in Parts 2 and 3 of *International Tables* A1 are due in part to their history of creation, and, in part to material grounds. The tables were created independently by different authors at different times and were only combined at a late stage. The differences in presentation and the corresponding reasons are explained in the Appendix of *International Tables* A1.

Example 3.8

The group–subgroup relation $P4_2/mbc \to Cccm$ requires a cell transformation and an origin shift. In *International Tables* A1, Part 3, we find the transformation of the basis vectors in the column 'Axes':

$$\mathbf{a}' = \mathbf{a} - \mathbf{b},\ \mathbf{b}' = \mathbf{a} + \mathbf{b},\ \mathbf{c}' = \mathbf{c}$$

The transformation matrix is therefore:

$$\boldsymbol{P} = \begin{pmatrix} 1 & 1 & 0 \\ -1 & 1 & 0 \\ 0 & 0 & 1 \end{pmatrix}$$

In the column 'Coordinates' the listed coordinate transformations are $\frac{1}{2}(x-y)+\frac{1}{4}$, $\frac{1}{2}(x+y)+\frac{1}{4}$, z. Therefore, the origin shift is $\boldsymbol{p}'^{\mathrm{T}} = (\frac{1}{4}, \frac{1}{4}, 0)$ in the coordinate system of $Cccm$. The corresponding origin shift in the coordinate system of $P4_2/mbc$ is:

$$\begin{pmatrix} x_p \\ y_p \\ z_p \end{pmatrix} = \boldsymbol{p} = -\boldsymbol{P}\boldsymbol{p}' = -\begin{pmatrix} 1 & 1 & 0 \\ -1 & 1 & 0 \\ 0 & 0 & 1 \end{pmatrix} \begin{pmatrix} \frac{1}{4} \\ \frac{1}{4} \\ 0 \end{pmatrix} = \begin{pmatrix} -\frac{1}{2} \\ 0 \\ 0 \end{pmatrix}$$

For the same relation $P4_2/mbc \to Cccm$, the same basis transformation is listed in Part 2 of *International Tables* A1, but a different origin shift, $(0, \frac{1}{2}, 0)$.

3.6.7 Transformation of further crystallographic quantities

Any transformation of the basis vectors entails changes for all quantities that depend on the setting of the basis. Without proof, we list some of them. Since all mentioned quantities are vectors or vector coefficients, the changes are independent of any origin shift; the transformations require only the 3×3 matrices $\boldsymbol{P}$ and $\boldsymbol{P}^{-1}$.

$(h',k',l') = (h,k,l)\,\boldsymbol{P}$

The Miller indices h,k,l of lattice planes are transformed in the same way as the basis vectors. The new indices h',k',l' result from the same transformation matrix $\boldsymbol{P}$ as for the basis vectors **a**, **b**, **c**.

$V = abc \times \sqrt{1-\cos^2\alpha-\cos^2\beta-\cos^2\gamma+2\cos\alpha\cos\beta\cos\gamma}$

The reciprocal lattice vectors $\mathbf{a}^*$, $\mathbf{b}^*$, $\mathbf{c}^*$ are oriented perpendicular to the planes (1 0 0), (0 1 0), and (0 0 1). Their lengths are

$$a^* = 1/d_{100} = bc\sin\alpha/V,\ b^* = 1/d_{010} = ac\sin\beta/V,\ c^* = 1/d_{001} = ab\sin\gamma/V$$

$$\begin{pmatrix}\mathbf{a}^{*\prime}\\ \mathbf{b}^{*\prime}\\ \mathbf{c}^{*\prime}\end{pmatrix} = \boldsymbol{P}^{-1}\begin{pmatrix}\mathbf{a}^{*}\\ \mathbf{b}^{*}\\ \mathbf{c}^{*}\end{pmatrix}$$

with V being the volume of the unit cell and d_{100} = distance between adjacent planes (1 0 0). They are transformed in the same way as the coordinates by the inverse matrix $\boldsymbol{P}^{-1}$.

$$\begin{pmatrix}u'\\ v'\\ w'\end{pmatrix} = \boldsymbol{P}^{-1}\begin{pmatrix}u\\ v\\ w\end{pmatrix}$$

The coefficients u,v,w of a translation vector $\mathbf{t} = u\mathbf{a}+v\mathbf{b}+w\mathbf{c}$ are also transformed by the inverse matrix $\boldsymbol{P}^{-1}$.

Exercises

Solutions on page 321.

3.1. Zircon ($ZrSiO_4$), anatase (TiO_2), many rare-earth phosphates, arsenates, vanadates, and others crystallize in the space group $I4_1/amd$, space group number 141. In *International Tables A*, taking origin choice 2, we find among others the following coordinate triplets under the heading 'positions':

(8) $\bar{y}+\frac{1}{4},\bar{x}+\frac{1}{4},\bar{z}+\frac{3}{4}$ (10) $x+\frac{1}{2},y,\bar{z}+\frac{1}{2}$

Formulate these coordinate triplets as:

(a) mappings that transfer the point X with the coordinates x,y,z onto the point $\tilde{X}$ with the coordinates $\tilde{x},\tilde{y},\tilde{z}$;

(b) matrix–column pairs;

(c) 4×4 matrices.

(d) Apply eqn (3.6), page 22, consecutively to the given 4×4 matrices. Does the result depend on the sequence?

(e) Convert the results back to coordinate triplets and compare these with the listing in *International Tables A*, space group $I4_1/amd$, origin choice 2.

3.2. For some physical problem it is necessary to refer to space group $I4_1/amd$ in a primitive basis. This is chosen to be:

$$\mathbf{a}_P = \mathbf{a},\ \mathbf{b}_P = \mathbf{b},\ \mathbf{c}_P = \tfrac{1}{2}(\mathbf{a}+\mathbf{b}+\mathbf{c})$$

If this basis were chosen for the description of the space group in *International Tables A*, the data of *International Tables A*, origin choice 2, would have to be changed, as described in Section 3.6.

(a) What is the matrix of the basis transformation?

(b) How do the point coordinates transform?

(c) The symmetry operations (8) and (10) are mentioned in Exercise 3.1; symmetry operation (15) is $\bar{y}+\frac{3}{4},\bar{x}+\frac{1}{4},z+\frac{3}{4}$. What do the symmetry operations (8), (10), (15), and (15) + $(\frac{1}{2},\frac{1}{2},\frac{1}{2})$ look like in the new basis?

(d) What would the corresponding entries be in *International Tables A*, if this primitive basis were used? Take into account the standardization, that is translations are converted to values of 0 to <1 by the addition of integral numbers.

3.3. A subgroup of the space group $P\bar{6}m2$ is $P\bar{6}2m$ with the basis vectors $\mathbf{a}' = 2\mathbf{a}+\mathbf{b}$, $\mathbf{b}' = -\mathbf{a}+\mathbf{b}$, $\mathbf{c}' = \mathbf{c}$ and an origin shift $\boldsymbol{p}^{\mathrm{T}} = (\frac{2}{3}, \frac{1}{3}, 0)$. How many times has the unit cell of $P\bar{6}2m$ been enlarged? How do the coordinates transform?

3.4. The coordinate system of a (body-centred) tetragonal space group is to be transformed first to an orthorhombic coordinate system with the basis vectors $\mathbf{a}' = \mathbf{a}+\mathbf{b}$, $\mathbf{b}' = -\mathbf{a}+\mathbf{b}$, $\mathbf{c}' = \mathbf{c}$ and an origin shift $\boldsymbol{p}_1^{\mathrm{T}} = (0, \frac{1}{2}, 0)$, followed by a second transformation to a monoclinic system with $\mathbf{a}'' = \mathbf{a}'$, $\mathbf{b}'' = -\mathbf{b}'$, $\mathbf{c}'' = -\frac{1}{2}(\mathbf{a}' + \mathbf{c}')$ and an origin shift $\boldsymbol{p}_2^{\mathrm{T}} = (-\frac{1}{8}, \frac{1}{8}, -\frac{1}{8})$ (referred to the orthorhombic coordinate system). What are the transformations of the basis vectors and the coordinates from the tetragonal to the monoclinic coordinate system? What is the origin shift? Does the volume of the unit cell change?

3.5. The group–subgroup relation $Fm\bar{3}c \rightarrow I4/mcm$ (retaining the c axis) requires a basis transformation and an origin shift. In *International Tables* A1, Part 3, we find the basis transformation $\frac{1}{2}(\mathbf{a}-\mathbf{b})$, $\frac{1}{2}(\mathbf{a}+\mathbf{b})$, $\mathbf{c}$ and the coordinate transformation $x-y+\frac{1}{2}$, $x+y$, z. What is the corresponding origin shift in the coordinate system of $Fm\bar{3}c$? Compare the result with the origin shift of $\frac{1}{4}, \frac{1}{4}, 0$ given in Part 2 of the tables.

Basics of crystallography, part 2

4

4.1 The description of crystal symmetry in *International Tables A*: Positions

Three different kinds of description of the symmetry operations of crystals are used in *International Tables A*:

1. By one or more diagrams of the symmetry elements, see Section 6.4.1.
2. By one diagram of the 'general position', see Section 6.4.4.
3. By the coordinate triplets of the 'general positions', see Section 3.1.1. As shown there, the coordinate triplets not only specify the coordinates of the image points, but can also be regarded as *descriptions of the mappings*; see also Section 6.4.3.

In *International Tables A*, the coordinate triplets in the upper block of the 'Positions', such as shown at the beginning of Section 3.1.1 for the space group $I4_1/amd$, are a kind of shorthand notation of eqn (3.1) (page 20):

- the left part ($\tilde{x} =, \tilde{y} =, \tilde{z} =$) has been omitted;
- all components with the coefficients $W_{ik} = 0$ and $w_i = 0$ have been omitted.

The term (2) $\bar{x}+\frac{1}{2},\bar{y}+\frac{1}{2},z+\frac{1}{2}$ therefore means:

$$\boldsymbol{W} = \begin{pmatrix} -1 & 0 & 0 \\ 0 & -1 & 0 \\ 0 & 0 & 1 \end{pmatrix}, \quad \boldsymbol{w} = \begin{pmatrix} \frac{1}{2} \\ \frac{1}{2} \\ \frac{1}{2} \end{pmatrix}$$

The term (3) $\bar{y},x+\frac{1}{2},z+\frac{1}{4}$ is a shorthand notation of the matrix–column pair:

$$\boldsymbol{W} = \begin{pmatrix} 0 & -1 & 0 \\ 1 & 0 & 0 \\ 0 & 0 & 1 \end{pmatrix}, \quad \boldsymbol{w} = \begin{pmatrix} 0 \\ \frac{1}{2} \\ \frac{1}{4} \end{pmatrix}$$

This way, *International Tables A* present the analytic-geometrical tools for the description of crystal symmetry.

4.2 Crystallographic symmetry operations

By definition, crystallographic symmetry operations are always isometries; however, not every isometry can be a crystallographic symmetry operation.

This is due to the periodicity of crystals and to the restriction that the periodicity lengths may not be arbitrarily short.

Choose a primitive basis for a crystal, Definition 2.7 (page 14). Then every symmetry operation corresponds to a matrix–column pair $(\boldsymbol{W}, \boldsymbol{w})$. This includes an infinity of translations $(\boldsymbol{I}, \boldsymbol{w})$; every triplet $\boldsymbol{w}$ of integers represents a translation. However, the number of matrix parts $\boldsymbol{W}$ is finite, as follows from the following considerations.

Theorem 4.1 For every space group, represented by the matrix–column pairs $(\boldsymbol{W}, \boldsymbol{w})$ of its symmetry operations, there exist only a finite number of matrices $\boldsymbol{W}$.

The matrix $\boldsymbol{W}$ maps the basis $(\mathbf{a}_1, \mathbf{a}_2, \mathbf{a}_3)$ onto the vectors $(\tilde{\mathbf{a}}_1, \tilde{\mathbf{a}}_2, \tilde{\mathbf{a}}_3)$. For the set of all $\boldsymbol{W}$s we have a set of base images. Every basis vector is a lattice vector; therefore, its image vector is also a lattice vector, since the lattice has to be mapped onto itself under $\boldsymbol{W}$, otherwise $\boldsymbol{W}$ would not correspond to a symmetry operation. The set of all image vectors of a basis vector, say $\mathbf{a}_i$, have their endpoints on a sphere of radius a_i. If there were infinitely many image vectors $\tilde{\mathbf{a}}_i$, their endpoints would have to concentrate around at least one point on this sphere with an infinite density. For any two lattice vectors their difference also is a lattice vector; then there would have to exist arbitrarily short lattice vectors. As a consequence, only a finite number of image vectors $\tilde{\mathbf{a}}_i$ can exist for any i, and thus only finitely many matrices $\boldsymbol{W}$.

This conclusion is evidently independent of the dimension d of space. However, the maximal possible number of different matrices increases markedly with d: It is 2 for $d = 1$, 12 for $d = 2$, 48 for $d = 3$, and 1152 for $d = 4$.

A second restriction concerns the possible orders N of rotations. According to the aforementioned consideration of the matrices $\boldsymbol{W}$, these have to consist of integral numbers if a primitive basis has been chosen, because in that case all lattice vectors are integral. On the other hand, the matrix of any rotation with reference to an appropriate orthonormal basis, can be represented by

$$\boldsymbol{W} = \begin{pmatrix} \cos\varphi & -\sin\varphi & 0 \\ \sin\varphi & \cos\varphi & 0 \\ 0 & 0 & 1 \end{pmatrix}$$

The trace of the matrix is therefore $\mathrm{tr}(\boldsymbol{W}) = n$, n integral, as well as $\mathrm{tr}(\boldsymbol{W}) = 1 + 2\cos\varphi$.

The trace is independent of the basis. As a consequence, for crystallographic symmetry operations, we have:

$$1 + 2\cos\varphi = n, \quad n \text{ integral}$$

This restricts the possible values of φ to:

$$\varphi = 0^\circ,\ 60^\circ,\ 90^\circ,\ 120^\circ,\ 180^\circ,\ 240^\circ,\ 270^\circ, \text{ and } 300^\circ$$

Therefore, the order N is restricted to the values $N = 1, 2, 3, 4$, and 6. This is also valid for rotoinversions, since any rotation can be coupled with an inversion.

In crystallography, the types of crystallographic symmetry operations are designated by their **Hermann–Mauguin symbols** (Figs 4.1 and 4.2). These are:

- 1 for the identical mapping.
- $\overline{1}$ ('one bar') for the inversion.
- Rotations: A number N, $N = 2, 3, 4, 6$. This corresponds to the order of the rotation. If needed, the power of the rotation is mentioned; for example, $6^{-1} = 6^5$ for a rotation by $-60° = 300°$.
- Screw rotations: N_p designates a screw rotation consisting of a rotation N coupled with a translation parallel to the axis of rotation by p/N of the shortest lattice distance in this direction. The possible symbols are: 2_1 ('two sub one'), 3_1, 3_2, 4_1, 4_2, 4_3, 6_1, 6_2, 6_3, 6_4, and 6_5.
- Rotoinversions: $\overline{3}$, $\overline{4}$, and $\overline{6}$.
- Reflections: m (like *mirror*). m is identical to $\overline{2}$.

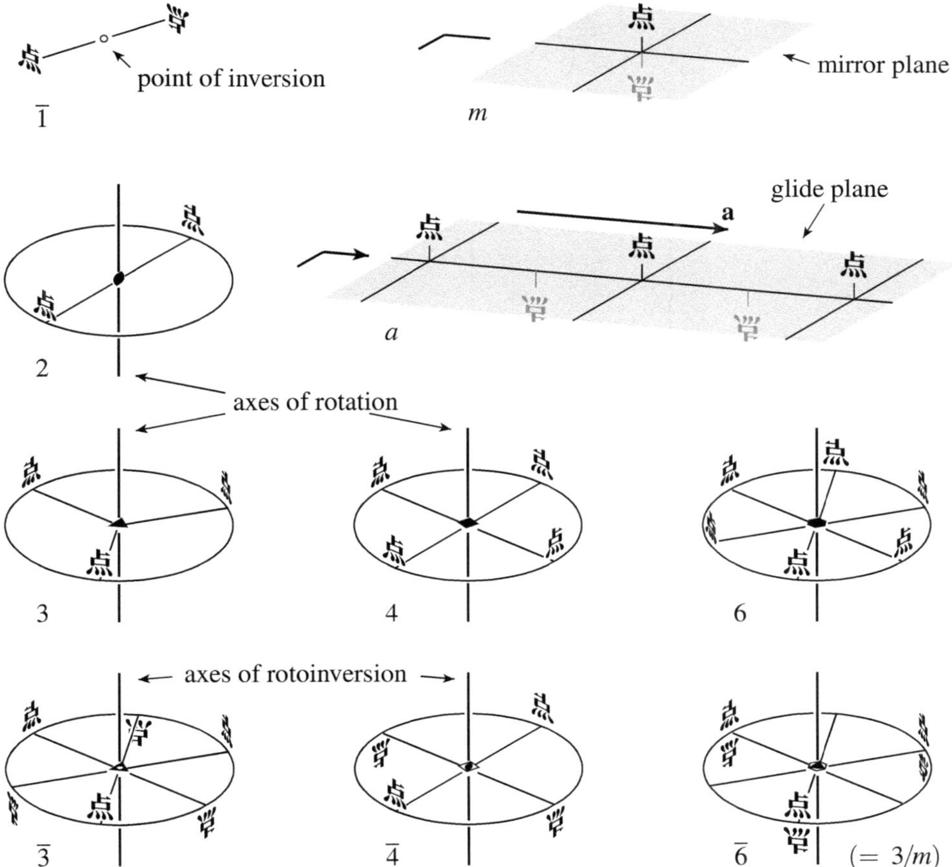

Fig. 4.1 The effect of different symmetry operations on the point 点 (Chinese symbol for point, pronounced diǎn in Chinese, ten in Japanese). The symmetry operations are designated by their Hermann–Mauguin symbols and by their graphical symbols.

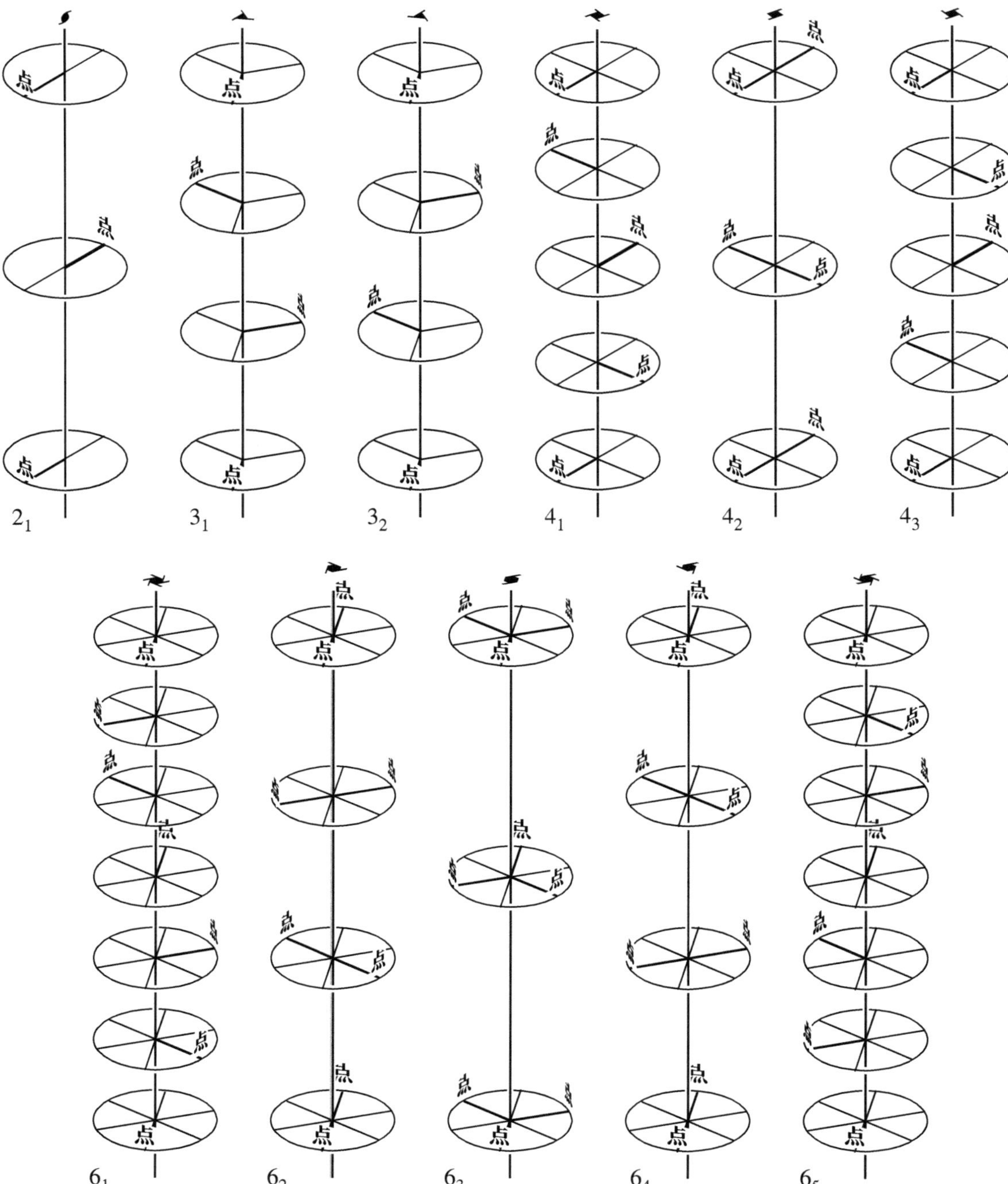

Fig. 4.2 The crystallographic screw axes with their Hermann–Mauguin and graphical symbols. The axes 3_1, 4_1, 6_1, and 6_2 are right-handed, 3_2, 4_3, 6_5, and 6_4 left-handed. One length of translation in the axis direction is depicted.

- Glide reflections: The letter m is replaced by a symbol that expresses the glide vector $\mathbf{v}$. The vector $\mathbf{v}$ is parallel to the glide plane; its length is one-half of a lattice vector. The main symbols among the conventional settings are:

  a, b, or c if the glide vector is $\frac{1}{2}\mathbf{a}$, $\frac{1}{2}\mathbf{b}$, or $\frac{1}{2}\mathbf{c}$, respectively; for plane groups it is g;

  n for glide vectors $\frac{1}{2}(\pm\mathbf{a}\pm\mathbf{b})$, $\frac{1}{2}(\pm\mathbf{b}\pm\mathbf{c})$, $\frac{1}{2}(\pm\mathbf{c}\pm\mathbf{a})$, or $\frac{1}{2}(\pm\mathbf{a}\pm\mathbf{b}\pm\mathbf{c})$;

  d for glide vectors $\frac{1}{4}(\pm\mathbf{a}\pm\mathbf{b})$, $\frac{1}{4}(\pm\mathbf{b}\pm\mathbf{c})$, $\frac{1}{4}(\pm\mathbf{c}\pm\mathbf{a})$, or $\frac{1}{4}(\pm\mathbf{a}\pm\mathbf{b}\pm\mathbf{c})$.

  e designates a glide plane having two mutually perpendicular glide vectors $\frac{1}{2}\mathbf{a}$, $\frac{1}{2}\mathbf{b}$, or $\frac{1}{2}\mathbf{c}$.

  For the special symbols g_1 and g_2 among non-conventional settings of tetragonal space groups see Section 10.3.3.

The mentioned symbols for single crystallographic symmetry operations give no indication about the orientation of the rotation or rotoinversion axis. As explained in more detail in Section 6.3.1, the orientation of an axis is expressed in a Hermann–Mauguin symbol by its position in the symbol.

In the plane, the same orders exist for rotations, since the traces of the matrices yield the same equation $2\cos\varphi = n$ with the same solutions for φ and N. However, only one type of symmetry operation exists in the plane for $\det(\boldsymbol{W}) = -1$, the reflection or glide reflection at a line, which is represented in an appropriate basis by:

$$\boldsymbol{W} = \begin{pmatrix} -1 & 0 \\ 0 & 1 \end{pmatrix}$$

A rotation axis belongs to every rotation. Among the symmetry rotations of crystals, these axes can only adopt certain mutual angles, otherwise their compositions would produce rotations with non-integral traces. This is another way to understand why for a crystal there can only exist a finite number of matrices $\boldsymbol{W}$. The different possible sets $\{\boldsymbol{W}\}$ of compatible matrices $\boldsymbol{W}$, the 32 crystal classes, can be derived in this way.

4.3 Geometric interpretation of the matrix–column pair $(\boldsymbol{W},\boldsymbol{w})$ of a crystallographic symmetry operation

Consider a crystallographic symmetry operation W, represented by the matrix–column pair $(\boldsymbol{W}, \boldsymbol{w})$, referred to a given coordinate system (the geometric interpretation would be impossible without the coordinate system).

Essentially, the following procedure can be applied to general isometries, without restrictions to the orders (see also Section 3.5).

First, we analyse the matrix part $\boldsymbol{W}$:

- $\det(\boldsymbol{W}) = +1$: rotation; $\det(\boldsymbol{W}) = -1$: rotoinversion;
- angle of rotation φ from $\cos\varphi = \frac{1}{2}(\pm\mathrm{tr}(\boldsymbol{W}) - 1)$ (4.1);
  the $+$ sign refers to rotations, the $-$ sign to rotoinversions.

This can be summarized in a table:

	$\det(\boldsymbol{W}) = +1$					$\det(\boldsymbol{W}) = -1$				
$\mathrm{tr}(\boldsymbol{W})$	3	2	1	0	−1	−3	−2	−1	0	1
type	1	6	4	3	2	$\overline{1}$	$\overline{6}$	$\overline{4}$	$\overline{3}$	$\overline{2} = m$
order	1	6	4	3	2	2	6	4	6	2

Characterization of the crystallographic symmetry operations

Every symmetry operation, translations excepted, is related to a *symmetry element*. This is a point, a straight line, or a plane that retains its position in space when the symmetry operation is performed.

1. Type 1 or $\overline{1}$:
   - 1: identity or translation with $\boldsymbol{w}$ as a column of the translation vector;
   - $\overline{1}$: inversion; the symmetry element is the *point of inversion F* (inversion centre):

$$\boldsymbol{x}_F = \frac{1}{2}\boldsymbol{w} \tag{4.2}$$

2. All other operations have a fixed axis (axis of rotation or rotoinversion); its direction can be calculated from $\boldsymbol{W}\boldsymbol{u} = \boldsymbol{u}$ (rotations) or $\boldsymbol{W}\boldsymbol{u} = -\boldsymbol{u}$ (rotoinversions). For a reflection and a glide reflection the symmetry element is not the axis, but the mirror plane or the glide plane (short for glide-reflection plane); the direction of the axis is normal to the plane.

3. The *coefficient of a screw rotation* or the *coefficient of a glide reflection* $\frac{1}{k}\boldsymbol{t}$ can be calculated from the corresponding matrix of rotation $\boldsymbol{W}$, which has order k, that is $\boldsymbol{W}^k = \boldsymbol{I}$:

$$\frac{1}{k}\boldsymbol{t} = \frac{1}{k}\left(\boldsymbol{W}^{k-1} + \boldsymbol{W}^{k-2} + \ldots + \boldsymbol{W} + \boldsymbol{I}\right)\boldsymbol{w} \tag{4.3}$$

If $\boldsymbol{t} = \boldsymbol{o}$, we have a rotation or a reflection. If $\boldsymbol{t} \neq \boldsymbol{o}$, we have a screw rotation or a glide reflection. In this case we obtain the *reduced operation*:

$$(\boldsymbol{I}, -\frac{1}{k}\boldsymbol{t})(\boldsymbol{W}, \boldsymbol{w}) = (\boldsymbol{W}, \boldsymbol{w} - \frac{1}{k}\boldsymbol{t}) = (\boldsymbol{W}, \boldsymbol{w}') \tag{4.4}$$

The column $\frac{1}{k}\boldsymbol{t}$ is called the *screw component* or the *glide component* of the column $\boldsymbol{w}$. The column $\boldsymbol{w}' = \boldsymbol{w} - \frac{1}{k}\boldsymbol{t}$ determines the *position* in space of the corresponding symmetry element. Therefore, $\boldsymbol{w}'$ is also called the positional component of $\boldsymbol{w}$. If $\boldsymbol{W}$ has only main diagonal coefficients (i.e. only $W_{ii} \neq 0$) then $W_{ii} = \pm 1$ holds and $\boldsymbol{w}_i$ is a screw or a glide coefficient for $W_{ii} = +1$ and a positional coefficient for $W_{ii} = -1$.

4. The fixed points of the isometry are determined by solution of eqn (3.14):

$$\boldsymbol{W}\boldsymbol{x}_F + \boldsymbol{w} = \boldsymbol{x}_F$$

This equation has no solution for screw rotations and glide reflections. The position of the screw axis or glide plane rather results from the reduced operation, eqn (4.4):

$$\boldsymbol{W}\boldsymbol{x}_F + \boldsymbol{w}' = \boldsymbol{x}_F \tag{4.5}$$

The conventional pairs $(\boldsymbol{W}, \boldsymbol{w})$ are listed in *International Tables A* in shorthand notation as 'general positions', see Section 3.1.1. Their geometric meaning can be found in the tables of the space groups under the heading 'symmetry operations'. They have been numbered in the same sequence as the coordinate triplets. More explanations follow in Section 6.4.3.

4.4 Derivation of the matrix–column pair of an isometry

The matrix–column pair $(\boldsymbol{W}, \boldsymbol{w})$ consists of 12 coefficients. To determine them, the coordinates of four non-coplanar image points have to be known. The most straightforward procedure is to take the image points of the origin and of the three endpoints of the basis vectors.

1. If $\tilde{O}$ is the image point of the origin O, we obtain:

$$\tilde{\boldsymbol{o}} = \boldsymbol{W}\boldsymbol{o} + \boldsymbol{w} = \boldsymbol{w} \tag{4.6}$$

Therefore, the coordinates $\tilde{\boldsymbol{o}}$ of $\tilde{O}$ are the coefficients of $\boldsymbol{w}$.

2. After having determined $\boldsymbol{w}$, the matrix $\boldsymbol{W}$ is obtained from the images of the points X_o, Y_o, and Z_o with

$$\boldsymbol{x}_o = \begin{pmatrix} 1 \\ 0 \\ 0 \end{pmatrix}, \quad \boldsymbol{y}_o = \begin{pmatrix} 0 \\ 1 \\ 0 \end{pmatrix}, \quad \boldsymbol{z}_o = \begin{pmatrix} 0 \\ 0 \\ 1 \end{pmatrix}$$

using the relations:

$$\tilde{\boldsymbol{x}}_o = \boldsymbol{W}\boldsymbol{x}_o + \boldsymbol{w}, \quad \tilde{\boldsymbol{y}}_o = \boldsymbol{W}\boldsymbol{y}_o + \boldsymbol{w}, \quad \tilde{\boldsymbol{z}}_o = \boldsymbol{W}\boldsymbol{z}_o + \boldsymbol{w} \quad \text{or} \tag{4.7}$$

$$\tilde{\boldsymbol{x}}_o = \begin{pmatrix} W_{11} \\ W_{21} \\ W_{31} \end{pmatrix} + \boldsymbol{w}, \quad \tilde{\boldsymbol{y}}_o = \begin{pmatrix} W_{12} \\ W_{22} \\ W_{32} \end{pmatrix} + \boldsymbol{w}, \quad \tilde{\boldsymbol{z}}_o = \begin{pmatrix} W_{13} \\ W_{23} \\ W_{33} \end{pmatrix} + \boldsymbol{w} \tag{4.8}$$

With the coordinates of $\tilde{\boldsymbol{x}}_o, \tilde{\boldsymbol{y}}_o, \tilde{\boldsymbol{z}}_o$ we obtain the matrix $\boldsymbol{W}$.

To check the result, calculate fixed points, the trace, the determinant, the order, and/or other known values. If the images of $\tilde{O}, \tilde{X}_o, \tilde{Y}_o$, or $\tilde{Z}_o$ are difficult to determine, use the images of other points; the calculations then become more complicated.

Exercises

Solutions on page 323.

4.1. Give the geometric interpretations for the symmetry operations (8), (10), (15), $(15)+(\frac{1}{2},\frac{1}{2},\frac{1}{2})=(15)_2$, and $(15)_{2n}$ mentioned in Exercises 3.1 and 3.2 (page 40; $(15)_{2n}$ means standardized). Take the matrices you obtained with Exercise 3.1 and apply the procedure explained in Section 4.3. For each of the mentioned operations, derive:

(a) the determinant $\det(\boldsymbol{W})$ and the trace $\mathrm{tr}(\boldsymbol{W})$;
(b) from that, the type of symmetry operation;
(c) the direction of the axis of rotation or of the normal to the plane;
(d) the screw and glide components;
(e) the position of the symmetry element;
(f) the Hermann–Mauguin symbol of the symmetry operation.
(g) What operations yield fixed points?

Compare the results with the figures and listings of the symmetry operations in *International Tables A*, space group $I4_1/amd$, origin choice 2.

Group theory

5

5.1 Two examples of groups

Consider two sets:

(1) The set $\mathbb{Z}$ of the integral numbers.
The set $\mathbb{Z} = \{0, \pm 1, \pm 2, \ldots\}$ is infinite. The result of the addition of two numbers z_1 and z_2 again is an integral number $z_3 = z_1 + z_2$. There exists the number 0 with the property $z + 0 = z$ for any z. For every z there exists another number $-z$ with the property $z + (-z) = 0$. The relation $z_1 + z_2 = z_2 + z_1$ always holds.

(2) The symmetry $\mathcal{G}$ of a square.
The set $\mathcal{G}$ consists of the eight mapping elements $g_1, g_2, \ldots, g_8$. These are (see Fig. 5.1) the rotations *4* counter-clockwise by 90°, *2* by 180°, and $\mathit{4}^{-1}$ by −90° (equivalent to $\mathit{4}^3$ by 270°); the reflections m_x, m_y, m_+, and m_- through the lines m_x, m_y, m_+ and m_-; and finally the identity mapping *1*, which maps every point onto itself. Each one of these mappings maps the square onto itself; any composition (sequence) of two mappings again yields a mapping of the figure onto itself. The composition of any mapping g with *1* reproduces g, and for any mapping g there exists the inverse mapping g^{-1}, such that g composed with g^{-1} yields the identity mapping. Contrary to $\mathbb{Z}$ the composition of two elements does not always yield the same element: *4* and m_x yield m_- if *4* is performed first, but m_+ if m_x is performed first.

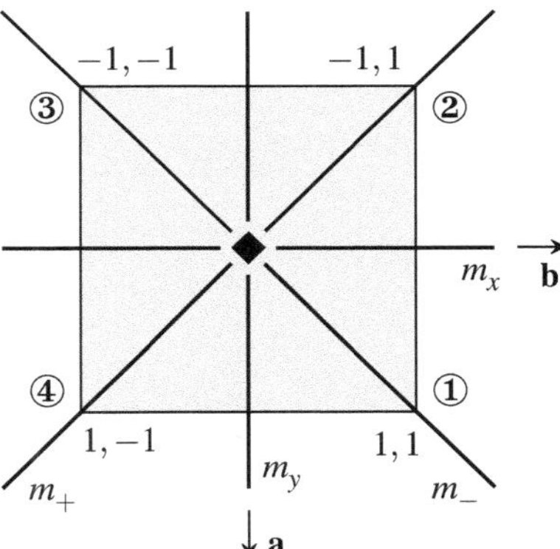

Fig. 5.1 The square and its symmetry elements (point of rotation ◆ and four mirror lines).

The two sets have another important property in common, with reference to the chosen kinds of composition, addition, and sequential mapping. Taking three elements z_1, z_2, z_3 or g_1, g_2, g_3, it makes no difference which elements are composed first as long as the sequence is not changed:

$$(z_1 + z_2) + z_3 = z_4 + z_3 = z_6$$

always yields the same result as

$$z_1 + (z_2 + z_3) = z_1 + z_5 = z_6$$

In the same way

$$(g_1 \circ g_2) \circ g_3 = g_4 \circ g_3 = g_6 = g_1 \circ (g_2 \circ g_3) = g_1 \circ g_5 = g_6$$

holds, where the sequential execution is expressed by the symbol $\circ$.[1] The sets $\mathbb{Z}$ and $\mathcal{G}$ are said to be *associative* with respect to the selected compositions.

The sets $\mathbb{Z}$ and $\mathcal{G}$ share the mentioned properties with many other sets. The term *group* has been coined for all sets having these properties.

[1] $g_1 \circ g_2$ reads 'g_1 followed by g_2'.

Table 5.1 Symmetry operations of a square and the corresponding permutations of its vertices, see Fig. 5.1.

1	(1)(2)(3)(4)
2	(13)(24)
4	(1234)
4^{-1}	(1432)
m_x	(12)(34)
m_y	(14)(23)
m_+	(2)(4)(13)
m_-	(1)(3)(24)

Table 5.2 Matrices of the symmetry operations of the square.

$$\boldsymbol{W}(1)=\begin{pmatrix}1&0\\0&1\end{pmatrix}\quad \boldsymbol{W}(2)=\begin{pmatrix}\bar{1}&0\\0&\bar{1}\end{pmatrix}$$

$$\boldsymbol{W}(4)=\begin{pmatrix}0&\bar{1}\\1&0\end{pmatrix}\quad \boldsymbol{W}(4^{-1})=\begin{pmatrix}0&1\\\bar{1}&0\end{pmatrix}$$

$$\boldsymbol{W}(m_x)=\begin{pmatrix}\bar{1}&0\\0&1\end{pmatrix}\quad \boldsymbol{W}(m_y)=\begin{pmatrix}1&0\\0&\bar{1}\end{pmatrix}$$

$$\boldsymbol{W}(m_+)=\begin{pmatrix}0&\bar{1}\\\bar{1}&0\end{pmatrix}\quad \boldsymbol{W}(m_-)=\begin{pmatrix}0&1\\1&0\end{pmatrix}$$

An example of a non-associative composition of elements of the set $\mathbb{Z}$ is subtraction. For example, $(5-3)-2=2-2=0$, but $5-(3-2)=5-1=4\neq 0$. However, we do not have to be concerned about associativity: mappings are associative, and symmetry deals with groups of mappings.

Let us consider the example of the group $\mathcal{G}$ in some more detail. The sequential execution of mappings is not always a simple matter. You may find it difficult to answer the question, 'What mapping (symmetry operation) results when a cube is first rotated anticlockwise about the direction [111] (body diagonal) by 120° and then about [010] (an edge) by 270°?' The replacement of mappings by analytical tools helps to keep track of the operations. Such tools are the permutations of the vertices ①, ②, ③, and ④; other tools are the matrices of the mappings. Exactly one permutation and one matrix corresponds to each mapping. It is convenient to choose the origin to be in the centre of the square; then the column $\boldsymbol{w}$ of the matrix–column pair $(\boldsymbol{W},\boldsymbol{w})$ is a zero column and it is sufficient to consider the matrix $\boldsymbol{W}$. The composition of mappings then corresponds to a sequence of permutations or to multiplication of the matrices, see Tables 5.1 and 5.2. The notation in Table 5.1 has the following meaning: (3) means the vertex ③ keeps its position; (13) means the vertices ① and ③ interchange their positions; (1234) means cyclic interchange of the vertices, ①→②→③→④→①.

Using permutations, we obtain the result of the above-mentioned sequential execution of *4* and m_x:

First *4*, then m_x

$$\begin{array}{ccccc} & ① & ② & ③ & ④\\ 4 & \downarrow & \downarrow & \downarrow & \downarrow\\ & ② & ③ & ④ & ①\\ m_x & \downarrow & \downarrow & \downarrow & \downarrow\\ & ① & ④ & ③ & ②\end{array}$$

This corresponds to the permutation (1)(3)(24) which is m_-

First m_x, then *4*

$$\begin{array}{ccccc} & ① & ② & ③ & ④\\ m_x & \downarrow & \downarrow & \downarrow & \downarrow\\ & ② & ① & ④ & ③\\ 4 & \downarrow & \downarrow & \downarrow & \downarrow\\ & ③ & ② & ① & ④\end{array}$$

This corresponds to the permutation (2)(4)(13) which is m_+

Using matrices, the first operation has to be written on the right side, since it is applied to the coordinate column, which is placed on the right side in the formula $\tilde{\boldsymbol{x}}=\boldsymbol{W}\boldsymbol{x}$:

First *4*, then m_x

$$\underset{\boldsymbol{W}(m_x)}{\begin{pmatrix}\bar{1}&0\\0&1\end{pmatrix}}\underset{\boldsymbol{W}(4)}{\begin{pmatrix}0&\bar{1}\\1&0\end{pmatrix}}=\begin{pmatrix}0&1\\1&0\end{pmatrix}=\boldsymbol{W}(m_-)$$

First m_x, then *4*

$$\underset{\boldsymbol{W}(4)}{\begin{pmatrix}0&\bar{1}\\1&0\end{pmatrix}}\underset{\boldsymbol{W}(m_x)}{\begin{pmatrix}\bar{1}&0\\0&1\end{pmatrix}}=\begin{pmatrix}0&\bar{1}\\\bar{1}&0\end{pmatrix}=\boldsymbol{W}(m_+)$$

If the group contains not too many elements, the results of the compositions can be presented in a table, the *group multiplication table*, see Table 5.3. In the top line, each column is labelled with a symmetry operation (group element), and so is each row in the left-most column. The top line refers to the operation performed first, the left column to the second operation. The entry at the intersection of a column and a row corresponds to the composition of the operations. Check the results obtained from *4* and m_x.

Table 5.3 Multiplication table of the symmetry group of the square.

	1	2	4	4^{-1}	m_x	m_+	m_y	m_-	← first symmetry operation
1	1	2	4	4^{-1}	m_x	m_+	m_y	m_-	
2	2	1	4^{-1}	4	m_y	m_-	m_x	m_+	
4	4	4^{-1}	2	1	m_+	m_y	m_-	m_x	
4^{-1}	4^{-1}	4	1	2	m_-	m_x	m_+	m_y	
m_x	m_x	m_y	m_-	m_+	1	4^{-1}	2	4	
m_+	m_+	m_-	m_x	m_y	4	1	4^{-1}	2	
m_y	m_y	m_x	m_+	m_-	2	4	1	4^{-1}	
m_-	m_-	m_+	m_y	m_x	4^{-1}	2	4	1	

↑
second symmetry operation

5.2 Basics of group theory

The observations made in Section 5.1 can be formalized with the so-called *group axioms* (*group postulates*) [53]. These are:

(1) *Closure*: A *group* is a set $\mathcal{G}$ of elements g_i, for which a composition law is defined, such that the composition of any two elements $g_i \circ g_k$ yields exactly one element $g_j \in \mathcal{G}$:

$$g_j = g_i \circ g_k$$

Remarks

- The notation $g_j \in \mathcal{G}$ means, g_j is an element of the set $\mathcal{G}$.
- We use: calligraphic letters such as $\mathcal{G}, \mathcal{H}$ for groups; slanted sans-serif lower-case letters or ciphers such as $g, h, 4$ for group elements; italic upper-case letters such as A, B for arbitrary sets; and italic lower-case letters such as a, b for their elements. Groups and sets are also designated by curly braces such as $\{g_1, g_2, \dots\} = \mathcal{G}$ or $\{a_1, a_2, \dots\} = A$. $\{\boldsymbol{W}\}$ means a group consisting of mapping matrices $\boldsymbol{W}_1, \boldsymbol{W}_2, \dots$.
- The composition of elements is often called '*multiplication*' and the result the '*product*', even if the composition is of some other kind. The composition sign $\circ$ is usually omitted.
- In most cases the kind of composition is clear. However, sometimes a specification is needed, for example, if the composition in $\mathbb{Z}$ should be addition or multiplication.
- The convention in crystallography is: When mappings of points (or coordinates) are expressed by matrices, then in a sequence of mappings the first matrix is written on the right, the second one on the left, see Section 5.1.
- The symmetry operations are the elements of a symmetry group. The symmetry elements (points of inversion, rotation axes, rotoinversion axes, screw axes, and reflection and glide planes) are not the elements of the group. This unfortunate terminology has historical reasons.

(**2**) The composition of elements is associative, see Section 5.1.

(**3**) One element of the group, the *identity element*, $g_1 = e$ has the property

$$eg_i = g_i e = g_i \text{ for all } g_i \in \mathcal{G}$$

(**4**) Every $g \in \mathcal{G}$ has an *inverse element* (or reciprocal element) $x \in \mathcal{G}$, such that $xg = gx = e$ holds. Usually, the inverse element is termed g^{-1}. The element inverse to g^{-1} is g. If $g \neq g^{-1}$, there are pairs g and g^{-1}.

For the following, some more basic terms are needed.

(**a**) The number $|\mathcal{G}|$ of elements in a group $\mathcal{G}$ is called the *order of the group*. If $|\mathcal{G}|$ is not finite (as in $\mathbb{Z}$), $\mathcal{G}$ is called a *group of infinite order* or an *infinite group*.

(**b**) Let $g \in \mathcal{G}$. Then, according to axiom 1, $gg = g^2$, $ggg = g^3, \dots$ are also elements of $\mathcal{G}$. If $|\mathcal{G}|$ is finite, there must exist a smallest number k, such that $g^k = e$ holds. This number k is called the *order of the element* g (not to be confused with the order of the group). If $|\mathcal{G}|$ is infinite, k can be infinite. For example, the group $\mathbb{Z}$ from Section 5.1 only has elements of infinite order (the 0 excepted).

(**c**) If $g_i g_k = g_k g_i$ holds for all element pairs $g_i, g_k \in \mathcal{G}$, $\mathcal{G}$ is called a *commutative* or *Abelian group*.[2]

[2] After the mathematician NIELS ABEL, 1802–1829.

(**d**) An arbitrary subset $A = \{a_1, a_2, a_3, \dots\} \subseteq \mathcal{G}$ is called a *complex* from $\mathcal{G}$. A complex usually does not fulfil the group axioms. $g_i A$ refers to the set $\{g_i a_1, g_i a_2, g_i a_3, \dots\}$ of the products of g_i with the elements of A; Ag_i designates the set $\{a_1 g_i, a_2 g_i, a_3 g_i, \dots\}$. If $B = \{b_1, b_2, b_3, \dots\} \subseteq \mathcal{G}$ is also a complex from $\mathcal{G}$, then AB is the set of all products $a_i b_k$, $AB = \{a_1 b_1, a_2 b_1, a_3 b_1, \dots, a_1 b_2, \dots, a_i b_k, \dots\}$; BA is the set of all products $b_i a_k$.

(**e**) A complex $H \subseteq \mathcal{G}$ that fulfils the group axioms is called a *subgroup*, written $\mathcal{H} \leq \mathcal{G}$. If $\mathcal{G}$ contains elements which do not occur in $\mathcal{H}$, that is, if $\mathcal{H}$ (as a set) is smaller than $\mathcal{G}$, then $\mathcal{H}$ is called a *proper subgroup* of $\mathcal{G}$, $\mathcal{H} < \mathcal{G}$. By analogy, $\mathcal{G} \geq \mathcal{H}$ is called a *supergroup* of $\mathcal{H}$ and $\mathcal{G} > \mathcal{H}$ a *proper supergroup* of $\mathcal{H}$. $\mathcal{G} \leq \mathcal{G}$ as a subgroup of itself and the unity element e (that always forms a group by itself) are the *trivial subgroups* of $\mathcal{G}$. If we regard e as being a group, we write $\{e\}$.

Definition 5.1 $\mathcal{H} < \mathcal{G}$ is called a *maximal subgroup* of $\mathcal{G}$ if there exists no intermediate group $\mathcal{Z}$ for which $\mathcal{H} < \mathcal{Z} < \mathcal{G}$ holds. If $\mathcal{H}$ is a maximal subgroup of $\mathcal{G}$, then $\mathcal{G} > \mathcal{H}$ is called a *minimal supergroup* of $\mathcal{H}$.

Example 5.1
Generation of the symmetry group of the square from the generator $\{4, m_x\}$:

$4 \circ m_x \to m_-$; $m_x \circ m_- \to 4^{-1}$;
$4 \circ m_- \to m_y$; $4 \circ m_y \to m_+$;
$m_x \circ m_y \to 2$; $4 \circ 4^{-1} \to 1$

(**f**) A complex of elements $g_1, g_2, \dots$ is called a set of *generators of* $\mathcal{G}$ if $\mathcal{G}$ (i.e. all elements of $\mathcal{G}$) can be generated by repeated compositions from the generators.
For example, the symmetry group of the square, among others, can be generated from $\{4, m_x\}$ or $\{m_x, m_+\}$ or $\{1, 4^{-1}, m_x, m_y\}$ or $\{4, 2, 4^{-1}, m_-, m_+\}$.

(**g**) A group $\mathcal{G}$ is called *cyclic* if it can be generated from one of its elements $a \in \mathcal{G}$ (a and a^{-1} in the case of infinite groups). For finite cyclic groups, the order of a is the group order. The order of $(\mathbb{Z}, +)$ is infinite, see above, letter (**b**); generators are $a = 1$ and $a^{-1} = -1$.

(**h**) A group of small order can be clearly represented by its group multiplication table, see the example in Table 5.3. The composition ab of a and b is placed in the intersection of the column of a and the row of b. Each column and each row of the group multiplication table lists each of the elements of the group once and only once.

Question. What elements of a group multiplication table have order 2? (Reply on page 323.)

(**i**) Groups that have the same group multiplication table, apart from the names or symbols, and if necessary, after rearrangement of rows and columns, are called *isomorphic*. This definition becomes unwieldy for groups of large order and meaningless for infinite groups. However, the essential property 'the same group multiplication table' can be defined without reference to a multiplication table.

Let $\mathcal{G} = \{g_1, g_2, \ldots\}$ and $\mathcal{G}' = \{g_1', g_2', \ldots\}$ be two groups. 'The same group multiplication table' means that for corresponding elements g_i and g_i' or g_k and g_k' their products $g_i g_k$ and $g_i' g_k'$ also correspond to each other. This is the basis of the definition:

Definition 5.2 Two groups $\mathcal{G}$ and $\mathcal{G}'$ are isomorphic, $\mathcal{G} \cong \mathcal{G}'$, if

(i) there exists a reversible mapping of $\mathcal{G}$ onto $\mathcal{G}'$, $g_i \rightleftarrows g_i'$;

(ii) the product $g_i' g_k'$ of the images g_i' and g_k' is equal to the image $(g_i g_k)'$ of the product $g_i g_k$ for every pair $g_i, g_k \in \mathcal{G}$. More formally: $g_i' g_k' = (g_i g_k)'$. The image of the product is equal to the product of the images.

With the aid of isomorphism all groups can be subdivided into *isomorphism classes* (isomorphism types) of isomorphic groups. Such a class is also called an *abstract group*; the groups themselves are *realizations* of the abstract group. In the group-theoretical sense, there is no distinction between different realizations of the same abstract group. This makes it possible to replace a group of mappings by the corresponding group of matrices (or of permutations) and thus renders it possible to treat the geometrical group with analytical tools. This is exactly what is exploited in Section 5.1, where three realizations of the symmetry group of the square are used:

1. the group of the mappings $\{1, 4, \ldots, m\_\}$;
2. the permutation group, the elements of which are listed in Table 5.1;
3. the group of the matrices mentioned in Table 5.2.

5.3 Coset decomposition of a group

Let $\mathcal{G}$ be a group and $\mathcal{H} < \mathcal{G}$ a subgroup. A *coset decomposition* of $\mathcal{G}$ with respect to $\mathcal{H}$ is defined as follows:

1. The subgroup $\mathcal{H}$ is the first *coset*.
2. If $g_2 \in \mathcal{G}$, but $g_2 \notin \mathcal{H}$, then the complex $g_2\mathcal{H}$ is the second coset (left coset, since g_2 is on the left side).

No element of $g_2\mathcal{H}$ is an element of $\mathcal{H}$ and all elements of $g_2\mathcal{H}$ are different. Therefore, the cosets $\mathcal{H}$ and $g_2\mathcal{H}$ have $|\mathcal{H}|$ elements each.

3. If $g_3 \in \mathcal{G}$ but $g_3 \notin \mathcal{H}$ and $g_3 \notin g_2\mathcal{H}$, the complex $g_3\mathcal{H}$ is the third (left) coset. All elements of $g_3\mathcal{H}$ are different and no element of $g_3\mathcal{H}$ appears in $\mathcal{H}$ or $g_2\mathcal{H}$; all elements are new.

4. The procedure is continued until no element of $\mathcal{G}$ is left over. As a result, $\mathcal{G}$ has been decomposed to *left cosets* with respect to $\mathcal{H}$. In the same way, $\mathcal{G}$ can also be decomposed to right cosets $\mathcal{H}, \mathcal{H}g_2, \ldots$.

5. The number of right and left cosets is the same; it is called the *index* of $\mathcal{H}$ in $\mathcal{G}$.

It follows from items (1)–(5) that each element of $\mathcal{G}$ appears in exactly one coset and that every coset contains $|\mathcal{H}|$ elements. Only the first coset is a group since it is the only one that contains the identity element. From $h_j\mathcal{H} = \mathcal{H}$ it follows that every element $g_i h_j$ of the coset $g_i\mathcal{H}$ can be used to generate the coset.

Taking $h_i \in \mathcal{H}$ and $n = |\mathcal{H}|$, the elements $g_i \in \mathcal{G}$ are distributed among the left cosets in the following manner:

first coset $\mathcal{H} =$	second coset $g_2\mathcal{H} =$	third coset $g_3\mathcal{H} =$	$\ldots$	ith coset $g_i\mathcal{H} =$
$e = h_1$	$g_2 e$	$g_3 e$	$\ldots$	$g_i e$
h_2	$g_2 h_2$	$g_3 h_2$	$\ldots$	$g_i h_2$
h_3	$g_2 h_3$	$g_3 h_3$	$\ldots$	$g_i h_3$
$\vdots$	$\vdots$	$\vdots$	$\vdots$	$\vdots$
h_n	$g_2 h_n$	$g_3 h_n$	$\ldots$	$g_i h_n$

Total of i cosets. Each of them contains the same number of elements. No one contains elements of another coset

Example 5.2

A few coset decompositions of the symmetry group of the square, $\mathcal{G} = \{1, 2, 4, 4^{-1}, m_x, m_y, m_+, m_-\}$ (cf. Fig. 5.1 and Table 5.3):

Decomposition with respect to $\mathcal{H} = \{1, 2\}$

1st coset	2nd coset	3rd coset	4th coset
left cosets			
$1 \circ \mathcal{H} = \{1, 2\}$	$4 \circ \mathcal{H} = \{4, 4^{-1}\}$	$m_x \circ \mathcal{H} = \{m_x, m_y\}$	$m_+ \circ \mathcal{H} = \{m_+, m_-\}$
right cosets			
$\mathcal{H} \circ 1 = \{1, 2\}$	$\mathcal{H} \circ 4 = \{4, 4^{-1}\}$	$\mathcal{H} \circ m_x = \{m_x, m_y\}$	$\mathcal{H} \circ m_+ = \{m_+, m_-\}$

There are four cosets, the index amounts to 4. In addition, in this case, the left and right cosets are equal.

Decomposition with respect to $\mathcal{H} = \{1, m_x\}$			
1st coset	2nd coset	3rd coset	4th coset
left cosets			
$1 \circ \mathcal{H} = \{1, m_x\}$	$4 \circ \mathcal{H} = \{4, m_-\}$	$2 \circ \mathcal{H} = \{2, m_y\}$	$4^{-1} \circ \mathcal{H} = \{4^{-1}, m_+\}$
right cosets			
$\mathcal{H} \circ 1 = \{1, m_x\}$	$\mathcal{H} \circ 4 = \{4, m_+\}$	$\mathcal{H} \circ 2 = \{2, m_y\}$	$\mathcal{H} \circ 4^{-1} = \{4^{-1}, m_-\}$
Left and right cosets are different.			

The *Theorem of Lagrange*[3] follows directly from the coset decomposition of a finite group:

[3] JOSEPH-LOUIS LAGRANGE, 1736–1823.

Theorem 5.3 If $\mathcal{G}$ is a finite group and $\mathcal{H} < \mathcal{G}$, then the order $|\mathcal{H}|$ of $\mathcal{H}$ is a divisor of the order $|\mathcal{G}|$ of $\mathcal{G}$.

There exist i cosets having $|\mathcal{H}|$ elements each, every element of $\mathcal{G}$ appears exactly once, and therefore $|\mathcal{G}| = |\mathcal{H}| \times i$. The index is:

$$i = \frac{|\mathcal{G}|}{|\mathcal{H}|} \tag{5.1}$$

From Theorem 5.3 it follows that a group having the order of a prime number p can only have trivial subgroups. The symmetry group of the square, having an order of 8, can only have subgroups of orders 1 ($\{1\}$, trivial), 2, 4, and 8 ($\mathcal{G}$, trivial).

Equation (5.1) is meaningless for infinite groups. However, if the elements of an infinite group are arranged in a sequence, we can delete from this sequence, say, every other element. The number of the remaining elements then is 'half as many', even though their number is still infinite. After deletion of all odd numbers from the infinite group $\mathbb{Z}$ of all integral numbers, the group of the even numbers remains. This is a subgroup of $\mathbb{Z}$ of index 2. This is due to the fact that in this way $\mathbb{Z}$ has been decomposed into two cosets: the subgroup $\mathcal{H}$ of the even numbers and a second coset containing the odd numbers:

first coset $\mathcal{H} =$		second coset $1 + \mathcal{H} =$	
$e = 0$			$1 + 0 = 1$
-2	2	$1 + (-2) = -1$	$1 + 2 = 3$
-4	4	$1 + (-4) = -3$	$1 + 4 = 5$
$\vdots$	$\vdots$	$\vdots$	$\vdots$

Similar considerations apply to space groups. Space groups are infinite groups, consisting of an infinite number of symmetry operations. However, we can say, 'the subgroup of a space group of index 2 consists of half as many symmetry operations'. 'Half as many' is to be taken in the same sense as, 'the number of even numbers is half as many as the number of integral numbers'.

For the finite index between two infinite groups $\mathcal{G} > \mathcal{H}$ we write $i = |\mathcal{G} : \mathcal{H}|$.

5.4 Conjugation

The coset decomposition of a group $\mathcal{G}$ separates the elements of $\mathcal{G}$ into classes; every element belongs to one class, i.e. to one coset. The cosets have the same number of elements, but the elements of a coset are quite different. For example, the identity element is part of the elements of the subgroup (first coset), while the order of all other elements is larger than 1. In this section we consider another kind of separation of the elements of $\mathcal{G}$: the subdivision into conjugacy classes. Generally, conjugacy classes have different sizes (lengths), but the elements of one class have common features.

Definition 5.4 The elements g_i and g_j, $g_i, g_j \in \mathcal{G}$, are called *conjugate in* $\mathcal{G}$ if there exists an element $g_m \in \mathcal{G}$ such that $g_j = g_m^{-1} g_i g_m$ holds. The set of elements, which are conjugate to g_i when g_m runs through all elements of $\mathcal{G}$ is called the *conjugacy class* of g_i.

Another way to express this is to say that g_i can be *transformed* to g_j by g_m. There may exist several elements $g_m, g_n, \ldots \in \mathcal{G}$ that transform g_i to g_j. The calculation rule $g_j = g_m^{-1} g_i g_m$ is called conjugation.

Referred to symmetry groups, this means: Two symmetry operations of the symmetry group $\mathcal{G}$ are conjugate if they are transformed one to another by some other symmetry operation of the same group $\mathcal{G}$.

Example 5.3
In the symmetry group of the square (Fig. 5.1), the rotation *4* transforms the reflection m_+ to m_- and vice versa:

$$\underset{m_+}{\begin{pmatrix} 0 & \bar{1} \\ \bar{1} & 0 \end{pmatrix}} = \underset{4^{-1}}{\begin{pmatrix} 0 & 1 \\ \bar{1} & 0 \end{pmatrix}} \underset{m_-}{\begin{pmatrix} 0 & 1 \\ 1 & 0 \end{pmatrix}} \underset{4}{\begin{pmatrix} 0 & \bar{1} \\ 1 & 0 \end{pmatrix}}$$

The same transformation $m_+ \rightleftharpoons m_-$ can be achieved by the rotation 4^{-1} and by the reflections m_x and m_y, while the remaining symmetry operations of the square leave m_+ and m_- unchanged. The reflections m_+ and m_- are conjugate in the symmetry group of the square. Together, they form a conjugacy class.

Properties of conjugation:

(1) Every element of $\mathcal{G}$ belongs to exactly one conjugacy class.

(2) The number of elements in a conjugacy class (the *length* of the conjugacy class) is different; however, it is always a divisor of the order of $\mathcal{G}$.

(3) If $g_i \in \mathcal{G}$ and if the equation

$$g_m^{-1} g_i g_m = g_i \qquad (5.2)$$

holds for all $g_m \in \mathcal{G}$, then g_i is called *self-conjugate*. Since eqn (5.2) is equivalent to $g_i g_m = g_m g_i$, this is also expressed: 'g_i is *commutative* with all elements of $\mathcal{G}$'.

(4) For Abelian groups, it follows from eqn (5.2) that every element is self-conjugate, and thus forms a conjugacy class by itself. Similarly, the identity element $e \in \mathcal{G}$ of any group forms a class by itself.

(5) Elements of the same conjugacy class have the same order.

Definition 5.5 Subgroups $\mathcal{H}, \mathcal{H}' < \mathcal{G}$ are called *conjugate subgroups* in $\mathcal{G}$, if there exists an element $g_m \in \mathcal{G}$ such that $\mathcal{H}' = g_m^{-1}\mathcal{H}\,g_m$ holds. The set of subgroups that are conjugate to $\mathcal{H}$ when g_m runs through all elements of $\mathcal{G}$ forms a conjugacy class.

Example 5.4
As explained in Example 5.3, m_+ and m_- are conjugate elements of the symmetry group $\mathcal{G}$ of the square. The groups $\{1, m_+\}$ and $\{1, m_-\}$ are conjugate subgroups of $\mathcal{G}$ because $4^{-1}\{1, m_+\}4 = \{1, m_-\}$. These two groups form a conjugacy class. The conjugate subgroups $\{1, m_x\}$ and $\{1, m_y\}$ form another conjugacy class; $4^{-1}\{1, m_x\}4 = \{1, m_y\}$.

Theorem 5.6 Conjugate subgroups are isomorphic and thus have the same order.

By conjugation, the set of all subgroups of $\mathcal{G}$ is subdivided into *conjugacy classes of subgroups*. Subgroups of the same conjugacy class are isomorphic. The number of subgroups in such a class is a divisor of the order $|\mathcal{G}|$. Every subgroup of $\mathcal{G}$ belongs to exactly one conjugacy class. Different conjugacy classes may contain different numbers of subgroups.

Conjugate subgroups of space groups are the subject of Chapter 8 .

Definition 5.7 Let $\mathcal{H} < \mathcal{G}$. If $g_m^{-1}\mathcal{H}\,g_m = \mathcal{H}$ holds for all $g_m \in \mathcal{G}$, $\mathcal{H}$ is called a *normal subgroup* of $\mathcal{G}$, designated $\mathcal{H} \lhd \mathcal{G}$ (also called an *invariant subgroup* or *self-conjugate subgroup*).

The equation $g_m^{-1}\mathcal{H}\,g_m = \mathcal{H}$ is equivalent to $\mathcal{H}\,g_m = g_m\mathcal{H}$. Therefore, the coset decomposition of a normal subgroup yields the same right and left cosets. The normal subgroup can also be defined by this property. That implies the self-conjugacy in $\mathcal{G}$. In Example 5.2, $\{1, 2\}$ is a normal subgroup, $\{1, 2\} \lhd \mathcal{G}$, but not so $\{1, m_x\}$.

Every group $\mathcal{G}$ has two trivial normal subgroups: the identity element $\{e\}$ and itself ($\mathcal{G}$). All other normal subgroups are called *proper normal subgroups*.

5.5 Factor groups and homomorphisms

The cosets of a group $\mathcal{G}$ with respect to a normal subgroup $\mathcal{N} \lhd \mathcal{G}$ by themselves form a group which is called the *factor group* (or quotient group) $\mathcal{F} = \mathcal{G}/\mathcal{N}$. The cosets are considered to be the new group elements. They are connected by complex multiplication, see Section 5.2, letter **(d)**. Replacing group elements with cosets can be compared to the packing of matches into matchboxes: First, there are matches; after packing them, only the matchboxes (now filled) remain visible, which are now the elements to be handled.

$\mathcal{G}/\mathcal{N}$ is pronounced 'G modulo N'.

Example 5.5
The point group $3m$ consists of the elements $1, 3, 3^{-1}, m_1, m_2, m_3$. Its subgroup 3 consists of the elements $1, 3, 3^{-1}$; it is also the first coset of the coset decomposition of $3m$ with respect to 3. The complex $m_1\{1, 3, 3^{-1}\} = \{m_1, m_2, m_3\}$ is the second coset. Left and right cosets coincide; therefore, the subgroup 3 is a normal subgroup. The factor group $3m/3$ consists of the two elements $\{1, 3, 3^{-1}\}$ and $\{m_1, m_2, m_3\}$. Together, the rotations are now considered to be one group element, and the reflections in common form another group element. The normal subgroup, in this case $\{1, 3, 3^{-1}\}$, is the new identity element of the factor group.

Factor groups are important not only in group theory, but also in crystallography and in representation theory. The following rules hold:

The normal subgroup $\mathcal{N}$ of the group $\mathcal{G}$ is the identity element of the factor group $\mathcal{F} = \mathcal{G}/\mathcal{N}$.

The element inverse to $g_i\mathcal{N}$ is $g_i^{-1}\mathcal{N}$.

Example 5.6

Let $\mathcal{G}$ be the group of the square (cf. Section 5.1) and let $\mathcal{N}$ be the normal subgroup $\{1,2\} \lhd \mathcal{G}$. The group multiplication table of the factor group $\mathcal{F}$ is:

	$\{1,2\}$	$\{4,4^{-1}\}$	$\{m_x,m_y\}$	$\{m_+,m_-\}$
$\{1,2\}$	$\{1,2\}$	$\{4,4^{-1}\}$	$\{m_x,m_y\}$	$\{m_+,m_-\}$
$\{4,4^{-1}\}$	$\{4,4^{-1}\}$	$\{1,2\}$	$\{m_+,m_-\}$	$\{m_x,m_y\}$
$\{m_x,m_y\}$	$\{m_x,m_y\}$	$\{m_+,m_-\}$	$\{1,2\}$	$\{4,4^{-1}\}$
$\{m_+,m_-\}$	$\{m_+,m_-\}$	$\{m_x,m_y\}$	$\{4,4^{-1}\}$	$\{1,2\}$

The identity element is $\{1,2\}$. The multiplication table is equal to the multiplication table of the point group $mm2$; therefore, the factor group is isomorphic to $mm2$.

In the preceding example $\mathcal{F}$ is isomorphic to a subgroup of $\mathcal{G}$. This is not necessarily so; factor groups of space groups very often are not isomorphic to a subgroup. The main difference between the factor groups $\mathcal{F}$ and the subgroups $\mathcal{H}$ of a group $\mathcal{G}$ can be demonstrated by a geometric comparison:

A subgroup corresponds to a section through a body; the section only discloses a part of the body, not the whole body, but of this part all of its details. The factor group corresponds to a projection of a body onto a plane: every volume element of the body contributes to the image, but always a column of volume elements is projected onto an image element; the individual character of the elements of the body is lost.

Normal subgroups and factor groups are intimately related to the homomorphic mappings or homomorphisms.

Definition 5.8 A mapping $\mathcal{G} \to \mathcal{G}'$ is called *homomorphic* or a *homomorphism* if for all pairs of elements $g_i, g_k \in \mathcal{G}$ that are mapped, $g_i \to g_i'$ and $g_k \to g_k'$,

$$(g_i g_k)' = g_i' g_k' \tag{5.3}$$

also holds. The image of the product is equal to the product of the images.

This is the same condition as for an isomorphism, see Section 5.2, letter (**i**). However, an isomorphism has exactly one image element per starting element, so that the mapping can be reversed, whereas under a homomorphism there is no restriction as to how many elements of $\mathcal{G}$ are mapped onto one element of $\mathcal{G}'$. Therefore, isomorphism is a special kind of homomorphism.

Example 5.7
Let $\mathcal{G} = \{W_1, \ldots, W_8\}$ be the group of the mapping matrices of the square (cf. Table 5.2); their determinants have the values $\det(W_i) = \pm 1$. If we assign its determinant to each matrix, then this is a homomorphic mapping of the group $\mathcal{G}$ of the matrices onto the group of numbers $\mathcal{G}' = \{-1, 1\}$; -1 and 1 form a group with respect to the composition by multiplication. Equation (5.3) of Definition 5.8 holds since $\det(W_i W_k) = \det(W_i)\det(W_k)$. The mapping cannot be reversed because the determinant 1 has been assigned to four of the matrices and the determinant -1 to the other four.

Closer inspection reveals that a homomorphism involves a close relation between $\mathcal{G}$ and $\mathcal{G}'$, see, for example, the textbook [53].

Theorem 5.9 Let $\mathcal{G} \rightarrow \mathcal{G}'$ be a homomorphism of $\mathcal{G}$ onto $\mathcal{G}'$. Then a normal subgroup $\mathcal{K} \trianglelefteq \mathcal{G}$ is mapped onto e', the identity element of $\mathcal{G}'$, and the cosets $g_i\mathcal{K}$ are mapped onto the remaining elements $g_i' \in \mathcal{G}'$. Therefore, the factor group $\mathcal{G}/\mathcal{K}$ is isomorphic to $\mathcal{G}'$. The normal subgroup $\mathcal{K}$ is called the *kernel* of the homomorphism. The homomorphism is an isomorphism if $\mathcal{K} = \{e\}$, that is, $\mathcal{G}'$ is isomorphic to $\mathcal{G}$.

Theorem 5.9 is of outstanding importance in crystallography. Take $\mathcal{G}$ as the space group of a crystal structure, $\mathcal{K}$ as the group of all translations of this structure, and $\mathcal{G}'$ as the point group of the macroscopic symmetry of the crystal; then, according to this theorem, the point group of a crystal is isomorphic to the factor group of the space group with respect to the group of its translations. This will be dealt with in more detail in Section 6.1.2.

5.6 Action of a group on a set

In spite of their importance, groups are not of primary interest in crystal chemistry. Of course, they are needed, since isometry groups are used to describe the symmetry of crystals. They are the foundation for all considerations, and knowledge of how to deal with them is essential to all profound reflections on crystal-chemical topics. However, primary interest is focused on the crystal structures themselves, on their composition from partial structures of symmetry-equivalent particles, and on the interactions between particles of equal or different partial structures. The symmetry group of the crystal structure is behind this. Therefore, what is really of interest, is the influence of the group on the points (centres of the particles) in point space: Which points are symmetry equivalent, which are invariant under what symmetry operations, etc. The conception of the *action of a group on a set* deals with this kind of questions. However, it is much more general, since the groups and the sets can be of any kind. As for groups, certain *postulates* must be fulfilled:

Let $\mathcal{G}$ be a group with the elements $e = g_1, g_2, \ldots, g_i, \ldots$ and let M be a set with the elements $m_1, m_2, \ldots, m_i, \ldots$.

Definition 5.10 The group $\mathcal{G}$ acts on the set M if:

1. $m_i = g_i m$ is a unique element $m_i \in M$ for every $g_i \in \mathcal{G}$ and every $m \in M$.
2. $em = m$ is fulfilled for every $m \in M$ and the identity element $e \in \mathcal{G}$.
3. $g_k(g_i m) = (g_k g_i)m$ holds for every pair $g_i, g_k \in \mathcal{G}$ and every $m \in M$.

Definition 5.11 The set of the elements $m_i \in M$, which are obtained by $m_i = g_i m$ when g_i runs through all elements of the group $\mathcal{G}$, is called the $\mathcal{G}$-orbit of m or the *orbit* $\mathcal{G}m$.

Applied to crystals, this means the following: Let $\mathcal{G}$ be the space group of a crystal structure and let m be an atom out of the set M of all atoms. The orbit $\mathcal{G}m$ is the set of all atoms that are symmetry-equivalent to the atom m in the crystal. For crystals we formulate in a more general way:

Definition 5.12 The mapping of a point X_o by the symmetry operations of a space group yields an infinite set of points, which is termed the *orbit of* X_o *under* $\mathcal{G}$, or the *crystallographic point orbit of* X_o, or the *$\mathcal{G}$-orbit of* X_o or for short $\mathcal{G}X_o$.

A $\mathcal{G}$-orbit is independent of the point X_o chosen from its points. The different $\mathcal{G}$-orbits of point space have no points in common. Two $\mathcal{G}$-orbits would be identical if they had a common point. Therefore, the space group causes a subdivision of the point space into $\mathcal{G}$-orbits. Crystallographic point orbits are dealt with in more detail in Section 6.5.

Example 5.8
The space group of zinc blende is $F\bar{4}3m$ (Fig. 1.1, page 4). Let a zinc atom be situated at the point $X_{\mathrm{Zn}} = (\frac{1}{4}, \frac{1}{4}, \frac{1}{4})$. By the symmetry operations of $F\bar{4}3m$ further (symmetry-equivalent) zinc atoms are situated at the points $\frac{3}{4}, \frac{3}{4}, \frac{1}{4}$, $\frac{3}{4}, \frac{1}{4}, \frac{3}{4}$, and $\frac{1}{4}, \frac{3}{4}, \frac{3}{4}$ and, in addition, at infinitely many more points, which result from the mentioned points by addition of (q, r, s), with $q, r, s =$ arbitrary positive or negative integral numbers. The complete set of these symmetry-equivalent zinc positions makes up a crystallographic point orbit. Starting from $X_{\mathrm{S}} = (0, 0, 0)$, the sulfur atoms occupy another orbit.

Definition 5.13 The set of all $g_i \in \mathcal{G}$ for which $g_i m = m$ holds is called the stabilizer $\mathcal{S}$ of m in $\mathcal{G}$.

In a crystal the stabilizer of m in $\mathcal{G}$ is the set of all symmetry operations of the space group $\mathcal{G}$ that map the atom m onto itself. The stabilizer is nothing other than the site symmetry of the point X_o where the atom is situated (Section 6.1.1). For the example of zinc blende that means: The orbit of the Zn atoms as well as the orbit of the S atoms has the point group $\bar{4}3m$ as its stabilizer.

The stabilizer is a subgroup of $\mathcal{G}$, $\mathcal{S} \leq \mathcal{G}$. If the element $g_k \in \mathcal{G}$, $g_k \notin \mathcal{S}$ maps the element $m \in M$ onto $m_k \in M$, then this is also valid for the elements $g_k\mathcal{S} \subset \mathcal{G}$. It can be shown that *exactly* these elements map m onto m_k.

Theorem 5.14 If $|\mathcal{G}|$ is the order of the finite group $\mathcal{G}$ and $|\mathcal{S}|$ the order of the stabilizer $\mathcal{S}$ of the element $m \in M$, then $L = |G|/|S|$ is the *length of the orbit* $\mathcal{G}m$.

Among symmetry groups, 'length' means the 'number of symmetry-equivalent points'. For example, the symmetry group $\mathcal{G}$ of the square has order $|\mathcal{G}| = 8$. The vertex $m =$ ① of the square (Fig. 5.1) is mapped onto itself by the symmetry operations 1 and m_-; the stabilizer therefore is the group $\mathcal{S} = \{1, m_-\}$ with order $|\mathcal{S}| = 2$ (site symmetry group of the point ①). The length of the orbit $\mathcal{G}m$ is $L = 8/2 = 4$; there are four points that are symmetry equivalent to ①.

Exercises

Solutions on page 323.

5.1. How can it be recognized in a multiplication table whether the result of a composition is independent of the sequence of the elements?

5.2. What are the orders of the symmetry operations of the symmetry group of the square in Section 5.1? Do not confuse the orders of the symmetry operations with the order of the group.

5.3. List the elements of the symmetry group of the trigonal prism. What is the order of the group? Write down the permutation group for the vertices of the trigonal prism in the same way as in Table 5.1 (label the vertices 1, 2, 3 for the lower base plane and 4, 5, 6 for the upper plane, vertex 4 being on top of vertex 1). Perform subsequent permutations to find out what symmetry operation results when the rotoinversion $\overline{6}$ is performed first, followed by the horizontal reflection m_z. Continue in the same way for other compositions of symmetry operations and write down the multiplication table.

5.4. Perform the right and left coset decompositions of $\overline{6}m2$ (symmetry group of the trigonal prism) with respect to the subgroup $\{1, 3, 3^{-1}\}$. Compare the two decompositions; what do you observe? What is the index? Repeat this with respect to the subgroup $\{1, m_1\}$ ($m_1 =$ reflection that maps each of the vertices 1 and 4 onto itself). What is different this time?

5.5. Which cosets are subgroups of $\mathcal{G}$?

5.6. Why is a subgroup of index 2 always a normal subgroup?

5.7. The two-dimensional symmetry group $4mm$ of the square and its multiplication table are given in Section 5.1.

(**a**) What are the subgroups of $4mm$ and how can they be found? Which of them are maximal?

(**b**) Why can m_x and m_+ not be elements of the same subgroup?

(**c**) What subgroups are mutually conjugate and what does that mean geometrically?

(**d**) What subgroups are normal subgroups?

(**e**) Include all subgroups in a graph exhibiting a hierarchical sequence in the following way: $4mm$ is at the top; all subgroups of the same order appear in the same row; every group is joined with every one of its maximal subgroups by a line; conjugate subgroups are joined by horizontal lines.

5.8. Write down the coset decomposition of the group $\mathbb{Z}$ of the integral numbers with respect to the subgroup of the numbers divisible by 5, $\{0, \pm5, \pm10, \pm15, \ldots\}$. What is the index of the subgroup $\{0, \pm5, \pm10, \pm15, \ldots\}$? Is it a normal subgroup?

5.9. The multiplication table of the factor group $\mathcal{F} = \mathcal{G}/\{1, 2\}$ of the group $\mathcal{G}$ of the square is given in Example 5.6 (page 60). Is $\mathcal{F}$ an Abelian group?

Basics of crystallography, part 3

6

The topics that have been dealt with in Chapters 3 (mappings) and 5 (groups) are applied in this chapter to the symmetry of crystals. This includes the clarification of certain terms, because the unfortunate nomenclature that has historically evolved has caused some confusion.

6.1 Space groups and point groups

The expression 'point group' is used for two different terms which are intimately related, but not really identical:

1. the symmetry of a molecule (or some other arrangement of particles) or the surroundings of a point in a crystal structure (site symmetry);
2. the symmetry of an ideally developed macroscopic crystal.

First, we consider molecular symmetry.

6.1.1 Molecular symmetry

Consider a molecule (or a finite cluster of atoms or something similar). The set of all isometries that map the molecule onto itself is called the *molecular symmetry*.

Definition 6.1 The molecular symmetry forms a group which is called the *point group* $\mathcal{P}_{\mathrm{M}}$ *of the molecule.*

Point groups are designated by Hermann–Mauguin or by Schoenflies symbols as explained in Sections 6.3.1 and 6.3.2, pages 73 and 76.

The group $\mathcal{P}_{\mathrm{M}}$ is finite if the molecule consists of a finite number of atoms and is mapped onto itself by a finite number of isometries. However, the group is infinite for linear molecules like H_2 and CO_2 because of the infinite order of the molecular axis.

Polymeric molecules actually consist of a finite number of atoms, but it is more practical to treat them as sections of infinitely large molecules, in the same way as crystals are treated as sections of ideal infinite crystals (i.e. crystal patterns). If an (infinitely long) ideal molecule has translational symmetry in one direction, then its symmetry group is a *rod group*. If it forms a layer with translational symmetry in two dimensions, its symmetry group is a *layer group*. Rod and layer groups are the subject of Section 7.4.

Molecules always have an extension in three-dimensional space; they are always three-dimensional objects. When one- or two-dimensional molecules are addressed in the chemical literature, usually molecules are meant whose atoms are linked by *covalent chemical bonds* forming chains or layers. When physicists speak of one- or two-dimensional structures, they usually have in mind strongly anisotropic physical properties, for example the electric conductivity. Nevertheless, the structures are always three-dimensional; one- or two-dimensional structures do not exist, even if they are termed so. However, the projection of a layer onto a parallel plane is two-dimensional.

According to their equivalence, point groups are classified into *point-group types*.

Definition 6.2 Two point groups $\mathcal{P}_{M1}$ and $\mathcal{P}_{M2}$ belong to the same point-group type if, after selection of appropriate bases (with origins at the centres of gravity), the matrix groups of $\mathcal{P}_{M1}$ and $\mathcal{P}_{M2}$ coincide.

All symmetry operations of a finite molecule leave its centre of gravity unchanged. If this is chosen as the origin, all symmetry operations are represented by matrix–column pairs of the kind $(\boldsymbol{W}, \boldsymbol{o})$. In practical work that means that it is sufficient to consider only matrices $\boldsymbol{W}$, that is 3×3 matrices.

Any arbitrary point of a molecule (e.g. an atom centre) may be mapped onto itself by symmetry operations other than the identity.

Definition 6.3 A point in a molecule has a definite *site symmetry* $\mathcal{S}$ (*site symmetry group*). It consists of all those symmetry operations of the point group of the molecule that leave the point fixed.

The site symmetry group $\mathcal{S}$ is always a subgroup of the point group of the molecule: $\mathcal{S} \leq \mathcal{P}_M$. It corresponds to the stabilizer, see Definition 5.13.

Definition 6.4 A set of symmetrically equivalent points X of a molecule is at the *general position* if the site symmetry $\mathcal{S}$ of the points consists of nothing more than the identity, $\mathcal{S} = \mathcal{I}$. Otherwise, if $\mathcal{S} > \mathcal{I}$, the points are at a *special position*.

In this context, the term 'position' does not have the meaning of 'a certain place in space', but is rather an abbreviation for 'Wyckoff position' according to Definition 6.6. Every point group has only *one* general position, which usually consists of several symmetry-equivalent points. However, the point group may have several special positions.

The point group $\mathcal{P}_M$ of the molecule acts on the molecule in the way described in Section 5.6. By analogy to Definition 5.12, the set of points which are symmetry-equivalent to a point X is the *orbit of X under $\mathcal{P}_M$*. According to Theorem 5.14, the length of the orbit of a point X_g at a general position under $\mathcal{P}_M$ is $L = |\mathcal{P}_M|$, that is there exist $|\mathcal{P}_M|$ symmetrically equivalent points of a general position. For points X_s of a special position having a site symmetry of order $|\mathcal{S}|$, there exist $|\mathcal{P}|/|\mathcal{S}|$ symmetrically equivalent points. Usually, the length of an orbit is called its multiplicity.

Theorem 6.5 The *multiplicity* of a point at a general position in a molecule is equal to the group order $|\mathcal{P}_M|$. If $|\mathcal{S}|$ is the order of the site symmetry of a point at a special position, the product of the multiplicity Z_s of this orbit with $|\mathcal{S}|$ is equal to the multiplicity Z_g of a point at the general position:

$$|\mathcal{S}| \times Z_s = Z_g.$$

Example 6.1
The symmetry of the NH_3 molecule consists of the rotations $1, 3, 3^{-1}$ and three reflections m_1, m_2, m_3. It is a group of order 6. The atoms occupy two special positions:
N atom with $|\mathcal{S}_N| = 6$ and $Z_N = 1$; three H atoms with $|\mathcal{S}_H| = 2$ and $Z_H = 3$. The symmetry for every one of the hydrogen atoms is alike. In this case, there exist three kinds of orbits:

(1) Special position on the threefold rotation axis. There is only one particle (N atom).
(2) Special position on a mirror plane. There are three symmetrically equivalent particles (H atoms).
(3) General position anywhere else comprising six equivalent particles (not actually existing in this example).

Definition 6.6 Two $\mathcal{P}_M$-orbits O_1 and O_2 (orbits under $\mathcal{P}_M$) belong to the same *Wyckoff position* if, after having selected two arbitrary points $X_1 \in O_1$ and $X_2 \in O_2$, their site symmetries $\mathcal{S}_1$ and $\mathcal{S}_2$ are conjugate in $\mathcal{P}_M$. In other words, there exists a symmetry operation g of the point group $\mathcal{P}_M$ that fulfils the equation $\mathcal{S}_2 = g^{-1}\mathcal{S}_1 g \quad (g \not\subset \mathcal{S}_1, \mathcal{S}_2)$.

Example 6.2
The pattern of the atoms of a metal-porphirin complex has the same two-dimensional point group as the square (cf. Fig. 5.1, page 51). The atoms C^1, C^6, C^{11}, C^{16} are symmetrically equivalent; they make up one orbit. The atoms H^1, H^6, H^{11}, H^{16} make up another orbit. The atom C^1 is situated on the mirror line m_y; its site symmetry group is $\mathcal{S}(C^1) = \{1, m_y\}$. The site symmetry group of the atom H^6 is $\mathcal{S}(H^6) = \{1, m_x\}$. With the aid of the group multiplication table (Table 5.3, page 53) we obtain:

$$4^{-1}\{1, m_y\}4 = \{1, m_x\}$$

Therefore, the site symmetry of the atom C^1 is conjugate to that of H^6; the orbits of the atoms C^1 and H^6, that is, the points $C^1, C^6, C^{11}, C^{16}, H^1, H^6, H^{11}$, and H^{16} belong to the same Wyckoff position. The orbit comprising the four atoms N^1, N^2, N^3, and N^4 belongs to another Wyckoff position: N^1 is situated on the mirror line m_- having the site symmetry group $\mathcal{S}(N^1) = \{1, m_-\}$. $\{1, m_-\}$ and $\{1, m_y\}$ are different site symmetry groups; there is no symmetry operation g of the square which fulfils $m_- = g^{-1}m_y g$. All other C and H atoms have the site symmetry group $\{1\}$ (general position); they commonly belong to another Wyckoff position.

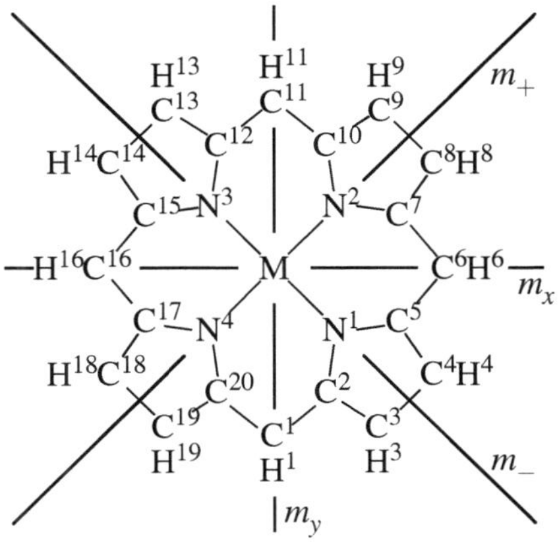

The example shows: The orbit of the symmetry-equivalent points C^1, C^6, C^{11}, C^{16} belongs to the same Wyckoff position as H^1, H^6, H^{11}, H^{16}. Atoms at special positions belong to the same Wyckoff position if they occupy symmetry elements that are equivalent by a symmetry operation of the point group. The mirror lines m_x and m_y are equivalent by fourfold rotation and thus are conjugate; all points on these mirror lines (their point of intersection excepted) belong to the same Wyckoff position. The point of intersection (position of the M atom) has another site symmetry and belongs to another Wyckoff position. The mirror lines m_+, m_- are not conjugate to m_x, m_y; points on them belong to two different Wyckoff positions.

Do not get irritated by the singular form of the terms 'general position', 'special position', and 'Wyckoff position'. Every such position may comprise many points (e.g. centres of atoms).

6.1.2 The space group and its point group

The 230 space-group types[1] were derived in 1891 by FEDOROV and by SCHOENFLIES, and somewhat later by BARLOW. The periodic structure of crystals was proven 21 years later (1912) when the first X-ray diffraction experiment was performed with a crystal by LAUE, FRIEDRICH, and KNIPPING. The periodicity had been assumed a long time before, but symmetry considerations had been possible only by observation of the shape of macroscopic real crystals. Like molecules, they belong to a finite point group. Nowadays we know that crystals are finite sections of periodic structures, and the problem of the macroscopic crystal symmetry is posed in a different way to how it is for molecules.

[1] Space group types, not space groups, cf. Sections 6.1.3 and 6.6.

Real crystals are formed by crystal growth, which involves a parallel advancement of crystal faces. The faces do not remain invariant, but the normals on them do. These are directions and thus have the character of vectors. Usually, the external shape of a crystal does not really correspond to its symmetry because of impairment during the growth process. The real symmetry of the ideally grown crystal is obtained from the normals: the bundle of vectors on the crystal faces is not susceptible to growth impairment.

Definition 6.7 The point group of a crystal structure is the symmetry group of the bundle of the normals on the crystal faces.

All kinds of faces that appear on a crystal species have to be considered, even very small faces that can be missing on individual crystals.

After having selected a coordinate system, the corresponding symmetry operations are not represented by matrix–column pairs $(\boldsymbol{W}, \boldsymbol{w})$, but only by the matrix parts $\boldsymbol{W}$, see Section 3.2, Theorem 3.5, page 25. Therefore, the group $\mathcal{P} = \{\boldsymbol{W}\}$ is finite, see Section 4.2, Theorem 4.1.

The set of *all* translations of a space group $\mathcal{G}$ forms a group $\mathcal{T}$, the translation group.

From a translation, the result of conjugation is always another translation, see eqn (6.1). It follows that:

Application of the rules of Section 3.2, eqns (3.31) and (3.12), yields (cf. page 23):

$$\mathbb{W}^{-1}\mathbb{T}\mathbb{W} = \left(\begin{array}{c|c} \boldsymbol{I} & \boldsymbol{W}^{-1}\boldsymbol{t} \\ \hline \boldsymbol{o}^{\mathrm{T}} & 1 \end{array}\right) = \left(\begin{array}{c|c} \boldsymbol{I} & \boldsymbol{t}' \\ \hline \boldsymbol{o}^{\mathrm{T}} & 1 \end{array}\right) = \mathbb{T}' \quad (6.1)$$

Theorem 6.8 The translation group $\mathcal{T}$ is not only a subgroup of the space group $\mathcal{G}$, but even a normal subgroup: $\mathcal{T} \lhd \mathcal{G}$.

What is the coset decomposition, see Section 5.3, of $\mathcal{G}$ with respect to $\mathcal{T}$? Let us consider an example:

Example 6.3
Left coset decomposition of the space group $Pmm2 = \{1, 2, m_x, m_y, t_1, t_2, t_3, \dots\}$ with respect to the translation group $\mathcal{T} = \{1, t_1, t_2, t_3, \dots\}$:

1st coset	2nd coset	3rd coset	4th coset
$1 \circ \mathcal{T} = 1 \circ 1,$	$2 \circ \mathcal{T} = 2 \circ 1,$	$m_x \circ \mathcal{T} = m_x \circ 1,$	$m_y \circ \mathcal{T} = m_y \circ 1,$
$1 \circ t_1,$	$2 \circ t_1,$	$m_x \circ t_1,$	$m_y \circ t_1,$
$1 \circ t_2,$	$2 \circ t_2,$	$m_x \circ t_2,$	$m_y \circ t_2,$
$1 \circ t_3,$	$2 \circ t_3,$	$m_x \circ t_3,$	$m_y \circ t_3,$
$\vdots$	$\vdots$	$\vdots$	$\vdots$

$\mathcal{T}$ is considered as an (infinite) column of the translations. The first coset is represented by the identity 1, represented by $(\boldsymbol{I}, \boldsymbol{o})$. Any other symmetry operation (in the example the twofold rotation 2), represented by $(\boldsymbol{W}_2, \boldsymbol{w}_2)$, is chosen as the representative of the second coset. The remaining symmetry operations of this coset are represented by $(\boldsymbol{I}, \boldsymbol{t}_i)(\boldsymbol{W}_2, \boldsymbol{w}_2) = (\boldsymbol{W}_2, \boldsymbol{w}_2 + \boldsymbol{t}_i)$, so that all elements of the second coset have the same matrix $\boldsymbol{W}_2$. There can be no elements with the matrix $\boldsymbol{W}_2$ that do not appear in the second column, etc.

Theorem 6.9 Every coset of the decomposition of $\mathcal{G}$ with respect to $\mathcal{T}$ contains exactly those elements which have the same matrix part. Every matrix $\boldsymbol{W}$ is characteristic for 'its' coset.

Therefore, the number of cosets of $\mathcal{G}/\mathcal{T}$ is exactly the same as the number of matrices $\boldsymbol{W}$. If every coset is considered to be a new (infinite) group element, then the group consisting of these elements is nothing other than the factor group $\mathcal{G}/\mathcal{T}$.

Multiplication of an element of the ith coset with one of the kth coset yields:

$$(\boldsymbol{W}_k, \boldsymbol{w}_k + \boldsymbol{t}_m)(\boldsymbol{W}_i, \boldsymbol{w}_i + \boldsymbol{t}_n) = (\boldsymbol{W}_k\boldsymbol{W}_i, \boldsymbol{w}_k + \boldsymbol{t}_m + \boldsymbol{W}_k\boldsymbol{w}_i + \boldsymbol{W}_k\boldsymbol{t}_n)$$

This is an element of the coset represented by $\boldsymbol{W}_j = \boldsymbol{W}_k\boldsymbol{W}_i$. In addition, we have:

$$(\boldsymbol{W}_i, \boldsymbol{w}_i)(\boldsymbol{W}_i^{-1}, \boldsymbol{w}_j) = (\boldsymbol{I}, \boldsymbol{W}_i\boldsymbol{w}_j + \boldsymbol{w}_i) = (\boldsymbol{I}, \boldsymbol{t}_k)$$

The cosets taken as the elements of the factor group $\mathcal{G}/\mathcal{T}$ thus have the same group multiplication table as the matrix parts that represent them (aside from their labelling).

Theorem 6.10 The factor group $\mathcal{G}/\mathcal{T}$ is isomorphic to the point group $\mathcal{P}$, or: The point group $\mathcal{P}$ is a homomorphic mapping of the space group $\mathcal{G}$ with the translation group $\mathcal{T}$ as its kernel, see Section 5.5, Theorem 5.9.

The point groups of crystals are classified like the point groups of molecules. The point groups of molecules act on points, and thus operate in point space; the point groups of crystals map vectors onto one another, they operate in vector space.

Definition 6.11 Two crystallographic point groups $\mathcal{P}_1$ and $\mathcal{P}_2$ belong to the same point group type, called the *crystal class*, if a basis can be found such that the matrix groups $\{\boldsymbol{W}_1\}$ of $\mathcal{P}_1$ and $\{\boldsymbol{W}_2\}$ of $\mathcal{P}_2$ coincide.

There are 32 crystal classes in space and 10 in the plane.

6.1.3 Classification of the space groups

A subdivision of a set into subsets is called a classification if every one of the subsets belongs to exactly one class. The classification of the crystallographic point groups also results in a classification of the space groups into 32 *crystal classes of space groups*. However, other classifications are more important. We discuss three of them: The classification of the space groups into seven crystal systems, into 219 affine space-group types, and into 230 crystallographic space-group types (or positive-affine space-group types). The latter are frequently called *the 230 space groups*, although they are not really space groups, but 230 classes of infinite numbers of equivalent space groups. Note that space group and space-group type have to be distinguished, as explained below (Definition 6.13; see also Section 6.6, page 86).

Among the 32 crystal classes there exist seven that correspond to point groups of lattices. These seven point groups are called *holohedries*. All point groups can uniquely be assigned to these holohedries. A point group $\mathcal{P}$ belongs to a holohedry $\mathcal{H}$ if:

1. $\mathcal{P} \leq \mathcal{H}$;
2. the index $|\mathcal{H}|/|\mathcal{P}|$ is as small as possible.

The assignment of space groups to point groups entails an assignment to the holohedries.

Definition 6.12 The seven holohedries assigned to the space groups are the seven *crystal systems of space groups*.

The crystal systems are: triclinic, monoclinic, orthorhombic, tetragonal, trigonal, hexagonal, and cubic. Contrary to the 'natural' sequence 'trigonal, tetragonal, hexagonal', 'trigonal' and 'hexagonal' are placed together due to their close relationship.

The crystal systems and the crystal classes of space groups form the basis for the sequence of the tables in *International Tables A*.

The classification of the space groups into crystal systems is more coarse than that into crystal classes. A finer subdivision, that is, a subdivision of the crystal classes, is also desirable; for example, among the crystal class 2 some space groups have 2_1 screw rotations (and no rotations 2), symbol $P2_1$, and others have only rotations, $P2$. Up to now, both of them belong to the same class, but they should be separated. This can be done as follows.

Consider every space group referred to an appropriate coordinate system, preferably the conventional crystallographic coordinate system. Then every space group is characterized by the set of its matrix–column pairs $\{(\boldsymbol{W}, \boldsymbol{w})\}$.

Definition 6.13 Two space groups $\mathcal{G}_1$ and $\mathcal{G}_2$ belong to the same *affine space-group type* or are called affine equivalent, if the sets $\{(\boldsymbol{W}_1, \boldsymbol{w}_1)\}$ and $\{(\boldsymbol{W}_2, \boldsymbol{w}_2)\}$ of their matrix–column pairs coincide, referred to an appropriate coordinate system. They belong to the same *crystallographic space-group type* if the sets $\{(\boldsymbol{W}_1, \boldsymbol{w}_1)\}$ and $\{(\boldsymbol{W}_2, \boldsymbol{w}_2)\}$ of their matrix–column pairs coincide, referred to an appropriate *right-handed* coordinate system.

In space, there exist 219 affine space-group types. The term reflects that the transformation from an appropriate coordinate system of $\mathcal{G}_1$ to that of $\mathcal{G}_2$ generally requires an affine transformation (i.e. with distortion of the lattice). $\mathcal{G}_1$ and $\mathcal{G}_2$ are two *different* space groups of the same space group type if their *lattice dimensions are different.*

The 230 crystallographic space-group types (often called, not quite correctly, 'the 230 space groups') are obtained if only right-handed coordinate systems are permitted. In chemistry, the distinction between right- and left-handed molecules can be essential, and in crystallography it is desirable to distinguish right-handed screw axes (e.g. 4_1) from left-handed ones (e.g. 4_3). This restriction causes a finer classification. Eleven affine space-group types split into enantiomorphic pairs of crystallographic space-group types (Table 6.1).

A convenient way to compare space groups for equivalence is to compare the space-group diagrams of *International Tables A*, see Section 6.4, or to compare the Hermann–Mauguin symbols.

Table 6.1 The 11 pairs of enantiomorphic space-group types (designated by their Hermann–Mauguin symbols, Section 6.3.1).

right-handed	— left-handed
$P3_1$	$P3_2$
$P3_121$	$P3_221$
$P3_112$	$P3_212$
$P4_1$	$P4_3$
$P4_122$	$P4_322$
$P4_12_12$	$P4_32_12$
$P6_1$	$P6_5$
$P6_2$	$P6_4$
$P6_122$	$P6_522$
$P6_222$	$P6_422$
$P4_132$	$P4_332$

6.2 The lattice of a space group

In crystallography, the vector lattice **T** is referred to a lattice basis or crystallographic basis $\mathbf{a}_1, \mathbf{a}_2, \mathbf{a}_3$, see Definition 2.5 (page 13). In this case all integral linear combinations $\mathbf{t} = t_1\mathbf{a}_1 + t_2\mathbf{a}_2 + t_3\mathbf{a}_3$ of the basis vectors are lattice vectors. A crystallographic basis can always be chosen such that *all* lattice vectors are integral linear combinations of the basis vectors; this is called a *primitive* basis, Definition 2.7. However, in crystallography, a *conventional crystallographic basis* is chosen, see Definition 2.6, such that the matrices of the symmetry operations become 'user friendly' and the metric tensor (cf. Definition 2.9, page 18) results in the simplest formulae for the calculation of distances and angles. This is achieved mainly by choosing basis vectors parallel to symmetry axes or perpendicular to symmetry planes (i.e. if they are symmetry adapted). As a consequence, the conventional basis is not always primitive; see the comments after Definition 2.7, page 14.

Definition 6.14 A lattice whose conventional basis is primitive is called a primitive lattice. The other lattices are called *centred lattices*.

A lattice is not 'primitive' or 'centred' as such, but (artificially) becomes so by the selection of the basis. The types of centring among the conventional bases in crystallography are the base centring A in the b–c plane, B in the a–c plane, C in the a–b plane, the face centring F of all faces, the body centring I (inner centring) in the middle of the cell, and the rhombohedral centring R (Fig. 6.1).

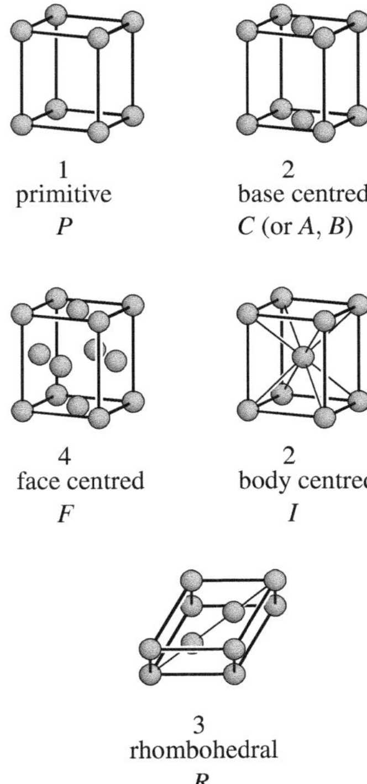

Fig. 6.1 Unit cells of centred bases and their symbols. The numbers specify how manifold primitive the respective cell is (i.e. by which factor the unit cell is enlarged relative to the corresponding primitive cell).

Referred to a primitive basis, the matrices of the translations have the form $(\boldsymbol{I}, \boldsymbol{t})$, with $\boldsymbol{t}$ being a column of *integral* numbers. Referred to a centred basis, the lattice vectors may have fractional numbers as coefficients.

The cell of a *primitive* lattice has no lattice vectors having their endpoints within the cell, the null vector (origin) excepted. The conventional *centred* lattices have centring vectors with the coefficients:

$A\ 0, \frac{1}{2}, \frac{1}{2}$ $\quad B\ \frac{1}{2}, 0, \frac{1}{2}$ $\quad C\ \frac{1}{2}, \frac{1}{2}, 0$ $\quad F\ 0, \frac{1}{2}, \frac{1}{2}\ \ \frac{1}{2}, 0, \frac{1}{2}\ \ \frac{1}{2}, \frac{1}{2}, 0$

$I\ \frac{1}{2}, \frac{1}{2}, \frac{1}{2}$ $\quad R\ \frac{2}{3}, \frac{1}{3}, \frac{1}{3}\ \ \frac{1}{3}, \frac{2}{3}, \frac{2}{3}$

The infinite set of all possible lattices is classified into Bravais types (also called Bravais lattices). The easiest way to envisage this classification is according to the space groups of their point lattices.

Definition 6.15 Two point lattices belong to the same *Bravais type* if their space groups belong to the same space-group type.

According to this definition, there exist 14 Bravais types, named after A. BRAVAIS, who was the first to derive them, in 1850. Since every point lattice also corresponds to a vector lattice, this includes a classification of the vector lattices (Table 6.2).

Table 6.2 The 14 Bravais types (Bravais lattices).

Name (abbreviation)	Metrics of the lattice	Centring
primitive triclinic (*aP*)	$a \neq b \neq c;\ \alpha \neq \beta \neq \gamma \neq 90°$	
primitive monoclinic (*mP*)	$a \neq b \neq c;\ \alpha = \gamma = 90°;\ \beta \neq 90°$	
base-centred monoclinic (*mC*)	$a \neq b \neq c;\ \alpha = \gamma = 90°;\ \beta \neq 90°$	$\frac{1}{2}, \frac{1}{2}, 0$
primitive orthorhombic (*oP*)	$a \neq b \neq c;\ \alpha = \beta = \gamma = 90°$	
base-centred orthorhombic (*oC*)	$a \neq b \neq c;\ \alpha = \beta = \gamma = 90°$	$\frac{1}{2}, \frac{1}{2}, 0$
face-centred orthorhombic (*oF*)	$a \neq b \neq c;\ \alpha = \beta = \gamma = 90°$	$0, \frac{1}{2}, \frac{1}{2};\ \frac{1}{2}, 0, \frac{1}{2};\ \frac{1}{2}, \frac{1}{2}, 0$
body-centred orthorhombic (*oI*)	$a \neq b \neq c;\ \alpha = \beta = \gamma = 90°$	$\frac{1}{2}, \frac{1}{2}, \frac{1}{2}$
primitive tetragonal (*tP*)	$a = b \neq c;\ \alpha = \beta = \gamma = 90°$	
body-centred tetragonal (*tI*)	$a = b \neq c;\ \alpha = \beta = \gamma = 90°$	$\frac{1}{2}, \frac{1}{2}, \frac{1}{2}$
primitive hexagonal (*hP*)	$a = b \neq c;\ \alpha = \beta = 90°;\ \gamma = 120°$	
rhombohedral trigonal (*hR*)*	$a = b \neq c;\ \alpha = \beta = 90°;\ \gamma = 120°$	$\frac{2}{3}, \frac{1}{3}, \frac{1}{3};\ \frac{1}{3}, \frac{2}{3}, \frac{2}{3}$
primitive rhombohedral (*rP*)*	$a = b = c;\ \alpha = \beta = \gamma \neq 90°$	
primitive cubic (*cP*)	$a = b = c;\ \alpha = \beta = \gamma = 90°$	
face-centred cubic (*cF*)	$a = b = c;\ \alpha = \beta = \gamma = 90°$	$0, \frac{1}{2}, \frac{1}{2};\ \frac{1}{2}, 0, \frac{1}{2};\ \frac{1}{2}, \frac{1}{2}, 0$
body-centred cubic (*cI*)	$a = b = c;\ \alpha = \beta = \gamma = 90°$	$\frac{1}{2}, \frac{1}{2}, \frac{1}{2}$

The symbol $\neq$ includes the possibility of a casual coincidence within experimental accuracy.
* *hR* and *rP* are identical, but with different settings of their basis vectors

6.3 Space-group symbols

Several kinds of symbols have been in use to designate space-group types. We deal with the Hermann–Mauguin symbols in detail and briefly with the Schoenflies symbols. In addition, especially in the Russian literature, the Fedorov symbols have been used [55]. For the historical development see Chapter 20. Since the Hermann–Mauguin symbols contain no information about the position of the chosen origin, although this is sometimes important, HALL developed correspondingly supplemented symbols [54].

6.3.1 Hermann–Mauguin symbols

The original version of the *Hermann–Mauguin symbols* is due to HERMANN[2] [56]; they were converted to an easy-to-use form by MAUGUIN[3] [57]. They are also called *international symbols*. Initially, the symbols were devised as a specification of a *system of generators of the space group*, see Section 5.2, letter **(f)**. This was not a system of as few generators as possible, but such that the space group could be generated in a most simple and clear way (for details, see *International Tables A*).[4]

[2] CARL HERMANN, 1898–1961.

[3] CHARLES MAUGUIN, 1878–1958.

[4] Section 1.4.3 in the sixth edition (2016) [15], Section 8.3.5 in previous editions [14].

Over the course of time, this view has changed: In *International Tables A*, a Hermann–Mauguin symbol designates the *symmetry* in outstanding directions, the *symmetry directions*. The symmetry in a symmetry direction **u** means the complete set $\{\mathsf{W}_i\}$ of symmetry operations W_i, whose rotation, screw, and rotoinversion axes or the normals on mirror or glide planes run parallel to **u**.

A direction **u** of non-trivial symmetry (i.e. higher than 1 or $\bar{1}$) is always a lattice direction, and a plane perpendicular to **u** is always a lattice plane. To perceive the '*symmetry in a symmetry direction* **u**', it is convenient to define a cell referred to **u** , that is, a symmetry-adapted cell.

Definition 6.16 A cell, defined by a shortest lattice vector in the direction of **u** and a *primitive basis* in the plane perpendicular to **u**, is called a *symmetry-adapted cell* or a *cell referred to* **u**.

If this cell is primitive, then the symmetry in the direction of **u** is *uniform*: All rotation or screw axes parallel to **u** are of the same kind; for example, only rotation axes 2 or only screw axes 2_1 or only screw axes 4_2; or only one kind of mirror or glide plane exists perpendicular to **u**; for example, only mirror planes m or only glide planes n.

If the cell is centred, rotations exist along with screw rotations and reflections along with glide reflections, or there jointly exist *different* kinds of screw rotations or of glide reflections. This is because the subsequent execution of a rotation and a centring translation results in a screw rotation, and a reflection and a translation results in a glide reflection. Examples: space group $C2$ has parallel 2 and 2_1 axes; space group $I4$ has 4 and 4_2 axes; space group $R3$ has 3, 3_1, and 3_2 axes; space group Cc has c and n glide planes.

Different symmetry directions can be symmetrically equivalent, for example the three fourfold axes parallel to the edges of a cube. Symmetry directions of this kind are combined to symmetry classes or *symmetry direction systems*. In crystals sometimes there are up to three classes of symmetrically equivalent directions of non-trivial symmetry. From each of these classes one *representative symmetry direction* is selected, the direction and its counter-direction being considered as *one* symmetry direction.

In *International Tables A*, the *full Hermann–Mauguin symbol* first designates the conventional lattice type (P, A, B, C, F, I, or R; Section 6.2).

The full Hermann–Mauguin symbol then specifies one system of generators of the symmetry group for every representative symmetry direction. If the normals to reflection or glide planes are parallel to rotation or screw axes, then both are separated by a fraction bar, for example, $2/c$, $6_3/m$ (however, $\bar{6}$ is used instead of $3/m$). Rotations are specified with priority over screw rotations, and reflections over glide reflections.

In the **full Hermann–Mauguin symbol** the *kind of symmetry* is specified by its *component* and the *orientation* of the symmetry direction by the *place in the symbol.* The sequence of the representative symmetry directions in the symbol depends on the crystal system. The crystal system and the number of representative symmetry directions is revealed by the Hermann–Mauguin symbol:

(1) *No* symmetry direction: *triclinic* (only $P1$ and $P\bar{1}$).

(2) *One* symmetry direction, twofold symmetry: *monoclinic.* The representative symmetry direction in crystallography is usually **b**, in physics and chemistry sometimes **c**, occasionally **a**. The full Hermann–Mauguin symbol of monoclinic space groups also states the symmetry in the **a**, **b**, and **c** directions (as in the orthorhombic system), each of the two non-symmetry directions being labelled by a '1'. Examples: $P211$ (**a** axis setting), $P121$ (**b** axis setting), $P112$ (**c** axis setting); $P1m1$ (**b** axis setting), $P11m$ (**c** axis setting); $C12/c1$ (**b** axis setting); $A112/a$ (**c** axis setting); $P12_1/c1$ (**b** axis setting), $P112_1/a$ (**c** axis setting). If a non-conventional setting is selected, the full symbol must be specified; the short symbol (see below) then is not sufficient.

(3) *Three* mutually perpendicular symmetry directions parallel to the coordinate axes, *only twofold* symmetries: *orthorhombic.* The sequence of the symmetry directions is **a**, **b**, **c**. Examples: $P222_1$, $I222$, $Cmc2_1$, $P2_1/n2_1/n2/m$, $F2/d2/d2/d$.

(4) *One* symmetry direction with *higher than twofold* symmetry in the point group (3, $\bar{3}$, 4, $\bar{4}$, $4/m$, 6, $\bar{6}$, or $6/m$): *trigonal*, *tetragonal*, or *hexagonal* space groups. The direction of this symmetry direction is **c**. The other representative symmetry directions are oriented perpendicular to **c** and have a maximal order of 2. The sequence of the symmetry directions in the Hermann–Mauguin symbol is **c**, **a**, **a** − **b**. Examples: $P31c$, $P3c1$, $P6_522$, $P\bar{6}2m$, $P\bar{6}m2$, $P6_3/m2/c2/m$. If there are no symmetry axes in both of the directions **a** and **a** − **b**, they are mentioned only in exceptional cases, that is, only the short Hermann–Mauguin symbol is given. Examples: $P4_1$, $P6/m$ (meaning $P4_1 11$ and $P6/m11$).

With a rhombohedral R lattice there are only two representative symmetry directions: **c** and **a** for a hexagonal setting of the coordinate system, corresponding to $[111]$ and $[1\bar{1}0]$ for a rhombohedral coordinate system (i.e. $a = b = c$; $\alpha = \beta = \gamma \neq 90°$). Examples: $R3$, $R\bar{3}$, $R32$, $R3m$, $R3c$, $R\bar{3}2/c$.

(5) One symmetry direction system of *four* directions having threefold axes parallel to the four body diagonals of the cube: *cubic*. A further symmetry direction system runs parallel to the cube edges and in some cases another symmetry direction system runs parallel to the six face diagonals. The sequence of the representative symmetry directions in the Hermann–Mauguin symbol is **a** (cube edge), **a** + **b** + **c** (body diagonal); if present, additionally **a** + **b** (face diagonal).
Note: Contrary to trigonal and rhombohedral space groups, cubic space groups do not have the component 3 or $\overline{3}$ directly after the lattice symbol, but in the third position.
Examples: $P23$ ($P321$ and $P312$ are trigonal), $Ia\overline{3}$, $P4_2 32$, $F\overline{4}3m$, $P4_2/m\overline{3}2/n$, $F4_1/d\overline{3}2/m$.

Particularities:

(1) The presence of points of inversion is mentioned only for $P\overline{1}$. In all other cases the presence or absence of points of inversion can be recognized as follows: they are present if and only if there are either rotoinversion axes of odd order or rotation axes of even order perpendicular to planes of reflection or glide reflection (e.g. $2/m$, $2_1/c$, $4_1/a$, $4_2/n$, $6_3/m$). Only the full symbol reveals this for some space groups.

(2) $P\overline{4}2m$ and $P\overline{4}m2$ are the Hermann–Mauguin symbols of different space-group types. $P3$, $P3_1$, $R3$, $R3c$ are correct symbols, but not so $P32$ or $P3c$. For the latter, the symmetry in the directions **a** *and* **a** − **b** *must* be mentioned: $P321$ and $P312$ are a pair of different space-group types.

(3) In two cases it is necessary to depart from the rule that rotations are mentioned with priority over screw rotations. Otherwise two space-group types would obtain the same Hermann–Mauguin symbol $I222$ and another two would obtain $I23$. These four space group types have twofold rotation axes that run parallel to the coordinate axes along with twofold screw axes, due to the I centring. For one space-group type each of the mentioned symbols is maintained (when non-parallel axes 2 *intersect* each other); the other two types are labelled $I2_1 2_1 2_1$ and $I2_1 3$ (they have non-intersecting axes).

In the **short Hermann–Mauguin symbol** the symmetry information of the full symbol is reduced, and the symbols are simpler, but remain sufficiently informative (the symbol contains at least one set of generators of the space group). The reflections and glide reflections of the full symbol are kept in the short symbol; only rotations and screw rotations are omitted. In the short symbols of centrosymmetric orthorhombic space groups only three planes are mentioned, for example $Pbam$ (full symbol $P2_1/b2_1/a2/m$); see Table 6.3.

Only C centrings and c glide reflections are used for the short symbols of monoclinic space groups. The full symbol must be given for other settings, for example $A112/m$, not $A2/m$; $P12_1/n1$, not $P2_1/n$ (although the symbol $P2_1/n$ abounds in the literature).

The very scarcely used *extended Hermann–Mauguin symbol* specifies nearly the entire symmetry of every symmetry direction. That includes additionally present screw axes and glide planes in the case of centred lattices. We do not deal here with the details; they can be looked up in *International Tables A*.[5]

[5] Section 1.5.4 in the sixth edition (2016) Section 4 in previous editions.

Table 6.3 Examples of short and full Hermann–Mauguin symbols.

short	full	short	full	short	full
Cm	$C1m1$	$Pbcm$	$P2/b2_1/c2_1/m$	$P\bar{3}c1$	$P\bar{3}2/c1$
$P2_1/c$	$P12_1/c1$	$Cmcm$	$C2/m2/c2_1/m$	$P6/mcc$	$P6/m2/c2/c$
$C2/c$	$C12/c1$	$P4_2/nmc$	$P4_2/n2_1/m2/c$	$Fd\bar{3}m$	$F4_1/d\bar{3}2/m$

The short Hermann–Mauguin symbol can be completed to the full symbol, and from this it is possible to derive the full set of symmetry operations of the space group. However, some familiarity with Hermann–Mauguin symbols is required before they can be handled securely in difficult cases. This is due to the fact that the Hermann–Mauguin symbols depend on the orientation of the space-group symmetry relative to the conventional basis. This property makes them more informative but also less easy to handle. Different symbols can refer to the same space-group type.

Example 6.4
$P2/m2/n2_1/a$, $P2/m2_1/a2/n$, $P2_1/b2/m2/n$, $P2/n2/m2_1/b$, $P2/n2_1/c2/m$, and $P2_1/c2/n2/m$ denote the same orthorhombic space group type, No. 53 ($Pmna$).

$P2_1/n2_1/m2_1/a$, $P2_1/n2_1/a2_1/m$, $P2_1/m2_1/n2_1/b$, $P2_1/b2_1/n2_1/m$, $P2_1/m2_1/c2_1/n$, and $P2_1/c2_1/m2_1/n$ are the (full) Hermann–Mauguin symbols of another orthorhombic space group type, No. 62 ($Pnma$).

The symbols mentioned first refer to the conventional settings. The other five are non-conventional settings, with differently oriented bases. More details on non-conventional settings are the subject of Section 10.3.

The point group symbol corresponding to a space group can be obtained from the Hermann–Mauguin symbol in the following way:

1. the lattice symbol is deleted (P, A, B, C, F, I or R);
2. all screw components are deleted (the subscript ciphers are deleted);
3. the letters for glide reflections (a, b, c, n, d, e) are replaced by m.

Examples: $C2/c \rightarrow 2/m$

$P2/m2/n2_1/a$ (short $Pmna$) $\rightarrow 2/m2/m2/m$ (short mmm)

$I\bar{4}2d \rightarrow \bar{4}2m$

$I4_1/a\bar{3}2/d$ (short $Ia\bar{3}d$) $\rightarrow 4/m\bar{3}2/m$ (short $m\bar{3}m$)

6.3.2 Schoenflies symbols

Schoenflies symbols were developed 35 years before the Hermann–Mauguin symbols. Compared to their original form, some of them have been slightly altered.

Rotoreflections are used instead of rotoinversions. A rotoreflection results from a coupling of a rotation with a reflection through a plane perpendicular to the rotation axis. Rotoreflections and rotoinversions state identical facts, but the orders of their rotations differ in pairs if they are not divisible by 4:

rotoreflection (Schoenflies)	S_1	S_2	S_3	S_6	S_4
rotoinversion (Hermann–Mauguin)	$\overline{2} = m$	$\overline{1}$	$\overline{6}$	$\overline{3}$	$\overline{4}$

In Section 6.1.2 the space groups are assigned to crystal classes according to their point groups. SCHOENFLIES[6] introduced symbols for these crystal classes (point-group types) in the following way:

[6] ARTHUR SCHOENFLIES, 1853–1928.

C_1 no symmetry.

C_i a centre of inversion is the only symmetry element.

C_s a plane of reflection is the only symmetry element.

C_N an N-fold rotation axis is the only symmetry element.

S_N an N-fold rotoreflection axis is the only symmetry element; only S_4 is used; for symbols replacing S_3 and S_6 see the following.

C_{Ni} there is an N-fold rotation axis (N odd) and a centre of inversion on the axis. Identical to S_M with $M = 2 \times N$.

D_N there are N twofold rotation axes perpendicular to an N-fold rotation axis.

C_{Nh} there is a vertical N-fold rotation axis and a horizontal reflection plane. C_{3h} is identical to S_3. There is an inversion centre if N is even.

C_{Nv} an N-fold vertical rotation axis is situated at the intersection line of N vertical reflection planes.

D_{Nh} there is an N-fold vertical rotation axis, N horizontal twofold rotation axes, N vertical reflection planes, and a horizontal reflection plane. There is an inversion centre if N is even.

D_{Nd} an N-fold vertical rotation axis contains a $2N$-fold rotoreflection axis and N horizontal twofold axes have bisecting directions between N vertical reflection planes. There is an inversion centre if N is odd. Identical to S_{Mv} with $M = 2 \times N$.

O_h symmetry of an octahedron and a cube.

O as O_h without reflection planes (rotations of an octahedron).

T_d symmetry of a tetrahedron.

T_h symmetry of an octahedron with twofold instead of fourfold axes.

T as T_d and T_h without reflection planes (rotations of a tetrahedron).

Special non-crystallographic point groups:

I_h symmetry of an icosahedron and pentagonal dodecahedron.

I as I_h without reflection planes (rotations of an icosahedron).

$C_{\infty v}$ symmetry of a cone.

$D_{\infty h}$ symmetry of a cylinder.

K_h symmetry of a sphere.

The space-group types belonging to a crystal class were simply numbered consecutively by SCHOENFLIES; they are identified by superscript numbers.

The sequence of the crystal classes has not always been kept the same in the space-group tables. Since 1952 the space-group types have been numbered in *International Tables* from 1 to 230, with the consequence that this sequence can hardly be changed.

Some Schoenflies symbols are compared with the corresponding Hermann–Mauguin symbols in Table 6.4.

Schoenflies space-group symbols have the advantage that they designate the space-group types in a unique way and independent of the selection (setting) of a basis. They have the disadvantage that they only give direct information about the point-group symmetry. They lack information about the lattice type, which is expressed only indirectly by the superscript number.

Schoenflies symbols are concise, but contain less information than Hermann–Mauguin symbols. Schoenflies symbols continue to be very popular in spectroscopy, quantum chemistry, and to designate the symmetry of molecules. In crystallography they are hardly used anymore.

Table 6.4 Comparison of Schoenflies and Hermann–Mauguin symbols of the crystallographic and some additional point-group types and examples for a few space-group types.

Schoen-flies	Hermann–Mauguin	Schoen-flies	Hermann–Mauguin	Schoen-flies	Hermann–Mauguin short	full
Point-group types						
C_1	1	C_i	$\bar{1}$	C_s	m	
C_2	2	C_{2h}	$2/m$	C_{2v}	$mm2$	
C_3	3	$C_{3h} = S_3$	$\bar{6} = 3/m$	C_{3v}	$3m$	
C_4	4	C_{4h}	$4/m$	C_{4v}	$4mm$	
C_6	6	C_{6h}	$6/m$	C_{6v}	$6mm$	
S_4	$\bar{4}$	$C_{3i} = S_6$	$\bar{3}$	$C_{\infty v}$	∞m	
D_2	222	$D_{2d} = S_{4v}$	$\bar{4}2m$	D_{2h}	mmm	$2/m2/m2/m$
D_3	32	D_{3h}	$\bar{6}2m$	D_{3d}	$\bar{3}m$	$\bar{3}2/m$
D_4	422	$D_{4d} = S_{8v}$	$\bar{8}2m$	D_{4h}	$4/mmm$	$4/m2/m2/m$
D_5	52	D_{5h}	$\overline{10}2m$	D_{5d}	$\bar{5}m$	$\bar{5}2/m$
D_6	622	$D_{6d} = S_{12v}$	$\overline{12}2m$	D_{6h}	$6/mmm$	$6/m2/m2/m$
				$D_{\infty h}$	∞/mm	$\infty/m2/m = \bar{\infty}2/m$
T	23	T_d	$\bar{4}3m$	T_h	$m\bar{3}$	$2/m\bar{3}$
		O	432	O_h	$m\bar{3}m$	$4/m\bar{3}2/m$
		I	235	I_h	$m\bar{3}\bar{5}$	$2/m\bar{3}\bar{5}$
Space-group types						
C_1^1	$P1$	C_i^1	$P\bar{1}$	C_s^1	Pm	$P1m1$
C_2^1	$P2$	C_2^2	$P2_1$	C_{2h}^5	$P2_1/c$	$P12_1/c1$
D_2^1	$P222$	C_{2v}^{12}	$Cmc2_1$	D_{2h}^{16}	$Pnma$	$P2_1/n2_1/m2_1/a$
C_{4h}^6	$I4_1/a$	D_{2d}^3	$P\bar{4}2_1m$	D_{4h}^9	$P4_2/mmc$	$P4_2/m2/m2/c$
C_{3i}^2	$R\bar{3}$	C_{6h}^2	$P6_3/m$	D_{6h}^4	$P6_3/mmc$	$P6_3/m2/m2/c$
T_d^2	$F\bar{4}3m$	O^3	$F432$	O_h^5	$Fm\bar{3}m$	$F4/m\bar{3}2/m$

6.4 Description of space-group symmetry in *International Tables A*

In most cases, the information concerning each space-group type can be found in *International Tables A* on two facing pages. The space-group symmetry is shown by diagrams, a list of the symmetry operations, and a table of the Wyckoff positions.

6.4.1 Diagrams of the symmetry elements

Consider as an example the space-group type $Pbcm$, No. 57. On the left page of the facing pages in *International Tables A* there are three diagrams, showing projections of the geometric sites of the symmetry elements of one unit cell (Fig. 6.2). As in the case of all orthorhombic space groups, every one of the three diagrams has two space-group symbols. The symbol mentioned in the heading and on top of the first diagram refers to the conventional setting. The other five symbols refer to non-conventional settings. If the book is turned such that the letters of a space-group symbol next to a diagram are upright, the **a** axis of the diagram points downwards and the **b** axis to the right. Beginning with the sixth edition of *International Tables A* (2016), the orientation of two axes has been marked with *a*, *b*, or *c* at each diagram; the third axis is perpendicular to the plane of the paper and points from the origin towards the reader, the origin being at the corner marked with a 0. These marks refer only to the conventional setting $Pbcm$.

Table 6.5 The most important graphic symbols for symmetry elements.

$\bar{1}$ ○					
Axes perpendicular to the paper plane					
2	2_1	$\bar{1}$ on 2	$\bar{1}$ on 2_1		
3	3_1	3_2	$\bar{3}$		
4	4_1	4_2	4_3	$\bar{1}$ on 4	$\bar{1}$ on 4_2
6	6_1	6_2	6_3	6_4	6_5
$\bar{1}$ on 6	$\bar{1}$ on 6_3	$\bar{4}$	$\bar{6}$		
Axes parallel to the paper plane					
2	2_1	4	$\bar{4}$	4_1	4_2
Axes inclined to the paper plane					
2	2_1	3	$\bar{3}$	3_1	3_2
Planes parallel to the paper plane; axes' directions **b**, **a**					
m	*a*	*b*	*n*	*e*	
Planes perpendicular to the paper plane; axes' directions **b**, **a**					
m	*b*	*c*	*n*	*d*	*e*

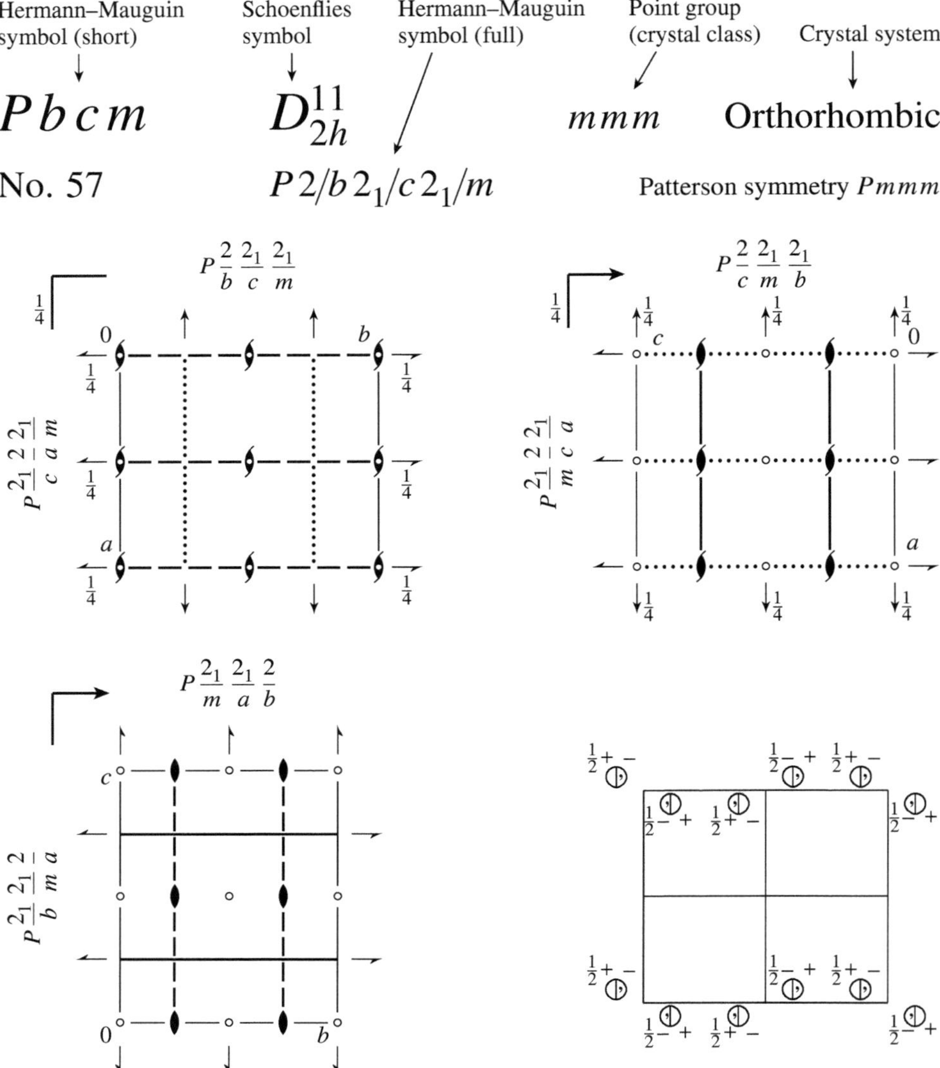

Fig. 6.2 Heading and diagrams from *International Tables A* for the orthorhombic space-group type $Pbcm$. Orientation of axes when the letters are upright: →**b** ↓**a**

Redrawn from [14] with permission of the International Union of Crystallography.

The symmetry elements are depicted by graphic symbols (Table 6.5). The kind of symmetry element and its orientation follow from the symbol. The heights z in the direction of **c** (direction of view) are specified for points of inversion and for axes and planes parallel to the paper plane as fractional numbers $0 < z < \frac{1}{2}$ if $z \neq 0$. All symmetry elements at height z occur again at height $z + \frac{1}{2}$.

Triclinic and monoclinic space groups are also shown by three projections along the three coordinate axes; all of them are referred to the conventional setting, which is mentioned in the heading. The axes are marked in the diagrams. Monoclinic space groups are treated with the two settings with symmetrically unique axis b and c. In those cases in which, in addition to this monoclinic axis, there is another special direction (centring vector, glide vector), three different cell choices are distinguished; therefore, the space-group types $C2$, Pc, Cm, Cc, $C2/m$, $P2/c$, $P2_1/c$, and $C2/c$ do not take the usual two pages, but eight pages per space-group type.

For each tetragonal, trigonal, hexagonal, and cubic space-group type only one diagram is shown in projection along **c**, having the standard orientation **a** downwards and **b** to the right. Trigonal space groups with rhombohedral lattice are described twice each, for a rhombohedral (primitive) cell and a hexagonal (rhombohedrally centred) cell; the diagrams for both are identical. The depiction of the symmetry elements for cubic space groups with an F-centred lattice only comprises one quarter of the cell, the contents of the other three quarters being the same.

6.4.2 Lists of the Wyckoff positions

For every space-group type there is a table of Wyckoff positions under the heading 'Positions', usually on the right page of the facing pages.[7] For the space-group type $Pbcm$ it reads:

Multiplicity, Wyckoff letter, Site symmetry			Coordinates			
8	e	1	(1) x,y,z	(2) $\bar{x},\bar{y},z+\frac{1}{2}$	(3) $\bar{x},y+\frac{1}{2},\bar{z}+\frac{1}{2}$	(4) $x,\bar{y}+\frac{1}{2},\bar{z}$
			(5) $\bar{x},\bar{y},\bar{z}$	(6) $x,y,\bar{z}+\frac{1}{2}$	(7) $x,\bar{y}+\frac{1}{2},z+\frac{1}{2}$	(8) $\bar{x},y+\frac{1}{2},z$
4	d	$..m$	$x,y,\frac{1}{4}$	$\bar{x},\bar{y},\frac{3}{4}$	$\bar{x},y+\frac{1}{2},\frac{1}{4}$	$x,\bar{y}+\frac{1}{2},\frac{3}{4}$
4	c	$2..$	$x,\frac{1}{4},0$	$\bar{x},\frac{3}{4},\frac{1}{2}$	$\bar{x},\frac{3}{4},0$	$x,\frac{1}{4},\frac{1}{2}$
4	b	$\bar{1}$	$\frac{1}{2},0,0$	$\frac{1}{2},0,\frac{1}{2}$	$\frac{1}{2},\frac{1}{2},\frac{1}{2}$	$\frac{1}{2},\frac{1}{2},0$
4	a	$\bar{1}$	$0,0,0$	$0,0,\frac{1}{2}$	$0,\frac{1}{2},\frac{1}{2}$	$0,\frac{1}{2},0$

The Wyckoff positions are numbered alphabetically from bottom to top by the Wyckoff letters in the second column. The Wyckoff position with the highest site symmetry is always placed in the bottom line and has the letter a. For any point belonging to a Wyckoff position, the number of symmetry-equivalent points within one unit cell is specified by the multiplicity in the first column. Usually, a Wyckoff position is labelled by its multiplicity and the Wyckoff letter, for example $4d$ for the second Wyckoff position in the preceding list.

The site symmetry in the third column is stated in an oriented way, having the same sequence of symmetry directions as in the space-group symbol (page 74). Those representative symmetry directions that have no symmetry higher than 1 or $\bar{1}$ are marked by points. For example, the site symmetry given for the Wyckoff position $4d$ is $..m$; therefore, its site symmetry is a reflection through a plane perpendicular to the third symmetry direction (**c**).

The general position is always the first one of the list of Wyckoff positions; it is labelled by the 'highest' necessary lower-case letter of the alphabet (for $Pbcm$ this is e), and it always has the site symmetry 1. Exceptionally, the general position of $Pmmm$ is labelled α, because $Pmmm$ has 27 Wyckoff positions, which is one more than the number of letters in the alphabet.

[7] The term 'positions' used here in *International Tables* is unfortunate, because what is meant is 'Wyckoff positions'. In the first (trilingual) edition of *International Tables* (1935) [26] the German term *Punktlage* had been used. It means the set of points that are symmetry-equivalent to a point in a space group. In the later English editions of *International Tables* the term *Punktlage* was translated to 'position'; it has been continued to be used with this meaning, although it differs from the common meaning of the word position. In every space group, there exists only *one* general position (in singular).

The coordinate triplets of the general position are numbered, for $Pbcm$ from (1) to (8). As explained in Section 4.1 (page 43), the corresponding symmetry operation can be derived from a coordinate triplet: The coordinate triplet is transcribed to a matrix–column pair and its geometric meaning is derived by the method explained in Section 4.3 (page 47):

$$(1): \begin{pmatrix} 1 & 0 & 0 \\ 0 & 1 & 0 \\ 0 & 0 & 1 \end{pmatrix}, \begin{pmatrix} 0 \\ 0 \\ 0 \end{pmatrix} \quad (2): \begin{pmatrix} -1 & 0 & 0 \\ 0 & -1 & 0 \\ 0 & 0 & 1 \end{pmatrix}, \begin{pmatrix} 0 \\ 0 \\ \frac{1}{2} \end{pmatrix} \quad \ldots \quad (8): \begin{pmatrix} -1 & 0 & 0 \\ 0 & 1 & 0 \\ 0 & 0 & 1 \end{pmatrix}, \begin{pmatrix} 0 \\ \frac{1}{2} \\ 0 \end{pmatrix}$$

The general position is the most important one, but it is not the only way to express the space-group symmetry. In addition, the matrix–column pairs of the corresponding symmetry operations are explicitly listed, see the next section.

6.4.3 Symmetry operations of the general position

The symmetry operations are listed under the heading 'Symmetry operations', usually on the left page of the facing pages. The symmetry operation corresponding to a coordinate triplet of the general position is specified after the number of the coordinate triplet. For the example of the space-group type $Pbcm$ the listing is:

(1) 1	(2) $2(0,0,\frac{1}{2})\quad 0,0,z$	(3) $2(0,\frac{1}{2},0)\quad 0,y,\frac{1}{4}$	(4) $2\quad x,\frac{1}{4},0$
(5) $\bar{1}\quad 0,0,0$	(6) $m\quad x,y,\frac{1}{4}$	(7) $c\quad x,\frac{1}{4},z$	(8) $b\quad 0,y,z$

This is to be interpreted in the following way:

The symmetry operation (1) is the identity;

(2) is a twofold rotation about the axis $0,0,z$ combined with a shift by $(0,0,\frac{1}{2})$, that is, it is a twofold screw rotation;

(3) is a twofold rotation about the axis $0,y,\frac{1}{4}$ combined with a shift by $(0,\frac{1}{2},0)$, that is, a twofold screw rotation;

(4) is a twofold rotation about the axis $x,\frac{1}{4},0$;

(5) is an inversion through the point $0,0,0$;

(6) is a reflection through the plane $x,y,\frac{1}{4}$;

(7) is a glide reflection with the glide direction c through a glide plane $x,\frac{1}{4},z$;

(8) is a glide reflection with the glide direction b through a glide plane $0,y,z$.

In general, an entry consists of the following data:

1. (n) number of the coordinate triplet.
2. Hermann–Mauguin symbol of the operation, for example 2 or c. The sense of rotation is marked by $^+$ or $^-$, for example $\bar{4}^+$ or $\bar{4}^-$. If a triplet of numbers in parentheses follows, for example $2(0,0,\frac{1}{2})$, this corresponds to the column of the screw or glide vector.
3. Parameterized representation of the symmetry element (point, axis, or plane), for example $0,0,z$ or $0,y,\frac{1}{4}$ or $x,\frac{1}{4},z$. For rotoinversions, the axis and the point of inversion are given, for example $\bar{4}^-\ 0,\frac{1}{2},z;\ 0,\frac{1}{2},\frac{1}{4}$.

The listing for space groups with centred lattices consists of several blocks, one for $0,0,0$ and one for each centring vector, which is mentioned on top of the block, for example 'For $(\frac{1}{2},\frac{1}{2},\frac{1}{2})+$ set'. Translations as symmetry operations occur only in the latter blocks, specified, for example, by $t(\frac{1}{2},\frac{1}{2},\frac{1}{2})$. Further exhaustive explanations can be found, as for all components of the space-group tables, in the instructions in Chapter 2.1 of *International Tables A* 2016 (Guide to the use of the space-group tables).

6.4.4 Diagrams of the general positions

Only one diagram per space group is shown for the general position. It has the standard orientation (**a** downward, **b** to the right; Fig. 6.2 bottom right). Monoclinic space groups have one diagram each for the two settings 'unique axis *b*' and 'unique axis *c*'. The outlines of the cell have been drawn as thin lines, as well as the lines $x,\frac{1}{2},0$ and $\frac{1}{2},y,0$ ($x=y$ for hexagonal cells). The starting point of the orbit is within the cell, close to the origin (upper left corner), slightly above the plane of the paper, which is expressed by a $+$ sign (for $+z$). The depicted points cover all points within the unit cell and points in the close vicinity of the unit cell; their heights are given as $\frac{1}{2}+$ (for $\frac{1}{2}+z$), $-$ (for $-z$), $\frac{1}{2}-$ (for $\frac{1}{2}-z$), etc. (y for monoclinic *b* settings).

The points are represented by circles. Each point corresponds to exactly one symmetry operation, which maps the starting point onto the considered point. Image points of symmetry operations of the second kind, that is, those with $\det(\boldsymbol{W})=-1$ for their matrix part $\boldsymbol{W}$, are marked by a comma in the middle of the circle. If the starting point were a right-hand glove, the points with a comma would correspond to left-hand gloves. If there are reflection planes parallel to the paper plane, the points of projection of equivalent points coincide. In this case the circle is subdivided by a vertical line, and exactly one of the semicircles contains a comma.

The diagrams for cubic space groups differ somewhat (if available, look up the page of a cubic space group in *International Tables A*). The points of the orbit are connected by lines forming a polyhedron around the origin and around its translationally equivalent points. Up to the fifth edition (2006), three diagrams form two pairs of stereoscopic views, allowing stereo views of the configurations (look at the left image with the left eye and at the central image with the right eye or at the central image with the left eye and at the right image with the right eye). These stereoscopic views have been discontinued in the sixth edition of *International Tables A* (2016). However, for the space-group types No. 221 ($Pm\overline{3}m$) to 230 ($Ia\overline{3}d$) there are more clearly arranged images with slightly inclined unit cells and shaded polyhedra.

6.5 General and special positions of the space groups

The *site symmetry group* $\mathcal{S}_X$ of a point X is defined for a space group $\mathcal{G}$ in the same way as for a molecular symmetry group $\mathcal{P}_M$: it is the subgroup of $\mathcal{G}$ consisting of those symmetry operations of $\mathcal{G}$ that leave X unchanged, see

Definition 6.3. Again, *general* and *special positions* are distinguished, Definition 6.4. However, in the case of space groups it is not immediately clear that the order $|\mathcal{S}|$ of $\mathcal{S}$ must be finite. This follows from the following:

Theorem 6.17 The matrix–column pairs $(\boldsymbol{W}_k, \boldsymbol{w}_k)$ of the elements $s_k \in \mathcal{S}$ have different matrix parts $\boldsymbol{W}_k$; each $\boldsymbol{W}_k$ can occur at most once.

If two group elements $s_m \in \mathcal{S}$ and $s_n \in \mathcal{S}$ had the same matrices, $\boldsymbol{W}_m = \boldsymbol{W}_n$, the following would hold (cf. eqns (3.10) and (3.12) in Section 3.2, page 23):

$$\mathbb{W}_m \mathbb{W}_n^{-1} = \left(\begin{array}{c|c} \boldsymbol{W}_m & \boldsymbol{w}_m \\ \hline \boldsymbol{o}^{\mathrm{T}} & 1 \end{array}\right) \left(\begin{array}{c|c} \boldsymbol{W}_n^{-1} & -\boldsymbol{W}_n^{-1}\boldsymbol{w}_n \\ \hline \boldsymbol{o}^{\mathrm{T}} & 1 \end{array}\right) = \left(\begin{array}{c|c} \boldsymbol{I} & \boldsymbol{w}_m - \boldsymbol{w}_n \\ \hline \boldsymbol{o}^{\mathrm{T}} & 1 \end{array}\right) = \mathbb{T}$$

That is a translation. However, translations have no fixed points and thus cannot be part of the site symmetry. The group $\{\boldsymbol{W}\}$ of *all* matrix parts of $\mathcal{G}$ is finite, see Section 4.2 (page 43), and, therefore, so is $\mathcal{S}$.

The Wyckoff positions of the space groups $\mathcal{G}$ yield the real base for a concise and complete description of a crystal structure. In Section 2.3 (page 13) two ways of putting together a crystal structure are described:

1. by lining up unit cells;
2. by interlacing particle lattices.

Now we can add:

(3) We start from the centre of gravity of a particle in the unit cell and add the centres of gravity of the corresponding (infinite) $\mathcal{G}$-orbit. We continue with the centre of gravity of a particle not yet considered, etc.; in this way the crystal structure is set up from a finite number of $\mathcal{G}$-orbits. Usually, several point lattices form part of a $\mathcal{G}$-orbit.

Whereas the coordinate triplets of the points of a point group designate individual points, the coordinate triplets of the Wyckoff positions of the space groups are representatives of their point lattices. The matrix–column pairs derived from the general position do not represent single mappings, but cosets of $\mathcal{G}$ with respect to $\mathcal{T}$. The multiplicity Z in the first column of the tables corresponds to the product of the order of the crystal class and the number of centring vectors, divided by the order $|\mathcal{S}|$ of the site-symmetry group $\mathcal{S}$. This is nothing other than the number of symmetry-equivalent points in the unit cell.

A Wyckoff position consists of infinitely many $\mathcal{G}$-orbits if the coordinate triplet of the representing point includes at least one free parameter, see the following example. If there is no free parameter, as in the case of the Wyckoff position $4b$ $\bar{1}$ $\frac{1}{2}$,0,0 of the space-group type $Pbcm$, the Wyckoff position consists of only one $\mathcal{G}$-orbit.

Example 6.5
In the space-group $Pbcm$ the $\mathcal{G}$-orbits $\mathcal{G}X_1$ for $X_1 = 0.094, \frac{1}{4}, 0$, and $\mathcal{G}X_2$ for $X_2 = 0.137, \frac{1}{4}, 0$ belong to the same Wyckoff position $4c$ $x, \frac{1}{4}, 0$ with the site symmetry 2. X_1 and X_2 belong to the *same* site symmetry group $\mathcal{S}$, consisting of the identity and a twofold rotation, but their orbits are different.

6.5.1 The general position of a space group

The importance of the general position of a space group $\mathcal{G}$ has been stressed repeatedly.[8] The (numbered) coordinate triplets listed in *International Tables A* in the upper block of the 'Positions' (cf. Section 6.4.2) can be interpreted as a shorthand notation of the matrix–column pairs of symmetry operations. They form a system of representatives of the cosets of $\mathcal{G}/\mathcal{T}$, that is, they contain exactly one representative of every coset. The choice of the representatives, in principle, is arbitrary. It is standardized such that $0 \leq w_i < 1$ holds for the coefficients of the columns $\boldsymbol{w}$. This way the coset belonging to $(\boldsymbol{W}, \boldsymbol{w})$ contains exactly all matrix–column pairs $(\boldsymbol{W}, \boldsymbol{w} + \boldsymbol{t})$, with $\boldsymbol{t}$ running through the coefficient columns of all translations. The number of representatives is equal to the order $|\mathcal{P}|$ of the point group, due to the isomorphism of the factor group $\mathcal{G}/\mathcal{T}$ with the point group $\mathcal{P}$.

For primitive lattices, $\boldsymbol{t}$ is a triplet of integral numbers; centred lattices additionally have rational numbers. This makes the standardization of the representatives ambiguous. In fact, the choice of the representatives has been changed in some cases. For example, for the space-group type $Cmma$ (termed $Cmme$ since 2002), in the 1952 edition *International Tables* we find $\frac{1}{2} - x, \bar{y}, z$, but $\bar{x}, \bar{y} + \frac{1}{2}, z$ since 1983. The reason is that the representatives and their sequences were selected according to different procedures in 1952 and 1983.

The components that make up a space group are such that they render it possible to cover the infinite set of symmetry operations by a finite number of specifications. The bases in *International Tables A* have been selected in such a way that all matrices consist of integral numbers.

[8] Remember that a space group has only one general position (in singular).

6.5.2 The special positions of a space group

The order of the site-symmetry group $\mathcal{S}$ of a special position is $|\mathcal{S}| > 1$. The set of the matrix parts $\boldsymbol{W}$ of the elements of $\mathcal{S}$ form a group which is isomorphic to the group of the matrix–column pairs $\{(\boldsymbol{W}, \boldsymbol{w})\}$ of $\mathcal{S}$. Since $\{\boldsymbol{W}\}$ is a subgroup of the matrix group of the point group $\mathcal{P}$ of $\mathcal{G}$, it follows that:

Theorem 6.18 Every site-symmetry group $\mathcal{S}$ of a space group $\mathcal{G}$ is isomorphic to a subgroup of the point group $\mathcal{P}$ of $\mathcal{G}$.

The site-symmetry group $\mathcal{S}$ of an arbitrary point X at a special position always has infinitely many conjugate groups $\mathcal{S}_i$. Due to the three-dimensional periodicity of the lattice, every point X has an infinite number of symmetry-equivalent points X_i, that is, points belonging to the orbit of X; their site-symmetry groups are conjugate. If $\mathcal{S}$ consists of rotations about an axis, all points of the orbit $\mathcal{G}X$, which are located on the rotation axis, have the *same* group $\mathcal{S}$. In addition, there exist an infinite bunch of parallel axes and thus an infinite number of groups $\mathcal{S}_i$. Similar considerations apply to the points located on a reflection plane; they have the same site-symmetry group, and there are an infinite number of these groups on the infinitely many parallel planes.

Generally, there are several site symmetry groups of the same kind. For example, in a centrosymmetric space group, the centres of inversion which are translation equivalent to the centre $\overline{1}$ at the origin are located at points X_k having integral coordinates. There are additional inversion centres at $\frac{1}{2},0,0$; $0,\frac{1}{2},0$; $0,0,\frac{1}{2}$; $\frac{1}{2},\frac{1}{2},0$; $\frac{1}{2},0,\frac{1}{2}$; $0,\frac{1}{2},\frac{1}{2}$; and $\frac{1}{2},\frac{1}{2},\frac{1}{2}$. They can be derived from eqn (4.2) (page 48), according to which the fixed points of inversion are at $\boldsymbol{x}_F = \frac{1}{2}\boldsymbol{w}$, with the translations $\boldsymbol{w} = (1,0,0)$; $(0,1,0)$; $(0,0,1)$; $(1,1,0)$; $(1,0,1)$; $(0,1,1)$; and $(1,1,1)$. Every centrosymmetric space group has eight inversion centres in a primitive unit cell; they are not translation equivalent, but some of them may become symmetry equivalent by symmetry operations other than translations.

Example 6.6
A glance at *International Tables A* shows that $Pmmm$, No. 47, has eight kinds of inversion centres (in this case the centres $\overline{1}$ are hidden in the eight Wyckoff positions of site symmetry mmm). $Pbcm$, No. 57, has only two Wyckoff positions with the site symmetry $\overline{1}$, each with a multiplicity of 4, because every four of the inversion centres are equivalent by reflections and rotations.

The coordinate triplets of special positions can only be interpreted as such, and no longer as descriptions of mappings. This is because the first representative is invariant not only under the identity mapping, but under $|\mathcal{S}|$ mappings, and it is converted by $|\mathcal{S}|$ mappings onto other representatives.

In practical work, two aspects of the special positions are of importance:

1. Special positions of higher-symmetry space groups often have multiplicities that correspond to the number of certain equivalent particles in the unit cell, whereas the multiplicity of the general position is too high for other particles. In this case, these other particles can be located only at a special position, in accordance with the chemical composition. For example, the unit cell of CaF_2 contains four Ca^{2+} ions (multiplicity of 4); then the F^- ions can only be situated at a Wyckoff position of multiplicity 8.
2. If a building block of a crystal structure is to occupy a special position, the symmetry of its surroundings cannot be higher than the proper symmetry of this building block. For example, the centre of gravity of a tetrahedral molecule cannot be placed at a special position whose site symmetry contains an inversion. This often restricts the possible positions to be taken.

6.6 The difference between space group and space-group type

Repeatedly, we have mentioned that a space group should not be confused with a space-group type. As explained in Section 6.1.3, a space-group type is a class of infinite many equivalent space groups. A space group as well as a space-group type are characterized by their symmetry, which is expressed by a space-group symbol and refers to symmetry matrices $\{(\boldsymbol{W},\boldsymbol{w})\}$.

The lattice of a space group, but not that of a space-group type, has well-defined lattice parameters that correspond to the appropriate coordinate system of a crystal structure. Atoms are located at specific places.

Consider as an example the structures of rutile and trirutile (Fig. 12.8, page 157). Both structures belong to the same space-group **type** $P4_2/mnm$, but the basis vector **c** and the number of atoms in the unit cell of trirutile are triplicated. The symmetries of both of the individual structures, rutile and trirutile, are also designated by the symbol $P4_2/mnm$. Rutile and trirutile have a specific lattice each, but their lattices do not coincide; in this case the symbol refers to two *different space groups*.

At first glance it may seem confusing that the same symbol is used both for a space-group type and for the individual space groups of this type. But in fact there is no confusion, because a space group never serves for anything other than to designate the symmetry of a specific crystal structure, including the specification of the lattice parameters and the atomic coordinates.

To put it another way: A space group is the group of symmetry operations of some specific crystal structure including its translations. There are an infinite number of possible space groups. A space-group type is one of 230 possible ways that crystallographic symmetry operations can be combined in space, without numerically fixed values for the translations.

The tables of *International Tables A*, in the first place, are the tables of space-group types, with arbitrary values for the lattice parameters, arbitrary occupation of Wyckoff positions, and arbitrary atomic coordinates. However, if the symmetry of a specific crystal structure is being described, with specific lattice parameters and atomic coordinates, the corresponding table describes an individual space group.

*International Tables A*1 are the tables of the subgroups of the space groups. Group–subgroup relations exist only between space groups, not between space-group types. No specific values of lattice parameters have been listed, but for every group–subgroup pair it is unequivocal how the lattice parameters and the atomic sites of the subgroup result from those of the original group.

For another juxtaposition of space group and space-group type see Table 6.6.

Table 6.6 Juxtaposition of space group and space-group type.

space group	**space-group type**
Geometric space (vector space, Euclidean space).	Algebraic space (point space, affine space).
Symmetry is regarded as a mapping of space onto itself. Permitted are translations and rotations as congruent mappings as well as reflections, inversion, and rotoinversion.	The affine mapping of vector space requires preservation and parallelism of straight lines and of ratios of distances between points on a straight lines, but otherwise arbitrary distortions are permitted.
Distances and angles can be measured in vector space if the lengths of basis vectors and the angles between them are known.	Lattice points in affine space are specified as coordinate triplets of natural numbers x, y, z, without fixation of metric relationships.
A table in *International Tables A* describes a space group only if the lattice parameters have been specified.	With no specification of the lattice parameters, that is, arbitrary values for the translations $t(1,0,0)$, $t(0,1,0)$ and $t(0,0,1)$, *International Tables A* describe the space-group types.

Exercises

Solutions on page 325.

6.1. Denominate the crystal systems corresponding to the following space groups:

$P4_1 32$; $P4_1 22$; $Fddd$; $P12/c1$; $P\bar{4}n2$; $P\bar{4}3n$; $R\bar{3}m$; $Fm\bar{3}$.

6.2. What is the difference between the space groups $P6_3 mc$ and $P6_3 cm$?

6.3. To what crystal classes (point groups) do the following space groups belong?
$P2_1 2_1 2_1$; $P6_3/mcm$; $P2_1/c$; $Pa\bar{3}$; $P4_2/m2_1/b2/c$?

Subgroups and supergroups of point and space groups

A structural relationship entails a symmetry relationship. Changes of symmetry occur during phase transitions or when an isotropic surrounding is replaced by anisotropic mechanical forces or by the action of electric or magnetic fields. Part of the symmetry of a molecule or crystal is then lost ('the symmetry is broken'). In the mentioned cases, there often exists a group–subgroup relation between the symmetry groups of the involved substances or phases. Therefore, it is useful to consider the foundations of such relations.

7.1 Subgroups of the point groups of molecules

In this section those molecular symmetries and their subgroups are considered that may occur as crystallographic point groups. A diagram of group–subgroup relations of non-crystallographic point groups can be found in *International Tables A*.[1]

The relations between a point group and its subgroups can be depicted by a graph. Two important aspects should be taken into account:

1. to include as many as possible of such relations in a graph;
2. to manage with graphs that are as few in number and as clear as possible.

Every crystallographic point group is either a subgroup of a cubic point group of type $4/m\bar{3}\,2/m$ (short symbol $m\bar{3}m$) of order 48 or of a hexagonal point group of type $6/m\,2/m\,2/m$ ($6/mmm$) of order 24. Therefore, only two graphs are needed to display all group–subgroup relations.

In the graphs of Figs 7.1 and 7.2 the symbol of every point group is connected with its *maximal* subgroups by lines. The order of the group mentioned on the left side corresponds to the height in the graph. The graphs shown are contracted graphs, in which subgroups of the same type are mentioned only once. For example, $4/m\bar{3}\,2/m$ actually has three subgroups of the type $4/m\,2/m\,2/m$, with their fourfold rotation axes aligned along x, y, and z, respectively, but $4/m\,2/m\,2/m$ is mentioned only once. The complete graphs, which would contain all subgroups (e.g. three times $4/m\,2/m\,2/m$), would take much more space; Fig. 7.1 would consist of 98 Hermann–Mauguin symbols with a bewildering number of connecting lines. Some point-group types appear in both graphs.

A direction is called *unique* if it is not equivalent by symmetry to any other direction, not even the counter-direction. The ten point groups framed in the

[1] Sixth edition (2016), Section 3.2.1.4, Fig. 3.2.1.6, [15]; fifth edition (2006), Section 10.1.4, Fig. 10.1.4.3 [14].

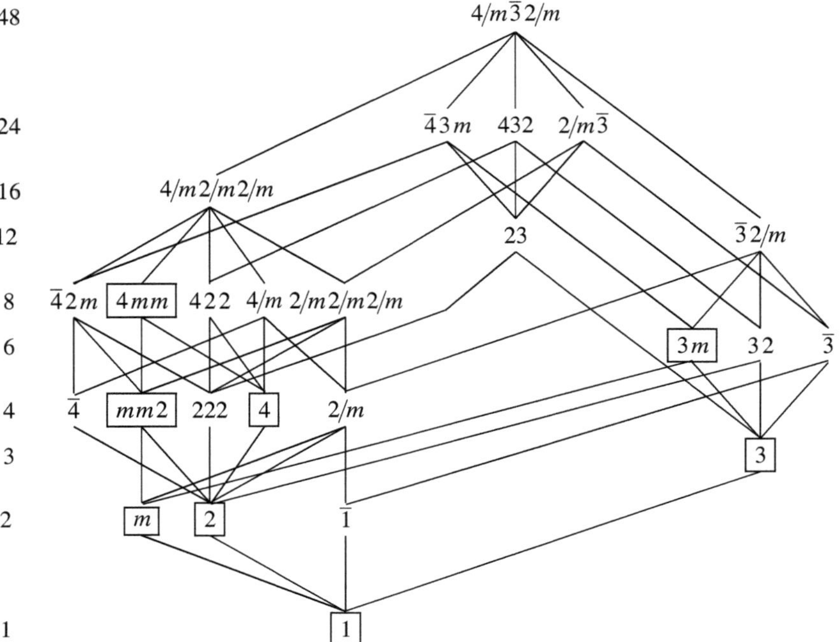

Fig. 7.1 Subgroup graph (contracted) of the point group $4/m\bar{3}2/m$ ($m\bar{3}m$). The order at the left side is given on a logarithmic scale. Polar groups are framed.

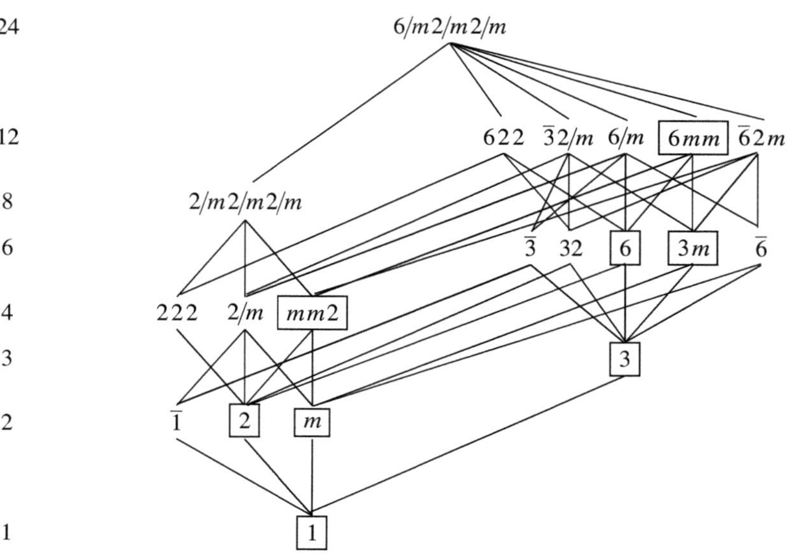

Fig. 7.2 Subgroup graph (contracted) of the point group $6/m2/m2/m$ ($6/mmm$). Polar groups are framed.

graphs of Figs 7.1 and 7.2 have at least one unique direction. They are called *polar point groups*. Certain properties, for example an electric dipole moment, require a polar point group.

Figures 7.1 and 7.2 only show symmetry relations. A symmetry reduction from a group $\mathcal{G}$ to a subgroup $\mathcal{H}$ also implies a change of the equivalence conditions within a molecule:

1. The site symmetries of the atoms are reduced, or
2. the point orbits of symmetrically equivalent atoms split into different orbits, or
3. both happen.

The symmetry reduction is often accompanied by a relaxed fixation of the atoms: Parameters (coordinates, parameters of thermal motion) with fixed or coupled values become independent.

7.2 Subgroups of the space groups

When crystal structures are related or when one of them is converted to another by a phase transition, keeping the general arrangement, the symmetries of the crystal structures are related. A few examples have been presented in Section 1.2, and many more are dealt with in Part II.

'Related symmetry' means:

1. the symmetry of one crystal structure is a subgroup of the symmetry of the other; or
2. both crystal structures have a common supergroup, that is, they have different partial symmetries of a higher symmetry; or
3. both crystal structures have a common subgroup, that is, they have part of their symmetries in common, but none is contained in the other. This case, however, requires particular prudence; in principle, it is always possible to find an infinite number of common subgroups and to use them to invent meaningless 'symmetry relations'.

On the other hand, symmetry relations often indicate the presence of structural relations, and it may be worth checking them. However, this should be done with caution, as shown by the example of the pair of structures CO_2 – FeS_2; both have the same space group type $Pa\overline{3}$, similar lattice parameters, and the same occupied Wyckoff positions, and yet the structural details are so different that they cannot be considered to belong to the same structure type (cf. Section 9.5, page 123).

For practical work, we look for groups that are intermediate between the starting space group $\mathcal{G}$ and the candidate subgroup $\mathcal{H}$. First, we look up the maximal subgroups $\mathcal{H}_{1i}$ of $\mathcal{G}$ (cf. Definition 5.1, page 54), then all maximal subgroups $\mathcal{H}_{2k}$ of $\mathcal{H}_{1i}$, etc., until the desired subgroup with the index $i = |\mathcal{G} : \mathcal{H}|$ has been reached. This way we obtain one or several chains of consecutive maximal subgroups from $\mathcal{G}$ to $\mathcal{H}$. The indices $|\mathcal{G} : \mathcal{H}_{1i}|$, $|\mathcal{H}_{1i} : \mathcal{H}_{2k}|$, must be divisors of the index $|\mathcal{G} : \mathcal{H}|$ (Lagrange's Theorem 5.3, page 57).

For this kind of procedure a theorem by C. HERMANN is of special value. First we define three special kinds of subgroups of space groups.

Let $\mathcal{G}$ be a space group with point group $\mathcal{P}_{\mathcal{G}}$ and normal subgroup of all translations $\mathcal{T}_{\mathcal{G}}$, and, correspondingly, let $\mathcal{H} < \mathcal{G}$ be a subgroup with point group $\mathcal{P}_{\mathcal{H}}$ and normal subgroup $\mathcal{T}_{\mathcal{H}}$.

Definition 7.1 $\mathcal{H} < \mathcal{G}$ is called a *translationengleiche subgroup* if $\mathcal{G}$ and $\mathcal{H}$ have the same group of translations, $\mathcal{T}_{\mathcal{H}} = \mathcal{T}_{\mathcal{G}}$; therefore, $\mathcal{H}$ belongs to a crystal class of lower symmetry than $\mathcal{G}$, $\mathcal{P}_{\mathcal{H}} < \mathcal{P}_{\mathcal{G}}$.[2]

[2] See margin note No. 1 in Chapter 11 (page 141) for comments referring to the terms *translationengleiche* and *klassengleiche*.

Definition 7.2 $\mathcal{H} < \mathcal{G}$ is called a *klassengleiche subgroup*, if $\mathcal{G}$ and $\mathcal{H}$ belong to the same crystal class, $\mathcal{P}_{\mathcal{H}} = \mathcal{P}_{\mathcal{G}}$; therefore, $\mathcal{H}$ has fewer translations than $\mathcal{G}$, $\mathcal{T}_{\mathcal{H}} < \mathcal{T}_{\mathcal{G}}$.[2]

Definition 7.3 A *klassengleiche* subgroup is called an *isomorphic subgroup* if $\mathcal{G}$ and $\mathcal{H}$ belong to the same affine space-group type.

Isomorphic subgroups are a special case of *klassengleiche* subgroups. An isomorphic subgroup has either the same standard Hermann–Mauguin symbol as the supergroup or that of the enantiomorphic partner.

Remark: The mentioned definition for isomorphic subgroups means that the space group and its subgroup have 'the same group multiplication table' (cf. Section 5.2). It does not include the case that a space group has two or more subgroups that are mutually isomorphic without being isomorphic to the space group. Occasionally, that causes misunderstandings.

Definition 7.4 $\mathcal{H}$ is called a *general subgroup*, if $\mathcal{T}_{\mathcal{H}} < \mathcal{T}_{\mathcal{G}}$ and $\mathcal{P}_{\mathcal{H}} < \mathcal{P}_{\mathcal{G}}$ hold.

A general subgroup is neither *translationengleiche* nor *klassengleiche*.

The remarkable *theorem of Hermann* [58] is then:

Theorem 7.5 A *maximal* subgroup of a space group is either *translationengleiche* or *klassengleiche*.

The proof of the theorem follows by construction of the intermediate group $\mathcal{Z}$, $\mathcal{G} \geq \mathcal{Z} \geq \mathcal{H}$, which consists of those cosets of $\mathcal{G}$ that occur in $\mathcal{H}$, possibly with fewer translations. $\mathcal{Z}$ is evidently a *translationengleiche* subgroup of $\mathcal{G}$ and a *klassengleiche* supergroup of $\mathcal{H}$. Either $\mathcal{Z} = \mathcal{G}$ or $\mathcal{Z} = \mathcal{H}$ must hold for a maximal subgroup.

Due to Hermann's theorem it is sufficient to consider only the *translationengleiche* and the *klassengleiche* subgroups. The maximal subgroups for every space-group type are listed in *International Tables*, Volume *A* 1983–2006 and Volume *A*1. In Volume *A* they can be found under the headings 'Maximal non-isomorphic subgroups' and 'Maximal isomorphic subgroups of lowest index'. However, among the *klassengleiche* subgroups with an enlarged conventional cell (mentioned under **IIb**) only the space-group types of the subgroups are listed, and not all of the individual subgroups themselves. In addition, Volume *A* lacks the important information about any necessary origin shifts. The complete listing of all subgroups can be found in Volume *A*1 [16].

A detailed description of the data and instructions for their use appear in Volume *A* (up to the fifth edition), Section 2.2.15, and in Volume *A*1, Chapters 2.1 and 3.1.

In addition to Hermann's theorem, further restrictions apply to the maximal subgroups of the space groups. Some of them, which are useful when setting up subgroup tables and in the practical application of group–subgroup relations, are mentioned in the following, without proofs. The proofs can be found in *International Tables* A1 in a theoretical chapter by G. NEBE.[3]

[3] Chapter 1.5 in the first edition, 2004; Chapter 1.3 in the second edition, 2010.

Theorem 7.6 Every space group $\mathcal{G}$ has an infinite number of maximal subgroups $\mathcal{H}$. They are space groups and their indices are powers of prime numbers p^1, p^2, or p^3; $p > 1$.

Remarks

1. A subgroup of, say, index 6 cannot be maximal.
2. Prime numbers p^1 apply to triclinic, monoclinic, and orthorhombic space groups $\mathcal{G}$;
   p^1 and p^2 to trigonal, tetragonal, and hexagonal $\mathcal{G}$;
   p^1, p^2, and p^3 to cubic $\mathcal{G}$.
3. Certain restrictions apply to the possible values of the prime numbers p, depending on $\mathcal{G}$ and the subgroup (e.g. only prime numbers of the kind $p = 6n + 1$ with n = integral). See Chapter 21.
4. An index of $p = 1$ is not permitted, it does not indicate a group–subgroup relation. It would mean $\mathcal{H} = \mathcal{G}$ (i.e. an isotypic structure, cf. Section 9.5).

Theorem 7.7 There are only a finite number of *maximal non-isomorphic subgroups* $\mathcal{H}$ of $\mathcal{G}$ because: If i is the index of $\mathcal{H}$ in $\mathcal{G}$, $i = |\mathcal{G} : \mathcal{H}|$, then $\mathcal{H}$ being non-isomorphic to $\mathcal{G}$ is only possible if i is a divisor of $|\mathcal{P}|$, the order of the point group $\mathcal{P}$ of $\mathcal{G}$.

Since the orders of the crystallographic point groups only contain the factors 2 and 3, maximal non-isomorphic subgroups can only have the indices 2, 3, 4, and 8. However, the index 8 is excluded. Actually, not all of these possibilities do occur. All maximal non-isomorphic subgroups of triclinic, monoclinic, orthorhombic, and tetragonal space groups have index 2; those of trigonal and hexagonal space groups have indices 2 or 3; indices 2, 3, and 4 only occur among cubic space groups.

Isomorphic subgroups may also have the mentioned indices, for example 2. A space group of type $P1$, for example, has seven subgroups of index 2, all of which are isomorphic.

Theorem 7.8 The number N of the subgroups of index 2 of a space group $\mathcal{G}$ is $N = 2^n - 1$, $0 \leq n \leq 6$.

The mentioned theorems show that a space group has infinitely many maximal subgroups. However, there are only a finite number which are non-isomorphic.

Table 7.1 Coset decomposition of the space group $\mathcal{G} = Pmm2$ with respect to the group of translations $\mathcal{T}$ (cf. Example 6.3, page 69). The group elements that are deleted upon symmetry reduction to the *translationengleiche* subgroup $P2$ and the *klassengleiche* subgroup $Pcc2$ with doubled basis vector **c** have been crossed out. For $Pcc2$ the elements of $\mathcal{T}$ are designated by $(p,q,0)$, $(p,q,1)$, ... which means the translations $p\mathbf{a}, q\mathbf{b}, 0\mathbf{c}$, $p\mathbf{a}, q\mathbf{b}, \pm 1\mathbf{c}$, ... with $p,q = 0, \pm1, \pm2, \ldots$; $(0,0,0)$ is the identity translation.

Translationengleiche subgroup $P2$

1st coset	2nd coset	3rd coset	4th coset
$1\circ 1$	$2\circ 1$	~~$m_x\circ 1$~~	~~$m_y\circ 1$~~
$1\circ t_1$	$2\circ t_1$	~~$m_x\circ t_1$~~	~~$m_y\circ t_1$~~
$1\circ t_2$	$2\circ t_2$	~~$m_x\circ t_2$~~	~~$m_y\circ t_2$~~
$1\circ t_3$	$2\circ t_3$	~~$m_x\circ t_3$~~	~~$m_y\circ t_3$~~
$1\circ t_4$	$2\circ t_4$	~~$m_x\circ t_4$~~	~~$m_y\circ t_4$~~
⋮	⋮	⋮	⋮

Klassengleiche subgroup $Pcc2$ ($\mathbf{c}' = 2\mathbf{c}$)

1st coset	2nd coset	3rd coset	4th coset
$1\circ(p,q,0)$	$2\circ(p,q,0)$	~~$m_x\circ(p,q,0)$~~	~~$m_y\circ(p,q,0)$~~
~~$1\circ(p,q,1)$~~	~~$2\circ(p,q,1)$~~	$m_x\circ(p,q,1)$	$m_y\circ(p,q,1)$
$1\circ(p,q,2)$	$2\circ(p,q,2)$	~~$m_x\circ(p,q,2)$~~	~~$m_y\circ(p,q,2)$~~
~~$1\circ(p,q,3)$~~	~~$2\circ(p,q,3)$~~	$m_x\circ(p,q,3)$	$m_y\circ(p,q,3)$
$1\circ(p,q,4)$	$2\circ(p,q,4)$	~~$m_x\circ(p,q,4)$~~	~~$m_y\circ(p,q,4)$~~
⋮	⋮	⋮	⋮

7.2.1 Maximal *translationengleiche* subgroups

The conditions concerning *translationengleiche* subgroups are the simplest ones. They can only be non-isomorphic subgroups because the (finite) point group has been decreased. All translations are kept; complete cosets of the decomposition of $\mathcal{G}$ with respect to $\mathcal{T}$ have been deleted (Table 7.1). The graphs of the point groups of Figs 7.1 and 7.2 can be applied, since the factor group $\mathcal{G}/\mathcal{T}$ is isomorphic to the point group $\mathcal{P}$. Space-group symbols replace the point-group symbols.

There are 10 cubic space-group types of the crystal class $m\bar{3}m$, and thus there are 10 graphs corresponding to Fig. 7.1. However, additional graphs are needed because, for example, $Pbcm$ does not appear among the *translationengleiche* subgroups of a space group of the crystal class $m\bar{3}m$. Figures 7.3 and 7.4 are examples of such graphs.

The *translationengleiche* subgroups $\mathcal{H}$ of a space group $\mathcal{G}$ are completely listed in the subgroup tables of *International Tables A* (up to the fifth edition, 2006) under **I**. Since every $\mathcal{H}$ contains complete cosets of $\mathcal{G}/\mathcal{T}$, $\mathcal{H}$ can be completely characterized by specifying the numbers (n) of the representatives of these cosets. Note, however, that the standard coordinate system of $\mathcal{H}$ may differ from that of $\mathcal{G}$. It may be necessary to perform a coordinate transformation (Section 3.6) to obtain the standard data of $\mathcal{H}$.

A *translationengleiche* subgroup always has a lower space group number. That is not so for *klassengleiche* subgroups.

7.2.2 Maximal non-isomorphic *klassengleiche* subgroups

Non-isomorphic *klassengleiche* subgroups of every space group can also be listed completely. However, in Volume *A* of *International Tables* (up to the fifth edition, 2006) this has been done only partly; the complete list can be found in Volume *A*1. *Klassengleiche* subgroups have a reduced $\mathcal{T}$ and thus also every coset in the factor group $\mathcal{G}/\mathcal{T}$ is reduced, but the number of cosets remains unchanged (Table 7.1). Two possibilities are distinguished for practical reasons (there is no group-theoretical reason):

1. The conventional cell remains unchanged, that is, only centring translations are lost (of course, only applicable to centred settings).
2. The conventional cell is enlarged.

Case 1 can be treated in the same way as the case of *translationengleiche* subgroups since the representatives of $\mathcal{G}$ (with or without centring translations) remain present in $\mathcal{H}$. For this reason, subgroups of this kind have been completely listed in Volume *A* (up to the fifth edition) under **IIa** and characterized in the same way as under **I**. Case 2, the subgroups with an increased conventional cell, would have different coordinate triplets than $\mathcal{G}$ due to the changed cell. Therefore, only the kinds of cell enlargements and the *types* of the subgroups have been listed in Volume *A* (up to the fifth edition) under **IIb**, but neither their numbers nor the actual representatives. For example, in Volume *A*, space-group type $Pmmm$, in the listing of the **IIb** subgroups, the entry $Pccm$ refers to two subgroups, the entry $Cmmm$ to four, and $Fmmm$ to eight

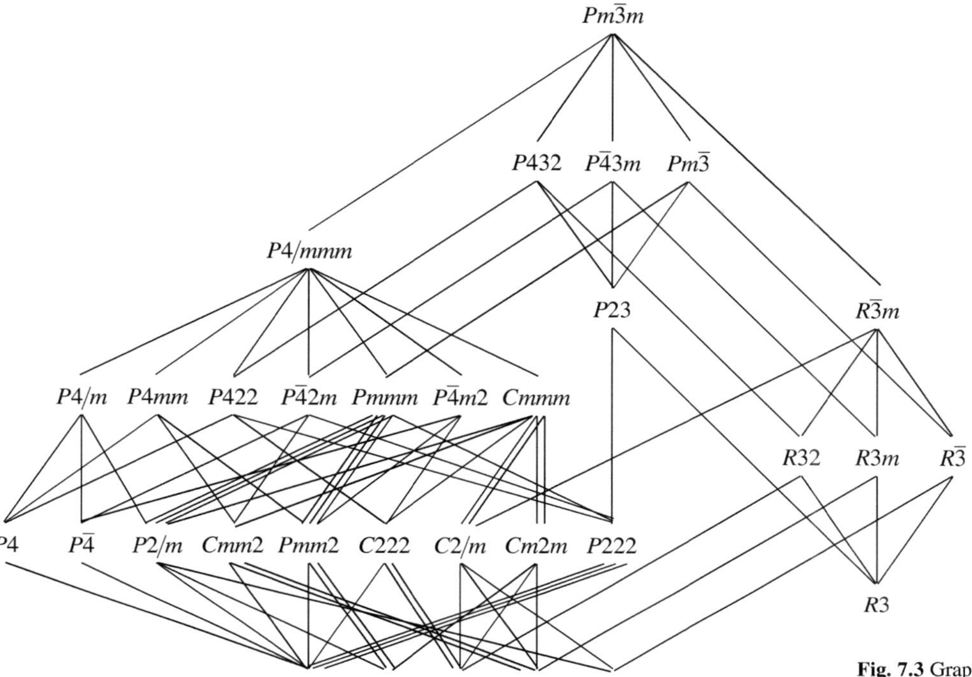

Fig. 7.3 Graph (contracted) of the *translationengleiche* subgroups of $Pm\bar{3}m$. Every conjugacy class of maximal subgroups is marked by one line. For example, there are three non-conjugate subgroups of $Pmmm$ of the type $Pmm2$, namely $Pmm2$, $Pm2m$, and $P2mm$, which are commonly designated by the conventional setting $Pmm2$. These may be conjugate in groups of higher order; in $Pm\bar{3}$ there is only one class of three conjugates of $Pmm2$. The trivial subgroup $P1$ is not mentioned.

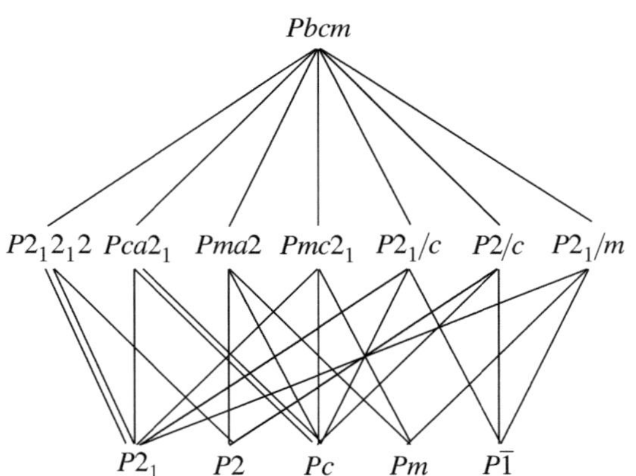

Fig. 7.4 Graph (contracted) of the *translationengleiche* subgroups of $Pbcm$. The kind of presentation is as in Fig. 7.3.

different subgroups. In Volume $A1$ all of these subgroups have been completely listed.

Klassengleiche subgroups can also be depicted in graphs, one for each crystal class. If the isomorphic subgroups are not considered (they exist always), 29 graphs are needed; some are rather simple, others more complicated. The crystal classes mmm, $mm2$, and $4/mmm$ have the most complicated ones. As examples, the graphs of the crystal classes $2/m$ and $m\bar{3}m$ are depicted in Figs 7.5 and 7.6. All 29 graphs of the *klassengleiche* subgroups can be found in *International Tables* $A1$.

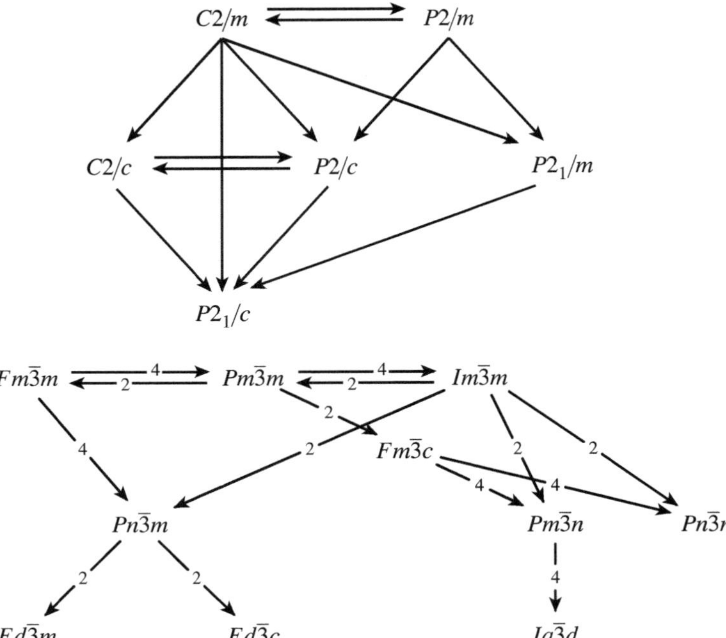

Fig. 7.5 Graph of the maximal *klassengleiche* subgroups of space groups of the crystal class $2/m$. Every space group $\mathcal{G}$ is connected with its maximal subgroups $\mathcal{H}$ by arrows; $\mathcal{H}$ is placed lower than $\mathcal{G}$ in the graph. If the type $\mathcal{G}$ can also occur as a subgroup of $\mathcal{H}$, both symbols are at the same height and the arrows point in the possible directions group $\to$ subgroup. All indices are 2.

Fig. 7.6 Graph of the maximal *klassengleiche* subgroups of space groups of the crystal class $m\bar{3}m$. The kind of presentation is as in Fig. 7.5. The indices are given in the arrows.

7.2.3 Maximal isomorphic subgroups

The number of maximal isomorphic subgroups are always infinite; therefore, it is only possible to list a small number of them individually. Only a few with the smallest indices have been included in Volume *A* of *International Tables* (up to the fifth edition, 2006), under the heading 'IIc Maximal isomorphic subgroups of lowest index'. In Volume *A*1 all up to index 4 are mentioned individually. In addition, series are mentioned that cover the complete infinite set of isomorphic subgroups with the aid of parameters. Concerning the possible values of the indices of isomorphic subgroups, see Chapter 21.

7.3 Minimal supergroups of the space groups

Whereas subgroups of space groups of finite index are always space groups, this restriction is not applicable to supergroups. However, this should not worry us as long as we deal with real crystal structures, whose symmetries can be described by space groups. Quasicrystals and incommensurately modulated structures, which are described with superspace groups in four- or five-dimensional space, are left out of consideration. We consider only three-dimensional space groups.

Neither do we consider another kind of supergroups, the black-and-white space groups (Shubnikov groups; magnetic space groups), where certain symmetry operations are connected with a colour inversion. In the Hermann–Mauguin symbol such symmetry operations are marked by a prime $'$, for example, $R\bar{3}c'$ (cf. *International Tables A* 2016, Chapter 3.6 [59] and the cited references).

By reversal of the definition for maximal subgroups of space groups, we define:

Definition 7.9 Let $\mathcal{H}$ be a maximal *translationengleiche*, *klassengleiche*, or isomorphic subgroup of the space group $\mathcal{G}$, $\mathcal{H} < \mathcal{G}$. Then $\mathcal{G}$ is a *translationengleiche*, *klassengleiche*, or isomorphic *minimal supergroup* of $\mathcal{H}$.

Even with restriction to space groups, supergroups are more manifold than subgroups. Subgroups result from the deletion of present symmetry operations; supergroups, however, result from the addition of symmetry operations.

Example 7.1
The deletion of the points of inversion from the space group $P\overline{1}$ results in *one translationengleiche*, maximal subgroup $P1$ of index 2. However, if points of inversion are added to $P1$, without changing the shape and size of the unit cell, there are an infinite number of possibilities where to place these points of inversion. Therefore, $P1$ has an infinity of minimal *translationengleiche* supergroups $P\overline{1}$ of index 2.

The tables of the supergroups in *International Tables A* (up to the fifth ed., 2006) and *A*1 are reversed listings of the subgroup tables. If a space group $\mathcal{H}$ appears as a maximal subgroup of a space group $\mathcal{G}$, then $\mathcal{G}$ is listed as a supergroup in the table of the space group $\mathcal{H}$. The table contains only *translationengleiche* and non-isomorphic *klassengleiche* supergroups. Not all of the individual supergroups have been listed; a listed space-group symbol may refer to several space groups of the same type, but only the corresponding conventional symbol is mentioned once. The tables contain information neither about the actual number of supergroups of the same type nor about origin shifts.

However, in Chapter 2.1 of the second edition of *International Tables A*1 (2010), 'Guide to the subgroup tables and graphs', instructions and examples have been included of how to derive the exact data of the minimal supergroups (only if the space group $\mathcal{H}$ is neither triclinic nor monoclinic and if the supergroup $\mathcal{G}$ is a space group).

A space group $\mathcal{G}$ that belongs to a different crystal system to the space group $\mathcal{H}$ can only be a supergroup $\mathcal{G} > \mathcal{H}$ if the lattice of $\mathcal{H}$ fulfils the metric conditions of the lattice of $\mathcal{G}$, or nearly so in practical work. For example, if $\mathcal{H}$ is orthorhombic and $\mathcal{G}$ is tetragonal, then $\mathcal{G}$ can only be a supergroup if the lattice parameters of $\mathcal{H}$ fulfil the condition $a = b$.

Example 7.2
In *International Tables*, Volumes *A* (up to the fifth edition) and *A*1, the following *translationengleiche* orthorhombic supergroups of the space group $P2_12_12$ are listed with their short Hermann–Mauguin symbols. The full symbols have been added here:

$Pbam$	$P2_1/b2_1/a2/m$	$Pccn$	$P2_1/c2_1/c2/n$	$Pbcm$	$P2/b2_1/c2_1/m$
$Pnnm$	$P2_1/n2_1/n2/m$	$Pmmn$	$P2_1/m2_1/m2/n$	$Pbcn$	$P2_1/b2/c2_1/n$

From the sequence of the 2_1 axes in the full symbols we can deduce:
$Pbam$, $Pccn$, $Pnnm$, and $Pmmn$ are supergroups with the same orientations of the coordinate systems. It is not mentioned that of these only $Pbam$

is a supergroup without an origin shift; to find this out, we have to look up the tables of the mentioned space groups in Volume *A*1 and look for the subgroup $P2_12_12$; it is there where the origin shifts are mentioned. $Pbcm$ and $Pbcn$ themselves are not actually supergroups, but only the conventional symbols of the space groups of four actual supergroups with exchanged axes and shifted origins; $Pbcm$ stands for the supergroups $P2_1/b2_1/m2/a$ and $P2_1/m2_1/a2/b$ without changed axes; $Pbcn$ stands for $P2_1/c2_1/n2/b$ and $P2_1/n2_1/c2/a$ (see Section 10.3 for Hermann–Mauguin symbols of non-conventional settings). The tables do not show directly that $Pbam$, $Pccn$, $Pnnm$, and $Pmmn$ refer to one supergroup each, while $Pbcm$ and $Pbcn$ refer to two each. However, this can be calculated by the procedure given in Section 2.1.7 of the second edition of *International Tables* A1 (2010).

In addition, four *translationengleiche* tetragonal supergroups are listed: $P42_12$, $P4_22_12$, $P\bar{4}2_1m$, and $P\bar{4}2_1c$. However, these are supergroups only if the space group $P2_12_12$ satisfies the condition $a = b$ (or nearly so in practice, i.e. $a \approx b$).

7.4 Layer groups and rod groups

Definition 7.10 The symmetry operations of an object in three-dimensional space form a *layer group* if it is periodical only in two dimensions, that is, if it has translational symmetry only in two dimensions; they form a *rod group* if it has translational symmetry only in one dimension.

The symmetry operations of an object in two-dimensional space form a *frieze group* if it has translational symmetry only in one dimension.

Layer, rod, and frieze groups are called *subperiodic groups*.

There are a few other terms for these groups (incomplete list):

Layer group: layer space group, net group, diperiodic group in three dimensions, two-dimensional group in three dimensions.

Rod group: stem group, linear space group, one-dimensional group in three dimensions.

Frieze group: ribbon group, line group in two dimensions.

The terms 'two-dimensional space group' instead of layer group and 'one-dimensional space group' instead of rod group, which can be found in the literature, are misleading and should not be used because these terms rather refer to plane groups and line groups.

Layer and rod groups were derived by C. HERMANN [60]. They have been compiled in *International Tables*, Volume *E* [61], in the same style as the space groups in Volume *A* (including their maximal subgroups).

As for space groups, the orders of rotations of layer groups are restricted to 1, 2, 3, 4, and 6. Therefore, there are only a finite number of layer-group types, namely 80. The order of rotations of rod groups parallel to the axis with translational symmetry is not restricted; therefore, there are an infinite number of rod-group types. If the orders of rotations are restricted to 1, 2, 3, 4, and 6, the number of rod-group types is 75. The maximal order of rotations of frieze groups is 2; there are seven types of frieze groups.

Layer groups are not to be confused with plane groups. For a plane group, space is restricted to two dimensions, which in principle is an infinitely thin plane. An (ideal) infinitely extended planar molecule like a graphene layer always has an extension in the third dimension (perpendicular to the molecular plane); its symmetry can only be designated by a layer group, in principle. The symmetry of the *pattern* of the graphene layer (i.e. its projection onto a parallel plane), on the other hand, can be designated by a plane group. The symmetry of layers having a thickness of several atoms, for example a silicate or a CdI_2 layer, cannot be described by such a projection.

The symbols used in *International Tables E* to denominate the symmetries of layer groups correspond to the Hermann–Mauguin symbols of space groups. The direction of **c** is considered to be perpendicular to the layer. Screw axes and glide vectors of glide planes can only occur perpendicular to **c**. The sole difference to the space-group symbols is the use of the lower case letters p and c instead of P and C for the first letter, which designates the lattice type (centring), for example, $p4/nmm$ (layer groups have no a, b, f, i, and r centrings). When using one of these symbols, it should be explicitly stated whether a layer or a plane group is meant, because the symbols in both cases begin with the same letters (p or c), and in some cases a layer group is designated with the same symbol as a plane group.

No centrings exist for rod groups. To distinguish them from layer groups, the symmetry symbol begins with a slanted $\mathscr{p}$ in script style, for example $\mathscr{p}4_2/mmc$. **c** is the direction having translational symmetry. Screw axes and glide vectors of glide planes can only occur parallel to **c**. Non-conventional settings, with translational symmetry along **a** or **b**, can be expressed by subscript letters a or b, for example, $\mathscr{p}_a 2_1 am$ (conventionally $\mathscr{p}mc2_1$).

Molecules of chain polymers tend to be entangled, in which case they have no overall symmetry; symmetry is then restricted to the local symmetry in the immediate surroundings of an atom. In crystalline polymers the chains are forced to align themselves and to adopt a symmetric conformation. This symmetry can be crystallographic, but often the symmetry of the single molecule within the crystal matrix is non-crystallographic.

Crystalline chain polymers often adopt helical molecular structures. In polymer science helices are designated by N/r ('N/r helix') where N is the number of repeating units[4] within one translation period and r is the corresponding number of helical 360° coil turns along the molecular chain. The corresponding Hermann–Mauguin screw axis symbol N_q can be calculated from [62, 63]:

$$Nn \pm 1 = rq \qquad (7.1)$$

The integers $n = 0, 1, 2, \ldots$ and q ($0 < q < N$) are to be chosen such that the equation is satisfied. In a helix designated by a Hermann–Mauguin symbol, all repeating units are symmetrically equivalent. For non-crystallographic screw axes with $N = 5$ or $N > 6$ that is fulfilled only approximately at the best, but in polymer chemistry they are treated as if they were symmetry equivalent,

[4] 'Repeating unit' and 'monomer unit' can be identical; however, in a case like polyethylene, $(CH_2)_\infty$, the repeating unit is CH_2 whereas the chemical monomer is C_2H_4.

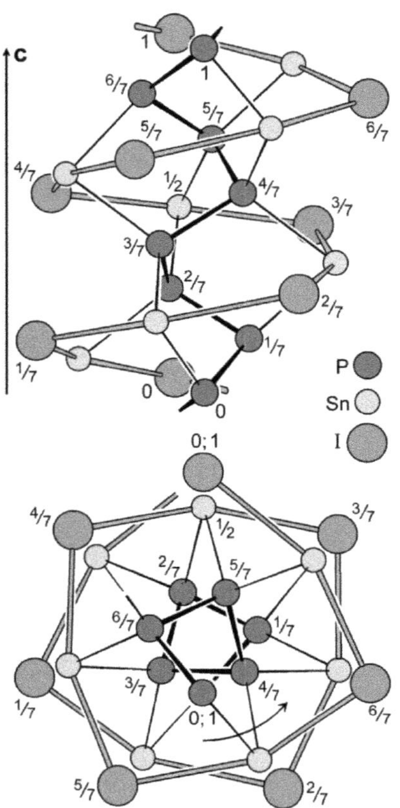

Fig. 7.7 Two views of the 7/2 *P*-helix of tin iodide phosphide SnIP having a 7_4 screw axis; rod group $\not{p}7_422$ [65]. The numbers are *z* coordinates. In the crystal, twofold crystallographic axes run perpendicular to the helix axis through the atoms at $z = 0$ and $z = \frac{1}{2}$.

according to the equivalence postulate [64]. The chemical handedness of the helix does not follow from the N/r symbol, but is specified by the additional letters *M* (minus; or *L*) and *P* (plus; or *R*) for left and right, respectively. The plus sign in equation (7.1) refers to *P*-helices, the minus sign to *M*-helices.

Example 7.3
Tin iodide phosphide consists of 7/2 helices of polyphosphide ions $(P^-)_\infty$ that have 7/2 helices $(SnI^+)_\infty$ winding around them (Fig. 7.7) [65]. There are 7 P atoms and 7 SnI repeating units per two helix turns. The racemic crystal contains right-handed and left-handed helices, which fulfil the non-crystallographic symmetries $\not{p}7_422$ and $\not{p}7_322$, respectively. Equation (7.1) is fulfilled if $7 \times 1 + 1 = 2 \times 4$ or $7 \times 1 - 1 = 2 \times 3$ (i.e. $q = 4$ or $q = 3$), and the corresponding Hermann–Mauguin symbol of the screw axis is 7_4 or 7_3, depending on handedness.
At the bottom right of Fig. 7.7 a turning angle to the right of $360°/7 = 51.4°$ is drawn; after a $\frac{1}{7}$ turn we find an atom at a height of $z = \frac{4}{7}$, which is symmetry equivalent to the P atom at $z = 0$. From that follows $q = 4$.

The N/r symbol cannot be deduced uniquely from the Hermann-Mauguin symbol, because not only a 7/2 *P*-helix, but also a 7/9 and any other 7/(2 modulo 7) *P* helix and any 7/(5 modulo 7) *M*-helix also has a 7_4 screw axis.

Crystallographic N_q screw axes are termed right-handed if $q < \frac{1}{2}N$ and left-handed if $q > \frac{1}{2}N$. This handedness does not have to be identical to the chemical *P* or *M* handedness. For example, a right-handed 3/1-*P* helix has a right-handed 3_1 screw axis, but a right-handed 3/2-*P* helix has a left-handed 3_2 screw axis.

In order to also have a uniform denomination for non-crystallographic screw axes of any order we define as follows:

Definition 7.11 A screw axis N_q of any order *N* is called right-handed if $q < \frac{1}{2}N$ and left-handed if $q > \frac{1}{2}N$.

That the handedness sometimes does not coincide is due to the different points of view: In chemistry the *P* or *M* handedness depends on the helical sequence of chemical bonds along the polymeric chain; for symmetry considerations the existence of chemical bonds is irrelevant. For example, the symmetry of the screw axis in Fig. 7.7 is 7_4 no matter where we draw the chemical bonds or whether we draw bonds at all.[5]

The crystallographic rod site symmetry in the crystal is called penetration rod group (Subsection 7.4.1). It is a common subgroup of the molecular rod group and the space group of the crystal. In the case of SnIP it is $\not{p}121$; that is a subgroup of $\not{p}7_422$ and of $\not{p}7_322$ and of the space group $P12/c1$ of the crystal.

Somewhat different symbols for layer and rod groups that had been in use before the publication of *International Tables E* in 2002 go back to BOHM and DORNBERGER-SCHIFF [67,68]. Hermann–Mauguin symbols of space groups had been used, the symmetry directions without translational symmetry being set in parentheses. Again, the unique direction is **c**. Examples: layer group

[5] Unfortunately, the International Union of Pure and Applied Chemistry (IUPAC) abandoned the well-established N/r terminology and now recommends to call a 7/2 helix a 7_2 helix, although the corresponding screw axis is not 7_2, but 7_3 or 7_4, depending on handedness [66].

$P(4/n)mm$ [short] or $P(4/n)\,2_1/m2/m$ [full symbol]; rod group $P4_2/m(mc)$ [short] or $P4_2/m(2/m2/c)$ [full].

A layer group is the subgroup of a space group that has lost all translations in the space dimension perpendicular to the layer. It corresponds to the factor group of the space group with respect to the group of all translations in this direction. For example, the layer group $pmm2$ is a subgroup of the space group $Pmm2$; it is isomorphic to the factor group $Pmm2/\mathcal{T}_z$, $\mathcal{T}_z$ being the group of translations having z components.

Example 7.4
The symmetry group of a single selected polymeric molecule of mercury oxide is the rod group $\mathscr{p}\,2/m2/c2_1/m$ ($\mathscr{p}\,mcm$ for short); $P(mc)m$ in Bohm–Dornberger-Schiff notation.

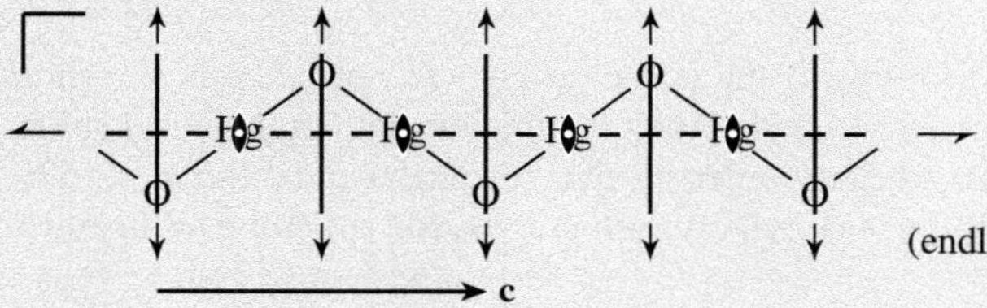

(endless chain)

Example 7.5
The symmetry group of a graphene layer is the layer group $p6/m2/m2/m$ ($p6/mmm$ for short; $P(6/m)mm$ in Bohm–Dornberger-Schiff notation).

7.4.1 Sectional layer groups and penetration rod groups

A point in a space group has a site symmetry, which is a subgroup of the space group. If the centre of gravity of a molecule is placed there, then the site symmetry is the maximal point symmetry that the molecule can adopt in the crystal. In other words, the symmetry of the molecule in the crystal is a common subgroup of the space group and the point symmetry of the isolated molecule.

The same is valid for layered molecules that have a certain layer group in the isolated state, and for chain-like polymeric molecules that have a certain rod group: the molecular symmetry in the crystal is a common subgroup of the space group and the layer group or rod group, respectively, of the single molecule. This common subgroup in turn is a layer or rod group. We can call it the layer-site symmetry and rod-site symmetry, respectively. The denominations of *International Tables* are *sectional layer group* and *penetration rod group*, respectively.

Definition 7.12 In a crystal, a sectional layer group consists of that subgroup of the symmetry operations of the crystal's space group that leave a transecting crystallographic plane invariant. A penetration rod group consists of that subgroup of the symmetry operations of the crystal's space group that leave a traversing crystallographic straight line invariant.

Sectional layer groups and penetration rod groups depend on the orientation and the location of the plane or the straight line relative to the space group. The location is specified by the coordinates of a point on the transecting plane or straight line.

Sectional layer groups of the space groups are listed in *International Tables E*, Parts 5 and 6 [61]. However, they cannot be found easily, because they are listed under the vague titles *Scanning of space groups* and *The scanning tables*. These denominations have their origin in the procedure of their derivation: A plane is scanned in parallel through the space group, whereby it fulfills one or the other layer group, depending on its location. Sectional layer groups serve to designate the layer-site symmetries not only of layered molecules in crystals, but also of interfaces in crystals. Sectional layer groups of the space groups can be obtained from the Bilbao Crystallographic Server (program `SECTIONS`; cf. Chapter 24).

Penetration rod groups of the space groups are not listed in *International Tables E*.[6] A procedure to derive the maximal crystallographic rod subgroups of rod groups with arbitrary order of their rod axes can be found at [63]. For example, $\mathscr{p}3_2 12$ is the maximal crystallographic rod subgroup of the rod group $\mathscr{p}9_5 12$.

[6] Since July 2023 penetration rod groups of the space groups can be retrieved with the program `RODS` of the Bilbao Crystallographic Server; at that time the program was still in the development stage. `https://www.cryst.ehu.es/#subperiodictop`

Exercises

Solutions on page 325.

For an exercise concerning supergroups see Exercise 10.5, page 138.

Exercises concerning *translationengleiche*, *klassengleiche*, and isomorphic subgroups are at the ends of Chapters 12, 13, and 14.

7.1. What layer or rod symmetry do the following polymeric molecules or ions have?

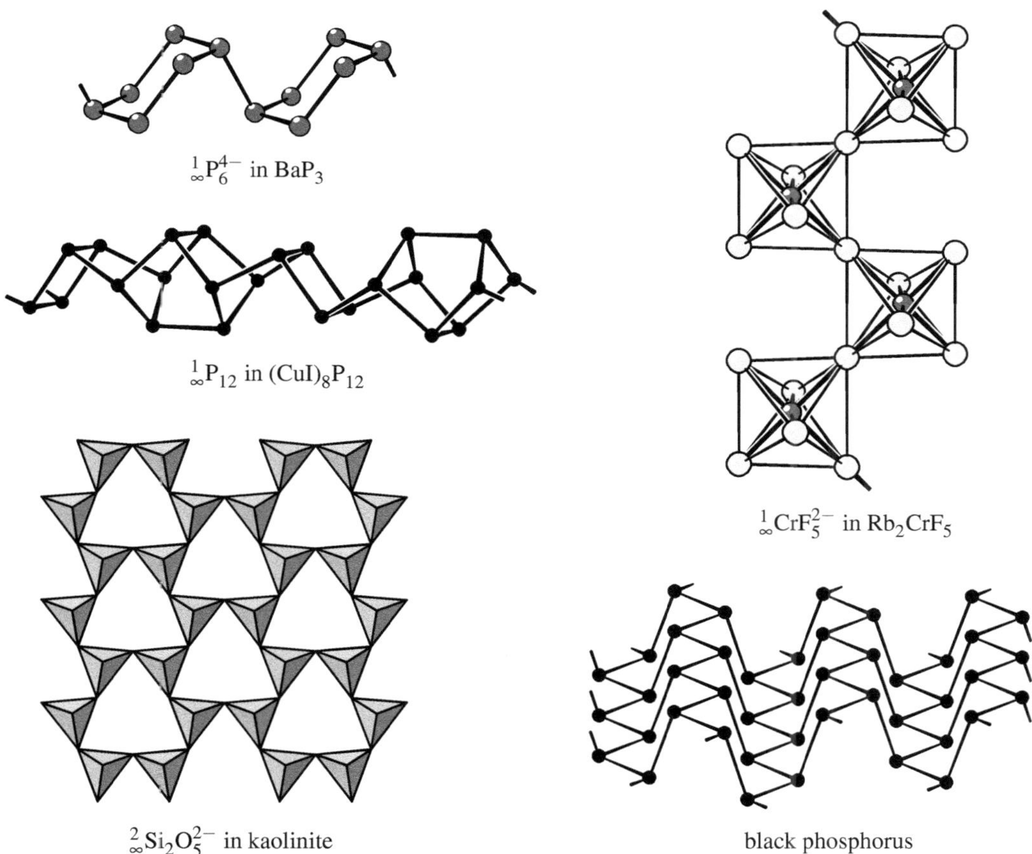

${}^{1}_{\infty}P_6^{4-}$ in BaP_3

${}^{1}_{\infty}P_{12}$ in $(CuI)_8P_{12}$

${}^{2}_{\infty}Si_2O_5^{2-}$ in kaolinite

${}^{1}_{\infty}CrF_5^{2-}$ in Rb_2CrF_5

black phosphorus

Conjugate subgroups and normalizers of space groups

This is a mathematically demanding chapter that provides a theoretical basis for the following chapter. Experts who are mainly concerned with the determination and description of crystal structures, as a rule, will get along without the theory and can skip the chapter.

8.1 Conjugate subgroups of space groups

Conjugate subgroups were defined on page 59 (Definition 5.5). In this section we consider some relations among conjugate subgroups of space groups by means of examples.

Let $\mathcal{G}$ be a space group and $\mathcal{H}$ a subgroup of $\mathcal{G}$, $\mathcal{H} < \mathcal{G}$. There can be groups conjugate to $\mathcal{H}$, which we designate by $\mathcal{H}'$, $\mathcal{H}''$, ... In common, $\mathcal{H}$, $\mathcal{H}'$, $\mathcal{H}''$, ... form a *conjugacy class*. Conjugate means: the groups $\mathcal{H}$, $\mathcal{H}'$, $\mathcal{H}''$, ... belong to the same space-group type, their lattices have the same dimensions, and they are equivalent by symmetry operations of $\mathcal{G}$. One says, '$\mathcal{H}$, $\mathcal{H}'$, $\mathcal{H}''$, ... are conjugate in $\mathcal{G}$', or '$\mathcal{H}'$, $\mathcal{H}''$, ... are conjugate to $\mathcal{H}$ in $\mathcal{G}$'.

Two kinds of conjugation of maximal subgroups can be distinguished:

1. Orientational conjugation. The conjugate subgroups have differently oriented unit cells. The orientations can be mapped one onto the other by symmetry operations of $\mathcal{G}$. Consider as an example the orthorhombic subgroups of a hexagonal space group (Fig. 8.1); the three unit cells are mutually rotated by

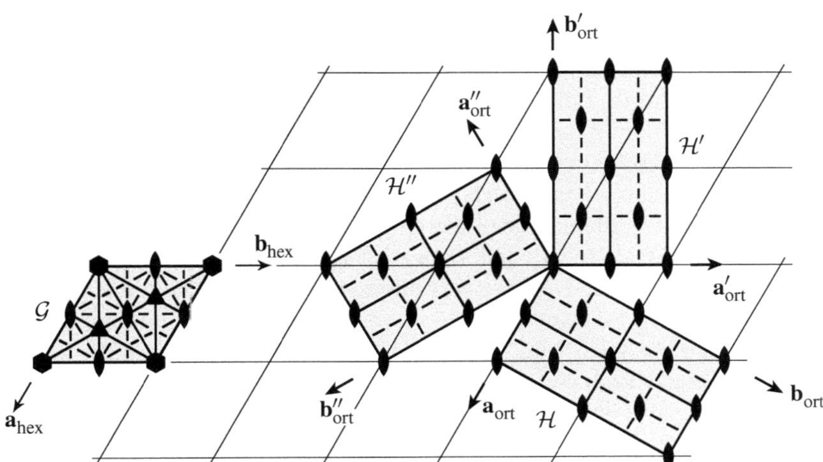

Fig. 8.1 Conjugate (C centred) orthorhombic subgroups ($\mathcal{H} = Cmm2$) of a hexagonal space group ($\mathcal{G} = P6mm$) with three different orientations.

120°, that is, they are equivalent by a rotation of order three of the hexagonal space group. The axis of order two which is contained in an axis of order six is maintained in the subgroups; the subgroups are conjugate in $\mathcal{G}$ by means of the lost axis of order three.

2. Translational conjugation. The primitive unit cell of the subgroups $\mathcal{H}$, $\mathcal{H}'$, $\mathcal{H}''$, ... must be enlarged by a (integral) factor ≥ 3 as compared to the primitive cell of $\mathcal{G}$. The conjugates differ in the selection of the symmetry elements of $\mathcal{G}$ that are being lost with the cell enlargement. If the unit cells of the conjugate subgroups $\mathcal{H}$, $\mathcal{H}'$, $\mathcal{H}''$, ... are set up according to the usual conventions, they differ in the positions of their origins; the positions of the origins of $\mathcal{H}$, $\mathcal{H}'$, $\mathcal{H}''$, ... are mapped onto each other by translation vectors of $\mathcal{G}$ that do not belong to $\mathcal{H}$.

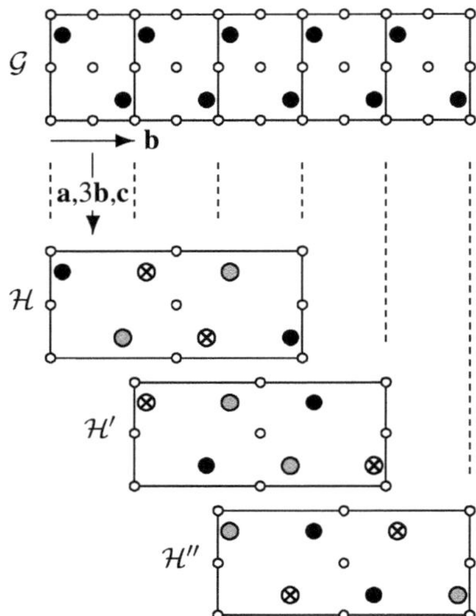

Fig. 8.2 Occurrence of conjugate subgroups by triplication of the unit cell.

Figure 8.2 shows an example where the unit cell of a centrosymmetric space group $\mathcal{G}$ (e.g. $P\bar{1}$) is enlarged by a factor of three. There are three subgroups $\mathcal{H}$, $\mathcal{H}'$, and $\mathcal{H}''$ that are conjugate in $\mathcal{G}$. Every one of them contains another third of the centres of inversion originally present. $\mathcal{H}$, $\mathcal{H}'$, and $\mathcal{H}''$ can be mapped onto each other by the translation vector **b** of $\mathcal{G}$. Instead of the initially symmetry-equivalent points ● we now have three kinds of points: ●, ⊗, and ◯. The three patterns of points are completely equivalent; the same pattern results if the 'colours' are interchanged.

A second example is shown in Fig. 8.3. In this case the unit cell of a centrosymmetric space group $\mathcal{G}$ is enlarged by a factor of four in two steps. The first step involves a doubling of the basis vector **b**, and one half of the inversion centres being lost. There are two subgroups $\mathcal{H}_1$ and $\mathcal{H}_2$ that are on a par; one contains one half of the original inversion centres, the other one contains the other half. However, $\mathcal{H}_1$ and $\mathcal{H}_2$ are not conjugate in $\mathcal{G}$, but belong to different conjugacy classes. $\mathcal{H}_1$ and $\mathcal{H}_2$ are not equivalent by a symmetry operation of $\mathcal{G}$; their origins are mutually shifted by $\frac{1}{2}$**b**. The patterns of the two inequivalent points ● and ⊗ are not alike and cannot be made equivalent by interchange of their colours. We return to this subject in Section 8.3, where non-conjugate subgroups of this kind are called *subgroups on a par* according to Definition 8.2.

The two unit cells depicted in Fig. 8.3 for the subgroup $\mathcal{H}_1$ differ in the positions of their origins, which are equivalent by the translation vector **b** of $\mathcal{G}$. And yet, in this case there are no conjugate subgroups; both unit cells comprise exactly the same selection of remaining inversion centres. For the description and distribution of the points ● and ⊗ it makes no difference whether the one or the other position of the origin is selected. $\mathcal{H}_1$ is a subgroup of $\mathcal{G}$ of index 2. This is an example for the universally valid statement that there never exist conjugate subgroups when the index is 2.

The situation is different when the unit cell is doubled a second time. At each of the steps $\mathcal{H}_1 \to \mathcal{H}_3$ and $\mathcal{H}_1 \to \mathcal{H}_3'$ again one half of the inversion centres are lost. $\mathcal{H}_3$ and $\mathcal{H}_3'$ are two different subgroups that are conjugate in $\mathcal{G}$ (but not in $\mathcal{H}_1$). Their unit cells shown in Fig. 8.3 are equivalent by the translation **b** of $\mathcal{G}$. Each one contains a different subset of one quarter of the inversion centres. Both patterns of distribution of the four kinds of points ●, ⊗, ◯, and ⊘ are absolutely equivalent, as can be recognized by interchange of the 'colours'. The index of $\mathcal{H}_3$ in $\mathcal{G}$ is 4.

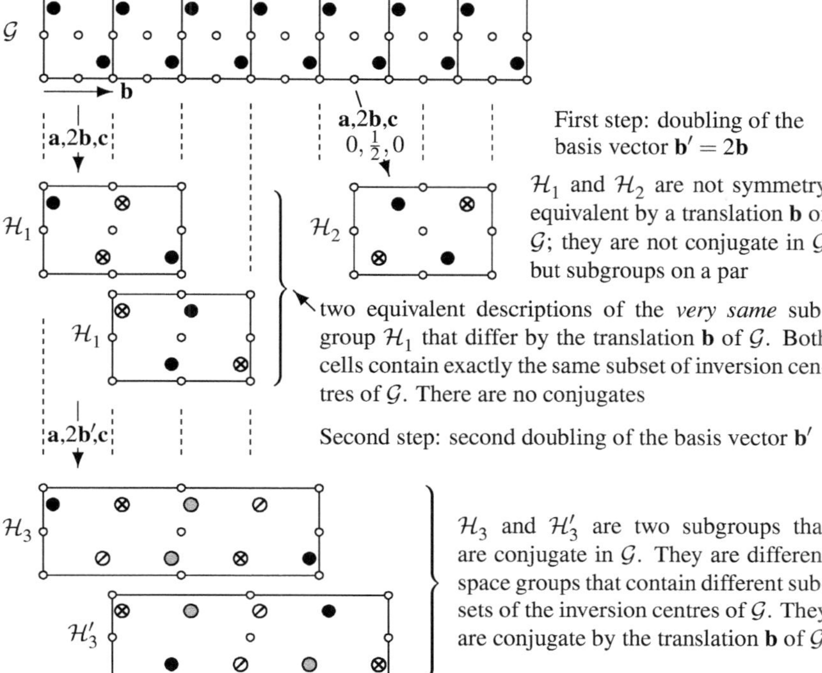

Fig. 8.3 Example for the occurrence of conjugate subgroups due to the loss of translational symmetry by enlargement of the unit cell by a factor of 4.

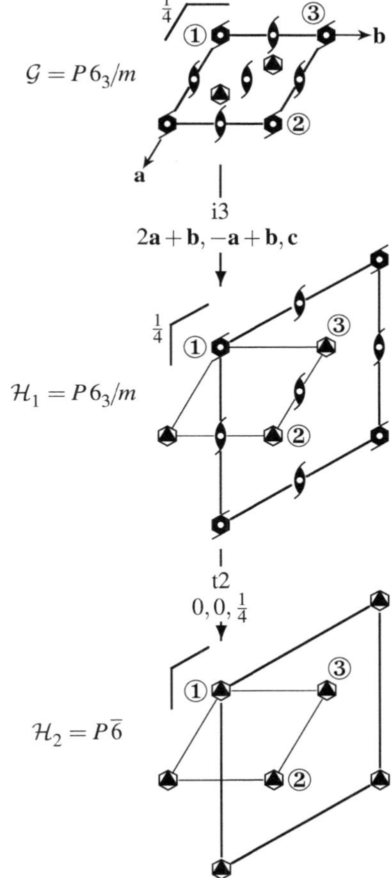

Fig. 8.4 The group $\mathcal{G}$ has three conjugate subgroups of type $\mathcal{H}_1$, but only one subgroup of type $\mathcal{H}_2$. The three conjugates to $\mathcal{H}_1$ differ in the positions of their origins. For $\mathcal{H}_2$ it makes no difference whether the origin is chosen at the position ①, ②, or ③; the result is the same subgroup in any case.

Cell enlargements do not always generate conjugate subgroups. If the cell of a space group is being enlarged in a direction in which the origin may float (i.e. is not fixed by symmetry), no conjugate subgroups result. For example, there are no isomorphic subgroups of $Pca2_1$ if the basis vector **c** is enlarged by an arbitrary integral factor, since the origin of $Pca2_1$ may float in the direction of **c**. In addition, there are some other cases where the enlargement of the unit cell by a factor ≥ 3 does not produce conjugate subgroups. An example is shown in Fig. 8.4.

Among *translationengleiche* maximal subgroups only orientational conjugation can occur; among *klassengleiche* and isomorphic maximal subgroups only translational conjugation.

8.2 Normalizers of space groups

According to Definition 5.5 two groups $\mathcal{H}$ and $\mathcal{H}'$ are conjugate subgroups in $\mathcal{G}$ if $\mathcal{H}$ can be mapped onto $\mathcal{H}'$ by an element $g_m \in \mathcal{G}$ by conjugation:

$$\mathcal{H}' = g_m^{-1}\mathcal{H}\, g_m \qquad g_m \notin \mathcal{H}$$

In addition, there always exist further elements $g_i \in \mathcal{G}$ that map $\mathcal{H}$ onto *itself.* These include at least the elements of $\mathcal{H}$ itself, but there may exist further elements with this property.

Definition 8.1 All elements $g_i \in \mathcal{G}$ that map a subgroup $\mathcal{H} < \mathcal{G}$ onto itself according to $\mathcal{H} = g_i^{-1}\mathcal{H}\,g_i$, taken by themselves, are the elements of a group. This group is called the *normalizer of* $\mathcal{H}$ *in* $\mathcal{G}$ and is designated by $\mathcal{N}_{\mathcal{G}}(\mathcal{H})$. Expressed mathematically:

$$\mathcal{N}_{\mathcal{G}}(\mathcal{H}) = \{g_i \in \mathcal{G} \mid g_i^{-1}\mathcal{H}\,g_i = \mathcal{H}\} \qquad (8.1)$$

The expression between the braces means: 'all elements g_i in $\mathcal{G}$ for which $g_i^{-1}\mathcal{H}\,g_i = \mathcal{H}$ holds'.

The normalizer is an intermediate group between $\mathcal{G}$ and $\mathcal{H}$: $\mathcal{H} \trianglelefteq \mathcal{N}_{\mathcal{G}}(\mathcal{H}) \leq \mathcal{G}$. It depends on $\mathcal{G}$ and $\mathcal{H}$. $\mathcal{H}$ is a normal subgroup of $\mathcal{N}_{\mathcal{G}}(\mathcal{H})$.

A special normalizer is the *Euclidean normalizer* of a space group. This is the normalizer of a space group $\mathcal{G}$ in the supergroup $\mathcal{E}$, the Euclidean group:

$$\mathcal{N}_{\mathcal{E}}(\mathcal{G}) = \{b_i \in \mathcal{E} \mid b_i^{-1}\mathcal{G}\,b_i = \mathcal{G}\}$$

The Euclidean group $\mathcal{E}$ comprises all isometries of three-dimensional space, that is, all distortion-free mappings. All space groups are subgroups of $\mathcal{E}$.

Consider as examples the images of the symmetry elements of the space groups in the left half of Fig. 8.5. A certain pattern can be recognized, the symmetry of which is higher than the symmetry of the space group itself. The symmetry operations that map equal symmetry elements of a space group $\mathcal{G}$ onto each other form a group (not necessarily a space group), which is nothing other than the Euclidean normalizer of $\mathcal{G}$. In Fig. 8.5 the Euclidean normalizers are shown on the right side. The Euclidean normalizer $\mathcal{N}_{\mathcal{E}}(\mathcal{G})$ so to speak describes the 'symmetry of the symmetry' of $\mathcal{G}$. A synonymous term is *Cheshire group* (so called after the 'Cheshire cat' from the children's story *Alice's Adventures in Wonderland*; first the cat appears grinning on the branches of a tree, later it disappears and nothing is left but its grin) [69].

In most cases $\mathcal{N}_{\mathcal{E}}(\mathcal{G})$ has a smaller unit cell than $\mathcal{G}$ (Fig. 8.5). For space groups whose origin floats in one or more directions (i.e. is not fixed by symmetry), the unit cell of the Euclidean normalizer is even infinitesimally small in the corresponding directions. For example, if $\mathcal{G}$ belongs to the crystal class $mm2$ and has the lattice basis $\mathbf{a}, \mathbf{b}, \mathbf{c}$, then the lattice basis of $\mathcal{N}_{\mathcal{E}}(\mathcal{G})$ is $\frac{1}{2}\mathbf{a}, \frac{1}{2}\mathbf{b}, \varepsilon\mathbf{c}$; the value of ε is infinitesimally small. In this case $\mathcal{N}_{\mathcal{E}}(\mathcal{G})$ is no longer a space group (the basis vectors of a space group may not be arbitrarily small); in the symbol of the normalizer this is expressed by a superscript ^{1}, ^{2}, or ^{3}, depending on the number of the corresponding directions.

In many cases of triclinic, monoclinic, and orthorhombic space groups, $\mathcal{N}_{\mathcal{E}}(\mathcal{G})$ also depends on the metric of the unit cell of $\mathcal{G}$. For example, the Euclidean normalizer of $Pbca$ is normally $Pmmm$ with halved lattice parameters, but it is $Pm\overline{3}$ if $a = b = c$ (Fig. 8.5). The Euclidean normalizer of $P2_1/m$ generally is $P2/m$ with halved lattice parameters; however, with a specialized metric of the cell it is $Pmmm$, $P4/mmm$, $P6/mmm$, or $Cmmm$ (Fig. 8.6). Normalizers with specialized metric have to be taken into account for *translationengleiche* subgroups; for example, a *translationengleiche* orthorhombic subgroup of a tetragonal space group still has the tetragonal cell metric $a = b \neq c$ (in practical work $a \approx b$).

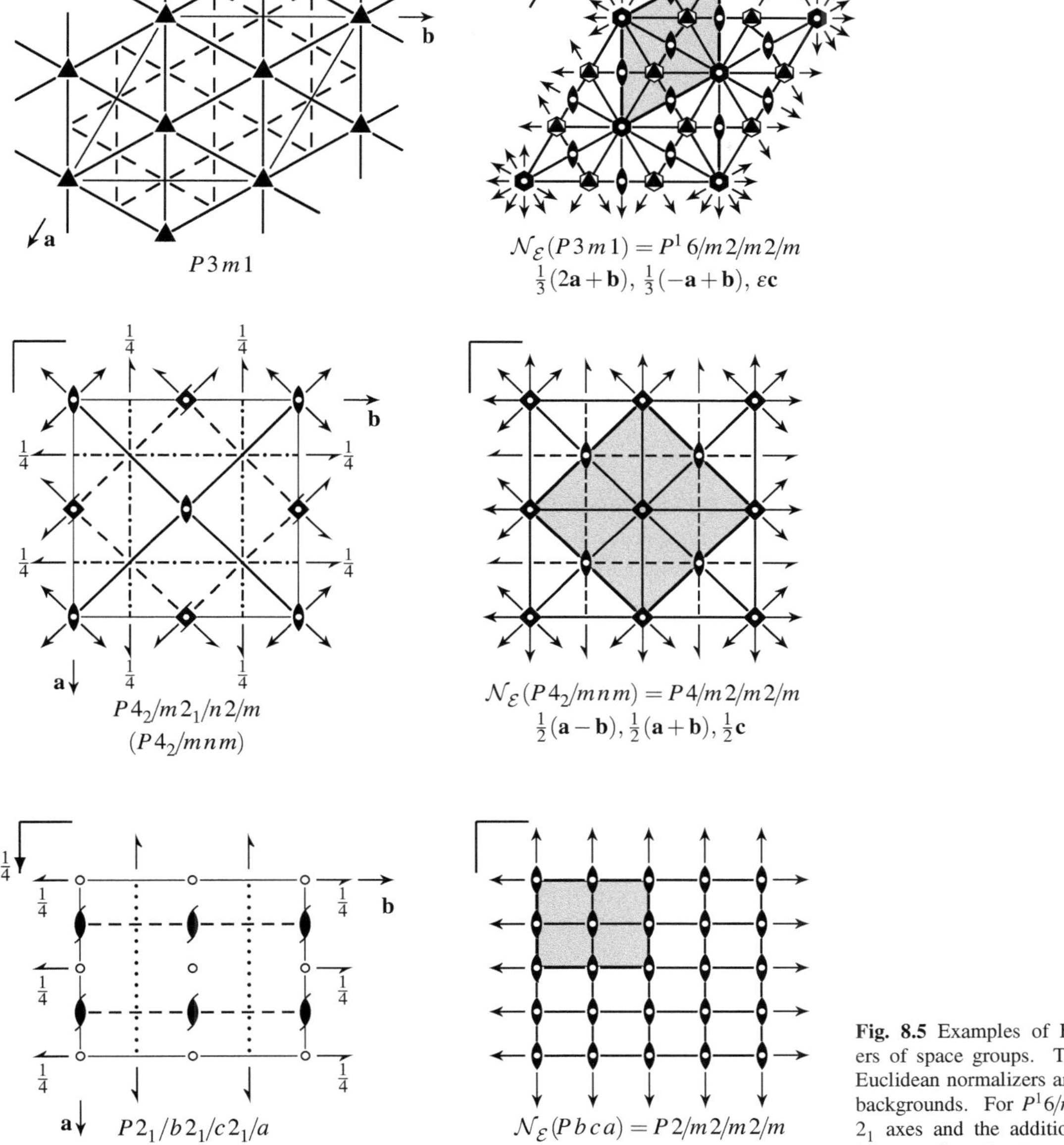

Fig. 8.5 Examples of Euclidean normalizers of space groups. The unit cells of the Euclidean normalizers are displayed by grey backgrounds. For $P^1 6/mmm$ the additional 2_1 axes and the additional glide planes in between the reflection planes have not been drawn for the sake of clarity. ε = infinitesimally small value.

Tables of Euclidean normalizers of all space groups can be found in *International Tables A*, beginning with the 1987 edition.[1] However, the Euclidean normalizers with specialized metric of the unit cell have been listed only since the fifth edition (2002); they can also be found in [70]. An extract from *International Tables A* is reproduced in Table 9.1 (page 117).

[1] Chapter 3.5 in the sixth edition (2016), Chapter 15 up to the fifth edition (2005).

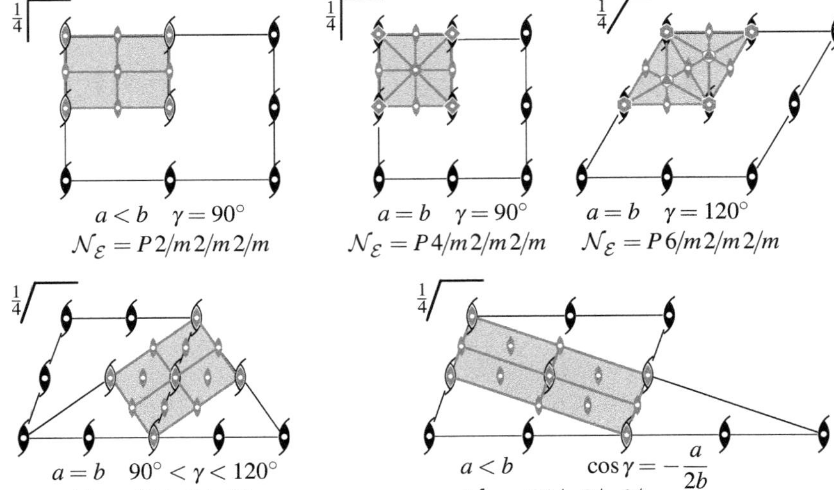

Fig. 8.6 Euclidean normalizers of the space group $P112_1/m$ with specialized metrics of the unit cell; **a** downwards, **b** to the right, monoclinic axis **c** perpendicular to the paper plane. The cells displayed by grey backgrounds correspond to the cells of the normalizers, for which only the symmetry axes and reflection planes parallel to **c** are shown, and whose third basis vector is $\frac{1}{2}\mathbf{c}$.

Finally, we mention two more normalizers. The *chirality-preserving Euclidean normalizer* $\mathcal{N}_{\mathcal{E}^+}(\mathcal{G})$ is the normalizer of a space group $\mathcal{G}$ in the chirality-preserving Euclidean group. That is the group of all isometries of three-dimensional space, but excluding all symmetry operations of the second kind (inversion, rotoinversion, reflection, glide reflection). The chirality-preserving normalizer is a subgroup of the Euclidean normalizer:

$$\mathcal{G} \le \mathcal{N}_{\mathcal{E}^+}(\mathcal{G}) \le \mathcal{N}_{\mathcal{E}}(\mathcal{G})$$

If $\mathcal{N}_{\mathcal{E}}(\mathcal{G})$ is centrosymmetric, $\mathcal{N}_{\mathcal{E}^+}(\mathcal{G})$ is the non-centrosymmetric subgroup of $\mathcal{N}_{\mathcal{E}}(\mathcal{G})$ of index 2 that is a supergroup of $\mathcal{G}$. If $\mathcal{N}_{\mathcal{E}}(\mathcal{G})$ is non-centrosymmetric, $\mathcal{N}_{\mathcal{E}^+}(\mathcal{G})$ and $\mathcal{N}_{\mathcal{E}}(\mathcal{G})$ are identical. Chirality-preserving Euclidean normalizers have been listed in *International Tables A* since the sixth edition (2016).

The *affine normalizer*, like the Euclidean normalizer, maps a space group onto itself, but in addition allows a distortion of the lattice. For example, only parallel 2_1 axes of the space group $P2_1/b2_1/c2_1/a$ ($a \neq b \neq c$) are mapped onto each other by the Euclidean normalizer; however, the affine normalizer also maps the differently oriented axes onto each other. The affine normalizer of the group $\mathcal{G}$ is the normalizer in the affine group, the group of all mappings (including distorting mappings). The affine normalizer is a supergroup of the Euclidean normalizer.

8.3 The number of conjugate subgroups. Subgroups on a par

According to eqn (8.1), the normalizer of $\mathcal{H}$ in $\mathcal{G}$, $\mathcal{N}_{\mathcal{G}}(\mathcal{H})$, is the group of all elements $g_m \in \mathcal{G}$ that map $\mathcal{H}$ onto itself by conjugation. These group elements are also elements of $\mathcal{N}_{\mathcal{E}}(\mathcal{H})$, the Euclidean normalizer of $\mathcal{H}$.

$\mathcal{N}_{\mathcal{G}}(\mathcal{H})$ is the largest common subgroup of $\mathcal{G}$ and $\mathcal{N}_{\mathcal{E}}(\mathcal{H})$; it consists of the intersection of the sets of symmetry operations of $\mathcal{G}$ and $\mathcal{N}_{\mathcal{E}}(\mathcal{H})$. This property makes it especially valuable; it renders it possible to derive the normalizer $\mathcal{N}_{\mathcal{G}}(\mathcal{H})$ from the tabulated Euclidean normalizers of the space groups.

The index j of $\mathcal{N}_{\mathcal{G}}(\mathcal{H})$ in $\mathcal{G}$ corresponds to the number of conjugate subgroups of $\mathcal{H}$ in $\mathcal{G}$ [71]. The arrows in the adjacent graph mark group–subgroup relations, which do not have to be maximal. In addition, space groups may coincide, $\mathcal{N}_{\mathcal{E}}(\mathcal{H}) = \mathcal{N}_{\mathcal{G}}(\mathcal{H})$ or $\mathcal{G} = \mathcal{N}_{\mathcal{G}}(\mathcal{H})$ or $\mathcal{N}_{\mathcal{G}}(\mathcal{H}) = \mathcal{H}$; the corresponding connecting arrow is then omitted. If $\mathcal{G} = \mathcal{N}_{\mathcal{G}}(\mathcal{H})$, then $j = 1$ and there are no subgroups conjugate to $\mathcal{H}$; this applies if $\mathcal{H}$ is a normal subgroup of $\mathcal{G}$.

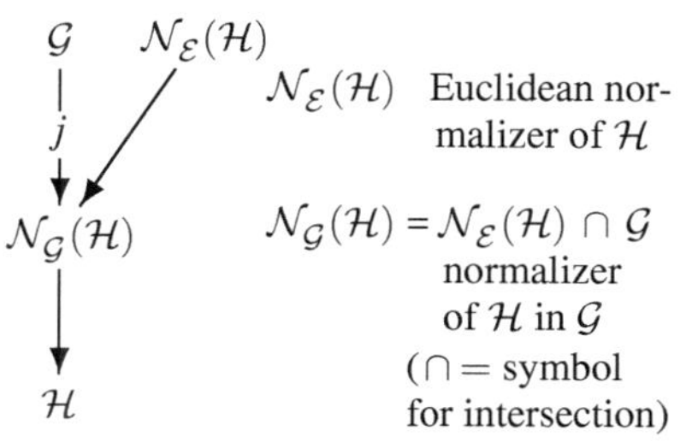

index j = number of conjugates of $\mathcal{H}$ in $\mathcal{G}$

Example 8.1

How many maximal conjugate subgroups of type $Cmcm$ does the space group $P6_3/mmc$ have?

The Euclidean normalizer of $Cmcm$ is $Pmmm$ with halved basis vectors (Table 9.1). Since $Cmcm$ is a maximal subgroup of $P6_3/mmc$, $\mathcal{N}_{\mathcal{G}}(\mathcal{H})$ must either be equal to $\mathcal{G}$ or equal to $\mathcal{H}$. In this case, $\mathcal{N}_{\mathcal{G}}(\mathcal{H}) = \mathcal{H}$ and the index 3 of $\mathcal{N}_{\mathcal{G}}(\mathcal{H})$ in $\mathcal{G}$ shows the existence of three conjugates of $Cmcm$ in $P6_3/mmc$. They have the three orientations as in Fig. 8.1. In the adjacent graph the basis vectors of every space group are given as vector sums of the basis vectors **a**, **b**, **c** of $\mathcal{G}$.

$Pmmm = \mathcal{N}_{\mathcal{E}}(\mathcal{H})$
$\frac{1}{2}\mathbf{a}$, $\frac{1}{2}(\mathbf{a}+2\mathbf{b})$, $\frac{1}{2}\mathbf{c}$

$\mathcal{G} = P6_3/mmc$
a, **b**, **c**

$Cmmm$
a, $\mathbf{a}+2\mathbf{b}$, $\frac{1}{2}\mathbf{c}$

3

$\mathcal{H} = Cmcm = \mathcal{N}_{\mathcal{G}}(\mathcal{H})$
a, $\mathbf{a}+2\mathbf{b}$, **c**

Among the subgroups of a space group $\mathcal{G}$ there may exist several conjugacy classes, $\mathcal{H}_1, \mathcal{H}_1', \mathcal{H}_1'', \ldots,\ \mathcal{H}_2, \mathcal{H}_2', \mathcal{H}_2'', \ldots,\ \ldots,$ all of which belong to the same space-group type and whose unit cells have the same dimensions.

Definition 8.2 Subgroups $\mathcal{H}_1, \mathcal{H}_2, \ldots < \mathcal{Z} \leq \mathcal{G}$ which are *not* conjugate in $\mathcal{G}$ but conjugate in one of the Euclidean normalizers $\mathcal{N}_{\mathcal{E}}(\mathcal{G})$ or $\mathcal{N}_{\mathcal{E}}(\mathcal{Z})$ are called *subgroups on a par* in $\mathcal{G}$.[2] They belong to different conjugacy classes, have the same lattice dimensions, and the same space-group type.

[2] We avoid the expression 'equivalent subgroups', which has different meanings in the literature; in addition, it should not be confused with 'symmetry equivalent'.

We already met subgroups on a par in Fig. 8.3 (page 107). There, $\mathcal{H}_1$ and $\mathcal{H}_2$ are subgroups on a par in $\mathcal{G}$; they are *not* symmetry equivalent by a symmetry operation of $\mathcal{G}$. $\mathcal{H}_3$ and $\mathcal{H}_3'$ are subgroups on a par of $\mathcal{H}_1$; they are not conjugate in $\mathcal{H}_1$, but they are conjugate in $\mathcal{G}$ and also in $\mathcal{N}_{\mathcal{E}}(\mathcal{H}_1)$. Examples of subgroups on a par are dealt with in the following Example 8.2 and in Section 12.2, page 151.

Subgroups on a par $\mathcal{H}_1, \mathcal{H}_2, \ldots$ that are *maximal* subgroups of $\mathcal{G}$ are conjugate in $\mathcal{N}_{\mathcal{E}}(\mathcal{G})$. Then one of the relations shown in Fig. 8.7 holds. The mentioned indices i and j show how many of these conjugacy classes occur and how many conjugates are contained in each of them. Every conjugacy class of the subgroups on a par contains the same number of conjugates. The number $i \times j$ refers to all conjugates in all conjugacy classes of the subgroups on a par; for this quantity the expression 'Euclidean equivalent subgroups' can be found in the literature [71, 76] which, however, has caused misunderstandings (it is only applicable if $\mathcal{H}_1, \mathcal{H}_2, \ldots$ are maximal subgroups of $\mathcal{G}$).

If $\mathcal{H}$ is not a maximal subgroup of $\mathcal{G}$, there may be subgroups on a par that are conjugate in $\mathcal{N}_{\mathcal{E}}(\mathcal{Z})$, the Euclidean normalizer of an intermediate group $\mathcal{Z}$, $\mathcal{G} > \mathcal{Z} > \mathcal{H}$. If $\mathcal{H}$ is a maximal subgroup of $\mathcal{Z}$, we can elucidate with one of the diagrams of Fig. 8.7 if there exist subgroups on a par (different conjugacy classes); for this purpose, $\mathcal{G}$ is to be replaced by $\mathcal{Z}$.

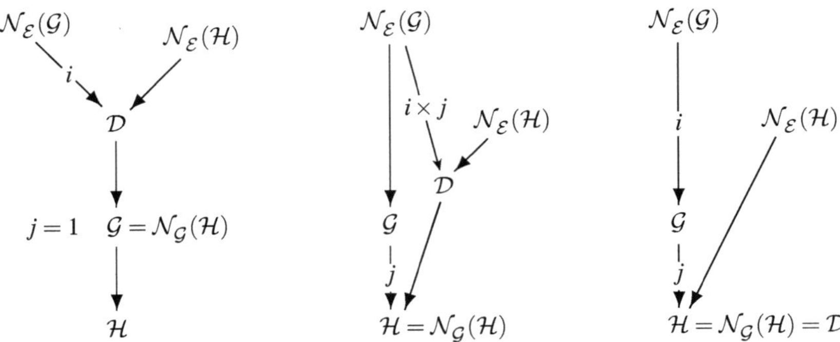

$\mathcal{D}=\mathcal{N}_\mathcal{E}(\mathcal{G}) \cap \mathcal{N}_\mathcal{E}(\mathcal{H})=$ largest common subgroup of $\mathcal{N}_\mathcal{E}(\mathcal{G})$ and $\mathcal{N}_\mathcal{E}(\mathcal{H})$

Fig. 8.7 Possible group–subgroup relations between the space group $\mathcal{G}$, its maximal subgroup $\mathcal{H}$, and their Euclidean normalizers. Except for $\mathcal{G}$ and $\mathcal{H}$, two or more of the groups may coincide; the connecting arrow is then omitted. The group–subgroup relations marked by the arrows, from $\mathcal{G}$ to $\mathcal{H}$ excepted, do not have to be maximal.
Index j of $\mathcal{N}_\mathcal{G}(\mathcal{H})$ in $\mathcal{G}$ = number of conjugates to $\mathcal{H}$ in $\mathcal{G}$ in a conjugacy class;
index i = number of conjugacy classes.

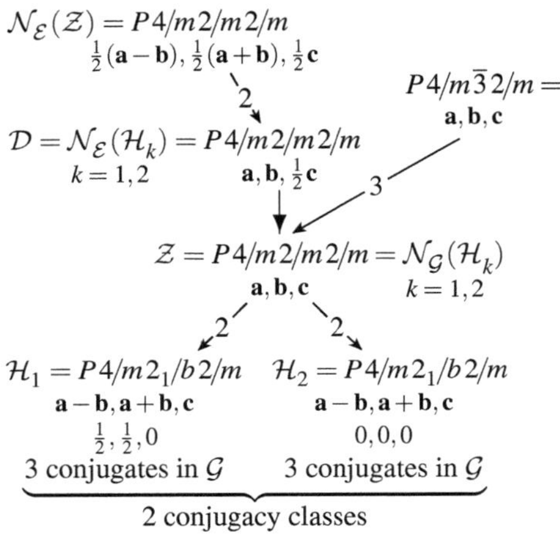

compare with the left diagram of Fig. 8.7, with $\mathcal{D}=\mathcal{N}_\mathcal{E}(\mathcal{H}_k)$, $i=2$, and inserting $\mathcal{Z}$ instead of $\mathcal{G}$

Example 8.2

Starting from the cubic space group of perovskite, $\mathcal{G}=Pm\bar{3}m$, symmetry can be reduced in two steps to the space group $P4/mbm$. As can be seen by the relations shown in the margin, there are two different subgroups on a par of the type $P4/mbm$, $\mathcal{H}_1$ and $\mathcal{H}_2$, which have the tetragonal **c** vector parallel to cubic **c** and the origin positions $\frac{1}{2},\frac{1}{2},0$ and $0,0,0$ in the coordinate system of $\mathcal{G}$. They are conjugate in $\mathcal{N}_\mathcal{E}(\mathcal{Z})$, the Euclidean normalizer of the intermediate group $\mathcal{Z}$, and belong to two conjugacy classes. The index of $\mathcal{N}_\mathcal{G}(\mathcal{H}_k)$ in $\mathcal{G}$ is 3. Therefore, the conjugacy classes represented by $\mathcal{H}_1$ and $\mathcal{H}_2$ consist of three groups each that are conjugate in $\mathcal{G}$; they have **c** parallel to **a**, **b**, and **c** of cubic $\mathcal{G}$, respectively. The existence of two subgroups on a par, $\mathcal{H}_1$ and $\mathcal{H}_2$, with **c** parallel to cubic **c** can be recognized by the index 2 of $\mathcal{D}$ in $\mathcal{N}_\mathcal{E}(\mathcal{Z})$. They render possible two different kinds of distortion of the perovskite structure (Fig. 8.8).

If we are interested in the subgroups of $\mathcal{G}$ in a tree of group–subgroup relations, in general only one representative of each conjugacy class needs to be considered, since all representatives are symmetry equivalent from the point of view of $\mathcal{G}$ and thus are absolutely equal. However, subgroups on a par, being non-conjugate, should all be considered, even though they also have the same space-group type and the same lattice dimensions. In the tree shown to the left of Example 8.2 only one representative is mentioned of each conjugacy class. The complete graph, which includes all conjugate subgroups, is depicted in Fig. 8.9.

In *International Tables A1* all maximal subgroups are listed for every space group. All conjugate maximal subgroups are mentioned in Part 2 of the tables; braces indicate which of them belong to the same conjugacy classes. In Part 3 of the Tables (Relations between the Wyckoff positions) only one representative is listed for every conjugacy class; however, if there exist conjugate subgroups with orientational conjugation, their basis transformations are mentioned. The relations between the Wyckoff positions for a group–subgroup pair are always the same for conjugate subgroups.

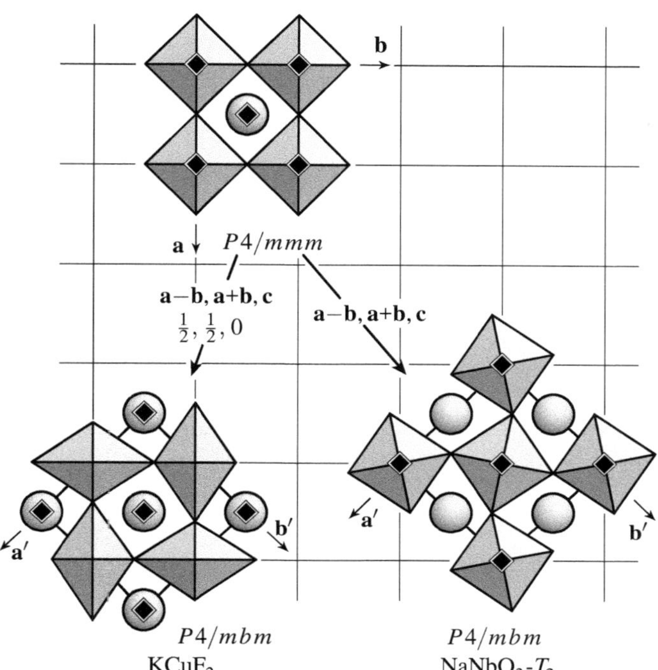

Fig. 8.8 Two variants of the perovskite structure with different kinds of distortion of the coordination octahedra in two different *klassengleiche* subgroups on a par. With the cell enlargement from $P4/mmm$ to $P4/mbm$, in one case the fourfold rotation axes running through the centres of the octahedra are lost; in the other case those axes running through the cations that are drawn as spheres are lost.

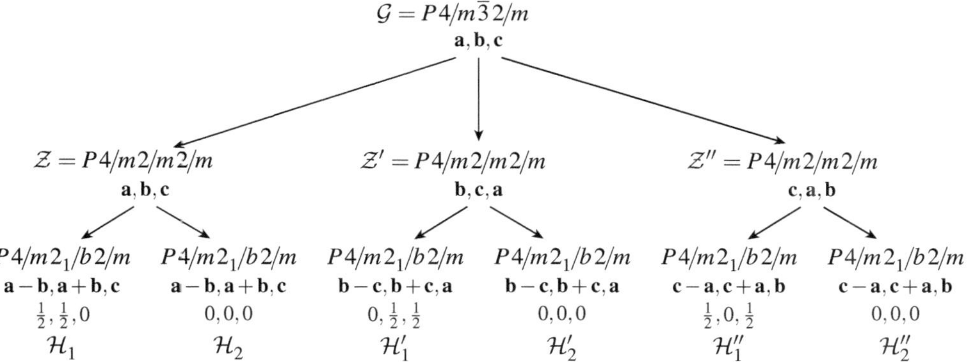

Fig. 8.9 Complete graph of the group–subgroup relations from Example 8.2 including all conjugate subgroups (without Euclidean normalizers). The basis vectors of the subgroups are given as linear combinations of the basis vectors **a**, **b**, and **c** of $\mathcal{G}$, and the origin positions refer to the coordinate system of $\mathcal{G}$. Subgroups that are conjugate in $\mathcal{G}$ are distinguished by primes $'$ and $''$.

Exercises

Normalizers are in Table 9.1. Some of the problems require access to *International Tables A* or A1.
Solutions on page 325.

8.1. Every cubic space group has four conjugate maximal rhombohedral subgroups. Is this orientational or translational conjugation?

8.2. Use the images of the symmetry elements in *International Tables A* to draw a diagram each with the symmetry elements of the 'symmetry of the symmetry' of the space groups $P2_1\,2_1\,2_1$, $Pbam$, $P4_1$. What are the Hermann–Mauguin symbols and the basis vectors of the Euclidean normalizers? Does any one of the orthorhombic space groups have a normalizer with enhanced symmetry if $a = b$?

8.3. Let $\mathcal{G} = R\overline{3}m$ and $\mathcal{H} = P\overline{3}m1$ with the same lattice parameters (hexagonal setting of the unit cell). Set up a tree of group–subgroup relations with the aid of *International Tables A* 2006 or A1 including $\mathcal{N}_{\mathcal{E}}(\mathcal{H})$ and $\mathcal{N}_{\mathcal{G}}(\mathcal{H})$. How many conjugate subgroups are there for $\mathcal{H}$ in $\mathcal{G}$?

8.4. Consider Fig. 8.4 and show with the aid of the normalizers why there are three conjugates of the space group $\mathcal{H}_1$ in $\mathcal{G}$ but only one of $\mathcal{H}_2$.

8.5. How many subgroups on a par $\mathcal{H} = P6_3/mmc$ in $\mathcal{G} = P6/mmm$ exist with doubled **c**? Are there any conjugates of $\mathcal{H}$ in $\mathcal{G}$?

Equivalent descriptions of crystal structures and chirality

9.1 Standardized description of crystal structures

At least the following data are needed for a unique description of a crystal structure:

1. the dimensions of the lattice, expressed by the lattice parameters a, b, c, α, β, γ;
2. the space group (Hermann–Mauguin symbol, perhaps its number according to *International Tables*), given the case with mention of the origin choice (origin choice 1 or 2);
3. for every point orbit occupied by atoms, the coordinates of one of its atoms.

These data can be specified in infinitely different ways, in principle. To be able to compare different crystal structures, to give them a systematic order, and to be able to handle them in databases, a uniform as possible description of the many structures is desirable. Therefore, in the course of time, rules have been developed for a standardized description, namely [72, 73]:

- The conventional setting according to *International Tables A* is chosen for the space group. If the tables allow for different choices, the preference is: origin at a centre of inversion, if present (origin choice 2); hexagonal axes for rhombohedral space groups; monoclinic axis b and 'cell choice 1' for monoclinic space groups.
- If there is still free scope to choose the unit cell, the *reduced cell* is chosen. Detailed instructions to establish such a cell can be found in *International Tables A*.[1] The most important ones are:

  $$a \leq b \leq c; \qquad |\cos\gamma| \leq \frac{a}{2b}; \qquad |\cos\beta| \leq \frac{a}{2c}; \qquad |\cos\alpha| \leq \frac{b}{2c}$$

  All angles $60° \leq \alpha,\ \beta,\ \gamma \leq 90°$ or all $90° \leq \alpha,\ \beta,\ \gamma \leq 120°$.
- Atomic coordinates for every representative atom of a point orbit such that $0 \leq x < 1,\ 0 \leq y < 1,\ 0 \leq z < 1$ and a minimum for $\sqrt{x^2 + y^2 + z^2}$.

[1] Chapter 3.1.3 in the sixth edition (2016) [15]; Chapter 9.2 in previous editions [14].

- Sequence of the atoms according to the sequence of the Wyckoff symbols in *International Tables A* (from top to bottom); if the Wyckoff symbols coincide, sequence with increasing coordinate values of x, then y, then z.

The program STRUCTURE TIDY [74] can be used to standardize any set of data this way.

9.2 Equivalent descriptions of crystal structures

Contraventions of the rules mentioned in the preceding section are frequent, not only from negligence or ignorance of the rules, but often for compelling reasons. For instance, in the case of molecules, the atomic coordinates are preferably chosen such that the atoms belong to the same molecule and are listed in the sequence of their linkage, even if this is in contravention of the rules.

Standardized structural data can result in two similar structures being documented differently, such that their relationship can hardly be recognized or is even obscured. In order to make relations evident it is often necessary not to observe standardized descriptions. Relations become most evident when the unit cells exhibit strict correspondence of their settings, dimensions, axes' ratios, and atomic coordinates. When comparing crystal structures, cell transformations and the involved coordinate transformations should be avoided, if possible, even if this calls for non-conventional settings of space groups.

Even if the rules are observed, there are almost always several possible ways to describe the very same crystal structure. For example, for the structure of rock salt (space group $Fm\bar{3}m$) the origin can be chosen to be at the centre of an Na^+ or a Cl^- ion, which results in two different sets of coordinates. In this case the equivalence of both descriptions is easy to recognize. However, in many other cases this is by no means a simple matter.

For all space groups, except $Im\bar{3}m$ and $Ia\bar{3}d$, several different sets of atomic coordinates can be chosen describing one and the same structure in the same space-group setting. The number of equivalent coordinate sets for the space group $\mathcal{G}$ is exactly i, i being the index of $\mathcal{G}$ in its Euclidean normalizer $\mathcal{N}_{\mathcal{E}}(\mathcal{G})$ [75, 76]. That is due to the fact that the coset decomposition of $\mathcal{N}_{\mathcal{E}}(\mathcal{G})$ with respect to $\mathcal{G}$ results in i cosets (cf. Section 5.3, page 55). Each one of them corresponds to one equivalent set of coordinates.

To obtain one equivalent set of coordinates from another, we make use of the Euclidean normalizers as listed in *International Tables A*. Table 9.1 contains an excerpt from these tables. The columns under *Euclidean normalizer* $\mathcal{N}_{\mathcal{E}}(\mathcal{G})$ contain the (space) group symbols of the Euclidean normalizers and their basis vectors (as vector sums of the basis vectors of $\mathcal{G}$). For space groups with floating origins (not fixed by symmetry), such as $I4$, the numbers of pertinent dimensions of space are specified by superscript numbers in the symbols of $\mathcal{N}_{\mathcal{E}}(\mathcal{G})$, for example, $P^1 4/mmm$. In addition, the basis vectors of $\mathcal{N}_{\mathcal{E}}(\mathcal{G})$ are infinitesimally small in these directions, which is expressed by the numerical factor ε. The last column contains the index of $\mathcal{G}$ in $\mathcal{N}_{\mathcal{E}}(\mathcal{G})$. The index value corresponds to the number of equivalent sets of coordinates that can be used to describe a structure in the space group $\mathcal{G}$. By application of the coordinate

Table 9.1 Selection of Euclidean normalizers of space groups. Excerpt from *International Tables A*, sixth edition (2016) [15], Tables 3.5.2.3–3.5.2.5; Tables 15.2.1.3 and 15.2.1.4 in the editions of 2002 and 2005 [14]; Table 15.3.2 in the editions of 1987–1998.

Space group $\mathcal{G}$	Euclidean normalizer $\mathcal{N}_{\mathcal{E}}(\mathcal{G})$ and chirality-preserving normalizer $\mathcal{N}_{\mathcal{E}^+}(\mathcal{G})$		Additional generators of $\mathcal{N}_{\mathcal{E}}(\mathcal{G})$			Index of $\mathcal{G}$ in $\mathcal{N}_{\mathcal{E}}(\mathcal{G})$
Hermann–Mauguin symbol	Symbol	Basis vectors	Translations	Inversion through a centre at	Further generators	
$P12_1/m1$*	$P12/m1$	$\frac{1}{2}\mathbf{a}, \frac{1}{2}\mathbf{b}, \frac{1}{2}\mathbf{c}$	$\frac{1}{2},0,0;\ 0,\frac{1}{2},0;\ 0,0,\frac{1}{2}$			$8\cdot1\cdot1$
$P12_1/m1$†	$Bmmm$	$\frac{1}{2}(\mathbf{a}+\mathbf{c}), \frac{1}{2}\mathbf{b}, \frac{1}{2}(-\mathbf{a}+\mathbf{c})$	$\frac{1}{2},0,0;\ 0,\frac{1}{2},0;\ 0,0,\frac{1}{2}$		z, y, x	$8\cdot1\cdot2$
$C12/m1$*	$P12/m1$	$\frac{1}{2}\mathbf{a}, \frac{1}{2}\mathbf{b}, \frac{1}{2}\mathbf{c}$	$\frac{1}{2},0,0;\ 0,0,\frac{1}{2}$			$4\cdot1\cdot1$
$P2_12_12_1$*	$Pmmm$	$\frac{1}{2}\mathbf{a}, \frac{1}{2}\mathbf{b}, \frac{1}{2}\mathbf{c}$	$\frac{1}{2},0,0;\ 0,\frac{1}{2},0;\ 0,0,\frac{1}{2}$	$0,0,0$		$8\cdot2\cdot1$
	$\mathcal{N}_{\mathcal{E}^+}(\mathcal{G})$: $P222$	$\frac{1}{2}\mathbf{a}, \frac{1}{2}\mathbf{b}, \frac{1}{2}\mathbf{c}$	$\frac{1}{2},0,0;\ 0,\frac{1}{2},0;\ 0,0,\frac{1}{2}$			$8\cdot1$
$Cmcm$	$Pmmm$	$\frac{1}{2}\mathbf{a}, \frac{1}{2}\mathbf{b}, \frac{1}{2}\mathbf{c}$	$\frac{1}{2},0,0;\ 0,0,\frac{1}{2}$			$4\cdot1\cdot1$
$Ibam$*	$Pmmm$	$\frac{1}{2}\mathbf{a}, \frac{1}{2}\mathbf{b}, \frac{1}{2}\mathbf{c}$	$\frac{1}{2},0,0;\ 0,\frac{1}{2},0$			$4\cdot1\cdot1$
$I4$	$P^1 4/mmm$	$\frac{1}{2}(\mathbf{a}-\mathbf{b}), \frac{1}{2}(\mathbf{a}+\mathbf{b}), \varepsilon\mathbf{c}$	$0,0,t$	$0,0,0$	y, x, z	$\infty\cdot2\cdot2$
	$\mathcal{N}_{\mathcal{E}^+}(\mathcal{G})$: $P^1 422$	$\frac{1}{2}(\mathbf{a}-\mathbf{b}), \frac{1}{2}(\mathbf{a}+\mathbf{b}), \varepsilon\mathbf{c}$	$0,0,t$		$y, x, \bar{z}$	$\infty\cdot2$
$P4/n$	$P4/mmm$	$\frac{1}{2}(\mathbf{a}-\mathbf{b}), \frac{1}{2}(\mathbf{a}+\mathbf{b}), \frac{1}{2}\mathbf{c}$	$\frac{1}{2},\frac{1}{2},0;\ 0,0,\frac{1}{2}$		y, x, z	$4\cdot1\cdot2$
$I422$	$P4/mmm$	$\frac{1}{2}(\mathbf{a}-\mathbf{b}), \frac{1}{2}(\mathbf{a}+\mathbf{b}), \frac{1}{2}\mathbf{c}$	$0,0,\frac{1}{2}$	$0,0,0$		$2\cdot2\cdot1$
	$\mathcal{N}_{\mathcal{E}^+}(\mathcal{G})$: $P422$	$\frac{1}{2}(\mathbf{a}-\mathbf{b}), \frac{1}{2}(\mathbf{a}+\mathbf{b}), \frac{1}{2}\mathbf{c}$	$0,0,\frac{1}{2}$			$2\cdot1$
$P\bar{4}2c$	$P4/mmm$	$\frac{1}{2}(\mathbf{a}-\mathbf{b}), \frac{1}{2}(\mathbf{a}+\mathbf{b}), \frac{1}{2}\mathbf{c}$	$\frac{1}{2},\frac{1}{2},0;\ 0,0,\frac{1}{2}$	$0,0,0$		$4\cdot2\cdot1$
$I\bar{4}2d$	$P4_2/nnm$	$\frac{1}{2}(\mathbf{a}-\mathbf{b}), \frac{1}{2}(\mathbf{a}+\mathbf{b}), \frac{1}{2}\mathbf{c}$	$0,0,\frac{1}{2}$	$\frac{1}{4},0,\frac{1}{8}$		$2\cdot2\cdot1$
$P3_221$	$P6_422$	$\mathbf{a}+\mathbf{b}, -\mathbf{a}, \frac{1}{2}\mathbf{c}$	$0,0,\frac{1}{2}$		$\bar{x}, \bar{y}, z$	$2\cdot2$
$P\bar{3}m1$	$P6/mmm$	$\mathbf{a}, \mathbf{b}, \frac{1}{2}\mathbf{c}$	$0,0,\frac{1}{2}$		$\bar{x}, \bar{y}, z$	$2\cdot1\cdot2$
$R\bar{3}m$ (hex.)	$R\bar{3}m$ (hex.)	$-\mathbf{a}, -\mathbf{b}, \frac{1}{2}\mathbf{c}$	$0,0,\frac{1}{2}$			$2\cdot1\cdot1$
$P\bar{6}$	$P6/mmm$	$\frac{1}{3}(2\mathbf{a}+\mathbf{b}), \frac{1}{3}(-\mathbf{a}+\mathbf{b}), \frac{1}{2}\mathbf{c}$	$\frac{2}{3},\frac{1}{3},0;\ 0,0,\frac{1}{2}$	$0,0,0$	y, x, z	$6\cdot2\cdot2$
$P6_3/m$	$P6/mmm$	$\mathbf{a}, \mathbf{b}, \frac{1}{2}\mathbf{c}$	$0,0,\frac{1}{2}$		y, x, z	$2\cdot1\cdot2$
$P6_3mc$	$P^1 6/mmm$	$\mathbf{a}, \mathbf{b}, \varepsilon\mathbf{c}$	$0,0,t$	$0,0,0$		$\infty\cdot2\cdot1$
$P6_3/mmc$	$P6/mmm$	$\mathbf{a}, \mathbf{b}, \frac{1}{2}\mathbf{c}$	$0,0,\frac{1}{2}$			$2\cdot1\cdot1$
$Pm\bar{3}m$	$Im\bar{3}m$	$\mathbf{a}, \mathbf{b}, \mathbf{c}$	$\frac{1}{2},\frac{1}{2},\frac{1}{2}$			$2\cdot1\cdot1$
$Fm\bar{3}m$	$Pm\bar{3}m$	$\frac{1}{2}\mathbf{a}, \frac{1}{2}\mathbf{b}, \frac{1}{2}\mathbf{c}$	$\frac{1}{2},\frac{1}{2},\frac{1}{2}$			$2\cdot1\cdot1$

* without specialized lattice metric.
† if $a = c$, $90° < \beta < 120°$.

transformations listed in the column *Additional generators of* $\mathcal{N}_{\mathcal{E}}(\mathcal{G})$ to a set of coordinates we can calculate the equivalent sets of coordinates. The symbol t means a translation of an arbitrary amount. A few particularities have to be watched in the case of chiral crystal structures (see next section).

Example 9.1

Rock salt crystallizes in space group $Fm\bar{3}m$. The Euclidean normalizer is $\mathcal{N}_{\mathcal{E}}(Fm\bar{3}m) = Pm\bar{3}m$ ($\frac{1}{2}\mathbf{a}$, $\frac{1}{2}\mathbf{b}$, $\frac{1}{2}\mathbf{c}$) with index $i = 2$. Therefore, there are two possible sets of coordinates. One is obtained from the other according to the *additional generators of* $\mathcal{N}_{\mathcal{E}}(\mathcal{G})$ in Table 9.1 by the addition of $\frac{1}{2}$, $\frac{1}{2}$, $\frac{1}{2}$:

Na	$4a$	$0,0,0$	and	Na	$4b$	$\frac{1}{2},\frac{1}{2},\frac{1}{2}$
Cl	$4b$	$\frac{1}{2},\frac{1}{2},\frac{1}{2}$		Cl	$4a$	$0,0,0$

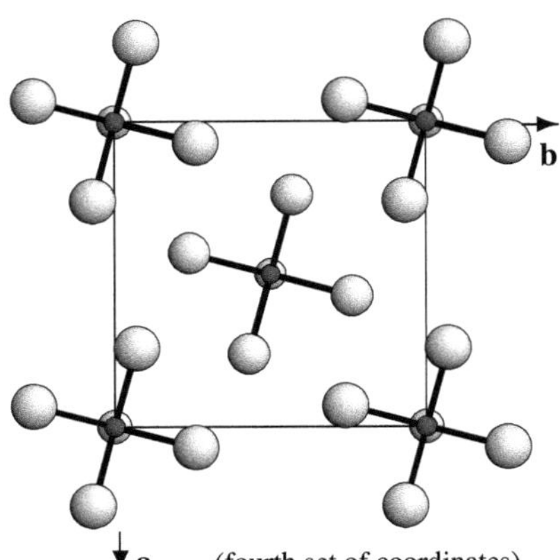

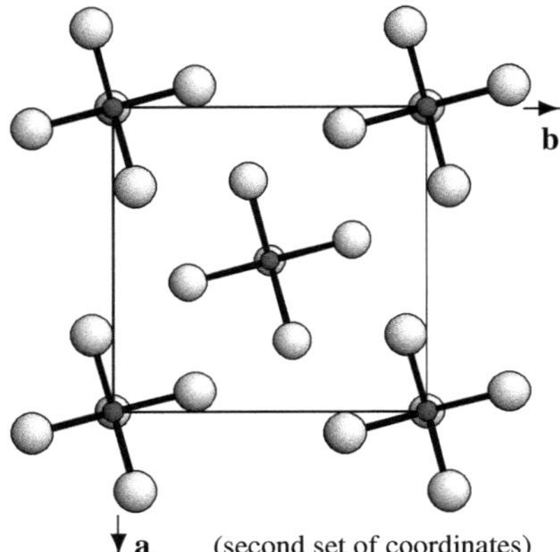

Fig. 9.1 Two equivalent descriptions of the crystal structure of $WOBr_4$. Two more result by inversion and infinitely many more by shifting the origin in the direction of **c**.

Example 9.2

$WOBr_4$ crystallizes in space group $I4$ with the atomic coordinates [77]:

	x	y	z
W	0	0	0.078
O	0	0	0.529
Br	0.260	0.069	0.0

The Euclidean normalizer of $I4$ is P^14/mmm with the basis vectors $\frac{1}{2}(\mathbf{a}-\mathbf{b})$, $\frac{1}{2}(\mathbf{a}+\mathbf{b})$, $\varepsilon\mathbf{c}$. The index of $I4$ in P^14/mmm is $\infty\cdot 2\cdot 2$ and thus is infinite (due to the infinitesimally small basis vector $\varepsilon\mathbf{c}$). By addition of $0,0,t$ to the coordinates of all atoms we obtain another equivalent set of coordinates. For this, there are infinitely many possibilities because t can adopt any arbitrary value. As expressed by the index $\infty\cdot 2\cdot 2$, in addition, there exist four equivalent sets of coordinates for every one of these infinitely many sets of coordinates. They are obtained by inversion at $0,0,0$ and by the transformation y, x, z. The equivalent sets of coordinates are thus:

(1)	W	0	0	$0.078+t$	(2)	0	0	$-0.078-t$
	O	0	0	$0.529+t$		0	0	$-0.529-t$
	Br	0.260	0.069	$0.0+t$		−0.260	−0.069	$0.0-t$
(3)	W	0	0	$0.078+t$	(4)	0	0	$-0.078-t$
	O	0	0	$0.529+t$		0	0	$-0.529-t$
	Br	0.069	0.260	$0.0+t$		−0.069	−0.260	$0.0-t$

with t = arbitrary. The situation is depicted in Fig. 9.1.

9.3 Chirality

Definition 9.1 An object is *chiral* if it cannot be superposed by pure rotation and translation on its image formed by inversion through a point.

The symmetry group of a chiral object contains no symmetry operations of the second kind, that is, no inversion, rotoinversion, reflection, or glide reflection. There exist two distinct enantiomorphic states that are interconverted by inversion. Further terms in this context are [78, 79]:

Absolute configuration	Spatial arrangement of atoms in a chiral molecule and its appropriate designation (e.g. by (*R*), (*S*) etc.)
Absolute (crystal) structure	Spatial arrangement of atoms in a chiral crystal and its description (lattice parameters, space group, atomic coordinates)
Enantiomorph	One out of a pair of objects of opposite handedness
Enantiomer	One molecule out of a pair of opposite handedness (special designation for enantiomorphic molecules)
Racemate	Equimolar mixture of a pair of enantiomers
Chirality sense (handedness)	Property that distinguishes enantiomorphs from one another; the two enantiomorphs of a pair have opposite handedness (left or right)
Achiral	Object that can be superposed by pure rotation and translation on its inverted image

We have to distinguish: the symmetry of a molecule, designated by its point group; the symmetry of a crystal, designated by its space group; and the symmetry of a space group, designated by its Euclidean normalizer.

Space groups are precisely chiral if their Euclidean normalizer has no symmetry operations of the second kind. These are the space groups with screw axes 3_1 or 3_2, 4_1 or 4_3, 6_1 or 6_5, as well as 6_2 or 6_4 if only one of these kinds is present. There are eleven pairs of enantiomorphic space group types (Table 6.1, page 71; the terms 'enantiomorphic space group' and 'chiral space group' are synonymous).

An enantiomorphic space group is a sufficient but not a necessary condition for a chiral crystal structure. Chirality is given also if the building blocks of a crystal in an achiral space group consist of only one kind of enantiomers or if the building blocks, without having to be chiral themselves, are arranged in a chiral manner in the crystal. For example, the single strands of molecules in $WOBr_4$ are not chiral (Example 9.2 and Fig. 9.1); the square-pyramidal molecules fulfil the point symmetry $4mm$, and the strands of the molecules fulfil the rod group $\not{p}4mm$. The crystal structure, however, is chiral.

Chiral crystal structures are compatible only with space groups that have no inversion centres, rotoinversion axes, reflection, and glide-reflection planes; these symmetry elements would generate the opposite enantiomers, and the compound would have to be racemic. Chiral crystal structures can adopt one out of 65 space-group types; they are called the *Sohncke space-group types* (after L. SOHNCKE who was the first to derive them).[2] The Sohncke space-group types comprise the 11 pairs of enantiomorphic space-group types and a further 43 achiral space-group types. For details see [79]. The enantiomorphic structures of a chiral crystal structure in an achiral space group are equivalent by the Euclidean normalizer of their space group. The total number of equivalent sets of coordinates, including the enantiomorphic pairs, can be determined as described in the preceding section. If we want to determine the equivalent sets of coordinates of a chiral crystal structure without inclusion of the enantiomorphs, the chirality-preserving Euclidean normalizer $\mathcal{N}_{\mathcal{E}^+}(\mathcal{G})$ has to be applied instead of the Euclidean normalizer.

The chirality-preserving Euclidean normalizer is identical to the Euclidean normalizer in the case of the 11 pairs of enantiomorphic space-group types. For the other 43 Sohncke space-group types it is a non-centrosymmetric subgroup of index 2 of the Euclidean normalizer.

[2] Many structural researchers are not aware of the difference between chiral space groups and Sohncke space groups; frequently, even in textbooks, the term chiral space group is used although a Sohncke space group is meant.

Example 9.3

The phosphorus atoms in NaP form helical chains that have the symmetry (rod group), not imposed by the crystal symmetry, $\not{p}4_322$ [80]. The chains wind around 2_1 axes parallel to **b** in the space group $\mathcal{G} = P2_12_12_1$.

The Euclidean normalizer $\mathcal{N}_{\mathcal{E}}(\mathcal{G})$ is $Pmmm$ with halved basis vectors (cf. Table 9.1); the chirality-preserving Euclidean normalizer $\mathcal{N}_{\mathcal{E}^+}(\mathcal{G})$ is its non-centrosymmetric subgroup $P222$. The index 16 of $\mathcal{G}$ in $\mathcal{N}_{\mathcal{E}}(\mathcal{G})$ shows the existence of 16 equivalent sets of coordinates to describe the crystal structure, which includes 8 enantiomorphic pairs. The index of $\mathcal{G}$ in $\mathcal{N}_{\mathcal{E}^+}(\mathcal{G})$ is 8. That corresponds to the eight equivalent sets of coordinates that are obtained by application of the translations

$$0,0,0;\ \tfrac{1}{2},0,0;\ 0,\tfrac{1}{2},0;\ 0,0,\tfrac{1}{2};\ \tfrac{1}{2},\tfrac{1}{2},0;\ \tfrac{1}{2},0,\tfrac{1}{2};\ 0,\tfrac{1}{2},\tfrac{1}{2};\ \text{and}\ \tfrac{1}{2},\tfrac{1}{2},\tfrac{1}{2}$$

keeping the chirality. The other eight, with the opposite handedness, follow from the same translations and additional inversion.
The inversion converts the left-handed $\mathscr{p}4_3 22$ helices to right-handed $\mathscr{p}4_1 22$ helices. $\mathscr{p}4_3 22$ and $\mathscr{p}4_1 22$ helices are enantiomorphic. The chirality is a property of the polymeric $(P^-)_\infty$ ions.
The space group $P2_1 2_1 2_1$ itself is not chiral, but it contains no symmetry operations of the second kind; it is a Sohncke space group. The left- as well as right-handed form of NaP can crystallize in the space group $P2_1 2_1 2_1$.

If the space group itself is chiral, that is, if it belongs to one of the 11 pairs of enantiomorphic space-group types, the enantiomorphic pair of structures are not equivalent by the Euclidean normalizer. Therefore, when determining the number of equivalent sets of coordinates with the aid of the Euclidean normalizer, we obtain only those with the same chirality sense. For example, quartz exhibits two enantiomorphic forms in the space groups $P3_1 21$ (left quartz) and $P3_2 21$ (right quartz).[3] As explained in Example 9.4, there are four equivalent sets of coordinates for right quartz. For left quartz there are also four equivalent sets of coordinates that cannot be generated by the Euclidean normalizer of right quartz. To obtain a coordinate set of the opposite enantiomorph of an enantiomorphic space group, the coordinates have to be inverted at 0, 0, 0 and the opposite enantiomorphic space group has to be chosen.

Whereas the non-chiral space group $P2_1 2_1 2_1$ is compatible with $\mathscr{p}4_3 22$ as well as $\mathscr{p}4_1 22$ helices of NaP, the helical components of right quartz are incompatible with the space group $P3_1 21$ of left quartz.

[3] That 'left quartz' has right-handed 3_1 screw axes has historical reasons. The terms 'left quartz' and 'right quartz' were coined by CHR. S. WEISS in the nineteenth century, based on the morphology of quartz crystals, at times when the theory of space groups had not been developed and the experimental crystal structure determination was not possible. Casually, left quartz rotates the plane of polarized light to the left.

Example 9.4

The crystal structure of quartz was determined approximately one hundred times. Right quartz crystallizes in the space group $P3_2 21$ with the atomic coordinates:[4]

	x	y	z
Si	0.470	0	$\frac{1}{6}$
O	0.414	0.268	0.286

The Euclidean normalizer is $P6_4 22$ (Table 9.1). The index 4 shows four equivalent sets of coordinates. The other three are obtained by the translation $0,0,\frac{1}{2}$, the transformation $-x, -y, z$, and by translation and transformation:

Si	0.470	0	$\frac{2}{3}$	–0.470	0	$\frac{1}{6}$	–0.470	0	$\frac{2}{3}$
O	0.414	0.268	0.786	–0.414	–0.268	0.286	–0.414	–0.268	0.786

The four possibilities are shown in Fig. 9.2.

[4] Quartz in the space group $P3_2 21$ has left-handed chemical 3/1 *M* helices $(Si–O–)_3$ running parallel to **c**, having three repeating units per translation period. $AlPO_4$ has the same structure, with left-handed chemical 3/2 *M* helices $(Al–O–P–O–)_3$ that have three repeating units in two coil turns in the doubled translation period 2**c**; the space group is $P3_1 21$. Whereas the chemical handedness coincides, the screw axis has the opposite handedness (cf. page 100).

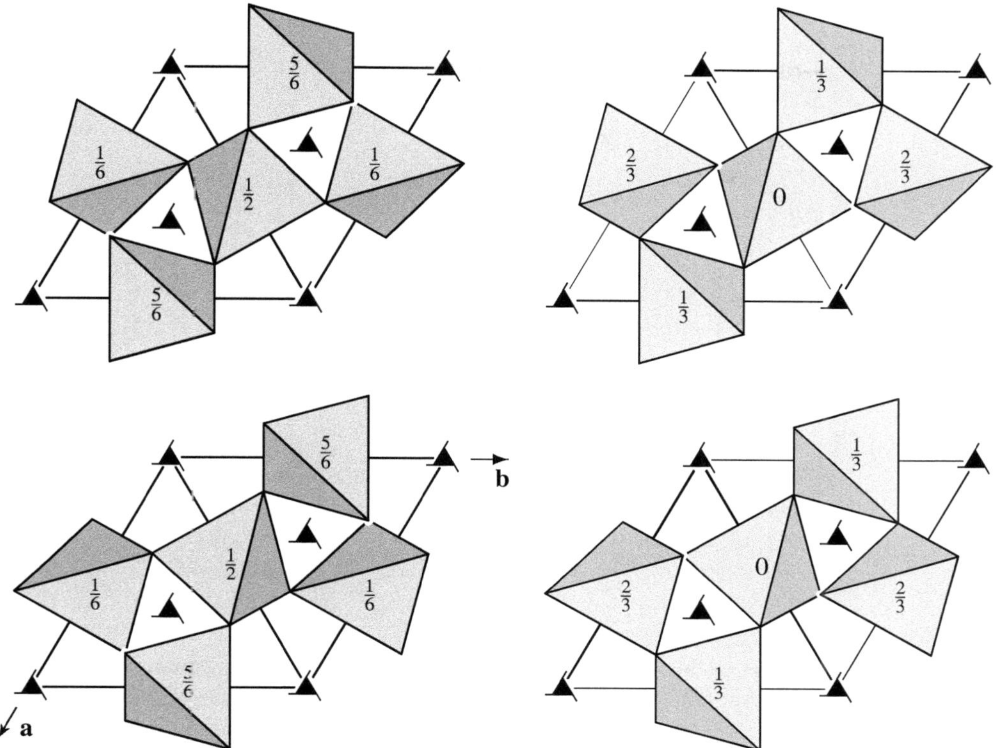

Fig. 9.2 The four equivalent descriptions for the crystal structure of right quartz. The numbers in the SiO_4 tetrahedra are the z coordinates of the Si atoms.

9.4 Wrongly assigned space groups

Due to the rapid collection of X-ray diffraction data and powerful computers and computer programs, the number of crystal structure determinations has greatly increased. Often computers have been used as 'black boxes' using preadjusted (default) routines, without the user worrying about what the computer has been doing. Correspondingly, the number of faulty structure determinations has increased. One of the most frequent errors is the choice of a wrong space group, in particular one of too low a symmetry [81–90]. This is always the case if the Euclidean normalizer $\mathcal{N}_{\mathcal{E}}(\mathcal{G})$ of the chosen space group $\mathcal{G}$ generates less than i equivalent sets of coordinates (i = index of $\mathcal{G}$ in $\mathcal{N}_{\mathcal{E}}(\mathcal{G})$). The correct space group is then an intermediate group between $\mathcal{G}$ and $\mathcal{N}_{\mathcal{E}}(\mathcal{G})$. Another reason for the wrong choice of space groups are twinned crystals, see Section 18.4.

Example 9.5
LaB_2C_2 was described in the space group $P\bar{4}2c$ with the set of coordinates mentioned first [91]. The Euclidean normalizer is $P4/mmm$ with $\frac{1}{2}(\mathbf{a}-\mathbf{b})$, $\frac{1}{2}(\mathbf{a}+\mathbf{b})$, $\frac{1}{2}\mathbf{c}$. The index of $P\bar{4}2c$ in $P4/mmm$ is 8. With the aid of Table 9.1 we supposedly obtain eight equivalent sets of coordinates:

	x	y	z	$\frac{1}{2}+x$	$\frac{1}{2}+y$	z	x	y	$\frac{1}{2}+z$	$\frac{1}{2}+x$	$\frac{1}{2}+y$	$\frac{1}{2}+z$
La	0	0	0	$\frac{1}{2}$	$\frac{1}{2}$	0	0	0	$\frac{1}{2}$	$\frac{1}{2}$	$\frac{1}{2}$	$\frac{1}{2}$
B	$\frac{1}{2}$	0.226	$\frac{1}{4}$	0	0.726	$\frac{1}{4}$	$\frac{1}{2}$	0.226	$\frac{3}{4}$	0	0.726	$\frac{3}{4}$
C	0.173	$\frac{1}{2}$	$\frac{1}{4}$	0.673	0	$\frac{1}{4}$	0.173	$\frac{1}{2}$	$\frac{3}{4}$	0.673	0	$\frac{3}{4}$
	$-x$	$-y$	$\frac{1}{2}-z$	$\frac{1}{2}-x$	$\frac{1}{2}-y$	$\frac{1}{2}-z$	$-x$	$-y$	$-z$	$\frac{1}{2}-x$	$\frac{1}{2}-y$	$-z$
La	0	0	$\frac{1}{2}$	$\frac{1}{2}$	$\frac{1}{2}$	$\frac{1}{2}$	0	0	0	$\frac{1}{2}$	$\frac{1}{2}$	0
B	$\frac{1}{2}$	−0.226	$\frac{1}{4}$	0	0.274	$\frac{1}{4}$	$\frac{1}{2}$	−0.226	$\frac{3}{4}$	0	0.274	$\frac{3}{4}$
C	−0.173	$\frac{1}{2}$	$\frac{1}{4}$	0.337	0	$\frac{1}{4}$	−0.173	$\frac{1}{2}$	$\frac{3}{4}$	0.337	0	$\frac{3}{4}$

In $P\bar{4}2c$ the following positions are symmetry equivalent [14, 15]:

$0,0,0$ and $0,0,\frac{1}{2}$ $\quad$ $\frac{1}{2},\frac{1}{2},0$ and $\frac{1}{2},\frac{1}{2},\frac{1}{2}$ $\quad$ $\frac{1}{2},y,\frac{3}{4}$ and $\frac{1}{2},-y,\frac{3}{4}$ $\quad$ $0,y,\frac{3}{4}$ and $0,-y,\frac{3}{4}$
$\frac{1}{2},y,\frac{1}{4}$ and $\frac{1}{2},-y,\frac{1}{4}$ $\quad$ $0,y,\frac{1}{4}$ and $0,-y,\frac{1}{4}$ $\quad$ $x,\frac{1}{2},\frac{3}{4}$ and $-x,\frac{1}{2},\frac{3}{4}$ $\quad$ $x,0,\frac{3}{4}$ and $-x,0,\frac{3}{4}$
$x,\frac{1}{2},\frac{1}{4}$ and $-x,\frac{1}{2},\frac{1}{4}$ $\quad$ $x,0,\frac{1}{4}$ and $-x,0,\frac{1}{4}$

Therefore, the atoms of the sets of coordinates mentioned one beneath the other belong to the same set of coordinates in $P\bar{4}2c$. There are really only four independent sets of coordinates. The symmetry of the space group $P\bar{4}2c$ is too low. The correct space group is a supergroup of $P\bar{4}2c$, namely $P4_2/mmc$ [90].

Errors of this kind can lead to unreliable or even wrong atomic coordinates and interatomic distances, which occasionally lead to grotesque misinterpretations or 'explanations'. The most frequent descriptions of structures with too low a symmetry concern space groups with floating origins or not recognized rhombohedral symmetry. The space group *Cc* has been wrongly assigned more often than any other [81, 82, 84].

Since the Euclidean normalizers also depend on the metric of the unit cell, this has to be paid attention to. If the correct space group has a different metric, the coordinates have to be transformed to its unit cell. Therefore, in the course of a crystal structure determination it should always be checked if the lattice can be transformed to that of a crystal system of higher symmetry. An example is given in Exercise 9.7 (page 127).

9.5 Isotypism and structure types

Definition 9.2 The crystal structures of two compounds are *isotypic* if their atoms are distributed in a like manner and if they have the same or the enantiomorphic space group.

Expressed more accurately:
The crystal structures of two compounds are *crystal-chemically isotypic* if they have the same or the enantiomorphic space-group type with comparable lattice dimensions, if their atomic positions are at equal Wyckoff positions with exactly or nearly the same coordinates, and if the geometrical and bonding interrelationships of the atoms are similar.

A *structure type* comprises a series of isotypic crystal structures. It is named after a *prototype*; that is an arbitrarily chosen compound with this structure that has usually been known for a long time.

The definition summarizes terms that were formulated in 1990 by a commission of the International Union for Crystallography [92]. The implementation of how to assign crystal structures to certain structure types in a database can be found at [93].

'Equal Wyckoff position' is tantamount to 'the same Wyckoff symbol', provided that the structural data of both compounds have been standardized in the same manner. A synonymous term is 'isopointal', but this term can be misleading, because the points belonging to the same Wyckoff position can be quite different and have different surroundings.

The occupation of certain Wyckoff positions is sometimes specified by a *Wyckoff sequence*. It consists of a sequence of the Wyckoff letters of the positions taken by atoms. For example, the Wyckoff sequence dc^4a in the space group $Pnma$ means that the Wyckoff positions d and a are occupied with one atom each per asymmetric unit, and the Wyckoff position c with four atoms that have different atomic coordinates. The values of the free parameters of the Wyckoff positions $8d$ (x, y, z) and $4c$ $(x, \frac{1}{4}, z)$ are not specified; they are only predetermined for $4a$ $(0, 0, 0)$.[5]

'Equal Wyckoff position' is a mandatory, but by no means a sufficient requirement for isotypism. If a Wyckoff position has free parameters for the atomic position, these may differ only slightly among the pair of matching atoms. Otherwise, the condition of 'similar geometrical interrelationships' would not be fulfilled. The surroundings around of every one of the matching atoms must be similar, i.e. the coordination polyhedra and their linkage must be essentially equal.

One of the isotypic structures can be conceptually generated from the other if atoms of an element are replaced by atoms of another element without changing their locations in the crystal structure (one-to-one correspondence for all atomic positions). The absolute values of the lattice dimensions and of the interatomic distances may differ,[6] but the angles between the basis vectors and the relative lattice dimensions (axes' ratios) must be similar.

An affine distortion is thus permitted. However, a strict definition of isotypism is not possible. It is not quite clear how large the deviations may be if there are striking differences among free parameters for atomic positions or axes' ratios. If all parameters are fixed by symmetry, the situation is clear. NaCl and MgO definitely are isotypic. However, pyrite and solid carbon dioxide cannot be considered to be isotypic, in spite of certain agreements.

Evidently, the parameters x deviate too much from one another. CO_2 has two O atoms placed on a threefold rotation axis at only 115 pm from a C atom; the corresponding Fe–S distance in pyrite amounts to 360 pm, whereas six other S atoms surround the Fe atom at only 226 pm. A C atom is bonded to only two other atoms, an Fe atom to six. The CO_2 crystal consists of CO_2 molecules; in the pyrite crystal there is a framework of nearly octahedrally coordinated Fe^{2+} ions linked via disulfide ions $[S–S]^{2-}$. If a disulfide ion is considered as a unit, this is a distorted NaCl-type structure. The interrelations of the atoms in CO_2 are fundamentally different from those of pyrite.

[5] If a Wyckoff position has free parameters, the location of the atom is geometrically unspecified. For example, e is the Wyckoff letter for the general position x, y, z of the space group $P2_1/c$; in this case the Wyckoff sequence e^4 means four kinds of atoms at arbitrary locations and is meaningless beyond that.

[6] It belongs to the essence of isotypism that the absolute values of distances play no role.

Pyrite [94] $Pa\bar{3}$					CO_2 [95] $Pa\bar{3}$				
$a = 542$ pm					$a = 562$ pm				
Fe	$4a$	0	0	0	C	$4a$	0	0	0
S	$8c$	x	x	x	O	$8c$	x	x	x
with $x = 0.384$					with $x = 0.118$				

Interatomic distances per atom $\leq$345 pm

S–S	1 × 218 pm	C=O	2 × 115 pm
Fe–S	6 × 226 pm	C···O	6 × 311 pm
S···S	6 × 307 pm	O···O	6 × 318 pm
	6 × 332 pm		6 × 345 pm

Table 9.2 Comparison of structural data of some representatives of the chalcopyrite type ABX_2, space group $I\bar{4}2d$ [98]. Wyckoff positions: A on $4b$ $(0,0,\frac{1}{2})$; B on $4a$ $(0,0,0)$; X on $8d$ $(x,\frac{1}{4},\frac{1}{8})$. The angle ψ specifies the turning of the BX_4 coordination tetrahedra about **c** relative to **a**. The deviation parameters Δ were calculated in relation to $CuGaTe_2$, which was chosen as the ideal representative. ICSD = Collection number of the Inorganic Chemistry Structural Database [3].

Formula	c/a	x	Bond angles X–A–X /°		Bond angles X–B–X /°		Bond lengths		ψ/°	Δ	ICSD
ABX_2			2×	4×	2×	4×	A–X /pm	B–X /pm			
$CuGaTe_2$	1.98	0.2434	110.7	108.9	109.2	109.6	262	258	0.2	–	74456
$CuFeS_2$	1.97	0.2426	111.1	108.7	109.5	109.5	230	226	0.9	0.006	2518
$InLiTe_2$	1.95	0.2441	111.5	108.5	110.3	109.1	277	273	0.3	0.016	59112
$AgGaS_2$	1.79	0.2092	119.5	104.7	111.1	108.7	256	228	4.3	0.133	23618
$LiPN_2$	1.56	0.1699	129.7	100.7	114.5	107.0	209	164	10.8	0.335	66007
$LiBO_2$	1.55	0.1574	130.2	100.0	113.4	107.5	196	148	12.8	0.355	34156
$NaPN_2$	1.40	0.1239	137.5	97.5	115.2	106.5	241	164	18.6	0.532	241818

To specify the degree of deviation in a quantitative manner, we can define *deviation parameters* [96–98]. If the data of two structures that are to be compared have been standardized in the same way, we can define a characteristic value $\Delta(x)$ in which the coordinate deviations of all atoms are combined:

$$\Delta(x) = \frac{\sum m\sqrt{(x_1-x_2)^2+(y_1-y_2)^2+(z_1-z_2)^2}}{\sum m} \tag{9.1}$$

x_1, y_1, z_1 are the coordinates of an atom of one structure and x_2, y_2, z_2 are those of the corresponding atom of the other structure. m is the corresponding multiplicity of the Wyckoff position. The sum is taken over all atoms of the asymmetric unit. A second characteristic value $\Delta(a)$ is used to relate the axes' ratios:

$$\Delta(a) = \frac{(b_1/a_1)(c_1/a_1)}{(b_2/a_2)(c_2/a_2)} \geq 1 \tag{9.2}$$

Both characteristic values can be combined into a deviation parameter Δ [98]:

$$\Delta = [\sqrt{2}\Delta(x)+1]\Delta(a) - 1 \tag{9.3}$$

The standardization of both structural data has to be performed such that the smallest possible value of Δ results.

Δ is equal to zero if there is complete coincidence of both structures. For the example pyrite – CO_2 we calculate $\Delta = 0.43$. Further examples are listed in Table 9.2, referring to structural data of representatives of the chalcopyrite type. In chalcopyrite ($CuFeS_2$) Cu and Fe atoms have a tetrahedral coordination by S atoms, and the tetrahedra are linked via common vertices. The compounds $LiPN_2$, $LiBO_2$, and $NaPN_2$ at the end of the list exhibit considerable deviations from the ideal values $c/a = 2.0$, $x = 0.25$ (free parameter of the X atoms), and $\psi = 0°$ (turning angle of the tetrahedra about **c**); the tetrahedra are strongly distorted. Nevertheless, the general linking of the atoms remains unchanged,

Table 9.3 Rules to determine, whether two crystal structures are isotypic.

	fulfilled for CO_2–pyrite	fulfilled for $CuFeS_2$–$InLiTe_2$	fulfilled for $CuFeS_2$–$LiBO_2$	fulfilled for α-$NaFeO_2$–$CuFeO_2$†
Matching chemical composition*	yes	yes	yes	yes
Same space group type	yes	yes	yes	yes
Same occupation of Wyckoff positions*	yes	yes	yes	yes
Similar axes' ratios	yes	yes	roughly	approximately
Nearly coincident atomic coordinates	no	yes	roughly	no
Small deviation parameter Δ	0.43	0.01	0.34	0.17
Similar geometric interrelations for all atoms (interatomic distances, angles and bonding to nearest neighbours, coordination polyhedra)	no	yes	roughly	no
	not isotypic	isotypic	approximately isotypic	not isotypic

* The chemical composition must be different, but the relative numbers of atoms in the chemical formulas must be the same. Arbitrary partial occupancy of Wyckoff positions is permitted.

† See page 196.

and these compounds could be considered to be isotypic with $CuFeS_2$. The sulfur atoms in chalcopyrite, taken by themselves, are arranged in a cubic-closest packing of spheres and have coordination number 12 relative to atoms of their kind. The N and O atoms in $LiPN_2$, $LiBO_2$, and $NaPN_2$, however, have only six adjacent atoms of the same kind. This has to do with the considerably shorter P–N and B–O bond lengths. In this and in similar cases it is advisable to use the term 'isotypism' with caution and to rather disclose the relationship by an appropriate circumscription. Parameters like the turning angle ψ or the deviation parameter Δ are expedient for this purpose. Another example of crystal structures that at a first glance could appear to be isotypic but definitely are not so is presented in Section 14.3.2, page 196; that example shows that Δ gives only a crude hint on whether two compounds are isotypic or not.

Rules to determine whether two crystal structures are isotypic are summarized in Table 9.3.

Definition 9.3 Two structures are *homeotypic* if they are similar, but the aforementioned conditions for isotypism are relaxed in that [92]:

(1) their space groups are different and are related by a group–subgroup relation;

(2) positions occupied by one kind of atoms in the structure are taken by several kinds of atoms in the other structure in an ordered manner (substitution derivatives); or

(3) the geometric conditions differ (significant deviations of axes ratios, angles, or atomic coordinates).

An example of substitution derivatives is: C (diamond) – ZnS (zinc blende) – $CuFeS_2$. The most appropriate method to work out relations of homeotypic structures whose space groups differ are crystallographic group–subgroup relations. Structures are also termed homeotypic if a single atom is replaced by a building block consisting of several atoms. Known examples are: the Nowotny phase Mn_5Si_3C as an analogue to the apatite structure $Ca_5(PO_4)_3F$ [99] and $K_2[PtCl_6]$ as an analogue to CaF_2 ($PtCl_6^{2-}$ ions at the Ca positions).

If two ionic compounds have the same structure type in such a way that the places of the cations of the one compound are taken by the anions of the other one and vice versa ('exchange of anions and cations'), they sometimes are called 'antitypes'. Example: Li_2O crystallizes in the 'anti-CaF_2 type', i.e. the Li^+ ions take the places of the F^- ions and the O^{2-} ions those of the Ca^{2+} ions.

Exercises

Solutions on page 326

9.1. Many tetraphenylphosphonium salts having square-pyramidal or octahedral anions crystallize in the space group $P4/n$. The coordinates for $P(C_6H_5)_4[MoNCl_4]$ are (origin choice 2, i.e. origin at a centre of inversion) [100]:

	x	y	z		x	y	z
P	$\frac{1}{4}$	$\frac{3}{4}$	0	Mo	$\frac{1}{4}$	$\frac{1}{4}$	0.121
C 1	0.362	0.760	0.141	N	$\frac{1}{4}$	$\frac{1}{4}$	–0.093
C 2	0.437	0.836	0.117	Cl	0.400	0.347	0.191

(values for the H atoms and C 3 to C 6 omitted)

How many equivalent sets of coordinates can be used to describe the structure? What are the corresponding coordinates?

9.2. The vanadium bronzes β'-$Cu_{0.26}V_2O_5$ [101] and β-$Ag_{0.33}V_2O_5$ [102] are monoclinic, both with space group $C2/m$ (the positions of the Cu and Ag atoms show partial occupancy). The coordinates are (without O atoms):

β'-$Cu_{0.26}V_2O_5$
$a = 1524$, $b = 361$,
$c = 1010$ pm, $\beta = 107.25°$

	x	y	z
Cu	0.530	0	0.361
V1	0.335	0	0.096
V2	0.114	0	0.120
V3	0.287	0	0.407

β-$Ag_{0.33}V_2O_5$
$a = 1539$, $b = 361$,
$c = 1007$ pm, $\beta = 109.7°$

	x	y	z
Ag	0.996	0	0.404
V1	0.117	0	0.119
V2	0.338	0	0.101
V3	0.288	0	0.410

Are the two structures isotypic or homeotypic (apart from the different partial occupancies of the Cu and Ag positions)?

9.3. Are the following three crystal structures isotypic?

$NaAg_3O_2$ [103] *Ibam*
$a = 616$, $b = 1044$, $c = 597$ pm

		x	y	z
Na	4*b*	$\frac{1}{2}$	0	$\frac{1}{4}$
Ag1	4*c*	0	0	0
Ag2	8*e*	$\frac{1}{4}$	$\frac{1}{4}$	$\frac{1}{4}$
O	8*j*	0.289	0.110	0

Na_3AlP_2 [104] *Ibam*
$a = 677$, $b = 1319$, $c = 608$ pm

		x	y	z
Al	4*a*	0	0	$\frac{1}{4}$
Na1	4*b*	$\frac{1}{2}$	0	$\frac{1}{4}$
Na2	8*j*	0.312	0.308	0
P	8*j*	0.196	0.101	0

Pr_2NCl_3 [105] *Ibam*
$a = 1353$, $b = 685$, $c = 611$ pm

		x	y	z
N	4*a*	0	0	$\frac{1}{4}$
Cl 1	4*b*	0	$\frac{1}{2}$	$\frac{1}{4}$
Cl 2	8*j*	0.799	0.180	0
Pr	8*j*	0.094	0.177	0

9.4. Are the following two crystal structures isotypic?

Na_6FeS_4 [106] $P6_3mc$
$a = 895$, $c = 691$ pm

		x	y	z
Na1	$6c$	0.146	–0.146	0.543
Na2	$6c$	0.532	0.468	0.368
Fe	$2b$	$\frac{1}{3}$	$\frac{2}{3}$	0.25
S1	$2b$	$\frac{1}{3}$	$\frac{2}{3}$	0.596
S2	$6c$	0.188	–0.188	0.143

Ca_4OCl_6 [107, 108] $P6_3mc$
$a = 907$, $c = 686$ pm

		x	y	z
Ca1	$2b$	$\frac{1}{3}$	$\frac{2}{3}$	0.427
Ca2	$6c$	0.198	–0.198	0.0
O	$2b$	$\frac{1}{3}$	$\frac{2}{3}$	0.106
Cl1	$6c$	0.136	–0.136	0.385
Cl2	$6c$	0.464	0.536	0.708

9.5. In 2001 it was announced that rambergite has an 'anti-wurtzite' structure, with the opposite absolute configuration to that of wurtzite [109]. Why is this nonsense?

wurtzite (ZnS) $P6_3mc$
$a = 382$, $c = 626$ pm

	x	y	z
Zn	$\frac{1}{3}$	$\frac{2}{3}$	0
S	$\frac{1}{3}$	$\frac{2}{3}$	0.375

rambergite (MnS) $P6_3mc$
$a = 398$, $b = 645$ pm

	x	y	z
Mn	$\frac{2}{3}$	$\frac{1}{3}$	0
S	$\frac{2}{3}$	$\frac{1}{3}$	0.622

9.6. The crystal data for two compounds are listed in the following. Decide whether the mentioned space groups are possibly wrong.

GeS_2-Ii [110] $I\bar{4}2d$
$a = 548$, $c = 914$ pm

	x	y	z
Ge	0	0	0
S	0.239	$\frac{1}{4}$	$\frac{1}{8}$

Na_2HgO_2 [111] $I422$
$a = 342$, $b = 1332$ pm

	x	y	z
Na	0	0	0.325
Hg	0	0	0
O	0	0	0.147

9.7. From the published lattice parameters of Na_4AuCoO_5 [112] it may be conjectured that the structure is not monoclinic (space group $P2_1/m$), but B-centred orthorhombic. What is the correct space group and what are the atomic coordinates?

$a = 555.7$, $b = 1042$, $c = 555.7$ pm, $\beta = 117.39°$

	x	y	z		x	y	z
Au	0	0	0	Co	0.266	$\frac{3}{4}$	0.266
Na1	0.332	0.000	0.669	O1	0.713	0.383	0.989
Na2	0.634	$\frac{3}{4}$	0.005	O2	0.989	0.383	0.711
Na3	0.993	$\frac{1}{4}$	0.364	O3	0.433	$\frac{1}{4}$	0.430

How to handle space groups

10

10.1 Wyckoff positions of space groups

The (infinite) set of symmetry-equivalent points in a space group $\mathcal{G}$ is called a *$\mathcal{G}$-orbit* or *crystallographic point orbit* (or point configuration), cf. Definitions 5.11 and 5.12 (page 62) [113–116]. If it is clear that the subject in question concerns points of a space group, for short we simply call it an orbit or point orbit. If the coordinates of a site are completely fixed by symmetry (e.g. $\frac{1}{4}, \frac{1}{4}, \frac{1}{4}$), then the orbit is identical to the corresponding *Wyckoff position* of the space group. However, if there are one or more freely variable coordinates (e.g. x in $x, \frac{1}{4}, 0$), the Wyckoff position comprises an infinity of possible orbits; they differ in the values of the variable coordinate(s) (cf. Definition 6.6 (page 67) and Section 6.5). For example, in space group $Pbcm$ the set of points that are symmetry equivalent to, say, $0.391, \frac{1}{4}, 0$, makes up one orbit. The set corresponding to $0.468, \frac{1}{4}, 0$ belongs to the same Wyckoff position $4c$ of $Pbcm$, but to another orbit (its variable coordinate x is different).

It is customary to designate a Wyckoff position of a space group by its *Wyckoff symbol*. It consists of the multiplicity and the Wyckoff letter, for example $4c$, see Section 6.4.2, page 81.

A consequence of this kind of designation is the dependence of the multiplicity on the size of the chosen unit cell. For example, the multiplicities of rhombohedral space groups are larger by a factor of 3 if the unit cell is not referred to rhombohedral, but to hexagonal axes.

Many space groups have different Wyckoff positions with the same type of point symmetry; combined they form a Wyckoff set (called *Konfigurationslage* by [113]). These Wyckoff positions are mapped onto one another by the affine normalizer of the space group. The *affine normalizer*, like the Euclidean normalizer (cf. page 110), maps a space group onto itself, but additionally permits an expansion or compression of the lattice.

Example 10.1
Space group No. 23, $I222$, has six Wyckoff positions with the site symmetry 2; together they make up one Wyckoff set:

$4e$ $(x, 0, 0)$ and $4f$ $(x, 0, \frac{1}{2})$ on twofold rotation axes parallel to **a**,
$4g$ $(0, y, 0)$ and $4h$ $(\frac{1}{2}, y, 0)$ on twofold rotation axes parallel to **b**,
$4i$ $(0, 0, z)$ and $4j$ $(0, \frac{1}{2}, z)$ on twofold rotation axes parallel to **c**.

However, if the lattice parameters are $a \neq b \neq c$, in the preceding example the positions $4e$, $4f$ cannot be considered to be equivalent to the positions $4g$, $4h$ and to $4i$, $4j$, being on differently oriented axes. On the other hand, the positions $4e$ and $4f$ are equivalent; they are mapped onto each other by the *Euclidean* normalizer (they are equivalent in the Euclidean normalizer).

As explained in Section 9.2, as a rule, there exist several equivalent sets of coordinates that describe the very same crystal structure; they can be interconverted with the aid of the Euclidean normalizer. The change from one to another coordinate set can imply an interchange of Wyckoff positions among the positions that are equivalent in the Euclidean normalizer. If the origin of space group $I222$ is shifted by $0, 0, \frac{1}{2}$, the atoms at the Wyckoff position $4e$ are shifted to $4f$ and vice versa. This is similar to the interchange of Na and Cl atoms in Example 9.1 (page 117).

With monoclinic space groups, different unit cells can be chosen for the same structure; their bases can be interconverted by transformations such as $\mathbf{a} \pm n\mathbf{c}, \mathbf{b}, \mathbf{c}$ ($n =$ integral number). This can also cause an interchange of Wyckoff symbols. The same applies to basis transformations of the space group $P\overline{1}$.

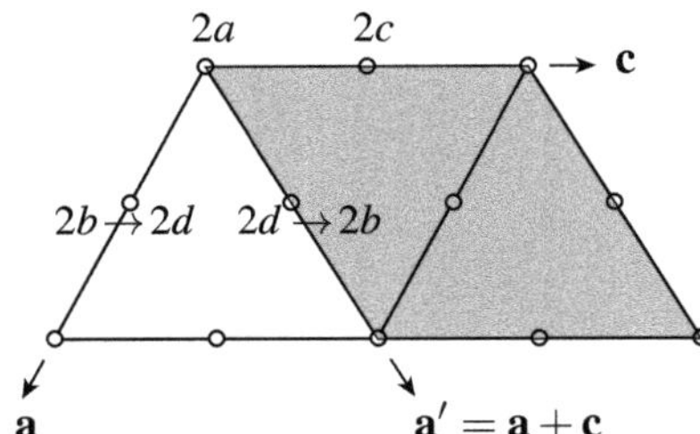

Example 10.2

The space group $P12_1/c1$ has four kinds of inversion centres:

$$2a,\ 0,0,0; \quad 2b,\ \tfrac{1}{2},0,0; \quad 2c,\ 0,0,\tfrac{1}{2}; \quad 2d,\ \tfrac{1}{2},0,\tfrac{1}{2}$$

By transformation to a cell with the basis $\mathbf{a}+\mathbf{c}, \mathbf{b}, \mathbf{c}$ (grey in the image in the margin), the Wyckoff positions $2a$ and $2c$ keep their Wyckoff symbols, whereas the other two are interchanged $2b \rightleftharpoons 2d$.

10.2 Relations between the Wyckoff positions in group–subgroup relations

For every group–subgroup relation $\mathcal{G} \to \mathcal{H}$ it is essential to keep track of which Wyckoff positions of the subgroup $\mathcal{H}$ result from those Wyckoff positions of the space group $\mathcal{G}$ that are occupied by atoms. A group–subgroup relation can only be correct and meaningful if there exist clear connections for all atoms.

The atomic positions show clearly how the symmetry is being reduced step by step in a sequence of group–subgroup relations. At first, as a rule, the atoms occupy special positions, that is, they are located on certain symmetry elements with fixed values for the coordinates and with some specific site symmetry. When proceeding from group to subgroup, the atomic positions experience the following changes one by one or jointly [117]:

1. some or all coordinates x, y, z fixed or coupled by symmetry become independent (i.e. the atoms can shift away from a special position);
2. the site symmetry is reduced;
3. the point orbit splits into different orbits.

If the index of the symmetry reduction is 2, either the site symmetry is kept and the orbit splits or there is no splitting and the site symmetry is reduced.

There is a one-to-one relation between the points of an orbit and the corresponding points of a subgroup. They comprise the same number of points in the same volume. The multiplicity of a Wyckoff position shows up in the multiplicities of the corresponding Wyckoff positions of the subgroup. If the unit cell selected to describe the subgroup has the same size, the sum of the multiplicities of the positions of the subgroup must be equal to the multiplicity of the position of the starting group. For example, from a position with multiplicity of 6, a position with multiplicity of 6 can result, or it can split into two positions of multiplicity of 3, or into two with multiplicities of 2 and 4, or into three with multiplicity of 2. If the unit cell of the subgroup is enlarged or reduced by a factor f, then the sum of the multiplicities must also be multiplied or divided by this factor f.

Between the positions of a space group and those of its subgroups there exist unique relations for given relative positions of their unit cells. Usually, there are several possible relative positions; the relations between the Wyckoff positions may differ for different (arbitrary choices of) relative positions of the origins of the group and the subgroup.

The relations between the Wyckoff positions of the space groups and their subgroups can be derived from the data of *International Tables A*. However, that is a cumbersome task. Therefore, it is recommended to use the tables of *International Tables A*1. Listed are all maximal subgroups of all space groups and which Wyckoff positions of a subgroup result from the Wyckoff positions of each space group. The listed relations in each case *are valid only for the given basis transformation and origin shift.* With other basis transformations or origin shifts the Wyckoff symbols of the subgroup may have to be interchanged among the positions that are equivalent in the Euclidean normalizer. In addition, the Bilbao Crystallographic Server offers the computer program `WYCKPLIT` [465]; it computes the relations after input of the space group, subgroup, basis transformation, and origin shift (cf. Section 24.2.3).

10.3 Non-conventional settings of space groups

As a rule, it is recommended to describe crystal structures with the conventional settings of the space groups and taking into account the standardization rules mentioned in Section 9.1. However, standardized settings may mean that related crystal structures have to be described differently, with the consequence that the similarities become less clear and may even be obscured. For the comparison of crystal structures, it is preferable to set up all unit cells in a most uniform way and to avoid cell transformations as far as possible, even if this requires the utilization of non-conventional settings of the space groups. In this section some instructions are given referring to such non-conventional settings.

A conventional setting is not to be confused with a standard setting. A standard setting fulfills the conditions mentioned in Section 9.1. A setting is conventional if it is completely tabulated in *International Tables A*, with all coordinate triplets, Wyckoff positions, etc. Conventional settings thus are all listed cell choices of monoclinic space groups with a monoclinic b and c, but not a axis, all tabulated origin choices and rhombohedral space groups with rhombohedral and with hexagonal bases. Not completely tabulated settings are non-conventional, especially in the case of orthorhombic space groups.

10.3.1 Orthorhombic space groups

Among orthorhombic space groups it is frequently suitable to choose settings that deviate from the listings in *International Tables A*. This is due to the fact that there exists no special preference for any of the axes directions **a**, **b**, and **c** in the orthorhombic system. Generally, there are six possible settings. All six are mentioned in *International Tables A*,[1] but only one of them is considered as the conventional setting and has been completely tabulated. The other five are obtained by interchange of the axes, namely:

[1] Table 1.5.4.4 in the sixth edition (2016) [118],
Table 4.3.2.1 in previous editions [119].

1. Conventional setting: **a b c**
2. Cyclic interchange 'forwards': **c a b**
3. Cyclic interchange 'backwards': **b c a**
4. Interchange of **a** and **b**: $\mathbf{b\,a\,\bar{c}}$ or $\mathbf{b\,\bar{a}\,c}$ or $\mathbf{\bar{b}\,a\,c}$
5. Interchange of **a** and **c**: $\mathbf{c\,b\,\bar{a}}$ or $\mathbf{c\,\bar{b}\,a}$ or $\mathbf{\bar{c}\,b\,a}$
6. Interchange of **b** and **c**: $\mathbf{a\,c\,\bar{b}}$ or $\mathbf{a\,\bar{c}\,b}$ or $\mathbf{\bar{a}\,c\,b}$

The notation **c a b** means: the original **a** axis is now in the position **b**, etc., or: convert **a** to **b**, **b** to **c**, **c** to **a**. The options 4 to 6 (interchange of two axes) require the reversal of the direction of one axis in order to retain a right-handed coordinate system.

The interchange of the axes has the following consequences:

1. The lattice parameters a, b, c must be interchanged.
2. In the Hermann–Mauguin symbol the sequence of the symmetry directions has to be changed.
3. The notations of the glide directions a, b, c in the Hermann–Mauguin symbol have to be interchanged. m, n, d, and e remain unaltered.
4. The notations of the centrings A, B, C have to be interchanged. P, F, and I remain unaltered.
5. In the coordinate triplets of the atomic positions the sequence and the notations have to be interchanged. If the direction of an axis is reversed, the signs of the corresponding coordinates have to be inverted. Usually, the Wyckoff symbols remain unaltered.[2] However, caution is required if the interchange of axes concerns space groups whose symbols do not reveal whether or how the axes have been interchanged (see Examples 10.5 and 10.6).
6. The notations of rotations and screw rotations do not change.

[2] The Wyckoff symbols and the coordinate triplets of the general and the special positions of non-conventional settings of the orthorhombic space groups can be retrieved with the program WYCKPOS at the BILBAO CRYSTALLOGRAPHIC SERVER (cf. Section 24.1).

Example 10.3

Two possibilities to interchange axes of the space group $P2/b2_1/c2_1/m$ ($Pbcm$, No. 57) and its Wyckoff position $4c$ ($x, \frac{1}{4}, 0$):

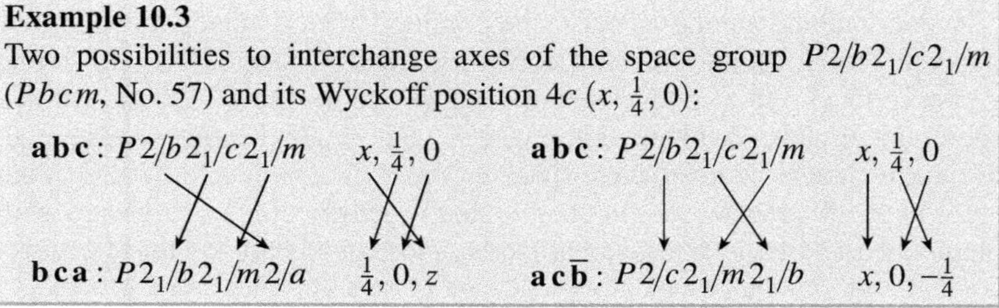

Example 10.4

Two possibilities to interchange axes of the space group $C2/m2/c2_1/m$ ($Cmcm$, No. 63) and its Wyckoff position $8g$ ($x, y, \frac{1}{4}$) with $x = 0.17$ and $y = 0.29$:

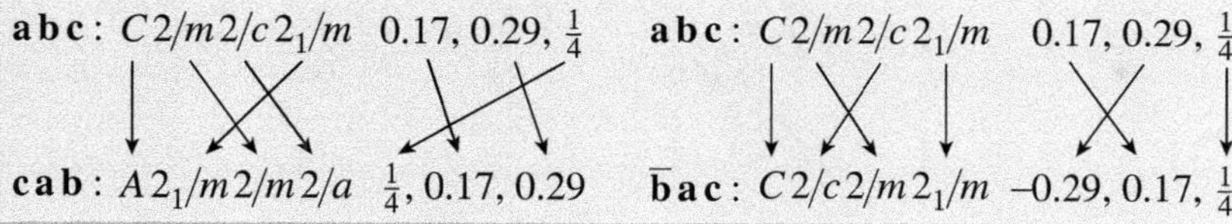

In most cases the non-conventional Hermann–Mauguin symbol shows in a unique way how the setting differs from the conventional setting. However, there are exceptions, as shown in the following two examples.

Example 10.5

In the space group $P222_1$ (No. 17) the Wyckoff position $2a$ (x, 0, 0) is situated on twofold rotation axes parallel to **a**, and the position $2c$ (0, y, $\frac{1}{4}$) on twofold rotation axes parallel to **b**. The following two possibilities to interchange axes result in the same non-conventional space-group symbol:

		$2a$	$2c$
abc:	$P222_1$	x, 0, 0	0, y, $\frac{1}{4}$
cab:	$P2_122$	0, y, 0	$\frac{1}{4}$, 0, z

		$2a$	$2c$
abc:	$P222_1$	x, 0, 0	0, y, $\frac{1}{4}$
$\mathbf{cb\bar{a}}$:	$P2_122$	0, 0, $\bar{z}$	$\frac{1}{4}$, y, 0

After cyclic interchange the twofold rotation axes of the Wyckoff position $2a$ are parallel to **b** at 0, y, 0. On the other hand, interchange of **a** and **c** yields twofold rotation axes parallel to **b** at $\frac{1}{4}$, y, 0. The non-conventional symbol $P2_122$ does not show how the basis vectors have been interchanged and where the rotation axes are situated. In such cases cyclic interchange should be preferred.

Example 10.6

The space group $Cmme$ ($C2/m2/m2/e$, No. 67) has a subgroup $Ibca$ ($I2_1/b2_1/c2_1/a$, No. 73) with doubled **c** vector. The Wyckoff position $4d$ (0, 0, $\frac{1}{2}$) of $Cmme$ becomes $8c$ (x, 0, $\frac{1}{4}$) of $Ibca$ with $x \approx 0$.

Starting from $Bmem$, a non-conventional setting (**b c a**) of $Cmme$, the same subgroup $Ibca$ is obtained by doubling **b**. The position $4d$ of $Bmem$ is (0, $\frac{1}{2}$, 0), from which we obtain (0, $\frac{1}{4}$, z) of $Ibca$ ($z \approx 0$). According to *International Tables A*, (0, $\frac{1}{4}$, z) is not the position $8c$, but $8e$ of $Ibca$; therefore, it has a denomination different to when starting from $Cmme$. However, if the non-conventional setting **b c a** is chosen for $Ibca$, the Wyckoff symbol $8c$ results for the position (0, $\frac{1}{4}$, z). The symbol $Ibca$ does not reveal this because cyclic interchange of the axes does not change the symbol. In this case the interchanged axes and Wyckoff symbols must explicitly be mentioned to avoid confusion.

	$4d$		$4d$		$4d$
$C2/m2/m2/e$	$0,\ 0,\ \frac{1}{2}$	$B2/m2/e2/m$	$0,\ \frac{1}{2},\ 0$	$B2/m2/e2/m$	$0,\ \frac{1}{2},\ 0$
$\mathbf{a}, \mathbf{b}, 2\mathbf{c}$ ↓	↓ ↓ ↓	$\mathbf{a}, 2\mathbf{b}, \mathbf{c}$ ↓	↓ ↓ ↓	$\mathbf{a}, 2\mathbf{b}, \mathbf{c}$ ↓	↓ ↓ ↓
$I2_1/b2_1/c2_1/a$	$x,\ 0,\ \frac{1}{4}$	$I2_1/b2_1/c2_1/a$	$0,\ \frac{1}{4},\ z$	$I2_1/b2_1/c2_1/a$	$0,\ \frac{1}{4},\ z$
	$8c$	$(\mathbf{a}'\mathbf{b}'\mathbf{c}')$	$8e$	$(\mathbf{b}'\mathbf{c}'\mathbf{a}')$	$8c$

10.3.2 Monoclinic space groups

The interchange of axes of monoclinic space groups has the same consequences as for orthorhombic space groups. In addition, the interchange of the angles α, β, γ has to be considered. *International Tables* A list the settings with **b** and **c** as the monoclinic axis. Settings with monoclinic **a** axis can be obtained from these settings in the same way as described in the preceding section.

A few additional particularities have to be taken into account in the case of monoclinic space groups with centrings or with glide planes. Three cell choices have each been listed for these space groups (cell choices 1, 2, and 3). The direction of the monoclinic axis and the cell choice can be uniquely recognized from the full Hermann–Mauguin symbol (Fig. 10.1). If a setting other than the standard setting is chosen ($P1\,2_1/c\,1$ in Fig. 10.1), the full symbol must be stated. A change of the cell choice entails a change not only of the Hermann–Mauguin symbol, but also of the monoclinic angle.

The interchange of axes of the space groups Cc and $C2/c$ requires special care. An interchange of the axes **a** and **c** (keeping the cell and $-\mathbf{b}$ as monoclinic axis), results in the following change of the space-group symbol:

$$C1\,2/c\,1 \text{ (cell choice 1)} \quad \longrightarrow \quad A1\,2/a\,1$$

$A1\,2/a\,1$ is not a conventional setting. One of the settings listed in *International Tables* is $A1\,2/n\,1$ (cell choice 2). $A1\,2/n\,1$ and $A1\,2/a\,1$ refer to the same space

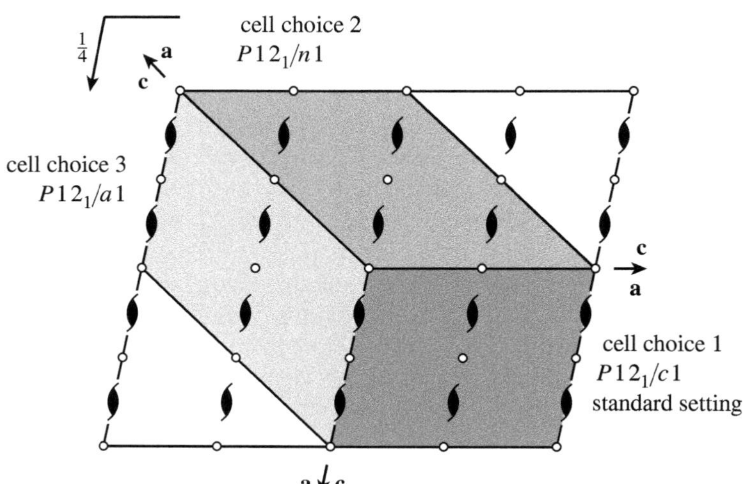

Fig. 10.1 The three cell choices for space group No. 14, $P2_1/c$, with monoclinic b axis.

group with the same cell, but with shifted origin positions. The space group has a as well as n glide planes; the one mentioned in the symbol is located at $y = 0$, the other one at $y = \frac{1}{4}$. The change of the setting from $A12/a1$ to $A12/n1$, keeping the axes directions, involves an origin shift by $0, \frac{1}{4}, \frac{1}{4}$ (Fig. 10.2).

Cyclic interchange of the axes $\mathbf{abc} \to \mathbf{cab}$ yields a change of the monoclinic axis from **b** to **c**. $C12/c1$ is then converted to $A112/a$. If **b** and **c** are interchanged, $\mathbf{abc} \to \mathbf{ac\bar{b}}$, the position of the origin must be watched. $C12/c1$ is then converted to $B112/b$ with glide planes b at $z = 0$ and n at $z = \frac{1}{4}$. $B112/b$ had been the setting used for the monoclinic **c** axis in the old editions of *International Tables* (up to 1969). However, since the 1983 edition it has been $B112/n$ (monoclinic c, cell choice 2), with glide planes n at $z = 0$ and b at $z = \frac{1}{4}$. $B112/b$ and $B112/n$ differ in their origin positions by $\frac{1}{4}, 0, \frac{1}{4}$. The denominations of the Wyckoff symbols in the old editions of *International Tables* (up to 1969) differ from those in the newer editions.

It is advisable to use non-conventional monoclinic settings that result from cyclic interchange of axes. In that case the coordinate triplets have to be interchanged correspondingly, but there are no origin shifts. The Wyckoff letters remain unchanged. In addition, the monoclinic angle keeps its value (β for monoclinic **b** axis). If **a** and **c** are interchanged, β only keeps its value if the direction of **b** is reversed, otherwise β has to be replaced by $180° - \beta$.

Sometimes it is useful to choose a non-conventional centred setting. For example, the space group $Cmcm$ has a subgroup $P112_1/m$ that can also be set as $C112_1/m$. This setting avoids a cell transformation:

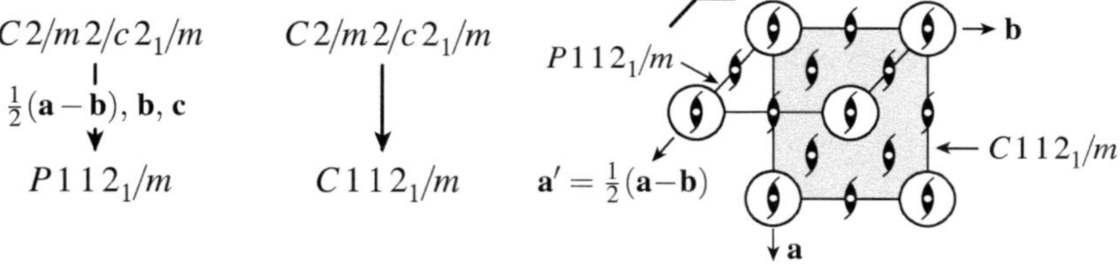

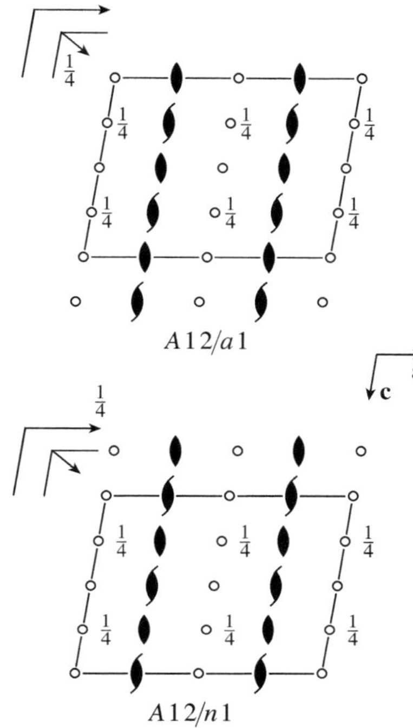

Fig. 10.2 The two settings $A12/a1$ and $A12/n1$ of the same space group with the same axes directions differ in the positions of their origins.

A disadvantage of non-conventional centrings is the lack of tables of the coordinate triplets for the Wyckoff positions. Wyckoff symbols are only unique if the corresponding coordinate triplets are explicitly stated.

10.3.3 Tetragonal space groups

Sometimes it is useful to choose C-centred instead of primitive settings or face-centred instead of body-centred settings of tetragonal space groups. Their $\mathbf{a}'$ and $\mathbf{b}'$ axes run diagonal to the axes of the conventional setting (Fig. 10.3). For example, this should be considered for a relation from a face-centred cubic space group to a tetragonal subgroup, which can then be described with unchanged axes and an F instead of a (conventional) I cell. This is advisable if a further subgroup follows which again has the metric of the original F cell; in this way two cell transformations can be avoided.

In the Hermann–Mauguin symbol of the non-conventional centred setting the symmetry elements referring to the directions **a** and $\mathbf{a} - \mathbf{b}$ are interchanged. In addition to the letters for the centrings, the following letters for glide planes have to be changed:

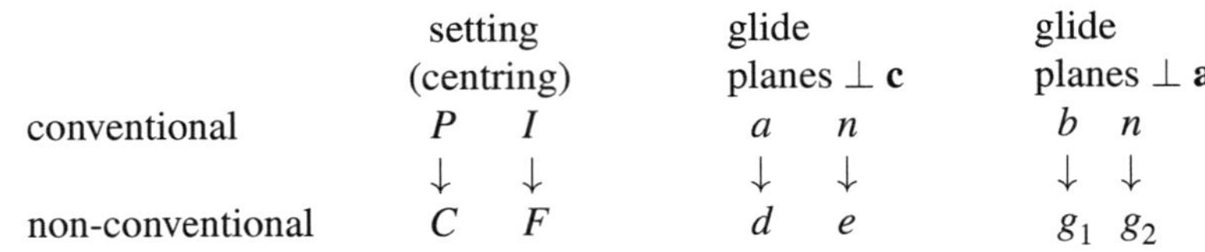

	setting (centring)		glide planes ⊥ **c**		glide planes ⊥ **a**	
conventional	P	I	a	n	b	n
	↓	↓	↓	↓	↓	↓
non-conventional	C	F	d	e	g_1	g_2

The glide planes g_1 and g_2 run perpendicular to $\mathbf{a}' - \mathbf{b}'$ and $\mathbf{a}' + \mathbf{b}'$ (in the non-conventional setting) and have glide components of $\frac{1}{4}, \frac{1}{4}, 0$ and $\frac{1}{4}, \frac{1}{4}, \frac{1}{2}$, respectively.

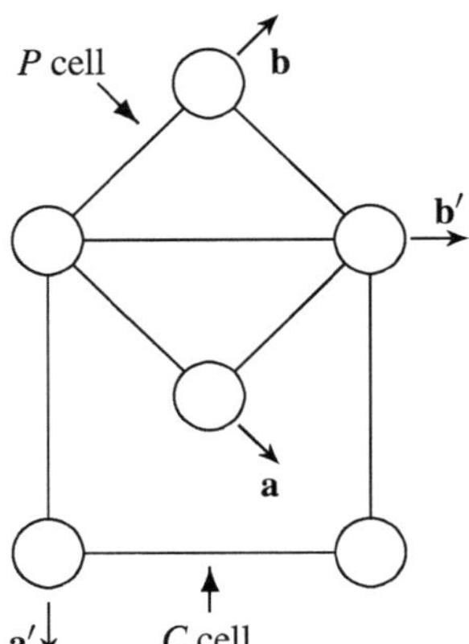

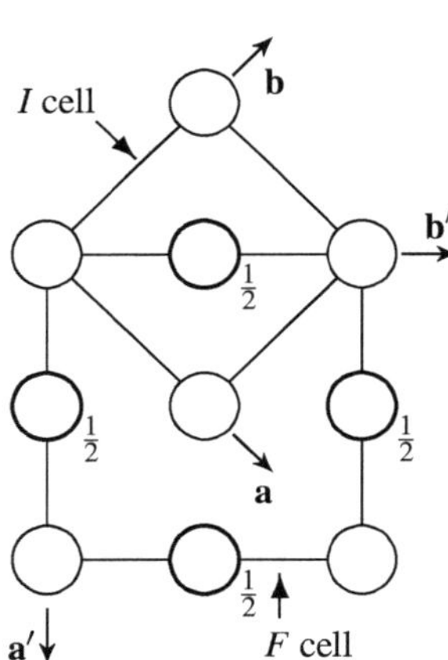

$\mathbf{a}' = \mathbf{a} - \mathbf{b} \quad \mathbf{b}' = \mathbf{a} + \mathbf{b}$

Fig. 10.3 Relative orientations of C and F centred tetragonal unit cells.

Example 10.7

Transformation of two tetragonal space groups to unit cells with doubled volume and diagonal **a** and **b** axes:

Example 10.8

Above 47 °C, $NiCr_2O_4$ has the cubic spinel structure. Upon cooling, the structure experiences a slight distortion and becomes tetragonal [120]. If the tetragonal structure is not described with a body-centred but with a face-centred setting, the small amount of the distortion concerning the coordinates of the O atoms becomes much more apparent (tree of group–subgroup relations as explained in Chapter 11):

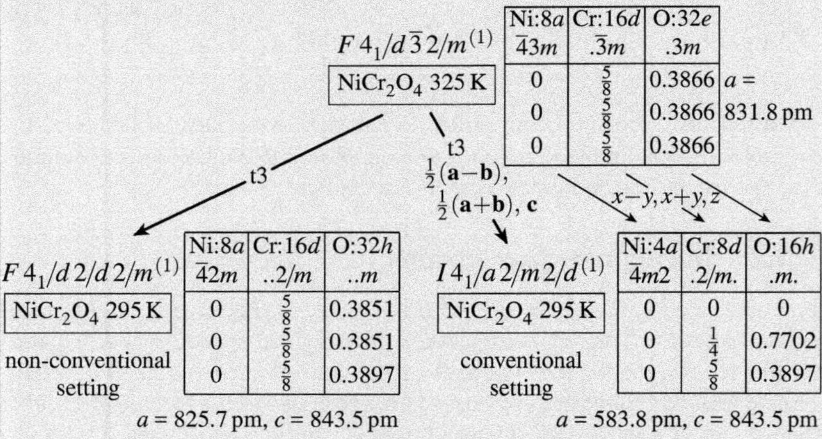

Transformation $I\,4_1/a\,2/m\,2/d \rightarrow F\,4_1/d\,2/d\,2/m$ according to $\mathbf{a}+\mathbf{b}, -\mathbf{a}+\mathbf{b}, \mathbf{c}$ and $\frac{1}{2}(x+y), \frac{1}{2}(-x+y), z$

Note the sequence in the Hermann–Mauguin symbols, also at the site symmetry of the Ni atom, and the halved multiplicities at the Wyckoff symbols of the conventional setting $I\,4_1/a\,2/m\,2/d$ (because of the halved size of the unit cell).

A disadvantage as for all non-conventional centrings is the lack of tables of the coordinates of the Wyckoff positions. Special care is required when space groups with d glide planes are involved. For instance, for the space group $I4_1/amd$ ($I4_1/a2/m2/d$) mentioned in Example 10.8, it makes a difference whether the transformation $I4_1/amd \rightarrow F4_1/ddm$ is performed according to $\mathbf{a}+\mathbf{b}, -\mathbf{a}+\mathbf{b}, \mathbf{c}$ or $\mathbf{a}-\mathbf{b}, \mathbf{a}+\mathbf{b}, \mathbf{c}$, because the d glide directions are different. The latter transformation results in wrong glide directions unless an origin shift of $-\frac{1}{4},-\frac{1}{4},-\frac{1}{4}$ is included in the group–subgroup relation. The transformation of space groups that allow two origin choices can involve different origin shifts depending on the origin choice.

10.3.4 Rhombohedral space groups

Basis transformations concerning rhombohedral space groups can be rather complex and are notoriously prone to errors. In addition, it has to be watched whether the setting refers to 'rhombohedral' or to 'hexagonal' axes.

Every cubic space group has rhombohedral subgroups that have their threefold axes parallel to one of the four directions of the body diagonals of the cube. Only in the case of primitive cubic space groups does the conventional cell of a maximal rhombohedral subgroup have the same unit cell, provided that rhombohedral axes have been chosen. The maximal rhombohedral subgroups of F- and I-centred cubic space groups can be derived without cell transformations if non-conventional F- or I-centred settings are chosen with rhombohedral axes (i.e. with $a = b = c$, $\alpha = \beta = \gamma \approx 90°$). The space-group symbol then begins with $F\overline{3}$, $F3$, $I\overline{3}$, or $I3$, for example $F\overline{3}m$ instead of $R\overline{3}m^{(\mathrm{rh})}$ (not to be confounded with $Fm\overline{3}$).

'Hexagonal' axes offer two possible settings called 'obverse' and 'reverse'; obverse is the conventional one. The settings differ in the kind of centring of the hexagonal cell, namely $\pm(\frac{2}{3}, \frac{1}{3}, \frac{1}{3})$ for obverse and $\pm(\frac{1}{3}, \frac{2}{3}, \frac{1}{3})$ for reverse. Sometimes cell transformations can be avoided if the reverse setting is chosen. For example, when a rhombohedral space group results from another rhombohedral space group by doubling of the (hexagonal) c axis, either the directions of **a** and **b** have to be reversed or the directions can be kept if one of the two space groups is chosen with a reverse setting. Since this cannot be expressed by the Hermann–Mauguin symbol, it has to be explicitly mentioned, preferably by a superscript $^{(\mathrm{rev})}$ after the corresponding Hermann–Mauguin symbol.

10.3.5 Hexagonal space groups

In former times, hexagonal space groups were described with an H cell or a C cell. These result from the conventional cell according to:

Basis vectors of the H cell: $2\mathbf{a}+\mathbf{b}, -\mathbf{a}+\mathbf{b}, \mathbf{c}$

Basis vectors of the C cell: $2\mathbf{a}+\mathbf{b}, \mathbf{b}, \mathbf{c}$ or $\mathbf{a}, \mathbf{a}+2\mathbf{b}, \mathbf{c}$

The C cell (orthohexagonal cell) corresponds to the cell of orthorhombic subgroups of hexagonal space groups. The H cell is centred in the positions $\pm(\frac{2}{3}, \frac{1}{3}, 0)$. Both kinds of cell are usually of little advantage and are used only occasionally.

Exercises

Solutions on page 327.

10.1. The cell of space group $P\bar{1}$ is to be converted to another cell with the basis vectors **a**, **b**, **a**+**b**+**c**. How do the Wyckoff symbols have to be interchanged? Set up the transformation matrices $\boldsymbol{P}$ and $\boldsymbol{P}^{-1}$. The Wyckoff symbols are: $1a\,(0,0,0)$; $1b\,(0,0,\frac{1}{2})$; $1c\,(0,\frac{1}{2},0)$; $1d\,(\frac{1}{2},0,0)$; $1e\,(\frac{1}{2},\frac{1}{2},0)$; $1f\,(\frac{1}{2},0,\frac{1}{2})$; $1g\,(0,\frac{1}{2},\frac{1}{2})$; $1h\,(\frac{1}{2},\frac{1}{2},\frac{1}{2})$; $2i\,(x,y,z)$.

10.2. What are the Hermann–Mauguin symbols of space group $P2_1/n2_1/m2_1/a$ for the axes settings **c**, **a**, **b** and **b**, $\bar{\mathbf{a}}$, **c**? What are the corresponding coordinates of the point 0.24, $\frac{1}{4}$, 0.61?

10.3. What has to be considered when the axes **a** and **b** of the space group $C\,1\,2/c\,1$ are interchanged?

10.4. What is the Hermann–Mauguin symbol of the space group $P4_2/n2_1/c2/m$ with a C-centred setting?

10.5. Among others, the following supergroups of space group $P2_1/c$ ($P\,1\,2_1/c\,1$, No. 14) are listed in *International Tables* with their standard symbols; the full symbols have been added here in parentheses:

$Pnna$ ($P2/n2_1/n2/a$); $Pcca$ ($P2_1/c2/c2/a$); $Pccn$ ($P2_1/c2_1/c2/n$); $Cmce$ ($C2/m2/c2_1/e$).

Only the standard symbols are listed for supergroups, even if this requires a basis transformation that is not mentioned. What are the Hermann–Mauguin symbols of the supergroups if their axes are oriented in the same way as for $P\,1\,2_1/c\,1$? Is it necessary, in some of the cases, to choose some other setting of $P\,1\,2_1/c\,1$ in order to be able to keep the orientation of the axes for the supergroup? What conditions must be met by the cell of $P\,1\,2_1/c\,1$ so that the mentioned space groups can actually be supergroups?

Part II

Symmetry Relations between Space Groups as a Tool to Disclose Connections between Crystal Structures

The group-theoretical presentation of crystal-chemical relationships

11

The consequent application of group theory in crystal chemistry yields a conclusive confirmation of the validity of the symmetry principle mentioned in Chapter 1. Particularly aspect 2 of the symmetry principle emerges in an impressive way. Disturbances like covalent bonds, lone electron pairs, or the Jahn–Teller effect, as a rule, cause symmetry reductions as compared to ideal models. However, the group-theoretical analysis shows that the symmetry reduction often corresponds to the smallest possible step, that is, the space group of a crystal structure is a *maximal* subgroup of a related conceivable or actually existent higher-symmetry structure.

If the symmetry reduction is such that all translations are being retained, the maximal subgroup $\mathcal{H}$ is called a *translationengleiche* subgroup of the space group $\mathcal{G}$. CARL HERMANN called these subgroups *zellengleiche* subgroups. However, this term has been replaced by *translationengleiche* because of possible misinterpretations. On page 155 it is explained why *zellengleiche* is prone to misinterpretations.[1]

If the symmetry reduction involves a loss of translations, the maximal subgroup $\mathcal{H}$ is either a *klassengleiche* or an *isomorphic* subgroup of the space group $\mathcal{G}$; isomorphic is an important special case of *klassengleiche*. The loss of translations is tantamount to an enlargement of the primitive unit cell, either by enlargement of the conventional unit cell or by the loss of centrings. In the outset of the development, the misleading terms 'equivalent' and 'isosymbolic' (with the same Hermann–Mauguin symbol) were used instead of 'isomorphic'. Enantiomorphic subgroups like $P3_1$ and $P3_2$ are isomorphic, although they appear as different, not 'isosymbolic' space-group types in *International Tables*.

With the aid of the mentioned terms of space-group theory it is possible to present symmetry relations between two crystal structures in a concise manner with a *Bärnighausen tree*. If the space group of the lower-symmetry structure is a maximal subgroup of the space group of the higher-symmetry structure, there is only one step of symmetry reduction. If it is not a maximal subgroup, we resolve the total symmetry reduction into a chain of sequential steps, each step representing the transition to a maximal subgroup. Therefore, we have to discuss only one of these steps in detail.

[1] German *translationengleiche* means 'with the same translations'; *zellengleiche* means 'with the same cell'; *klassengleiche* means 'of the same (crystal) class'. Of the different German declension endings only the form with terminal *-e* is commonly used in English. Native English-speaking experts of a commission of the International Union of Crystallography could not agree upon apt English terms with exactly the same meanings and officially decided to adopt the German terms. The abbreviations '*t*-subgroup' and '*k*-subgroup' are unfortunate because with '*t*-subgroup' it is not obvious if it is the loss or the conservation of the translations that is meant. Some non-English-speaking authors have used the expressions *translation-equivalent* and *class-equivalent*, which do not reflect the exact meaning. A few American authors use the terms *equi-translational* and *equi-class*.

For two structures that we want to interrelate, we place their space-group symbols one below the other and indicate the direction of the symmetry reduction by an arrow pointing downwards. See the example in the white field of the scheme on the opposite page 143.

Since they are more informative, it is advisable to use only the full Hermann–Mauguin symbols. In the middle of the arrow we insert the kind of maximal subgroup and the index of symmetry reduction, using the abbreviations t for *translationengleiche*, k for *klassengleiche*, and i for *isomorphic*. If the symmetry reduction involves a change of the size or setting of the unit cell, we also insert the new basis vectors expressed as vector sums of the basis vectors of the higher-symmetry cell.

For the sake of clarity it is recommended to avoid cell transformations whenever possible. If necessary, it is much better to fully exploit the possibilities offered by the Hermann–Mauguin symbolism and to choose space-group settings that do not correspond to the conventional settings of *International Tables* (cf. Section 10.3, page 131).

If the transition to a subgroup requires an origin shift of the unit cell, because otherwise the conventional position of the origin of a space group would have to be abandoned, we insert this in the arrow as a triplet of numbers that express the coordinates x_p, y_p, z_p of the new origin referred to the (old) coordinate system of the higher-symmetry cell. Origin shifts also tend to obscure relations, and they can be rather irksome. Nevertheless, there is no point in deviating from the standard origin settings of *International Tables*, because otherwise much additional information would be required for an unequivocal description. Note: The coordinate triplet specifying the origin shift in the group–subgroup arrow refers to the coordinate system of the higher-symmetry space group, whilst the corresponding changes of the atomic coordinates refer to the coordinate system of the subgroup and therefore are different. The new coordinates do not result from the addition of x_p, y_p, z_p to the old coordinates; see Sections 3.6.3, page 32, and 3.6.6, page 39.

Any change of the basis vectors and the origin is essential information that should never be omitted if there is a change.

Note: In *International Tables* A1 origin shifts given in Part 2 refer to the higher-symmetry space group. In Part 3 (tables of the relations of the Wyckoff positions) they are given only as parts of the coordinate transformations, and therefore, in the coordinate systems of the subgroups; to convert them to origin shifts referred to the coordinate system of the higher-symmetry space group, a conversion according to eqn (3.43), page 39, is necessary. *This calculation has to be performed*; one may not simply take the origin shift listed in Part 2, because unfortunately in Parts 2 and 3, as a rule, the same origin shift (out of several possible choices) has not been chosen for the same group–subgroup pair.

International Tables offer two origin choices for some space groups (‘origin choice 1’ and ‘origin choice 2’). The choice is specified by a superscript $^{(1)}$ or $^{(2)}$ after the space-group symbol, for example $P4/n^{(2)}$. Whether the chosen setting of rhombohedral space groups refers to rhombohedral or hexagonal axes is specified by superscript $^{(\mathrm{rh})}$ or $^{(\mathrm{hex})}$. Occasionally it may be useful to use a non-conventional rhombohedral ‘reverse’ setting, that is, with centring

Scheme of the formulation of the smallest step of symmetry reduction connecting two related crystal structures

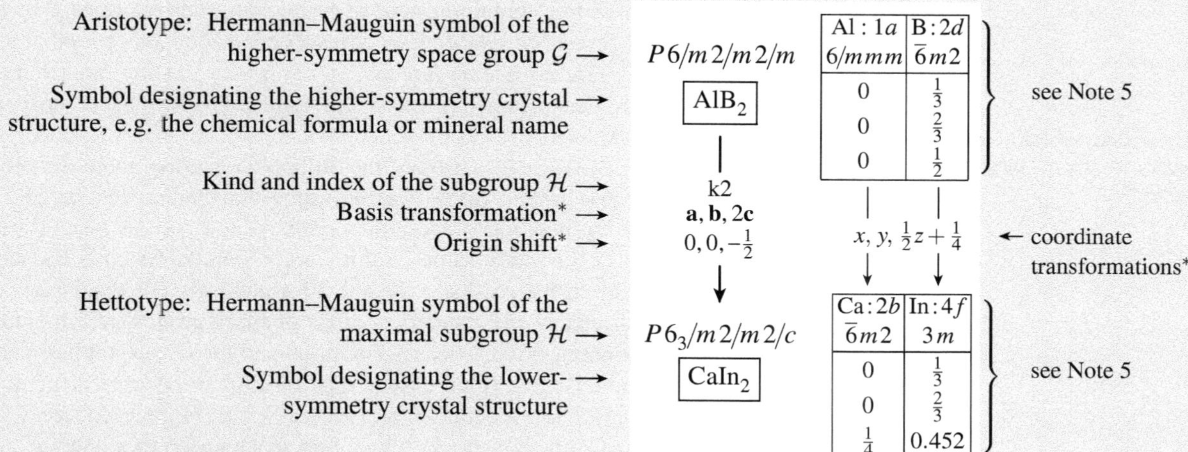

* mentioned only if there is a change

Explanatory notes

(1) Possible kinds of maximal subgroups $\mathcal{H}$ of a given space group $\mathcal{G}$:

Symbol	Term	Meaning
t	*translationengleiche*	$\mathcal{G}$ and $\mathcal{H}$ have the same translations; the crystal class of $\mathcal{H}$ is of lower symmetry than that of $\mathcal{G}$
k	*klassengleiche*	$\mathcal{G}$ and $\mathcal{H}$ belong the same crystal class; $\mathcal{H}$ has lost translational symmetry, its primitive unit cell is larger than that of $\mathcal{G}$
i	isomorphic	$\mathcal{G}$ and $\mathcal{H}$ belong to the same or the enantiomorphic space group type; $\mathcal{H}$ has lost translational symmetry, its unit cell is larger than that of $\mathcal{G}$

(2) The index i of a subgroup is the number of cosets of $\mathcal{H}$ in $\mathcal{G}$; $i > 1$. The number of symmetry operations of $\mathcal{H}$ is $1/i$ of those of $\mathcal{G}$ (in the same way as the number of even numbers is half the number of all integer numbers).

(3) Basis transformation: The three basis vectors of $\mathcal{H}$ are expressed as linear combinations of the basis vectors **a**, **b**, **c** of $\mathcal{G}$. Never omit this information if there is a change of the basis vectors.

(4) Origin shift: The coordinate triplet of the origin of $\mathcal{H}$ is given in the coordinate system of $\mathcal{G}$. Never omit this information if there is an origin shift.

(5) Additional information (only if there is enough space; otherwise this information must be listed in a separate table). The atomic positions are given in a box next to the space-group symbol in the following way:

element symbol: Wyckoff label site symmetry
x y z

The coordinates are given for one atom in the asymmetric unit. If a numeric value is fixed by symmetry, it is stated as 0 or as a fraction (e.g. 0, $\frac{1}{4}$, $\frac{1}{2}$). Free parameters are stated as decimal numbers (e.g. 0.0, 0.25, 0.53). If possible, align the site-symmetry symbols in one line with the space-group symbol.

vectors $\pm(\frac{1}{3}, \frac{2}{3}, \frac{1}{3})$ instead of 'obverse' with $\pm(\frac{2}{3}, \frac{1}{3}, \frac{1}{3})$; this is specified by superscript $^{(\text{rev})}$, for example $R\overline{3}^{(\text{rev})}$. Since obverse and reverse settings always refer to hexagonal axes, an additional $^{(\text{hex})}$ is obsolete in this case.

In a Bärnighausen tree containing several group–subgroup relations, the *vertical distances between the space-group symbols should be kept proportional to the logarithms of the corresponding indices*. This way all subgroups that are at the same hierarchical distance (i.e. with the same index from the aristotype), appear on the same line.[2]

[2] Isotypic structures may not be included in a Bärnighausen tree as 'subgroups' of index 1 (i.e. t1 or i1).

The program SUBGROUPGRAPH of the Bilbao Crystallographic Server is useful to produce first graphic drafts of group–subgroup chains (Section 24.2.2). However, it does not position space-group symbols at distances proportional to the logarithms of the index values and it cannot include the basis transformations and origin shifts, so that essential information is not displayed. In addition, it is restricted to the standard settings of space groups, which entails an unnecessary amount of basis transformations and the corresponding coordinate transformations; that often causes relationships to become difficult to manage. The program is not suited to plot correctly a Bärnighausen tree.

If several paths can be constructed from one space group to a general subgroup, via different intermediate space groups, usually there is no point in depicting all of them. There is no general recipe indicating which of several possible paths should be preferred. However, crystal-chemical and physical aspects should be used as a guide, and it should not be left to a computer to take the choice. First of all, the chosen intermediate groups should be:

1. Space groups having actually known representatives.
2. Among different modifications of the same compound: Space groups that disclose a physically realizable path for the symmetry reduction. Actually observed phase transitions should be given high priority. For phase transitions that are driven by certain vibrational modes (soft modes), those intermediate space groups should be considered that are compatible with these lattice modes (i.e. irreducible representations).
3. In the case of substitution derivatives: Space groups showing a splitting of the relevant Wyckoff position(s). These intermediate groups allow for substitution derivatives, even if no representative is yet known.

Group–subgroup relations are of little value if the usual crystallographic data are not given for every considered structure. The mere mention of the space-group symbols is insufficient. The atomic coordinates of all atoms in an asymmetric unit are of special importance. It is also important to present all structures in such a way that their relations become apparent. In particular, the coordinates of all atoms of the asymmetric units should exhibit strict correspondence, so that their positional parameters can immediately be compared. Of course, all examples dealt with in the following chapters have been documented along these lines, so that they give an impression of how close the relations actually are.

For all structures, the same coordinate setting and, among several symmetry-equivalent positions for an atom, the same location in the unit cell, should be chosen, if possible. In order to obtain the necessary correspondence between all structures that are to be compared, it is often necessary to transform coordinates. For nearly all space groups, several different equivalent sets of coordinates can be chosen, describing one and the same structure. It is by no means a simple matter to recognize whether two differently documented structures are alike or not (the literature abounds with examples of 'new' structures that really had been well known). For more details see Section 9.2, page 116.

If space permits, it is useful to list the Wyckoff symbols, the site symmetries, and the coordinates of the atoms next to the space-group symbols in the Bärnighausen tree. If there is not enough space, this information must be provided in a separate table.

Note: A Bärnighausen tree is an open tree and not a closed graph with branches that meet again. From the mathematical point of view the branches could be continued, resulting in a graph with a common subgroup of all space groups. However, in crystal chemistry we examine symmetry reductions due to distortions and/or atomic substitutions; a continuation to a common subgroup of all space groups would be meaningless.

Symmetry relations between related crystal structures

12

In this chapter, using a selection of simple examples, we point out the different kinds of group–subgroup relations that are important among related (homeotypic) crystal structures and how to set up Bärnighausen trees.

12.1 The space group of a structure is a *translationengleiche* maximal subgroup of the space group of another structure

The relation between pyrite and PdS_2

The space group $Pbca$ of PdS_2 is a *translationengleiche* maximal subgroup of $Pa\bar{3}$, the space group of pyrite (FeS_2). The threefold axes of the cubic space group have been lost; the index of the symmetry reduction is therefore 3. The twofold screw axes parallel to and the glide planes perpendicular to the cube edges have been retained. This could be formulated in the following manner:

$$P2_1/a\bar{3} \;\text{—t3}\rightarrow\; P2_1/a\bar{1}$$

However, the second symbol does not correspond to the conventions for the orthorhombic system. The cube edges are no longer equivalent in the orthorhombic system, so that $P2_1/a\bar{1}$ has to be replaced by $P2_1/b2_1/c2_1/a$ or the short Hermann–Mauguin symbol $Pbca$. As shown in Fig. 12.1, the atomic coordinates have not changed much. However, the two structures differ widely, the c axis of PdS_2 being strongly stretched. This is due to the tendency of bivalent palladium towards square planar coordination (electron configuration d^8), whereas the iron atoms in pyrite have octahedral coordination.

Strictly speaking, in the mathematical sense, the space groups of FeS_2 and PdS_2 are not really *translationengleiche* because of the different lattice parameters. In the strict sense, however, FeS_2 at 25.0 °C and at 25.1 °C would not have the same space group either, due to thermal expansion. Such a strict treatment would render it impossible to apply group-theoretical methods in crystal chemistry and physics in the first place. Therefore, certain abstraction from the strict mathematical viewpoint is necessary. We use the concept of the *parent clamping approximation* [121], that is, we act as if the lattices of the two homeotypic structures were the same or were 'clamped' the one with the other, and then we permit relaxation of the clamp. In other words: We permit an

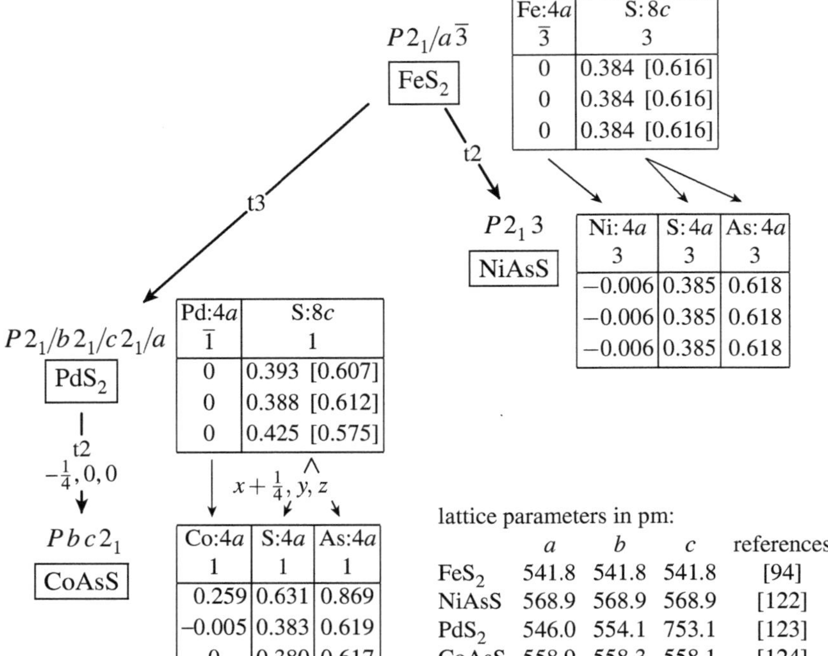

Fig. 12.1 Tree of group–subgroup relations for the family of structures of pyrite. Coordinates in brackets (not stated normally) refer to symmetry-equivalent positions.

affine distortion. The term 'isotypic', which we use to express that the crystal structures of NaCl, KCl, and MgO are of the same kind, with coinciding space groups, expressly permits such an affine distortion (but neither a change of the occupied Wyckoff positions nor larger deviations of the atomic coordinates).

There is no general rule indicating up to what degree deviations are tolerable. They should be judged with crystallographic, chemical, and physical expertise (see also the comments on isotypism in Section 9.5). Anything that helps to increase the understanding of structural relations is acceptable. However, it is recommended to be cautious with any tolerance.

Upon transition from $Pa\bar{3}$ to $Pbca$ none of the occupied Wyckoff positions split, but their site symmetries are reduced. Without the symmetry reduction from $\bar{3}$ to $\bar{1}$ the square coordination of the Pd atoms would not be possible.

Ternary derivatives of the pyrite type

If the positions of the sulfur atoms of pyrite and PdS_2 are replaced by two different kinds of atoms in an ordered 1 : 1 ratio, this enforces further symmetry reductions. The corresponding subgroups may only be subgroups in which the point orbits of the sulfur atoms split into symmetry-independent orbits. In the chosen examples NiAsS (gersdorffite) and CoAsS (cobaltite) the symmetry reductions consist in the loss of the inversion centres of $Pa\bar{3}$ and $Pbca$; the index is thus 2 in each case.

In both examples the site symmetries of the splitting point orbits are being kept (point group 3 for NiAsS, 1 for CoAsS). This always holds for subgroups

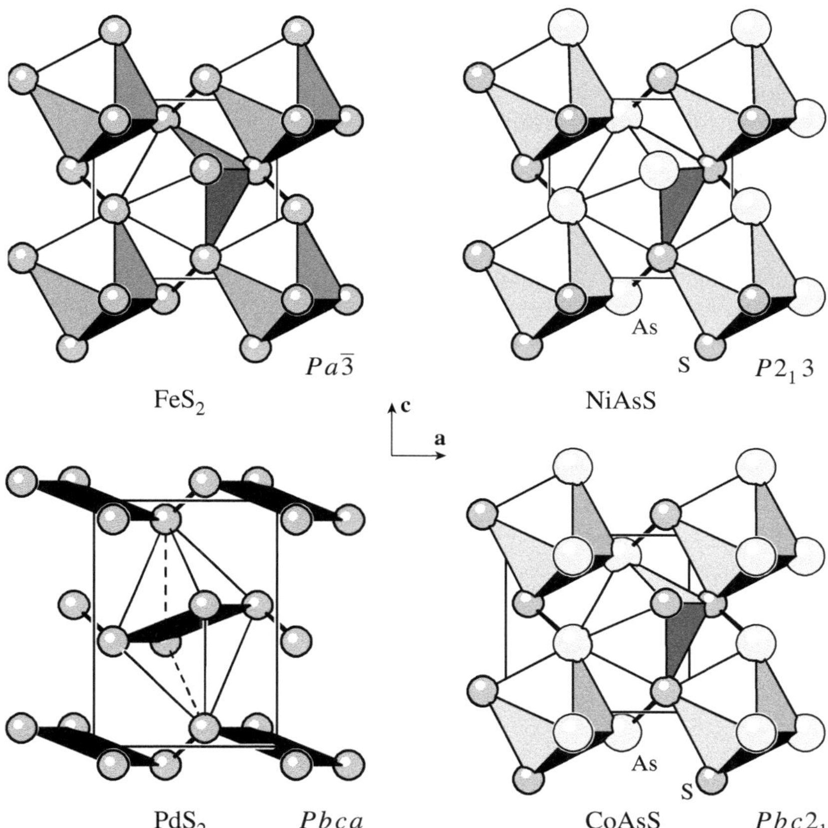

Fig. 12.2 Top views of the unit cells of pyrite, NiAsS, PdS$_2$, and CoAsS.

of index 2: Either the point orbit splits or there is a reduction of the site symmetry. Coordinate changes are not necessary, but may occur depending on the site symmetries. In our examples there are minor coordinate changes. In the case of NiAsS atomic shifts may only occur along the threefold rotation axes due to the site symmetry 3.

The relations between FeS$_2$, PdS$_2$, NiAsS, and CoAsS are summarized in Fig. 12.1 according to the instructions given in the scheme of Chapter 11 (page 143). $Pbc2_1$ is the non-conventional setting of the space group $Pca2_1$ after interchange of the axes a and b. A basis transformation from PdS$_2$ to CoAsS is avoided with this setting. Mind the origin shift; in the conventional description of $Pca2_1$, and therefore also of $Pbc2_1$, the origin is situated on one of the 2_1 axes and thus differs from that of $Pbca$. The origin shift of $-\frac{1}{4}, 0, 0$ in the coordinate system of $Pbca$ involves a change of atomic coordinates by $+\frac{1}{4}, 0, 0$ (i.e. with opposite sign).

The substitution derivatives NiAsS and CoAsS can only by connected by the common supergroup $P2_1/a\overline{3}$. A direct group–subgroup relation from $P2_13$ to $Pbc2_1$ does not exist, since $P2_13$ has no glide planes. Therefore, the mentioned structures belong to different branches of the pyrite tree. NiAsS and CoAsS differ in the distribution of the atoms (Fig. 12.2).

The crystal-chemical relation between α- and β-tin

Grey tin (α-Sn) has a cubic structure of diamond type. Under high pressure it is converted to tetragonal white tin (β-Sn). At first glance, according to the lattice parameters of the two modifications (Fig. 12.3), we would not expect a structural relation. However, we can imagine the structure of β-tin to be formed from that of α-tin by compression along one of the cube edges, without change of the coordinate triplet x, y, z of the tin atom. The distortion is enormous: the lattice parameter c decreases from 649 pm to 318 pm, and in this direction the atoms approach each other to this distance. The coordination number of a tin atom is increased from 4 to 6. Simultaneously, the lattice is widened in the a–b plane. The cell volume decreases to 79% of its original value.

In group-theoretical terms the transformation of α-tin to β-tin corresponds to the transition to a *translationengleiche* maximal subgroup of index 3 according to the notation:

$$F\,4_1/d\,\bar{3}\,2/m \;\text{—t3}\rightarrow\; F\,4_1/d\,\bar{1}\,2/m$$

Because of the compression along **c** the crystal system changes from cubic to tetragonal, and hence the space-group symbol has to be changed according to the conventions of the tetragonal system, namely:

$$F\,4_1/d\,\bar{3}\,2/m \;\text{—t3}\rightarrow\; F\,4_1/d\,2/d\,2/m$$

Instead of a face-centred tetragonal cell we can always choose a body-centred cell with half the volume. The new axes $\mathbf{a}'$ and $\mathbf{b}'$ of the small cell run along the face diagonals of the original cell and have half their lengths (cf. Fig. 10.3, page 136). Commonly β-tin is described with the smaller cell, which also requires a change of the space-group symbol to $I\,4_1/a\,2/m\,2/d$ (for short

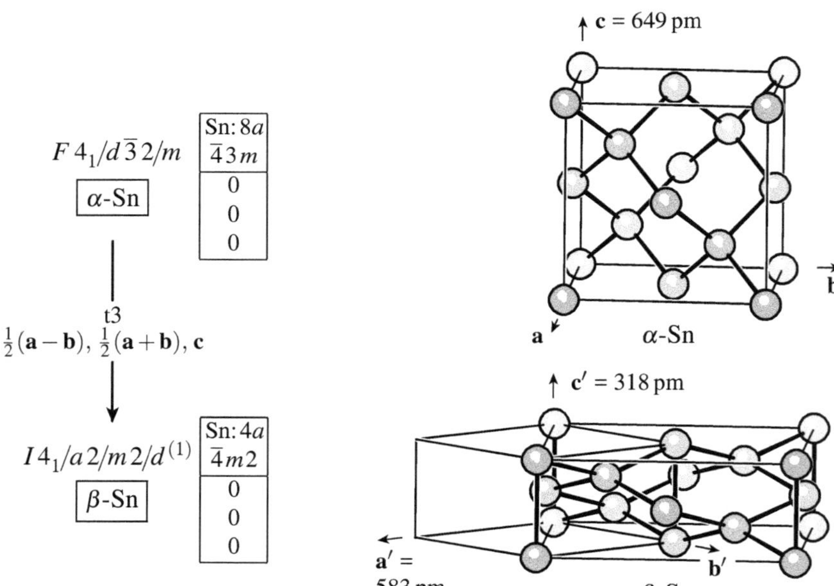

Fig. 12.3 The relation between α- and β-tin and their unit cells. The conventional cell of β-tin ($\mathbf{a}'$, $\mathbf{b}'$, $\mathbf{c}'$) is less appropriate for the comparison than the face-centred cell with $\mathbf{a} = \sqrt{2}\mathbf{a}'$.

$I4_1/amd$), as given in Fig. 12.3. The group–subgroup relation is depicted according to the scheme of page 143 in the left part of Fig. 12.3.

The group-theoretical relation does not justify the inference of a mechanism for the phase transition. No continuous deformation of an α-tin single crystal to a β-tin single crystal takes place under pressure. Rather, nuclei of β-tin are formed in the matrix of α-tin, which then grow at the expense of the α-tin. The structural rearrangement takes place only at the phase boundaries between the shrinking α-tin and the growing β-tin.

12.2 The maximal subgroup is *klassengleiche*

Two derivatives of the AlB_2 type

Consider two derivatives of the AlB_2 type as an example of *klassengleiche* subgroups. AlB_2 has a simple hexagonal structure in the space group $P6/mmm$. In the **c** direction, aluminium atoms and sheets of boron atoms alternate; the boron atom sheets are planar, like in graphite (Fig. 12.4) [125]. The ZrBeSi type has a similar structure [126], but the sheets consist of alternating Be and Si atoms (the sheet is like a sheet in hexagonal boron nitride BN). As a consequence, the inversion centres in the middles of the six-membered rings cannot be retained; the Zr atoms that replace the Al positions remain on inversion

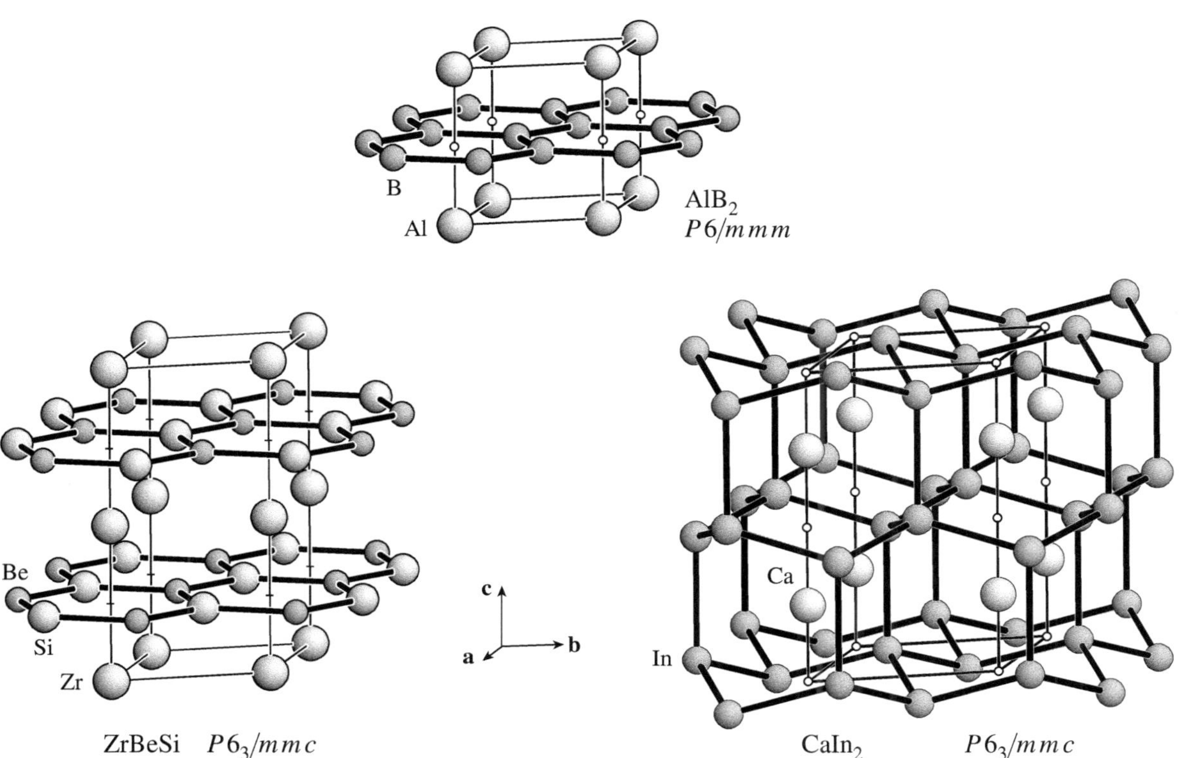

Fig. 12.4 The structures of AlB_2, ZrBeSi, and $CaIn_2$. The mirror planes perpendicular to **c** in $P6_3/mmc$ are at $z = \frac{1}{4}$ and $z = \frac{3}{4}$.

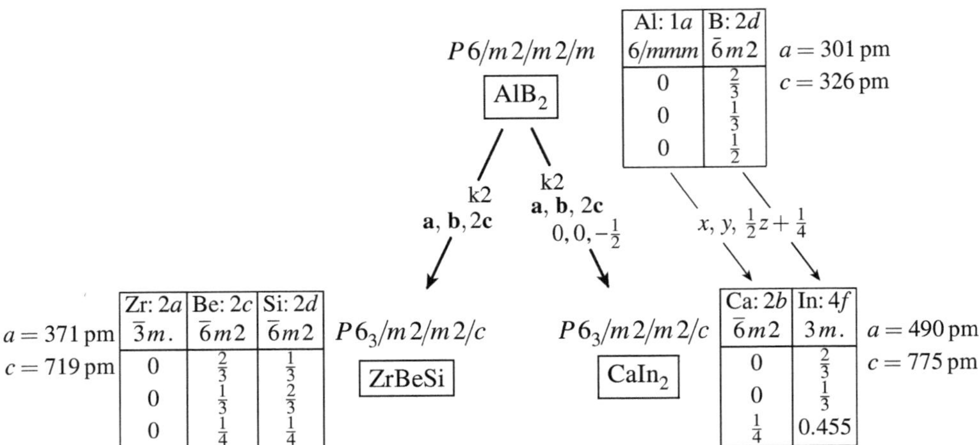

Fig. 12.5 Both hettotypes of the AlB_2 type have the same space-group type and a doubled c axis, but the space groups are different due to different origin positions relative to the origin of the aristotype. Because of the doubling of the c axis the z coordinates are halved. The origin shift of $0, 0, -\frac{1}{2}$ in the right branch requires that $\frac{1}{4}$ be added to the z coordinates.

centres. This enforces a symmetry reduction to the *klassengleiche* space group $P6_3/mmc$ with doubled **c** vector.

The doubling of c is the essential aspect in the symmetry reduction from the AlB_2 to ZrBeSi type. The index is 2: half of all translations are lost, together with half of the inversion centres, half of the symmetry axes perpendicular to **c**, and half of the mirror planes perpendicular to **c**. Instead of the mirror planes perpendicular to [210] (last m in the Hermann–Mauguin symbol $P6/mmm$) now there are glide planes c. The Wyckoff position $2d$ of the boron atoms of AlB_2 splits into two symmetry-independent positions $2c$ and $2d$ of the subgroup (Fig. 12.5, left), rendering possible occupation by atoms of two different species.

Figures 12.4 and 12.5 show another peculiarity. $P6/mmm$ has two *different klassengleiche* maximal subgroups of the same type $P6_3/mmc$ with doubled basis vector **c**. The second one corresponds to $CaIn_2$ [127, 128]. Here the graphite-like sheets of the AlB_2 type have become puckered layers of indium atoms, similar to those in grey arsenic. The indium atoms of adjacent layers have shifted parallel to **c** and have come close to each other in pairs, so that the result is a network as in lonsdaleite (hexagonal diamond). The alternating shift of the atoms no longer permits the existence of mirror planes in the layers; however, neighbouring layers are mutually mirror-symmetrical. The calcium atoms are on the mirror planes, but no longer on inversion centres. The difference between the two subgroups $P6_3/mmc$ consists of the selection of the symmetry elements that are being lost with the doubling of **c**.

The conventional description of the space groups according to *International Tables* requires an inversion centre to be at the origin of space group $P6_3/mmc$. The position of the origin at an Al atom of the AlB_2 type can be kept when the symmetry is reduced to that of ZrBeSi (i.e. Zr at the origin). The symmetry reduction to $CaIn_2$, however, requires an origin shift to the centre of one of the six-membered rings. In the coordinate system of the aristotype that is a shift by $0, 0, -\frac{1}{2}$, as marked in the middle of the group–subgroup arrow in Fig. 12.5. For the new atomic coordinates (in the coordinate system of the subgroup), the origin shift results in the addition of $+\frac{1}{4}$ to the z coordinates, that is, with the *opposite* sign of the shift stated in the group–subgroup arrow. In addition, due to the doubling of c, the z coordinates of the aristotype have to be halved. Therefore, the new z coordinate of the In atom is approximately $z' \approx \frac{1}{2}z + \frac{1}{4} = \frac{1}{2} \times \frac{1}{2} + \frac{1}{4} = \frac{1}{2}$. It cannot be exactly this value, because then there would have been no symmetry reduction and the space group would still be $P6/mmm$. The symmetry reduction requires the atom shift to $z' = 0.455$.

For two structures to be considered as related, the atom shift may not be arbitrarily large. There is no general rule indicating up to what amount atom shifts are permitted. As in the case of tolerable deviations of the lattices, they should be judged with crystallographic, chemical, and physical expertise.

In the relation $AlB_2 \rightarrow ZrBeSi$, the site symmetry $\overline{6}m2$ of the boron atoms is retained and the Wyckoff position splits. In the relation $AlB_2 \rightarrow CaIn_2$ it is the other way; the position does not split, the atoms remain symmetry equivalent, but their site symmetry is reduced to $3m1$ and the z coordinate becomes variable.

Among *klassengleiche* subgroups of index 2 there often exist two or four, and sometimes even eight subgroups of the same type with different origin positions. It is important to choose the correct one, with the correct origin shift. All of these subgroups are listed in *International Tables*, Volume $A1$, but not so in Volume A. They are subgroups on a par, that is, they are not conjugate, but belong to different conjugacy classes (cf. Section 8.3, page 110).

What we have dealt with here is a widespread phenomenon. For many crystal structures so-called superstructures are known; they have reduced translational symmetry. In X-ray diffraction patterns they have additional weak 'superstructure reflections' in between the main reflections of the basic structure. Whereas the term superstructure gives only a qualitative, informal outline of the facts, the group-theoretical approach permits a precise treatment.

The relation between $In(OH)_3$ and $CaSn(OH)_6$

Aside from the enlargement of the unit cell as in the preceding examples, the loss of translational symmetry is also possible without enlargement of the conventional unit cell if centring translations are lost. In any case, the primitive unit cell is enlarged. Consider as an example the relation of indium hydroxide [129, 130] and $CaSn(OH)_6$, which is known as the mineral burtite [131, 132]. All important data of the structures are given in Fig. 12.6. In the lower part of the figure the symmetry elements of both space groups are shown in the style of *International Tables*. Pay attention to how symmetry elements are eliminated by the symmetry reduction $I2/m\overline{3}$ —k2→ $P2/n\overline{3}$.

The structure of $In(OH)_3$ is intimately related to the structure of skutterudite, $CoAs_3$ [133, 134]. Its structure consists of vertex-sharing $CoAs_6$ octahedra that are mutually tilted in such a way that four octahedron vertices each come close to one another forming a slightly distorted square, corresponding to the formula $Co_4^{3+}(As_4^{4-})_3$. In $In(OH)_3$, $(OH^-)_4$ squares take the places of the As_4^{4-}

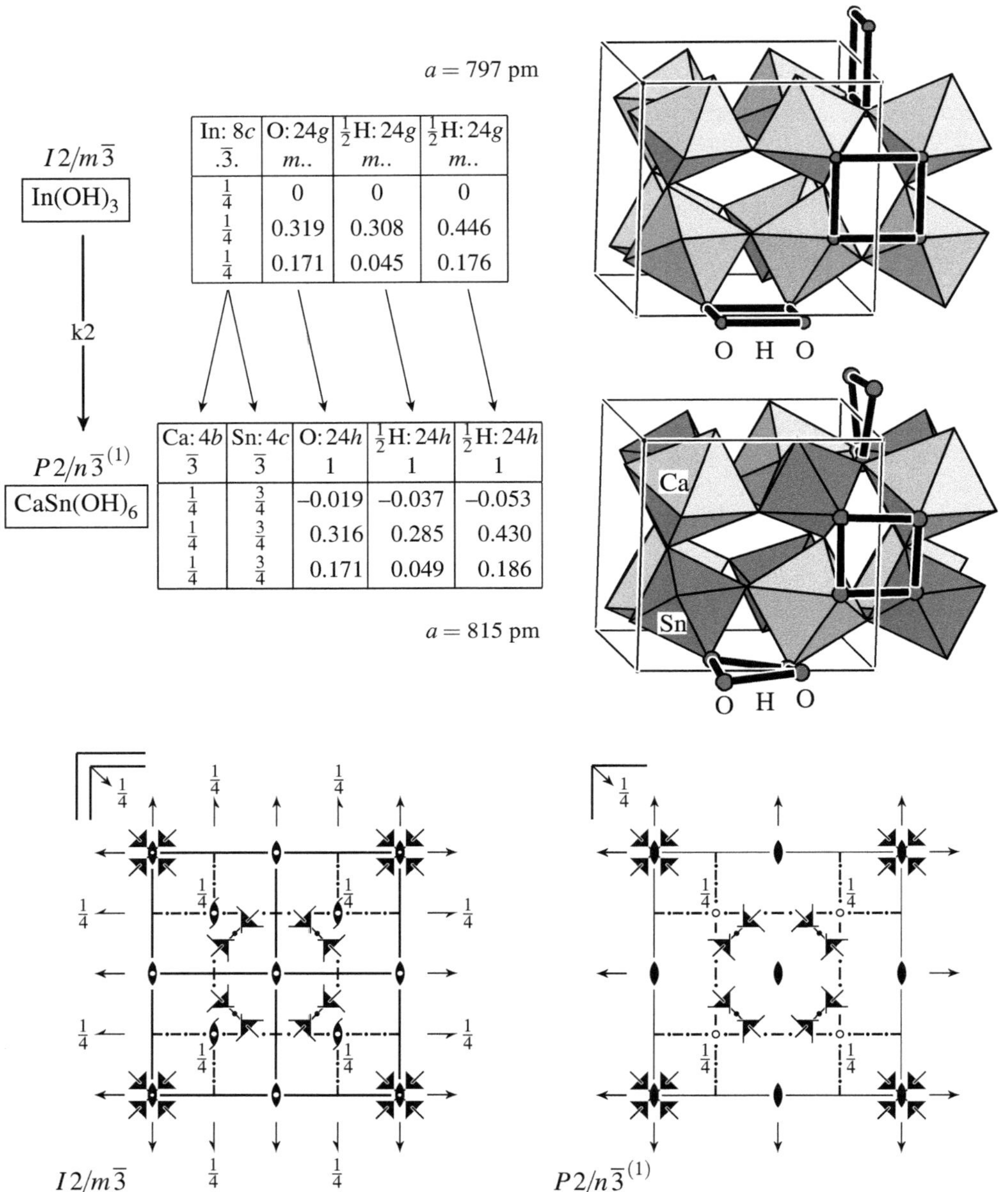

In: 8*c*	O: 24*g*	$\frac{1}{2}$H: 24*g*	$\frac{1}{2}$H: 24*g*
.$\bar{3}$.	*m*..	*m*..	*m*..
$\frac{1}{4}$	0	0	0
$\frac{1}{4}$	0.319	0.308	0.446
$\frac{1}{4}$	0.171	0.045	0.176

Ca: 4*b*	Sn: 4*c*	O: 24*h*	$\frac{1}{2}$H: 24*h*	$\frac{1}{2}$H: 24*h*
$\bar{3}$	$\bar{3}$	1	1	1
$\frac{1}{4}$	$\frac{3}{4}$	−0.019	−0.037	−0.053
$\frac{1}{4}$	$\frac{3}{4}$	0.316	0.285	0.430
$\frac{1}{4}$	$\frac{3}{4}$	0.171	0.049	0.186

Fig. 12.6 The relation between $In(OH)_3$ and $CaSn(OH)_6$. The H atoms exhibit misorder among two close sites in between two neighbouring O atoms, O–H···O and O···H–O, and therefore occupy their positions with a mean probability of one half (expressed by the formulation $\frac{1}{2}$H; data from neutron diffraction; cf. margin note No.2, page 213, for a remark on the term misorder). Three (distorted) $(OH^-)_4$ squares are shown by black bonding lines in the images to the right.
Remark for those who look meticulously: due to the doubling of the translation vectors in the [111], [11$\bar{1}$], [1$\bar{1}$1], and [$\bar{1}$11] directions, 3_1 axes are converted to 3_2 axes and vice versa.

squares, with O atoms at the vertices and hydrogen bonds on the edges. As a consequence of the loss of the body centring, the positions of the indium atoms of $In(OH)_3$ split into two independent positions, which are occupied by the calcium and tin atoms in burtite. In addition, the mirror planes on which the O and H atoms are placed in $In(OH)_3$ are lost, resulting in an additional degree of freedom. This renders possible two different sizes of the coordination octahedra in the subgroup $P2/n\overline{3}$, which are occupied by the Ca and Sn atoms. The $(OH^-)_4$ squares become slightly twisted.

The example shows that the old term '*zellengleiche*' (with the same cell) instead of '*translationengleiche*' (with the same translations) can be misinterpreted. The unit cells of $In(OH)_3$ and $CaSn(OH)_6$ have nearly the same size, and yet the subgroup is not a '*zellengleiche*' subgroup, because one of the cells is centred and the other one is not. The volumes of the primitive cells differ by a factor of 2; $CaSn(OH)_6$ has half as many translations as $In(OH)_3$.

12.3 The maximal subgroup is isomorphic

Isomorphic subgroups comprise a special category of *klassengleiche* subgroups. Every space group has an infinite number of isomorphic maximal subgroups. The index i agrees with the factor by which the unit cell is being enlarged. The indices are prime numbers $p \neq 1$; squares of prime numbers p^2 may occur in the case of tetragonal, hexagonal, and trigonal space groups, and for cubic space groups only cubes p^3 ($p \geq 3$) of prime numbers are possible. For many space groups not all prime numbers are permitted. The prime number 2 is often excluded, and for certain subgroups additional restrictions may apply (e.g. only prime numbers $p = 6n+1$) [135, 136]; for details see Chapter 21. Usually, in accordance with the symmetry principle, only small index values are observed. However, seemingly curious values like 13, 19, 31, or 37 do occur (see Exercise 14.2 for two examples).

In *International Tables A* (1983–2006 editions), isomorphic subgroups are listed only for the smallest possible indices. Looking there, it may happen that an isomorphic subgroup is missed. On the other hand, it may happen that one believes that a not-mentioned subgroup of some higher index is possible, although its index is a forbidden prime number. Therefore, it is safer to consult *International Tables A*1, where all possibilities are listed completely.

Since isomorphic subgroups have the same Hermann–Mauguin symbol (except for enantiomorphic space groups), it is of special importance to keep in mind the difference between space groups and space-group types. In the following examples the group and the subgroup belong to the same space-group type, but they are different space groups with different lattices. The volume of the unit cell of the subgroup is enlarged by the factor i.

The relation between CuF_2 and VO_2

Consider the pair of structures CuF_2–VO_2 as an example for the transition to an isomorphic subgroup. The details are shown in Fig. 12.7. The z coordinates are being halved with the switch-over from CuF_2 to VO_2, due to the doubling

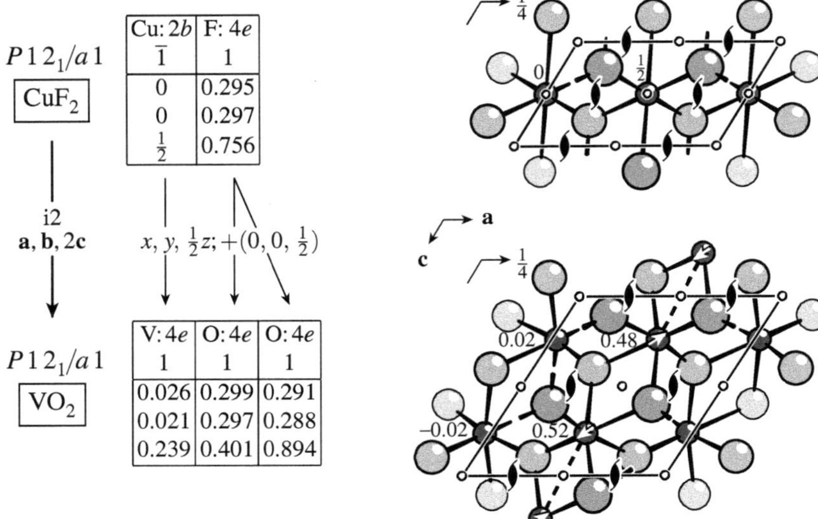

Fig. 12.7 The relation between CuF_2 and VO_2. Numbers in the images are y coordinates of the metal atoms. The white arrows on top of the four vanadium atoms in the bottom image mark the directions of their shifts.

Table 12.1 Crystal data of the compounds of Section 12.3.

	Lattice parameters a/pm	b/pm	c/pm	β/°	Space group	Z	References
CuF_2	536.2	456.9	330.9	121.1	$P12_1/a1$	2	[137]
VO_2	538.3	453.8	575.2	122.7	$P12_1/a1$	4	[138]
TiO_2 (rutile)	459.4	459.4	295.9		$P4_2/mnm$	2	[139]
$CoSb_2O_6$ (trirutile)	465.0	465.0	927.6		$P4_2/mnm$	2	[140]

of the basis vector **c**. Atoms from the neighbouring unit cell of CuF_2 are being included in the doubled cell, whereby the original coordinates $x, y, z+1$ are converted approximately, but by no means exactly, to the coordinates $x, y, \frac{1}{2}z+\frac{1}{2}$; if they were exactly at these values, the symmetry would not have been reduced. For the vanadium atoms, the atom at $\sim(x, y, \frac{1}{2}z+\frac{1}{2})$ does not have to be listed. This is due to the fact that the metal atom position does not split; the Cu positions at $0, 0, \frac{1}{2}$ and $0, 0, \frac{3}{2}$ become the V pair 0.026, 0.021, 0.239 and −0.026, −0.021, 0.761, whose atoms are symmetry equivalent by the inversion at $0, 0, \frac{1}{2}$. The multiplicities of the Wyckoff labels show that the numbers of all atoms in the unit cell are doubled.

The crystal data are listed in Table 12.1. The images of the unit cells in Fig. 12.7 show the elimination of one half of all inversion centres and screw axes, resulting from the doubling of the lattice parameter c. Whereas the Cu atoms in CuF_2 are located on inversion points in the centres of the coordination octahedra, the V atoms (electron configuration d^1) in VO_2 have been shifted along **c** forming pairs, with alternating V···V distances of 262 and 317 pm due to spin pairing at the shorter of these distances (two of these V–V contacts are marked in Fig. 12.7 by dotted lines). In addition, there is one shortened V=O bond (176 pm); the remaining V–O bonds have lengths of 186 to 206 pm. (Upon heating, the spin pairing is cancelled at 68°C, the V···V distances become equal, and VO_2 becomes metallic in the rutile type; cf. Exercise 12.6).

The rutile–trirutile relation

Another simple example of a relation between isomorphic space groups concerns trirutile [141]. As shown in Table 12.1 and Fig. 12.8, the c axis of trirutile is tripled as compared to rutile. This way it is possible to obtain an ordered distribution of two kinds of cations in a ratio of 1 : 2, keeping the space-group type $P4_2/mnm$. This is possible because the space group $P4_2/mnm$ has an isomorphic subgroup of index 3, which is nicely reflected by the term 'trirutile'. For the sake of a clear terminology, this term should only be used if it has this kind of group-theoretical foundation. The index 3 is the smallest possible index for an isomorphic subgroup of $P4_2/mnm$. A 'dirutile' with this space group type cannot exist; therefore, it is recommended not to use this term, even if lower-symmetry, non-tetragonal structures with doubled c axis can be developed from the rutile type [142, 143].

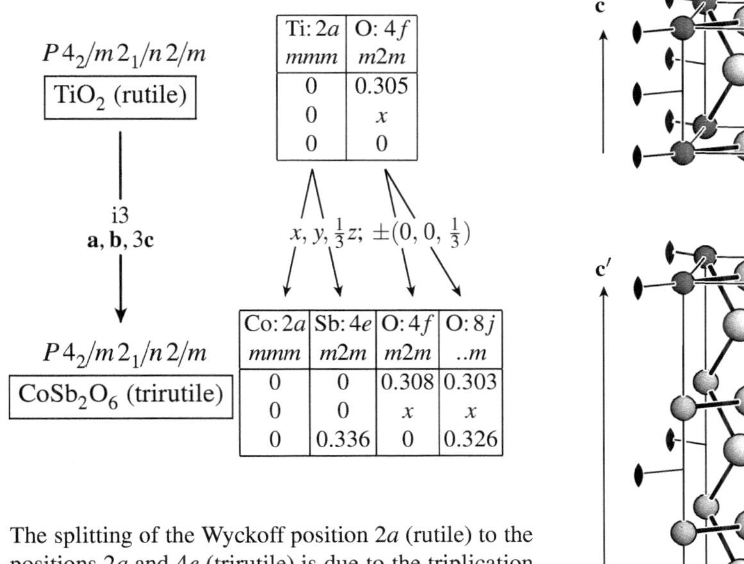

Ti: $2a$	O: $4f$
mmm	$m2m$
0	0.305
0	x
0	0

Co: $2a$	Sb: $4e$	O: $4f$	O: $8j$
mmm	$m2m$	$m2m$	$..m$
0	0	0.308	0.303
0	0	x	x
0	0.336	0	0.326

The splitting of the Wyckoff position $2a$ (rutile) to the positions $2a$ and $4e$ (trirutile) is due to the triplication of the unit cell. The translation-equivalent positions 0, 0, 0; 0, 0, 1; 0, 0, 2 of rutile are converted to 0, 0, 0; 0, 0, $\sim\frac{1}{3}$; 0, 0, $\sim\frac{2}{3}$ of trirutile

Fig. 12.8 Group–subgroup relation between rutile and trirutile. To clarify the elimination of symmetry elements, the twofold rotation axes have been included in the unit cells.

12.4 The subgroup is neither *translationengleiche* nor *klassengleiche*

Subgroups that are neither *translationengleiche* nor *klassengleiche* are called *general subgroups*. They cannot be maximal subgroups; there must be at least one intermediate group. Starting from a high-symmetry aristotype, it frequently occurs that the symmetry is first reduced to a *translationengleiche* subgroup, followed by a *klassengleiche* subgroup.

The relation between NiAs and MnP

The phase transition of the NiAs type to the MnP type has been the subject of thorough studies with a number of compounds (e.g. VS, MnAs) [144, 145]. The symmetry reduction involves two steps (Fig. 12.9). In the first step the hexagonal symmetry is lost; for this a slight distortion of the lattice would be sufficient. The orthorhombic subgroup has a *C*-centred cell. Due to the centring, the cell is *translationengleiche*, although its size is twice as big. The centring is removed in the second step, half of the translations being lost; therefore, it is a *klassengleiche* reduction of index 2.

The images in Fig. 12.9 show which symmetry elements are eliminated with the two steps of symmetry reduction. Among others, half of the inversion centres are being lost. Here we have to watch out: the eliminated inversion centres

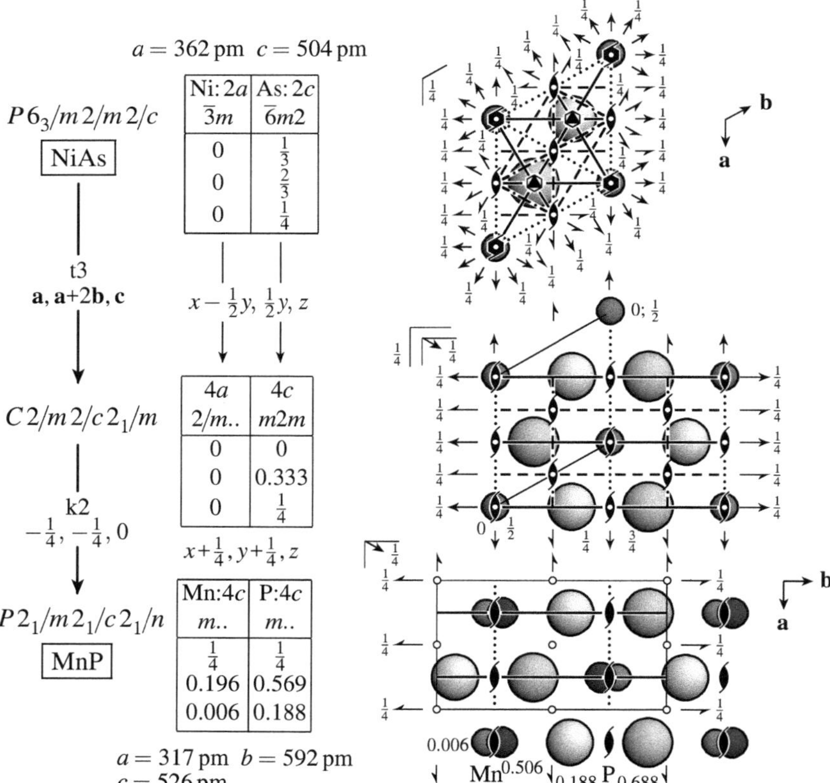

Fig. 12.9 The relation between NiAs and MnP. The numbers in the images to the right are z coordinates [146].

of the space group $Cmcm$ ($C2/m2/c2_1/m$) are those of the Wyckoff positions $4a$ (0, 0, 0) and $4b$ ($\frac{1}{2}$, 0, 0), while those of the Wyckoff position $8d$ ($\frac{1}{4}$,$\frac{1}{4}$,0) are retained. Since the subgroup $Pmcn$ ($P2_1/m2_1/c2_1/n$) should have its origin on a point of inversion, an origin shift is required. The shift of $-\frac{1}{4}, -\frac{1}{4}$,0 entails an addition of $\frac{1}{4}, \frac{1}{4}$,0 to the coordinates.

After addition of $\frac{1}{4}, \frac{1}{4}$,0 to the coordinates listed in Fig. 12.9 for the space group $Cmcm$, we obtain ideal values for an undistorted structure in the space group $Pmcn$. However, due to the missing distortion, the symmetry would still be $Cmcm$. The space group $Pmcn$ is attained only after the atoms have been shifted away from the ideal positions. The deviations concern mainly the y coordinate of the Mn atom (0.196 instead of $\frac{1}{4}$) and the z coordinate of the P atom (0.188 instead of $\frac{1}{4}$). These are significant deviations, but they are small enough to consider MnP as being a distorted variant of the NiAs type.

12.5 The space groups of two structures have a common supergroup

Two crystal structures can be intimately related even when there is no direct group–subgroup relation between their space groups. Instead, there may exist a common supergroup. The structures of NiAsS and CoAsS, presented in Section 12.1, offer an example. In this case, the pyrite type corresponds to the common supergroup. Even if there is no known representative, it may be worthwhile to look for a common supergroup.

The relation between $RbAuCl_4$ and $RbAuBr_4$

The crystal structures of the rubidium-halidoaurates $RbAuCl_4$ and $RbAuBr_4$ are monoclinic, but have different unit cells and space groups. If the settings given in Table 12.2 are chosen, the metric conditions suggest a close relation. The main difference obviously concerns the lattice parameter c, which is approximately twice as large for $RbAuCl_4$ as for $RbAuBr_4$.

With the aid of *International Tables* (volumes *A* up to 2006 or *A*1) it is easy to find out that both $I12/c1$ and $P12_1/a1$ can be derived from the common supergroup $C12/m1$ by *klassengleiche* symmetry reductions (Fig. 12.10; in Volume *A*1, Part 3, $I12/a1$ is listed as subgroup of $C12/m1$; $I12/c1$ is the same subgroup with interchanged axes a and c).

The symmetry elements of the three space groups are depicted in Fig. 12.11 in the style of *International Tables*. It is recommended to study the images thoroughly in order to get acquainted with the details of the different kinds of elimination of symmetry elements due to the loss of translations.

Uncritical inspection of the subgroup tables could lead to the assumption that the space group of $RbAuBr_4$ could be a maximal subgroup of that of $RbAuCl_4$, because $P12_1/c1$ is mentioned as a subgroup of $I12/a1$ in *International Tables* *A*1; after exchange of a and c these are the same as $P12_1/a1$ and $I12/c1$, respectively. However, this is wrong: The relation $I12/c1 - \text{k}2 \rightarrow P12_1/a1$ is valid only for an unchanged size of the unit cell (the loss of translations results from the loss of the centring).

Table 12.2 Crystal data of $RbAuCl_4$ and $RbAuBr_4$.

	$RbAuCl_4$	$RbAuBr_4$
a/pm	976.0	1029.9
b/pm	590.2	621.4
c/pm	1411.6	743.6
β/°	120.05	121.33
space group	$I12/c1$	$P12_1/a1$
references	[147]	[148]

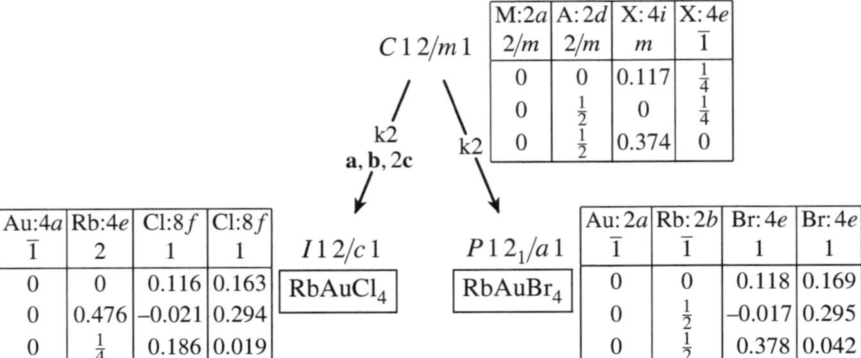

$C1\,2/m\,1$	M:2*a*	A:2*d*	X:4*i*	X:4*e*
	$2/m$	$2/m$	m	$\bar{1}$
x	0	0	0.117	$\frac{1}{4}$
y	0	$\frac{1}{2}$	0	$\frac{1}{4}$
z	0	$\frac{1}{2}$	0.374	0

$I1\,2/c\,1$ ($RbAuCl_4$)	Au:4*a*	Rb:4*e*	Cl:8*f*	Cl:8*f*
	$\bar{1}$	2	1	1
x	0	0	0.116	0.163
y	0	0.476	−0.021	0.294
z	0	$\frac{1}{4}$	0.186	0.019

$P1\,2_1/a\,1$ ($RbAuBr_4$)	Au:2*a*	Rb:2*b*	Br:4*e*	Br:4*e*
	$\bar{1}$	$\bar{1}$	1	1
x	0	0	0.118	0.169
y	0	$\frac{1}{2}$	−0.017	0.295
z	0	$\frac{1}{2}$	0.378	0.042

Fig. 12.10 The symmetry relation between $RbAuCl_4$ and $RbAuBr_4$.

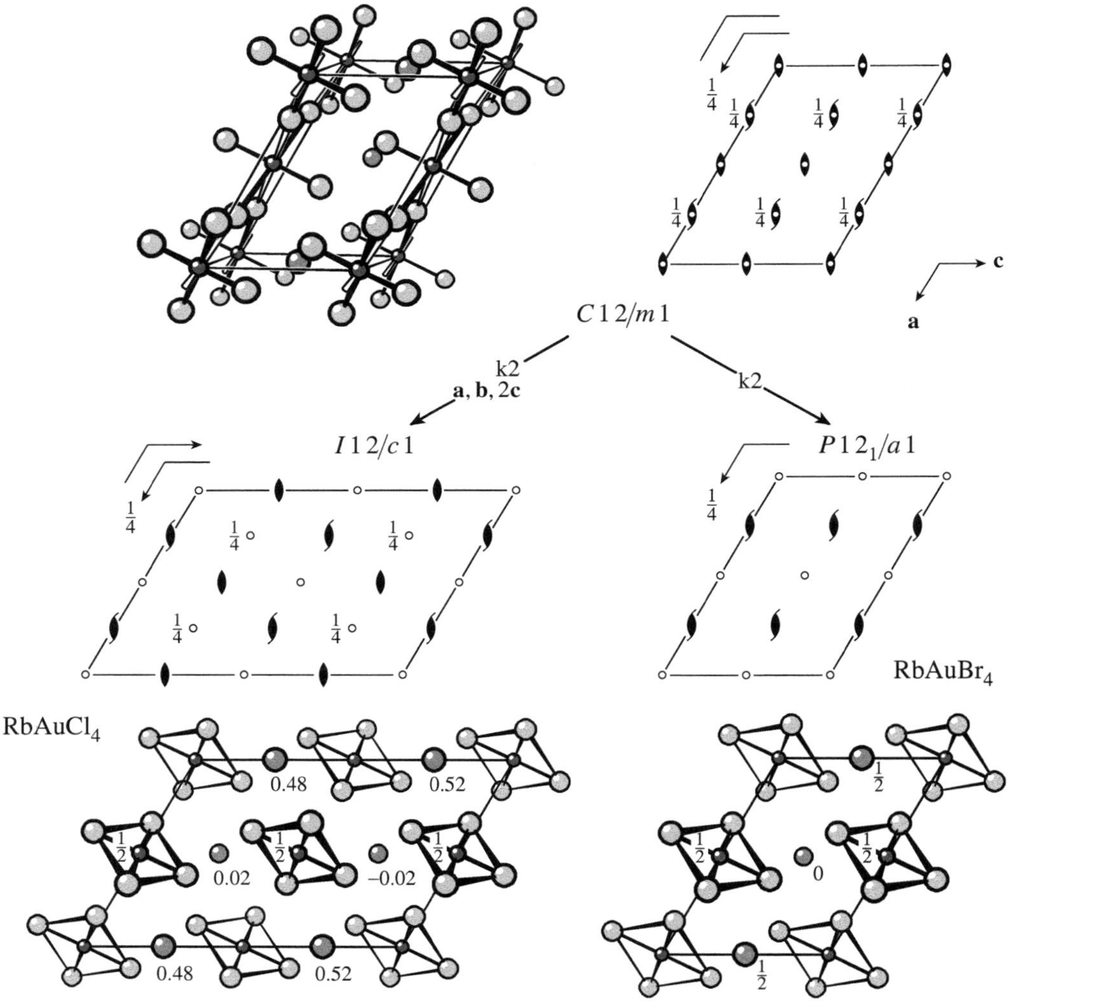

Fig. 12.11 Bottom: Projections of the structures of $RbAuCl_4$ and $RbAuBr_4$ along **b**. Centre: The corresponding symmetry elements. Top: Hypothetical aristotype in the common supergroup; in the perspective image the additional Au–halogen contacts are depicted by hollow bonds.

The lower part of Fig. 12.11 shows the crystal structures of both aurates. The structural similarity is evident. The positions of the Au and Rb atoms coincide almost completely. The difference concerns the relative orientations of the square AuX_4^- ions in their sequence along **c**. Whereas the $AuBr_4^-$ ions in $RbAuBr_4$ have the same tilting angle against the direction of view, the tilt of the $AuCl_4^-$ ions in $RbAuCl_4$ alternates with opposite angles.

Why don't the aurates crystallize in the common supergroup? The answer can be found by construction of the hypothetical structure in the space group $C12/m1$. Here each of the trivalent gold atoms would have to be coordinated by six halogen atoms, whereas their d^8 electron configuration requires a square coordination. At the top left of Fig. 12.11 the hollow bond lines show where the additional Au–halogen contacts would be; they would have the same lengths as the Au–halogen bonds in the *a*–*b* plane drawn in black. In addition, an undistorted octahedral coordination would require a considerable metric adjustment of the unit cell, with $a = b \approx 800$ pm and $a/b = 1$ instead of $a/b = 1.65$ (Fig. 12.12).

Now that the halidoaurates cannot adopt the structure in the supergroup $C12/m1$, the question remains if another compound could adopt this higher-symmetry structure. It would require a trivalent cation with a tendency towards octahedral coordination. Actually, no example is known as yet in the space group $C12/m1$. However, the pattern of vertex-sharing octahedra in the *a*–*b* plane corresponds to that of the $TlAlF_4$ type in the space group $P4/mmm$ [149] (cf. Exercise 12.7). In fact, the structure can be traced back to this structure type, requiring the metric adjustment to $a/b = 1$ and additionally a shearing deformation of the lattice such that the monoclinic angle is decreased from $\beta \approx 121°$ to 90°.

The settings of the space groups $I12/c1$ (standard: $C12/c1$) and $P12_1/a1$ (standard: $P12_1/c1$) were chosen because then the relation to the $TlAlF_4$ type does not require the interchange of axes.

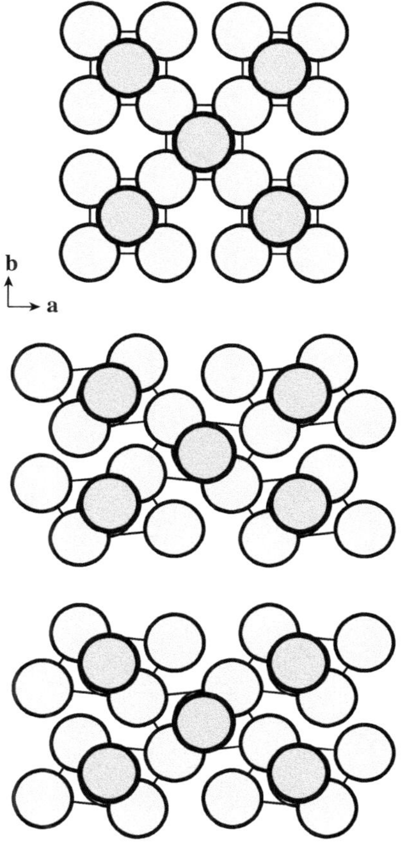

Fig. 12.12 Tetragonal MX_4^- layer as in $TlAlF_4$ and the distorted variant that occurs in $RbAuCl_4$ in two alternating orientations.

12.6 Large families of structures

For the sake of simplicity and clarity, the preceding sections have been restricted to simple examples. Large trees can be constructed using the modular way to put together Bärnighausen trees as set forth in the preceding sections. This way it is possible to introduce a systematic order to large areas of crystal chemistry using symmetry relations as the guiding principle. As an example, Figs 12.13 and 12.14 show structures that constitute the family of structures of the ReO_3 type.

The tree consists of two branches. The one shown in Fig. 12.13 contains hettotypes that result from substitutions of the metal atoms, including additional symmetry reductions due to distortions caused by the Jahn–Teller effect (Cu(II)-, Mn(III) compounds), covalent bonds (As–As bonds in $CoAs_3$), hydrogen bonds, and different relative sizes of the atoms. A section from this tree, namely the relation $CoAs_3$ (or $In(OH)_3$) $\rightarrow$ $CaSn(OH)_6$, is the subject of the preceding section 12.2 and Fig. 12.6. The relation $ReO_3 \rightarrow VF_3$ (FeF_3) is presented in more detail in Section 14.2.1 (page 183 and Fig. 14.4).

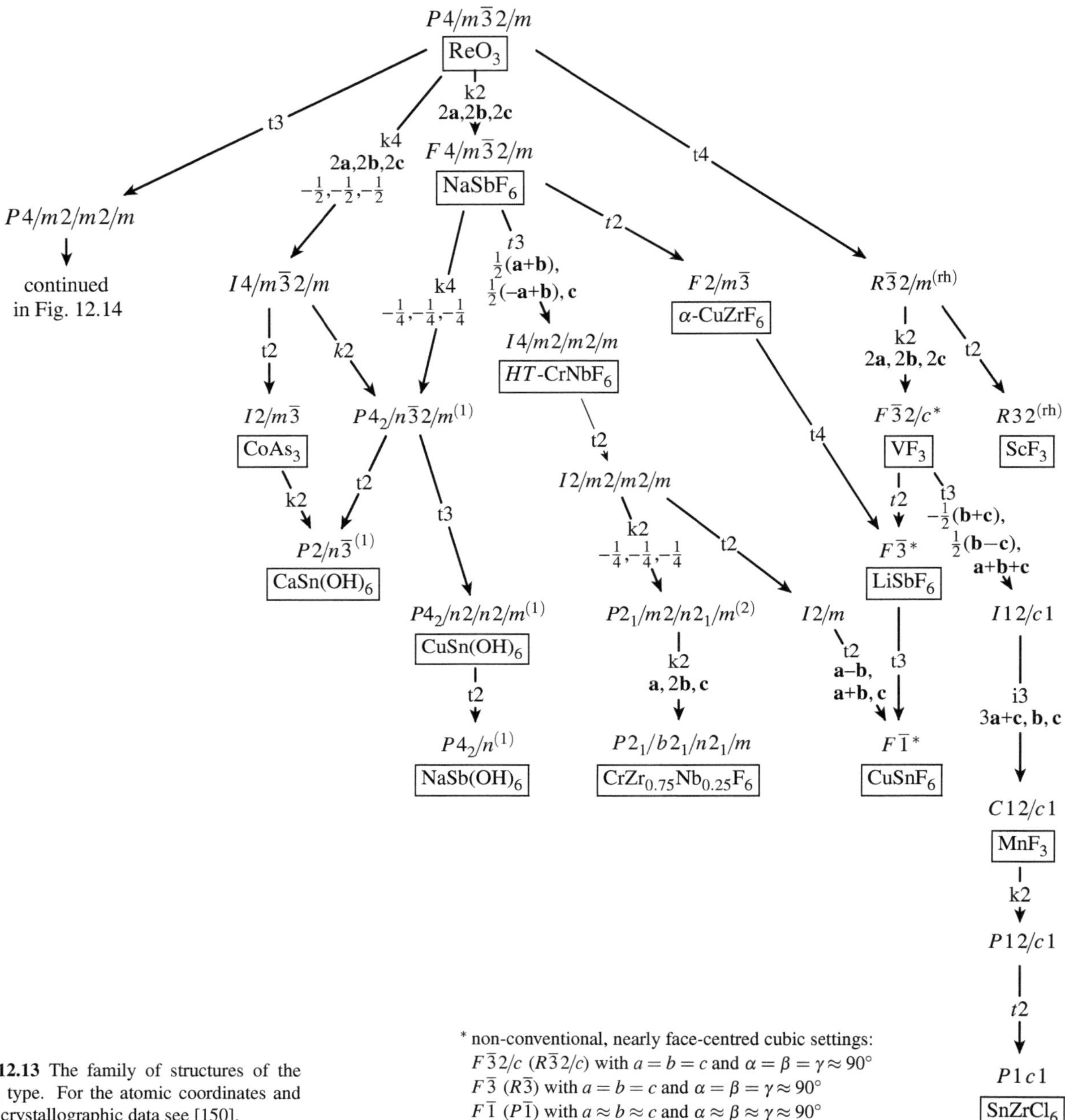

Fig. 12.13 The family of structures of the ReO_3 type. For the atomic coordinates and other crystallographic data see [150].

Of the second branch of the tree only the space group $P4/m2/m2/m$ is mentioned in Fig. 12.13; it is continued in Fig. 12.14. Only one compound is mentioned in this branch, WO_3. This is an example showing the symmetry relations among different polymorphic forms of a compound that are mutually transformed from one to another depending on temperature and pressure:

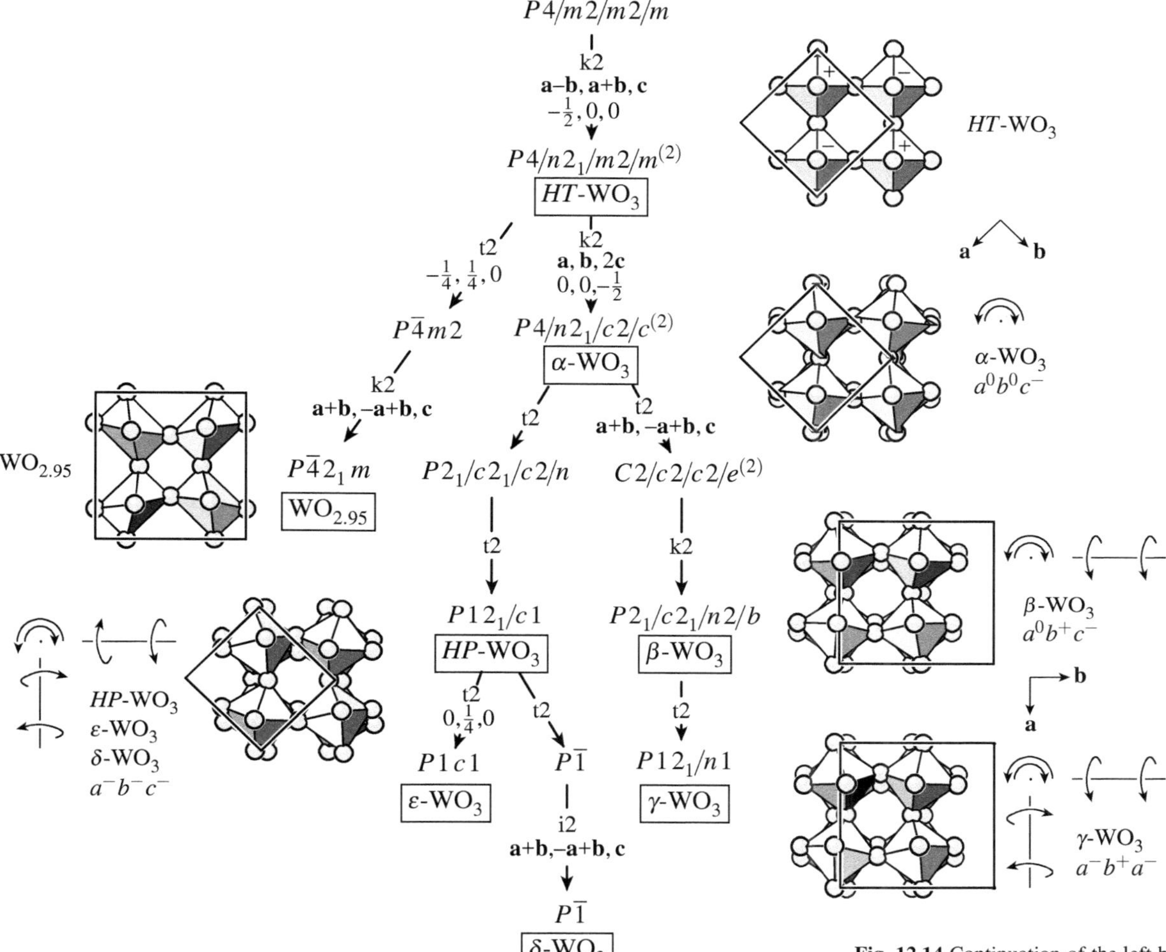

Fig. 12.14 Continuation of the left branch of Fig. 12.13 with polymorphic forms of WO_3. + and – in the image on the top right indicate in which direction the W atoms have been shifted from the octahedron centres. The bent arrows indicate how neighbouring octahedra have been turned as compared to the ReO_3 type.

$$
\begin{array}{ccccccccccc}
HT & \overset{1170\,\mathrm{K}}{\rightleftharpoons} & \alpha & \overset{990\,\mathrm{K}}{\rightleftharpoons} & \beta & \overset{600\,\mathrm{K}}{\rightleftharpoons} & \gamma & \overset{290\,\mathrm{K}}{\rightleftharpoons} & \delta & \overset{230\,\mathrm{K}}{\rightleftharpoons} & \varepsilon \\
 & & & & & & & & \upharpoonleft\downharpoonright{>}300\,\mathrm{MPa} & & \\
 & & & & & & & & HP & &
\end{array}
$$

Since all of these transformations exhibit strong hysteresis, $HT \rightleftharpoons \alpha$ and $\beta \rightleftharpoons \gamma$ excepted, the mentioned temperatures of transformation are only rough estimates. All mentioned modifications are derivatives of the ReO_3 type that result from three kinds of distortions, namely:

1. shifting of the W atoms out of the centres of the coordination octahedra;
2. mutual turning of the octahedra;
3. deformation of the octahedra.

(More modifications of WO_3 are known that are not derivatives of the ReO_3 type.)

Table 12.3 Crystal data of the modifications of tungsten trioxide.

	Space group	a/pm	b/pm	c/pm	$\alpha/°$	$\beta/°$	$\gamma/°$	References
HT-WO_3 (1200 K)	$P4/nmm$	530.3	530.3	393.5				[153]
α-WO_3 (1100 K)	$P4/ncc$	528.9	528.9	786.3				[153, 154]
HP-WO_3 (570 MPa)	$P1 2_1/c 1$	526.1	512.8	765		92.1		[155]
ε-WO_3 (230 K)	$P1c1$	527.8	516.2	767.5		91.7		[156]
β-WO_3 (623 K)	$Pcnb$	733.1	757.3	774.0				[154]
γ-WO_3 (573 K)	$P1 2_1/n 1$	732.7	756.4	772.7		90.5		[154]
δ-WO_3 (293 K)	$P\bar{1}$	731.3	752.5	768.9	88.8	90.9	90.9	[157]
$WO_{2.95}$	$P\bar{4}2_1m$	739	739	388				[158]

Even the most symmetrical modification, the high-temperature form *HT*-WO_3 (1200 K), is not cubic. It has W atoms that are shifted away from the octahedron centres parallel to **c**, combined with a slight elongation of the octahedra in this direction. This way, alternating short and long W–O bonds result in a strand of the vertex-sharing octahedra along **c**. In neighbouring strands the W atoms have been shifted in opposite directions. β-, γ-, and δ-WO_3 have their W atoms shifted towards an octahedron edge, ε-WO_3 and the high-pressure form *HP*-WO_3 towards an octahedron face.

In addition, the low-symmetry modifications have octahedra mutually turned in different ways. That is frequently designated by the Glazer notation [151, 152], which was developed for the 'double perovskites'. For example, in relation to the directions of **a**, **b**, and **c** of the eightfold unit cell of cubic perovskite or ReO_3, $a^+b^-c^0$ means: the consecutive octahedra in the direction of **a** have been turned about **a** to the same side, the consecutive octahedra in the direction of **b** have been turned about **b** to alternating sides, the consecutive octahedra in the direction of **c** have not been turned. For α-WO_3 there is only a mutual turning of the octahedra in opposite directions about **c**, $a^0b^0c^-$. β-WO_3 exhibits an additional turning of the octahedra about **b** to the same side, $a^0b^+c^-$, and γ-WO_3 also exhibits turning in opposite directions about **a** by the same amount as about **c**, $a^-b^+a^-$. *HP*-WO_3, ε-WO_3, and δ-WO_3 have mutually turned octahedra in opposite directions about **a**, **b**, and **c**. The previously described phase *HP*-WO_3 also appears at temperatures between 1070 K and 990 K [159].

The crystal data (Table 12.3) and the atomic coordinates of all atoms (Table 12.4) show that the modifications of WO_3 differ only slightly from one another.

The importance of group–subgroup relations during phase transitions is the subject of Chapter 16.

Further Bärnighausen trees of families of structures, some of them rather extensive, have been set up in the course of time. They include: Hettotypes of perovskite, containing a large variety of distortion and substitution derivatives [13, 160–163], the families of structures of rutile [142, 143, 164], of the CaF_2 type [142], of the AlB_2 type [165], of the $ThCr_2Si_2$ (or $BaAl_4$) type

Table 12.4 Atomic coordinates of the modifications of tungsten trioxide mentioned in Table 12.3.

	W			O			O		
	x	y	z	x	y	z	x	y	z
HT-WO_3	$\frac{1}{4}$	$\frac{1}{4}$	0.066	$\frac{1}{4}$	$\frac{1}{4}$	0.506	$\frac{1}{2}$	$\frac{1}{2}$	0
$x, y, \frac{1}{2}z+\frac{1}{4}$ ↓									
α-WO_3	$\frac{1}{4}$	$\frac{1}{4}$	0.283	$\frac{1}{4}$	$\frac{1}{4}$	0.503	0.525	0.475	$\frac{1}{4}$
HP-WO_3	0.256	0.268	0.288	0.255	0.173	0.512	0.558	0.454	0.301
$x, y-\frac{1}{4}, z$ ↓							−0.042	0.039	0.202
ε-WO_3*	0.255	0.030	0.287	0.252	−0.082	0.509	0.541	0.214	0.292
$\frac{1}{2}(x+y)$,	−0.256	0.490	−0.288	−0.254	0.566	0.494	0.451	0.294	−0.284
$\frac{1}{2}(-x+y), z$							−0.035	−0.206	0.216
↓							0.033	−0.279	−0.209
β-WO_3	0.252	0.029	0.283	0.220	−0.013	0.502	0.502	−0.032	0.279
							0.283	0.269	0.259
γ-WO_3	0.253	0.026	0.283	0.212	−0.002	0.500	0.498	−0.036	0.279
	0.246	0.033	0.781	0.277	0.028	0.000	0.000	0.030	0.218
							0.282	0.264	0.277
							0.214	0.257	0.742
δ-WO_3	0.257	0.026	0.285	0.210	−0.018	0.506	0.499	−0.035	0.289
$\frac{1}{2}(x+y)$,	0.244	0.031	0.782	0.288	0.041	0.004	0.001	0.034	0.211
$\frac{1}{2}(-x+y)-\frac{1}{4}, z$	0.250	0.528	0.216	6 additional O atoms			0.287	0.260	0.284
↓	0.250	0.534	0.719	at ca. $\frac{1}{2}-x, \frac{1}{2}+y, \frac{1}{2}-z$			0.212	0.258	0.729
$WO_{2.95}$	0.243	0.743	0.070	0.237	0.737	0.507	0.498	0.708	−0.024

* coordinates shifted by $x-0.245$, $y+\frac{1}{2}$, $z+0.037$ as compared to the literature [156].

[166,169] and other intermetallic compounds [167,168], of zeolites [170–172], Heusler alloys [173], antiperovskites [174], metal hydrides [175,176], hydroxide halides [177], and tetraphenylphosphonium salts [178, 179]. Many more citations can be found in the review article [179]. A comprehensive collection up to 2022 is presented by [180].

Exercises

To be solved, access to *International Tables A* and *A*1 is necessary. Solutions on page 328.

12.1. Prepare a (favourably enlarged) copy of Fig. 12.2. Insert all symmetry elements in the style of *International Tables* into the four unit cells. What symmetry elements are eliminated in each case?

12.2. The crystal data of low-temperature quartz are given in Example 9.4 (page 120); $a = 491$ pm, $c = 541$ pm. Upon heating above 573 °C it is converted to high-temperature quartz, space group $P6_222$, $a = 500$ pm, $c = 546$ pm, with the atomic coordinates Si, $\frac{1}{2}$, 0, $\frac{1}{2}$, and O, 0.416, 0.208, $\frac{2}{3}$ [181]. Derive the relation between the two structures. What kind of a relation is it? What has to be observed concerning the atomic coordinates? What additional degrees of freedom are present in the low-symmetry form?

12.3. The crystal data of α-$AlPO_4$ are [182]:

$P3_121$ $a = 494$ pm, $c = 1095$ pm

	x	*y*	*z*		*x*	*y*	*z*
Al	0.466	0	$\frac{1}{3}$	O1	0.416	0.292	0.398
P	0.467	0	$\frac{5}{6}$	O2	0.416	0.257	0.884

What is the symmetry relation with the structure of quartz? (cf. Exercise 12.2).

12.4. Most metals crystallize with one of the following packings of spheres.

cubic-closest packing of spheres (Cu type) $Fm\bar{3}m$

x	*y*	*z*
0	0	0

hexagonal-closest packing of spheres (Mg type) $P6_3/mmc$ $c/a = 1.633$

x	*y*	*z*
$\frac{1}{3}$	$\frac{2}{3}$	$\frac{1}{4}$

body-centred cubic packing of spheres (W type) $Im\bar{3}m$

x	*y*	*z*
0	0	0

How and from which of these packings are the structures of indium, α-mercury, protactinium, and α-uranium derived?

In $I4/mmm$
$a = 325.1$ $c = 494.7$

x	*y*	*z*
0	0	0

α-Hg $R\bar{3}m$
$a = 346.5$ $c = 667.7$
(hex. axes)

x	*y*	*z*
0	0	0

Pa $I4/mmm$
$a = 392.5$ $c = 324.0$

x	*y*	*z*
0	0	0

α-U $Cmcm$
$a = 285.4$ $b = 586.8$
$c = 495.8$ pm

x	*y*	*z*
0	0.398	$\frac{1}{4}$

12.5. Derive the structure of Tl_7Sb_2 [183] from one of the packings mentioned in Exercise 12.4.

$Im\bar{3}m$ $a = 1162$ pm

	x	*y*	*z*		*x*	*y*	*z*
Tl1	0	0	0	Tl3	0.350	0.350	0
Tl2	0.330	0.330	0.330	Sb	0.314	0	0

12.6. Normal VO_2 (the so-called M_1 phase) is reversibly converted at temperatures above 68 °C to the rutile structure (phase *R*). This includes a dramatic increase of its electric conductivity (transition from an insulator to a metal). There is another modification (the so-called M_2 phase [184]) if a small amount of the vanadium has been replaced by chromium. Take the data from Figs 12.7 and 12.8 to construct a Bärnighausen tree that relates these structures. Pay attention to the different settings of the space group No. 14 ($P12_1/a1$, $P12_1/n1$, $P12_1/c1$). The $CaCl_2$ type has to be considered as an intermediate group (cf. Fig. 1.2.)

Phase M_2 $A112/m$
$a = 452.6$ $b = 906.6$ $c = 579.7$ pm $\gamma = 91.9°$

	x	*y*	*z*		*x*	*y*	*z*
V1	0	0	0.281	O1	0.294	0.148	0.248
V2	0.531	0.269	$\frac{1}{2}$	O2	0.209	0.397	0
				O3	0.201	0.400	$\frac{1}{2}$

12.7. The crystal data of two modifications of $TlAlF_4$ at two different temperatures are given in the following [149]. Set up the group–subgroup relation between them. In what way do the modifications differ? Although the volumes of the cells differ by a factor of approximately 4, the index is only 2; how can that be?

$TlAlF_4$-$tP6$, 300 °C $P4/mmm$
$a = 364.9$ $c = 641.4$ pm

	x	y	z		x	y	z
Tl	$\frac{1}{2}$	$\frac{1}{2}$	$\frac{1}{2}$	F1	$\frac{1}{2}$	0	0
Al	0	0	0	F2	0	0	0.274

$TlAlF_4$-$tI24$, 200 °C $I4/mcm$
$a = 514.2$ $c = 1280.7$ pm

	x	y	z		x	y	z
Tl	0	$\frac{1}{2}$	$\frac{1}{4}$	F1	0.276	0.224	0
Al	0	0	0	F2	0	0	0.137

$TlAlF_4$-$tP6$ is known as the $TlAlF_4$ type.

12.8. Set up the symmetry relations between boehmite (γ-AlOOH) [185, 186] and the alkali metal hydroxide hydrates [187]. Do not consider the H atom positions. There are layers of edge-sharing octahedra parallel to the a–c plane. The lengths of the Al–O bonds in the (distorted) coordination octahedra are approximately 191 pm; the Rb–O and K–O distances are approximately 1.56 and 1.51 times longer. This is reflected in the a and c lattice parameters.

γ-AlOOH $Cmcm$
$a = 286.8$ $b = 1223.2$
$c = 369.5$ pm

	x	y	z
Al	0	0.179	$\frac{1}{4}$
O1	0	0.206	$\frac{3}{4}$
O2	$\frac{1}{2}$	0.083	$\frac{1}{4}$

$RbOH \cdot OH_2$ $Cmc2_1$
$a = 412.0$ $b = 1124.4$
$c = 608.0$ pm

	x	y	z
Rb	0	0.152	$\frac{1}{4}$
O1	0	0.163	0.75
O2	$\frac{1}{2}$	−0.032	0.146

$KOH \cdot OH_2$ $P112_1/a$
$a = 788.7$ $b = 583.7$
$c = 585.1$ pm $\gamma = 109.7°$

	x	y	z
K	0.075	0.298	0.254
O1	0.085	0.343	0.754
O2	0.237	−0.055	0.137

12.9. The space group $P4/ncc$ of α-WO_3 mentioned in Fig. 12.14 has additional n glide planes with the glide component $\frac{1}{2}(\mathbf{a}+\mathbf{b}) + \frac{1}{2}\mathbf{c}$ halfway between the c glide planes. It is a subgroup of $P4/nmm$ (HT-WO_3) that has glide planes with the glide component $\frac{1}{2}(\mathbf{a}+\mathbf{b})$, but the n glide planes seem to be missing. Since a symmetry reduction may only involve loss of and by no means addition of symmetry elements, there seems to be an error. Clarify the contradiction.

12.10. When the alloy $MoNi_4$ is quenched from 1200 °C, it crystallizes in the cubic Cu type with a random distribution of the atoms. If it is then annealed for hours at 840 °C, the atoms become ordered in a superstructure of the Cu type [188]. Set up the symmetry relations between the misordered and the ordered alloy.

$MoNi_4$, Cu type, $Fm\overline{3}m$
$a = 361.2$ pm

		x	y	z
Mo, Ni	$4a$	0	0	0

ordered $MoNi_4$, $I4/m$
$a = 572.0$ pm, $c = 356.4$ pm

		x	y	z
Mo	$2a$	0	0	0
Ni	$8h$	0.400	0.200	0

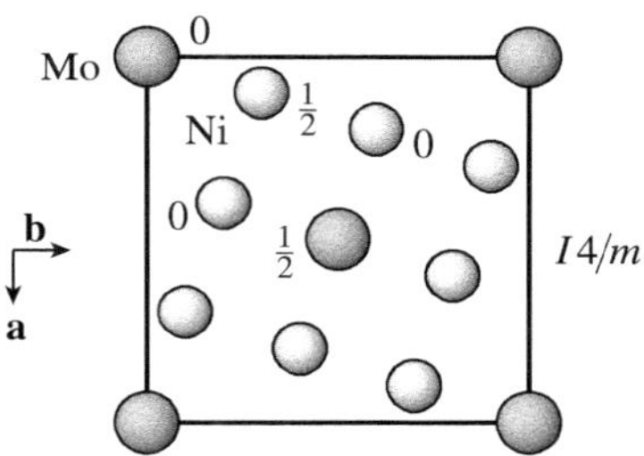

12.11. Misordered β-brass (CuZn) crystallizes in the W type: $I4/m\overline{3}2/m$, $a = 295.2$ pm, Cu and Zn statistically at position $2a$ (0, 0, 0). γ-brass (Cu_5Zn_8) is ordered: $I\overline{4}3m$, $a = 886.6$ pm, with the atomic coordinates [189]:

	x	y	z		x	y	z
Cu1	0.328	x	x	Zn1	0.608	x	x
Cu2	0.356	0	0	Zn2	0.312	x	0.037

What is the group–subgroup relation between β- and γ-brass? Comment: γ-brass has a vacant atom site as compared to β-brass.

Pitfalls when setting up group–subgroup relations

13

Unfortunately, the search for crystallographic group–subgroup relations is susceptible to pitfalls. Errors can easily be committed when noting down such relations uncritically.

Relations cannot be set up with space-group symbols alone. The lattice with given metric is an essential part of a space group. In crystal chemistry we compare crystal structures, and their characterization not only requires the space-group symbol, but also the lattice parameters and the coordinates of all atoms. Special attention must be paid to the sizes and orientations of the unit cells and the relative positions of their origins. In addition, the coordinates of all atoms of all subgroups must result straightforwardly from those of the aristotype. Moderate deviations of the lattice dimensions and of atomic locations are acceptable and often even necessary (cf. Sections 12.1 and 12.2).

Possible sources of errors are:

- Not taking into account necessary origin shifts.
- Wrong origin shifts.
- Wrong basis and/or coordinate transformations.
- Unnecessary basis transformations, for example just for the sake of clinging on to standard space-group settings.
- Errors handling non-conventional space-group settings.
- Lack of distinction between space groups and space-group types; group–subgroup relations exist only between space groups, not between space-group types. This implies that the volume of the primitive unit cell of a subgroup cannot have been decreased (apart from metric adjustments).
- Missing or wrong correspondence between the atomic positions of the group and the subgroup.
- Space-group or coordinate system settings among homeotypic structures that do not match.
- If the group–subgroup relations are correct, but origin shifts or basis transformations have not been stated, this can cause subsequent errors or misunderstandings.

In the following it is shown by means of examples how such errors can slip in.

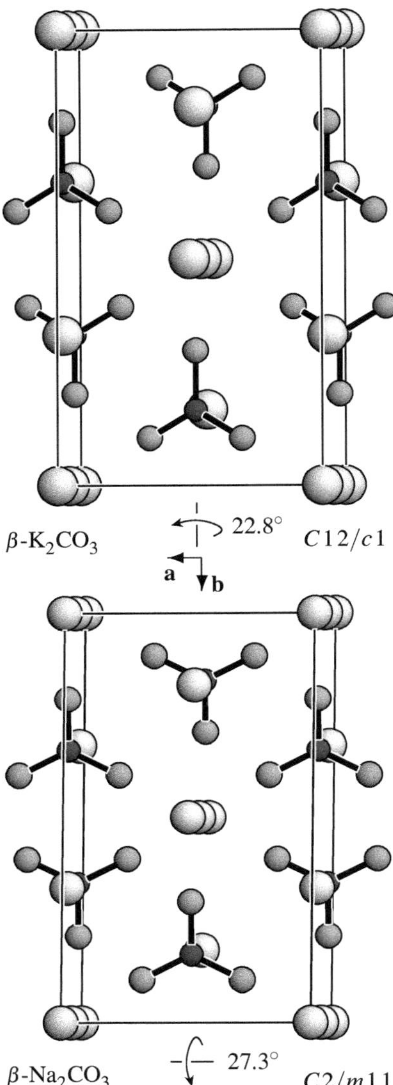

Fig. 13.1 Unit cells of β-K_2CO_3 and β-Na_2CO_3. The angles of tilt of the CO_3^{2-} ions are referred relative to a plane perpendicular to **c**.

13.1 Origin shifts

β-K_2CO_3 and β-Na_2CO_3 have similar structures and unit cells (Fig. 13.1) [190]. Along **c**, there are rows of alkali metal ions and rows in which alkali metal ions alternate with carbonate ions. The planes of the carbonate ions are not aligned exactly perpendicular to **c**. Compared to the perpendicular orientation, in the case of β-K_2CO_3, they are rotated about **b** and those of β-Na_2CO_3 are rotated about **a**. There is no group–subgroup relation between the space groups $C\,1\,2/c\,1$ and $C\,2/m\,1\,1$ of the two structures. However, in *International Tables A* 1983–2006 and *A*1 the space groups $Cmce$ and $Cmcm$ can be found as possible candidates for a common supergroup. In the listings of the supergroups the origin shifts are not mentioned in *International Tables*, neither in Volume *A* nor in Volume *A*1. Origin shifts are mentioned only in the listings of the subgroups in Volume *A*1. Therefore, the subgroups of $Cmcm$ and $Cmce$ have to be looked up in Volume *A*1. There we find that the relation $Cmce \longrightarrow C\,1\,2/c\,1$ requires an origin shift of $\frac{1}{4},\frac{1}{4},0$ (or $-\frac{1}{4},-\frac{1}{4},0$); all other relations ($Cmce \longrightarrow C\,2/m\,1\,1$, $Cmcm \longrightarrow C\,1\,2/c\,1$, $Cmcm \longrightarrow C\,2/m\,1\,1$) require no origin shifts. Since the values of the coordinates of all atoms of β-K_2CO_3 and β-Na_2CO_3 are nearly the same, there can be no origin shift. As a consequence, only $Cmcm$ and not $Cmce$ can be the common supergroup. Choosing $Cmce$ would be a gross mistake.

Another argument is in favour of the space group $Cmcm$: it points the way to a hexagonal aristotype in the space group $P6_3/mmc$ (Fig. 13.2). The images in Fig. 13.1 give the impression of a pseudohexagonal symmetry of the carbonates, and this agrees with the numerical ratio of $a\sqrt{3} \approx b$. In fact, the hexagonal aristotype is known in this case; α-K_2CO_3 ($> 420\,°C$) and α-Na_2CO_3 ($> 400\,°C$) are high-temperature modifications that crystallize in this space group. They have the CO_3^{2-} ions on threefold rotation axes and on mirror planes perpendicular to **c**.

We have chosen the non-conventional setting of $C\,2/m\,1\,1$ with monoclinic *a* axis for β-Na_2CO_3 to avoid a basis transformation (standard setting $C\,1\,2/m\,1$; *a* and *b* interchanged). This ensures a correspondence between the cells and coordinates of β-K_2CO_3 and β-Na_2CO_3 and makes their similarity obvious (Fig. 13.2).

Below 250 °C, K_2CO_3 forms another modification (γ-K_2CO_3) in the space group $P\,1\,2_1/c\,1$ that is also mentioned in Fig. 13.2. The relation $C\,1\,2/c\,1 \longrightarrow P\,1\,2_1/c\,1$ requires an origin shift by $-\frac{1}{4},-\frac{1}{4},0$, that is, $\frac{1}{4},\frac{1}{4},0$ has to be added to all coordinates of β-K_2CO_3; this is in accordance with the observed values of γ-K_2CO_3. If it were not so, the relation would be wrong.[1]

The maximal subgroups of every space group are listed in *International Tables A* 1983–2006. The corresponding origin shifts are missing in Volume *A*. They can only be found in *International Tables A*1.

However, *International Tables A*1 also has its shortcomings. In Volume *A*1 all space groups have been listed twice. In Part 2 of the volume (Maximal subgroups of the space groups), the origin shifts refer to the coordinate systems of the space groups. In Part 3 (Relations between the Wyckoff position), however, the origin shifts are given only as components of the coordinate transformations, and thus refer to the coordinate systems of the subgroup. In the

[1] The structure of sodium carbonate between 360 °C and −143 °C (γ-Na_2CO_3) has CO_3^{2-} ions, whose tilt angles vary from unit cell to unit cell following a sine wave, the periodicity of which is not commensurate with the periodicity of the lattice [195–198]. Structures of this kind are called incommensurately modulated structures. They cannot be adequately described by three-dimensional space groups, but require an extension to higher-dimensional superspace groups.

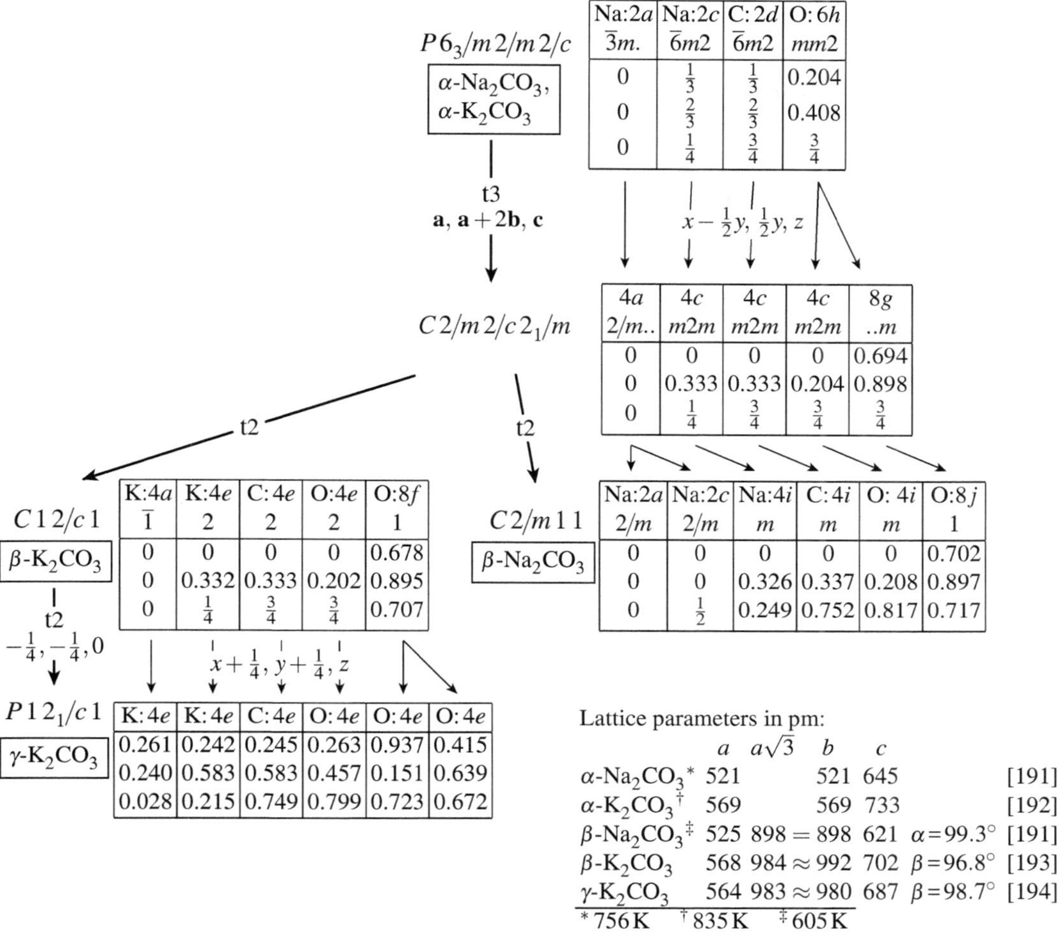

Fig. 13.2 Bärnighausen tree relating some modifications of the alkali metal carbonates K_2CO_3 and Na_2CO_3 (the O atoms of α-K_2CO_3 exhibit misorder). The position $8g$ of $C2/m2/c2_1/m$ results from the point –0.408, –0.204, $\frac{3}{4}$ which is a symmetry-equivalent point of the position $6h$ of $P6_3/m2/m2/c$.

group–subgroup arrow we state the origin shift with reference to the coordinate system of the high-symmetry space group; the corresponding transformations can be obtained from the data of Volume $A1$, Part 3, only after conversion to the coordinate system of the higher-symmetry space group. This is explained in Section 3.6.6 and Example 3.8, page 39. Given the case, this conversion *has to be* performed; one may not simply look up the corresponding transformation in Part 2 of Volume $A1$. The reason is that, as a rule, the origin shifts chosen for one and the same group–subgroup pair differ in Parts 2 and 3 of Volume $A1$ (often they have opposite directions). For example, the origin shift given in Fig. 13.2 for the relation $C12/c1 \rightarrow P12_1/c1$ corresponds to the data of space group $C12/c1$ in Volume $A1$, Part 3, whereas the origin shift mentioned in Part 2 is $\frac{1}{4}, \frac{1}{4}, 0$.

13.2 Subgroups on a par

The listings of the subgroups in *International Tables A* 1983–2006 are incomplete. Only among the *translationengleiche* subgroups and the *klassengleiche* subgroups of the kind '**IIa**' (*klassengleiche* with loss of centring) is the space group symbol of the subgroup repeated if there are several subgroups of the same type. They differ in the selection of the symmetry operations of the space group that are being retained in the subgroups; they are the ones specified by the listed generators. However, among the *klassengleiche* subgroups of the kind '**IIb**', with enlarged conventional unit cell, the individual subgroups are not listed as such, but only their space-group type is mentioned once, even if there are several subgroups of this type; in addition, the list of the retained generators is missing. The information missing in Volume *A* can be found in Volume *A*1; there, all maximal subgroups have been listed completely.

A space group may have several different subgroups of the same space-group type and with the same lattice dimensions; they can be either conjugate subgroups or subgroups on a par that belong to different conjugacy classes. See Definition 8.2 and the following text (page 111).

From the point of view of the aristotype, conjugate subgroups are symmetrically equivalent; only one of them needs to be considered, see Section 8.1, page 105, and the text after Example 8.2, page 112. However, an eye has to be kept upon the conjugacy classes (non-conjugate subgroups on a par). Subgroups on a par frequently occur among *klassengleiche* maximal subgroups of index 2. They have different positions of their origins. The examples of Fig. 8.8 (page 113) and Fig. 12.4 (page 151) show that structures whose space groups are subgroups on a par can be rather different. Therefore, it is important to choose the correct one out of several subgroups on a par. See also Fig. 15.1, page 201, where only the subgroup with the given origin shift results in the correct space group of KN_3.

The uncritical use of the program `MAXSUB` of the Bilbao Crystallographic Server can mislead the unwary user to choose only one subgroup of some type, although there are several in different conjugacy classes (cf. Section 24.2.1).

13.3 Wrong cell transformations

When searching for possible paths of symmetry reduction that will lead from a given aristotype to a given hettotype, the first thing to do is to write down all maximal subgroups of the aristotype, followed by their maximal subgroups, etc., until the hettotype is reached. It is important to keep track of all cell transformations and origin shifts for all of the involved group–subgroup relations. At the end all cell transformations and origin shifts must yield the correct lattice and origin of the hettotype.

The crystal structure of β-$IrCl_3$ can be regarded as a NaCl type in which two-thirds of the cation positions are vacant [199]. The dimensions of the unit cell correspond to:

$$a = \sqrt{2}a_{\mathrm{NaCl}} \quad b = 3\sqrt{2}b_{\mathrm{NaCl}} \quad c = 2c_{\mathrm{NaCl}} \quad (\text{with } a_{\mathrm{NaCl}} \approx 490.5 \text{ pm})$$

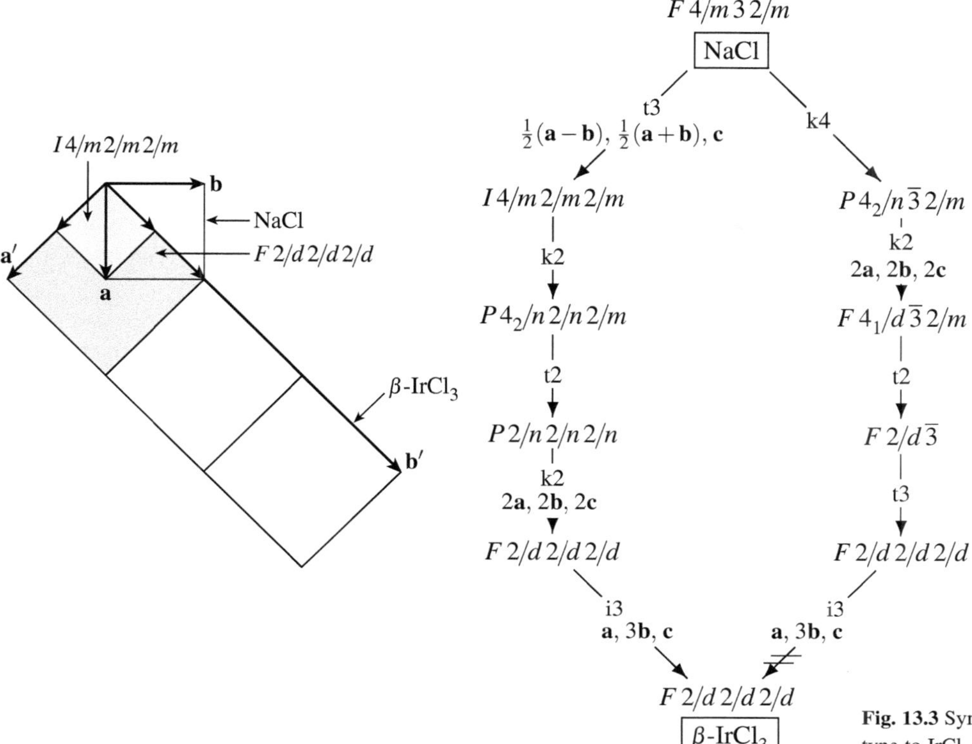

Fig. 13.3 Symmetry reduction from the NaCl type to $IrCl_3$ by removal of $\frac{2}{3}$ of the cations. The right branch is not correct. The image on the left shows the relations between the basis vectors of the space groups of the left branch (necessary origin shifts have not been marked).

The space group is $F\,d\,d\,d$. Two paths of symmetry reduction are depicted in Fig. 13.3. Of these, the right one is wrong; it does not yield the correct unit cell. That the two paths cannot yield the same result is also revealed by the indices; if all indices of a path are multiplied, the left path yields 72, the right one 144. That there must be an error could have been noticed immediately if the vertical distances between the space-group symbols in Fig. 13.3 were drawn proportional to the logarithms of the indices.

Multiplication of the transformation matrices of the subsequent basis transformations of Fig. 13.3 yields the total transformation matrix and reveals the wrong result of the right branch:[2]

$$\text{left branch:}\quad \begin{pmatrix} \frac{1}{2} & \frac{1}{2} & 0 \\ -\frac{1}{2} & \frac{1}{2} & 0 \\ 0 & 0 & 1 \end{pmatrix} \begin{pmatrix} 2 & 0 & 0 \\ 0 & 2 & 0 \\ 0 & 0 & 2 \end{pmatrix} \begin{pmatrix} 1 & 0 & 0 \\ 0 & 3 & 0 \\ 0 & 0 & 1 \end{pmatrix} = \begin{pmatrix} 1 & 3 & 0 \\ -1 & 3 & 0 \\ 0 & 0 & 2 \end{pmatrix}$$

$$\text{right branch:}\quad \begin{pmatrix} 2 & 0 & 0 \\ 0 & 2 & 0 \\ 0 & 0 & 2 \end{pmatrix} \begin{pmatrix} 1 & 0 & 0 \\ 0 & 3 & 0 \\ 0 & 0 & 1 \end{pmatrix} = \begin{pmatrix} 2 & 0 & 0 \\ 0 & 6 & 0 \\ 0 & 0 & 2 \end{pmatrix}$$

[2] Keep in mind that the matrices of consecutive basis transformations have to be multiplied in the sequence $\boldsymbol{P}_1\boldsymbol{P}_2\ldots$, whereas the inverse matrices needed for the coordinate transformations have to be multiplied in the reverse order $\ldots\boldsymbol{P}_2^{-1}\boldsymbol{P}_1^{-1}$ (or $\ldots\mathbb{P}_2^{-1}\mathbb{P}_1^{-1}$ if there are origin shifts); see Section 3.6.5, page 37.

13.4 Wrong settings of space groups

In a group–subgroup relation the setting of the subgroup has to be in accordance with the basis transformations. Gadolinium ferrate $GdFeO_3$ [200, 201] is one of the numerous variants of the perovskite type having mutually rotated coordination octahedra about the Fe atoms; these are connected by common vertices in three dimensions. The specified space group is usually *Pbnm* (non-conventional setting of *Pnma*), because then the direction of **c** can be kept as in the tetragonal perovskite variants.

The symmetry has to be reduced if the Fe atoms are to be substituted in an ordered way by two different kinds of atoms in an 1 : 1 ratio. With unchanged axes the two *translationengleiche* subgroups $P2_1/b\,1\,1$ and $P1\,2_1/n\,1$ come into question, both with a monoclinic angle of $\alpha \approx 90°$ or $\beta \approx 90°$, respectively. These are non-conventional settings of the space-group type $P1\,2_1/c\,1$. However, this may not lead to the assumption that the two subgroups are equal; they are two different subgroups, and they are not conjugate but belong to different conjugacy classes (they are not equivalent by a symmetry operation of *Pbnm*).

Of the symmetry elements of *Pbnm* ($P2_1/b2_1/n2_1/m$), for $P2_1/b\,1\,1$ the 2_1 axis along **a** is maintained as monoclinic axis, whereas for $P1\,2_1/n\,1$ it is the one along **b**. In addition, a resetting of $P1\,2_1/n\,1$ to $P1\,2_1/c\,1$ would result in a rather oblique monoclinic angle of $\beta \approx 144°$.

In both subgroups the positions of the Fe atoms are split into two independent positions. For the subgroup $P1\,2_1/n\,1$ more than 200 representatives are known as the $BaLaRuO_6$ type; the coordination octahedra about La are considerably larger than those about Ru. That is possible because in this space group every LaO_6 octahedron is linked with six RuO_6 octahedra and vice versa. In the subgroup $P2_1/b\,1\,1$ every octahedron is linked with four octahedra of the same kind and with two of the other kind, which is only possible if the octahedra have approximately equal sizes. Only a few perovskite variants are known to have the subgroup $P2_1/b\,1\,1$, and they only have equal atoms in all octahedra ($LaVO_3$ type). In such cases special caution is necessary when subgroups are retrieved with the program `MAXSUB` of the Bilbao Crystallographic Server (cf. Section 24.2.1). Further aspects concerning different settings of monoclinic space groups are dealt with in the following section.

13.5 Different paths of symmetry reduction

Frequently, it is possible to find several sequences of group–subgroup relations that will lead from an aristotype to a hettotype, that is, via different intermediate groups. If cell transformations or origin shifts are involved, then all of these sequences have to yield the same lattice, with the same cell size, cell orientation, and origin position. The sequential multiplication of all transformation matrices of each sequence must yield the same basis vectors and the same origin shift for the hettotype. If origin shifts are involved, use the 4×4 matrices, otherwise the 3×3 matrices are sufficient.

Silver oxide AgO is to be regarded as Ag(I)Ag(III)O_2. The crystal structure of its monoclinic modification is a derivative of the NaCl type, with distortions

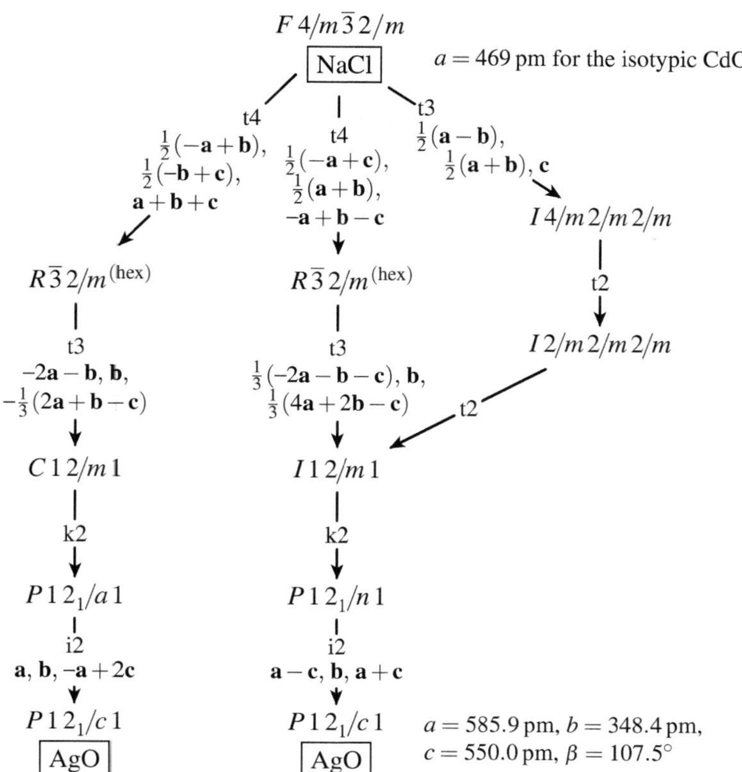

Fig. 13.4 Three sequences of symmetry reduction from the NaCl type to the monoclinic structure of AgO. The left sequence yields a differently oriented lattice of a conjugate subgroup. Note that $I\,4/m2/m2/m$ and $I\,2/m2/m2/m$ are in different lines to $R\bar{3}2/m$, reflecting the hierarchical distance from $F\,4/m\bar{3}2/m$.

to meet the coordination requirements of the silver atoms: linear for Ag(I) and square for Ag(III) [202–204]. Starting from the space group of the NaCl type ($F\,m\bar{3}\,m$), the symmetry can be reduced via a tetragonal or a rhombohedral intermediate group (Fig. 13.4 middle and right). The tetragonal c axis can be oriented along **a**, **b**, or **c** of $F\,m\bar{3}\,m$, while the c axis of the rhombohedral space group (hexagonal setting) can be along any one of the four body diagonals of the NaCl cell. In this case it is indeed cumbersome to find two sequences of group–subgroup relations that end with the same lattice of AgO. There are a total of 24 subgroups $P\,1\,2_1/c\,1$ that are conjugate in $F\,m\bar{3}\,m$ with six different orientations of the monoclinic b axis. That the central and the right branch of Fig. 13.4 result in the same subgroup can be recognized by multiplying the transformation matrices corresponding to the two paths down to the intermediate group $I\,1\,2/m\,1$, where the two paths via $R\bar{3}\,m$ and $I\,4/m\,m\,m$ meet. Multiplication of the matrices on the way via $R\bar{3}\,m$ yields:

$$\begin{pmatrix} -\frac{1}{2} & \frac{1}{2} & -1 \\ 0 & \frac{1}{2} & 1 \\ \frac{1}{2} & 0 & -1 \end{pmatrix} \begin{pmatrix} -\frac{2}{3} & 0 & \frac{4}{3} \\ -\frac{1}{3} & 1 & \frac{2}{3} \\ -\frac{1}{3} & 0 & -\frac{1}{3} \end{pmatrix} = \begin{pmatrix} \frac{1}{2} & \frac{1}{2} & 0 \\ -\frac{1}{2} & \frac{1}{2} & 0 \\ 0 & 0 & 1 \end{pmatrix}$$

This corresponds to the matrix for the only transformation on the way via $I\,4/mmm$. After further symmetry reduction down to $P\,1\,2_1/c\,1$ we obtain the transformation matrix for the conversion of the basis vectors from the NaCl type to AgO:

$$\begin{pmatrix} \frac{1}{2} & \frac{1}{2} & 0 \\ -\frac{1}{2} & \frac{1}{2} & 0 \\ 0 & 0 & 1 \end{pmatrix} \begin{pmatrix} 1 & 0 & 1 \\ 0 & 1 & 0 \\ -1 & 0 & 1 \end{pmatrix} = \begin{pmatrix} \frac{1}{2} & \frac{1}{2} & \frac{1}{2} \\ -\frac{1}{2} & \frac{1}{2} & -\frac{1}{2} \\ -1 & 0 & 1 \end{pmatrix}$$

The determinant of this matrix is 1, and the unit cell thus has the same volume as the NaCl cell. Another transformation is obtained for the left branch of Fig. 13.4:

$$\begin{pmatrix} -\frac{1}{2} & 0 & 1 \\ \frac{1}{2} & -\frac{1}{2} & 1 \\ 0 & \frac{1}{2} & 1 \end{pmatrix} \begin{pmatrix} -2 & 0 & -\frac{2}{3} \\ -1 & 1 & -\frac{1}{3} \\ 0 & 0 & -\frac{1}{3} \end{pmatrix} \begin{pmatrix} 1 & 0 & -1 \\ 0 & 1 & 0 \\ 0 & 0 & 2 \end{pmatrix} = \begin{pmatrix} 1 & 0 & -1 \\ -\frac{1}{2} & -\frac{1}{2} & -\frac{1}{2} \\ -\frac{1}{2} & \frac{1}{2} & -\frac{1}{2} \end{pmatrix}$$

The different results show the different orientations of the AgO cells. The determinant of the last matrix is also 1. By the multiplication of the first two of the preceding matrices we obtain the basis vectors of the intermediate groups $C1\,2/m\,1$ and $P1\,2_1/a\,1$:

$$\begin{pmatrix} 1 & 0 & 0 \\ -\frac{1}{2} & -\frac{1}{2} & -\frac{1}{2} \\ -\frac{1}{2} & \frac{1}{2} & -\frac{1}{2} \end{pmatrix}$$

The lengths of their basis vectors are obtained with the value a_{cub} of the cubic aristotype:

$$a_a = a(P2_1/a) = a_{\mathrm{cub}}\sqrt{1^2+(-\tfrac{1}{2})^2+(-\tfrac{1}{2})^2} = a_{\mathrm{cub}}\sqrt{\tfrac{3}{2}},$$
$$b_a = b(P2_1/a) = a_{\mathrm{cub}}\sqrt{(-\tfrac{1}{2})^2+(\tfrac{1}{2})^2} = a_{\mathrm{cub}}\sqrt{\tfrac{1}{2}},$$
$$c_a = c(P2_1/a) = a_{\mathrm{cub}}\sqrt{(-\tfrac{1}{2})^2+(-\tfrac{1}{2})^2} = a_{\mathrm{cub}}\sqrt{\tfrac{1}{2}}.$$

The angle β can be calculated with the aid of equation (2.2), page 17, and with $x_s = y_s = z_s = 0$:

$$\cos\beta(P2_1/a) = a_a^{-1}\cdot c_c^{-1}[1\cdot 0+(-\tfrac{1}{2})\cdot(-\tfrac{1}{2})+(-\tfrac{1}{2})\cdot(-\tfrac{1}{2})]a_{\mathrm{cub}}^2 = 0.5774;$$
$$\beta(P2_1/a) = 54.7^\circ.$$

The common cell for AgO results from this oblique unit cell at the cell doubling in the last step of the left branch of Fig. 13.4, $P1\,2_1/a\,1$ —i2→ $P1\,2_1/c\,1$. However, the corresponding basis transformation $\mathbf{a}, \mathbf{b}, -\mathbf{a}+2\mathbf{c}$ is not listed in *International Tables* A1 among the isomorphic subgroups of $P1\,2_1/a\,1$ of index 2. The mentioned isomorphic subgroups are $P1\,2_1/a\,1$ and $P1\,2_1/n\,1$. $P1\,2_1/n\,1$ ($\mathbf{a}_a$, $\mathbf{b}_a$, $2\mathbf{c}_a$) is identical to $P1\,2_1/c\,1$ ($\mathbf{a}_a$, $\mathbf{b}_a$, $-\mathbf{a}_a+2\mathbf{c}_a$), as shown in Fig. 13.5 (the indices *a*, *n*, and *c* designate the corresponding cell).

The example shows how complex matters can become due to the different possible settings of monoclinic space groups. In addition, care must be taken if rhombohedral space groups are involved; they often entail complicated basis transformations and errors may creep in.

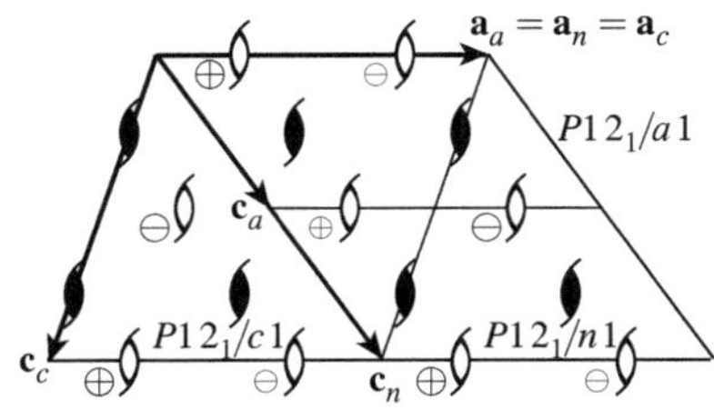

Fig. 13.5 Unit cells of the space group $P1\,2_1/a\,1$ and of its isomorphic subgroup with doubled cell in the settings $P1\,2_1/n\,1$ and $P1\,2_1/c\,1$. The hollow 2_1 axes fall away at the doubling of the $P1\,2_1/a\,1$ cell. ⊕ and ⊖ designate points that are equivalent by the glide plane, large and small points being symmetry equivalent only in the small $P1\,2_1/a\,1$ cell.

13.6 Forbidden addition of symmetry operations

A subgroup always has reduced symmetry and thus fewer symmetry operations. Erroneous adding of symmetry operations can be induced if symmetry operations present in the aristotype are first deleted in a sequence of group–subgroup relations and then reinstated at a later stage. This is not permitted, even if the hettotype itself is correct; in this case a different path must be found.

For example, starting from an aristotype that has twofold rotation axes, it is not permitted to choose a subgroup with 2_1 axes and then have the rotation axes reappear in a following subgroup. Consider the space group $Ccce$; it has two subgroups of the type $Pnna$ ($Pnna$ and $Pnnb$), and each one of these has a subgroup of type $P1\,2/n\,1$. Among the following four trees the first two are wrong:

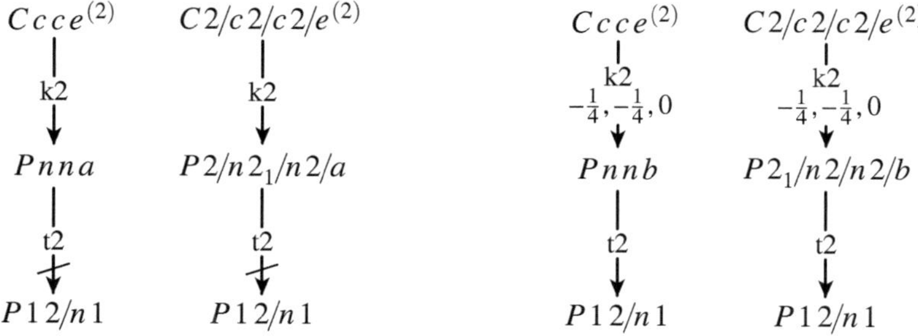

The first two trees represent the same relation, once with short, once with full Hermann–Mauguin symbols. The full symbols reveal the error: $P2/n2_1/n2/a$ does not have 2 but 2_1 axes parallel to **b**, which excludes rotation axes 2 in the subgroup $P1\,2/n\,1$. The two trees to the right side, which represent another relation twice, are correct ($Pnnb$ is a non-conventional setting of $Pnna$). The example shows why it is preferable to use full Hermann–Mauguin symbols.

Primitive subgroups of centred space groups often involve cell transformations with new basis vectors that result from linear combinations of fractions of the old basis vectors. For example, the cell transformation $\frac{1}{2}(\mathbf{a}-\mathbf{b}), \mathbf{b}, \mathbf{c}$ is required for the relation

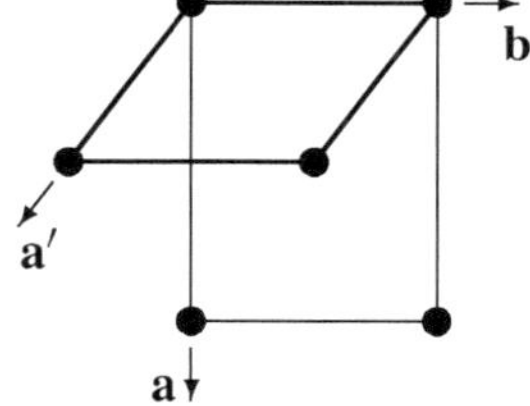

$$C2/m2/c2_1/m \;\text{—t2}\!\rightarrow\; P1\,1\,2_1/m$$

Such relations are frequent if the subgroup is monoclinic or triclinic, or if there are rhombohedral space groups. This must not lead to the belief that vector sums with fractional basis vectors are generally permitted. They are permitted only if they are compatible with corresponding centring vectors. By no means may new translations be added. The mentioned cell transformation is not permitted for the relation $P2/m2/c2_1/m$ —t2→ $P1\,1\,2_1/m$.

Exercises

Solutions on page 331.

13.1. Let two crystal structures have similar unit cells and similar atomic coordinates for all atoms. One of them crystallizes in the space group $P2_12_12_1$, the other one in $P112_1/a$. Use *International Tables A* 1983–2006 or *A*1 to find the two space groups that could be considered as common minimal supergroups. Only one of them is correct; which one is it?

13.2. What is the error among the cell transformations in the right branch of the tree in Fig. 13.3? Supplement the missing origin shifts and the missing relations between occupied positions in the left branch. Note: At the step $P2/n2/n2/n \rightarrow F2/d2/d2/d$ there are eight subgroups on a par with eight possible origin shifts.

NaCl type $Fm\bar{3}m$			
($a = 490.5$ pm)			
	x	y	z
M	0	0	0
Cl	0	0	$\frac{1}{2}$

β-$IrCl_3$ $Fddd^{(1)}$			
$a = 695$ $b = 2082$ $c = 981$ pm			
	x	y	z
Ir	$\frac{1}{4}$	0.167	$\frac{1}{4}$
Cl 1	0.220	$\frac{1}{2}$	$\frac{1}{2}$
Cl 2	0.247	0.162	0.488

Adapted from reference [199] by shifting from origin choice 2 to 1, and exchange of axes **b** and **c**.

13.3. Does the space group $Pm\bar{3}m$ have an isomorphic subgroup of index 8?

13.4. Is there any error among the following relations?

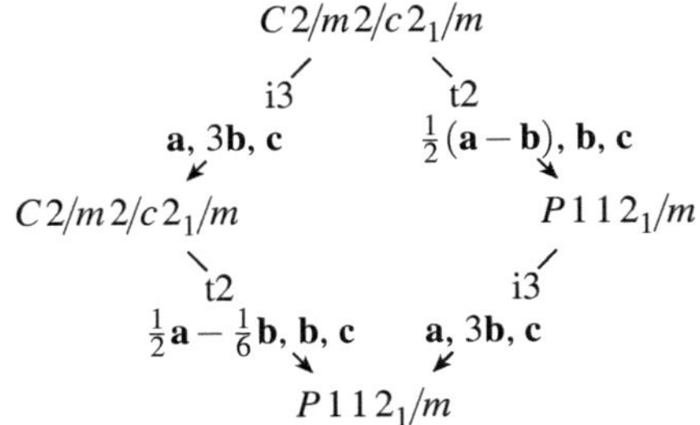

13.5. At the end of the preceding section it is claimed that the cell transformation $\frac{1}{2}(\mathbf{a}-\mathbf{b}), \mathbf{b}, \mathbf{c}$ is not compatible with the relation $P2/m2/c2_1/m$ —t2→ $P112_1/m$. Why is that so?

Derivation of crystal structures from closest packings of spheres

14.1 Occupation of interstices in closest packings of spheres

In the preceding chapters it was shown how to relate an aristotype and its substitution derivatives with the aid of group–subgroup relations. The removal of atoms from a structure to a defect structure and also the occupation of interstices in a structure can be treated in the same way. In other words, atoms are replaced by voids or interstices are filled with atoms. A known example of this kind of approach is the description of CdI_2 as a hexagonal-closest packing of iodine atoms in which one-half of the octahedral interstices are occupied by cadmium atoms.

The term 'interstitial compound' originates in this kind of view. These compounds include the transition metal hydrides and carbides. The direct occupation of voids by atoms can actually be performed in the case of the transition metal hydrides by reaction of the metals with hydrogen. This also applies to some other compounds, for example the electrode materials of lithium-ion batteries, where a reversible electrochemical intercalation of Li^+ ions takes place (e.g. $x\,Li^+ + Li_{1-x}CoO_2 + x\,e^- \rightleftharpoons LiCoO_2$). However, in most cases the intercalation of atoms into a given guest structure, keeping this structure, cannot actually be performed. The intercalation then only takes place in our imagination. In this sense, all relations that are dealt with in this chapter are purely mental or descriptive in a formal way. Nevertheless, the formal description is useful as it opens the view to many connections. Many more or less complicated crystal structures can be derived by the same procedure from well-known, simple structure types. In a formal way, the occupation of interstices is treated like a 'substitution' of real atoms for 'zero atoms'.

Essentially, the relations among substitution derivatives that are dealt with in the preceding chapters are also purely mental. Zinc blende cannot really be made from diamond by substitution of zinc and sulfur atoms for carbon atoms. However, we must be aware of the crystal-chemical and physical differences between the (mental) substitution of atoms and the filling of voids. In the case of substitution of atoms for atoms, the surroundings of the atoms and their linkage remain, but they experience drastic changes by the occupation of voids.

A large number of inorganic crystal structures can be derived from closest packings of spheres by the partial occupation of octahedral or tetrahedral interstices. In all closest packings of spheres the number of octahedral interstices is equal to the number of spheres, and the number of tetrahedral interstices is twice as large. The fraction of the voids to be occupied results from the chemical composition. In a pentahalide MX_5 whose halogen atoms form a closest packing of spheres and whose M atoms occupy octahedral voids, exactly one-fifth of the octahedral voids has to be occupied. The unit cell of the hexagonal-closest packing of spheres contains two octahedral interstices; the cell of the cubic-closest packing contains four octahedral interstices. To be able to occupy one-fifth of these interstices, first the unit cell has to be enlarged by a factor of 5 (or a multiple of 5). Enlargement of the unit cell is tantamount to removal of translational symmetry and elimination of further symmetry elements of the packing of spheres. That means that the space groups of the derivative structures have to be subgroups of the space group of the packing of spheres, and there must be at least one *klassengleiche* group–subgroup relation.

The packing of spheres itself can be chosen to be the aristotype; the initially symmetry-equivalent voids become non-equivalent positions if some of them are occupied by atoms. Just as well, the packing of spheres with completely filled voids can be chosen to be the aristotype; the derivative structures then result from partial vacation of these sites. Therefore, for example, the cubic-closest packing of spheres as well as the NaCl type can be chosen to be the aristotype.

Of course, in addition to the occupation of voids, the atomic positions of the packing of spheres themselves can be replaced by different atom species. See, for example, reference [205] to obtain an idea of how manifold these structures can be. In addition, to a certain degree, some of the spheres of the packing can be missing. An example is the perovskite structure ($CaTiO_3$), where the Ca and O atoms combined form a closest packing of spheres with Ti atoms in one-quarter of the octahedral voids. If the Ca position is vacant, this is the ReO_3 type.

14.2 Occupation of octahedral interstices in the hexagonal-closest packing of spheres

14.2.1 Rhombohedral hettotypes

The Bärnighausen tree in Fig. 14.1 shows how the structures of some compounds with rhombohedral symmetry can be derived from the hexagonal-closest packing of spheres. Figure 14.2 shows a corresponding section of the packing of spheres. A triplication of the (primitive) unit cell is necessary; the enlarged cell can be described by a primitive rhombohedral unit cell, but it is more common to choose the triply primitive hexagonal setting, the cell of which has a triplicated base in the *a–b* plane and a triple lattice constant *c* (this cell is larger by a factor of 9, but due to the centring at $\pm(\frac{2}{3},\frac{1}{3},\frac{1}{3})$ its primitive cell is only three times larger). In Fig. 14.1 the space groups are numbered from $\mathcal{G}_1$ to $\mathcal{G}_9$, which is referred to in Chapter 19.

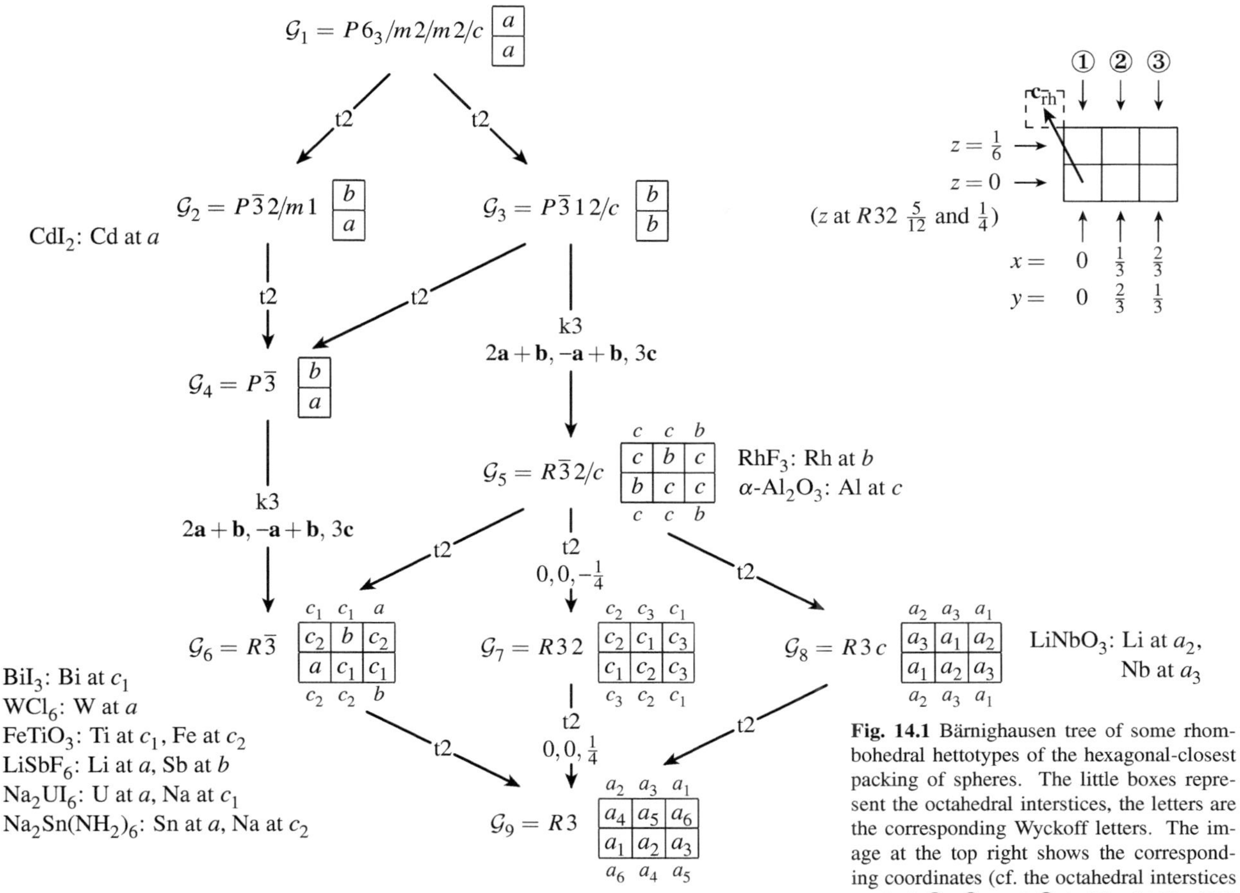

$\mathcal{G}_1 = P6_3/m2/m2/c$ [a / a]

t2 ↙ ↘ t2

$\mathcal{G}_2 = P\bar{3}2/m1$ [b / a]

CdI_2: Cd at a

$\mathcal{G}_3 = P\bar{3}12/c$ [b / b]

$\mathcal{G}_2$ → t2 → $\mathcal{G}_4$; $\mathcal{G}_3$ → t2 → $\mathcal{G}_4$

$\mathcal{G}_3$ → k3; **2a** + **b**, −**a** + **b**, 3**c** → $\mathcal{G}_5$

$\mathcal{G}_4 = P\bar{3}$ [b / a]

$\mathcal{G}_4$ → k3; 2**a** + **b**, −**a** + **b**, 3**c** → $\mathcal{G}_6$

$\mathcal{G}_5 = R\bar{3}2/c$

c	c	b
c	b	c
b	c	c
c	c	b

RhF_3: Rh at b
α-Al_2O_3: Al at c

$\mathcal{G}_5$ → t2 → $\mathcal{G}_6$; $\mathcal{G}_5$ → t2; $0, 0, -\frac{1}{4}$ → $\mathcal{G}_7$; $\mathcal{G}_5$ → t2 → $\mathcal{G}_8$

$\mathcal{G}_6 = R\bar{3}$

c_1	c_1	a
c_2	b	c_2
a	c_1	c_1
c_2	c_2	b

BiI_3: Bi at c_1
WCl_6: W at a
$FeTiO_3$: Ti at c_1, Fe at c_2
$LiSbF_6$: Li at a, Sb at b
Na_2UI_6: U at a, Na at c_1
$Na_2Sn(NH_2)_6$: Sn at a, Na at c_2

$\mathcal{G}_7 = R32$

c_2	c_3	c_1
c_2	c_1	c_3
c_1	c_2	c_3
c_3	c_2	c_1

$\mathcal{G}_8 = R3c$

a_2	a_3	a_1
a_3	a_1	a_2
a_1	a_2	a_3
a_2	a_3	a_1

$LiNbO_3$: Li at a_2, Nb at a_3

$\mathcal{G}_6$ → t2 → $\mathcal{G}_9$; $\mathcal{G}_7$ → t2; $0, 0, \frac{1}{4}$ → $\mathcal{G}_9$; $\mathcal{G}_8$ → t2 → $\mathcal{G}_9$

$\mathcal{G}_9 = R3$

a_2	a_3	a_1
a_4	a_5	a_6
a_1	a_2	a_3
a_6	a_4	a_5

①②③; $\mathbf{c}_{rh}$

$z = \frac{1}{6}$ →
$z = 0$ →
(z at $R32$ $\frac{5}{12}$ and $\frac{1}{4}$)

$x = 0 \quad \frac{1}{3} \quad \frac{2}{3}$
$y = 0 \quad \frac{2}{3} \quad \frac{1}{3}$

Fig. 14.1 Bärnighausen tree of some rhombohedral hettotypes of the hexagonal-closest packing of spheres. The little boxes represent the octahedral interstices, the letters are the corresponding Wyckoff letters. The image at the top right shows the corresponding coordinates (cf. the octahedral interstices marked ①, ②, and ③ in Fig. 14.2). Different point orbits of the same Wyckoff position are distinguished by subscript numbers (a_1, a_2, etc.). Wyckoff letters above and below the six little boxes mark neighbouring octahedral interstices (due to the rhombohedral symmetry the groupings of six octahedral voids are shifted sideways above each other, corresponding to the rhombohedral basis vector $\mathbf{c}_{rh}$ in the image at the top right) [206].

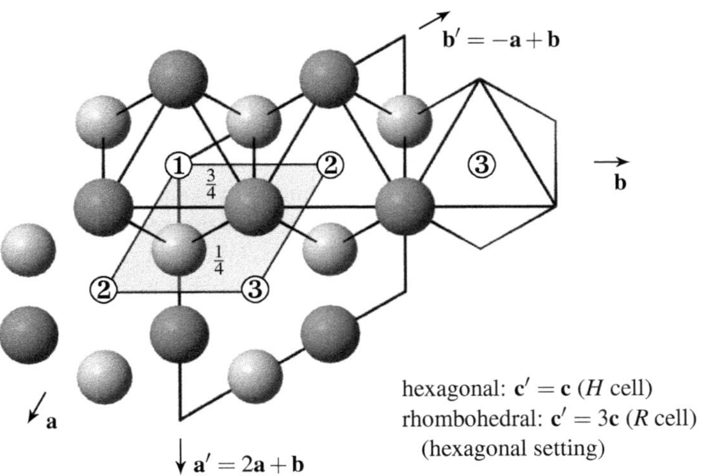

Fig. 14.2 Section from the hexagonal-closest packing. Cell with grey background: unit cell, space group $P6_3/m2/m2/c$. Large cell: triple cell with $\mathbf{c}' = \mathbf{c}$ for hexagonal and $\mathbf{c}' = 3\mathbf{c}$ for rhombohedral subgroups. The specified z coordinates of the spheres refer to $\mathbf{c}' = \mathbf{c}$. Six kinds of octahedral voids are marked by ①, ②, and ③; they are at $z = 0$ and $z = \frac{1}{2}$ (for $\mathbf{c}' = \mathbf{c}$) and $z = 0$ and $z = \frac{1}{6}$ (for $\mathbf{c}' = 3\mathbf{c}$).

The unit cell of the aristotype contains two spheres at the Wyckoff position $2d$, $\pm(\frac{2}{3}, \frac{1}{3}, \frac{1}{4})$ and two octahedral voids at $2a$ with the coordinates 0, 0, 0 and 0, 0, $\frac{1}{2}$. In the maximal *translationengleiche* subgroup $P\overline{3}2/m1$ the octahedral voids are no longer symmetry equivalent; if one of them is occupied and the other one remains empty, the result is the CdI_2 type. This step of symmetry reduction does not yet require an enlargement of the cell.

After triplication of the cell it contains six octahedral voids. They are represented in Fig. 14.1 by six little boxes and are marked by the corresponding Wyckoff letters; symmetry-equivalent octahedral interstices have the same label. If the little boxes are side by side, the corresponding octahedra have an edge in common; if they are one on top of the other, the octahedra have a face in common; if they are staggered with a common vertex, the octahedra share a vertex.

With the successive symmetry reduction, the voids successively become more and more symmetry independent, as can be seen by the increasing number of different Wyckoff letters. Finally, in the space group $R3$ they have all become independent. The atoms of the packing of spheres remain symmetry equivalent in all mentioned space groups, $R32$ and $R3$ excepted, which have two independent positions.

For compounds of the composition AX_3 (e.g. trihalides; X atoms form the packing of spheres) one-third of the six octahedral voids have to be occupied by A atoms (i.e. two of them are occupied and four remain vacant). We express this by the formulae $A_2\square_4X_6$ or $A\square_2X_3$. Two known structure types have this arrangement:

> BiI_3 in $R\overline{3}$. This is a layered structure whose occupied octahedra share common edges (position c_1 of $R\overline{3}$ in Fig. 14.1).
>
> RhF_3 in $R\overline{3}2/c$. All occupied octahedra share common vertices (position b of $R\overline{3}2/c$ in Fig. 14.1).

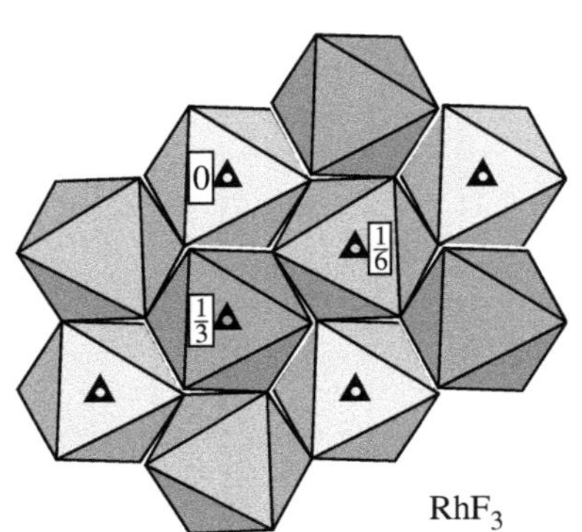

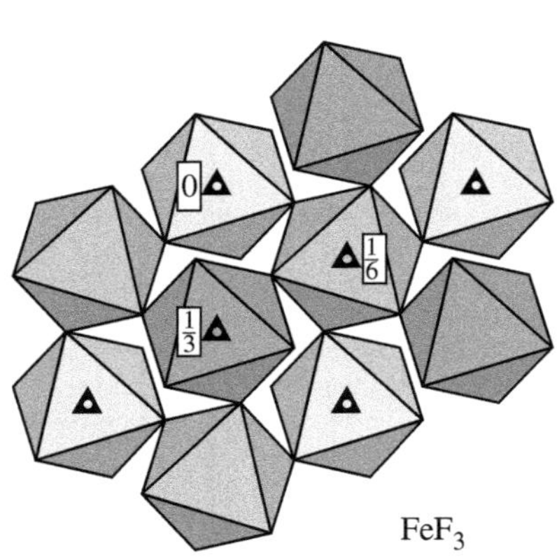

Fig. 14.3 Mutual rotation of occupied octahedra in the RhF_3 type about the threefold axes of the space group $R\overline{3}2/c$. The vertices of the cells are at the threefold axes running through the octahedra drawn in light grey at $z = 0$. The numbers are the z coordinates of the metal atoms. Only one half of the unit cell is shown along the direction of view (**c**).

If occupied and unoccupied octahedra of RhF_3 are interchanged, the space group remains $R\overline{3}2/c$. Now, four voids are occupied (position c of $R\overline{3}2/c$) and two are vacant, and the composition is $\square A_2X_3$. That corresponds to the structure of α-Al_2O_3 (corundum). Regarded this way, RhF_3 and α-Al_2O_3 are formally equivalent structures. From the point of view of crystal chemistry they are not: the occupied octahedra of α-Al_2O_3 share common faces and edges, whereas those of RhF_3 share only vertices. The crystal chemical difference also becomes apparent by comparison with the structures of further trifluorides. The symmetry of space group $R\overline{3}2/c$ permits a distortion of the packing of spheres in that the octahedra can be mutually rotated about the threefold axes (Fig. 14.3) [207]. The common vertices of the occupied octahedra in a trifluoride act as 'hinges', changing the A–F–A bond angles (Table 14.1). In an undistorted packing of spheres this angle amounts to 131.8°. In contrast, mutual rotation of octahedra of Al_2O_3 is hardly possible due to the rigid connection of the edge-sharing octahedra.

If the octahedra in a trifluoride are rotated until the A–F–A bond angle becomes 180°, the result is the ReO_3 type. It has a higher symmetry, namely cubic in the space group $Pm\overline{3}m$. That is a supergroup of $R\overline{3}2/c$. Therefore, the structures of the trifluorides can also be regarded as hettotypes of the ReO_3

Table 14.1 Some data concerning the structures of trifluorides AF_3 and related compounds.

Compound	Angle of rotation of the octahedra	Bond angle A—X—A	References
ReO_3	0°	180°	[208]
TiF_3	11.6°	170.6°	[211]
AlF_3	14.1°	157.0°	[209]
FeF_3	17.0°	152.1°	[210–212]
VF_3	19.1°	149.1°	[213]
CoF_3	19.3°	148.8°	[214]
GaF_3	20.4°	146.8°	[215]
InF_3	20.6°	146.8°	[216]
CrF_3	21.8°	144.8°	[217]
AlD_3	24.2°	140.7°	[218]
MoF_3	24.4°	141.0°	[219]
TeO_3	25.9°	137.9°	[220]
RhF_3	27.7°	135.4°	[221]
$TmCl_3$-III	29.3°	132.8°	[222]
IrF_3	30.3°	131.6°	[214]
$CaCO_3$	40.1°	(118.2°)	[223]

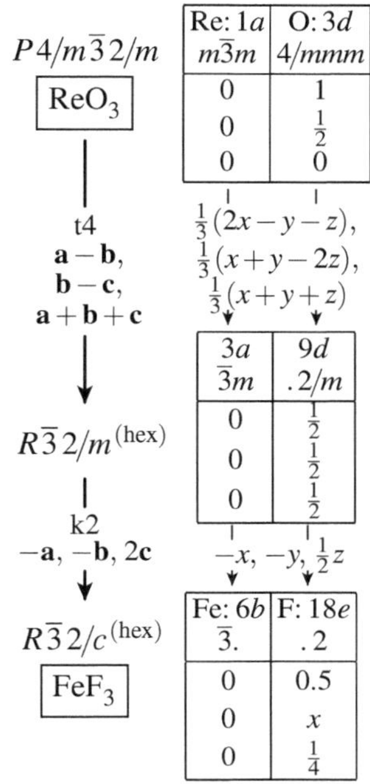

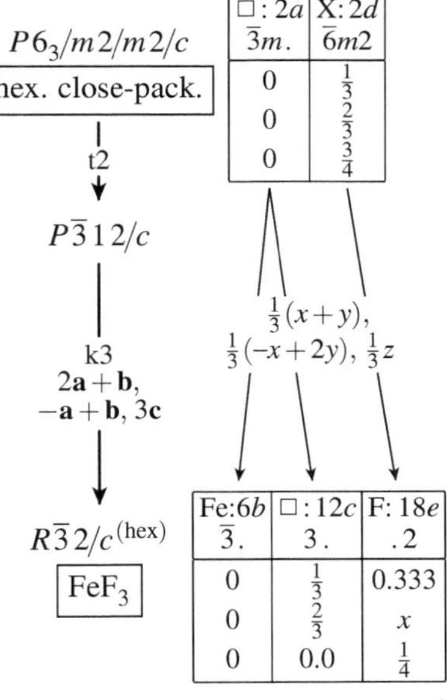

Fig. 14.4 Derivation of the FeF_3 structure either from the ReO_3 type or from the hexagonal-closest packing of spheres. The coordinates for FeF_3 given in the boxes are ideal values calculated from the aristotypes assuming no distortions. A y coordinate given as x means $y = x$. The Schottky symbol □ designates an unoccupied octahedral void.

type, from which they differ the more the A–F–A bond angles deviate from 180°. ReO_3 contains large cavities that have a cuboctahedral surrounding of 12 oxygen atoms. Upon rotation of the octahedra, the surrounding atoms move into the cavity and it becomes smaller. When the hexagonal-closest packing has been reached, this cavity becomes an octahedral interstice. The mutual rotation of the octahedra can be continued to a 'superdense' sphere packing. This corresponds to calcite ($CaCO_3$); calcite contains C atoms in the middles of certain octahedron faces; the C atom pulls up three O atoms each in a carbonate ion and squeezes them.

The actual rotation of the octahedra of FeF_3 (VF_3 type) is halfway between the two extremes. The group–subgroup relations are shown twice in Fig. 14.4, once with the ReO_3 type as the aristotype, and once with the hexagonal-closest packing of spheres as the aristotype. The space group $R\bar{3}2/c$ of FeF_3 is deliberately not given as a common subgroup of the space groups of the two aristotypes $P4/m\bar{3}2/m$ and $P6_3/m2/m2/c$, but two separate trees have been drawn; the two different points of view should not be mixed up. The tree with $P6_3/m2/m2/c$ as the aristotype corresponds to the non-executable, mental derivation of the FeF_3 structure from the packing of spheres by occupation of octahedral voids. On the other hand, the tree with the ReO_3 type as the aristotype involves no change of the chemical composition, but a distortion of the structure. This distortion can actually be performed. When FeF_3 is subjected to high pressures, the coordination octahedra experience a mutual rotation as described; at 9 GPa it is a nearly undistorted hexagonal-closest packing of fluorine atoms with the $c/(a\sqrt{3})$ ratio and the x coordinate of the F atom close to

Table 14.2 Observed lattice parameters, x coordinates of the F atoms, and angles of rotation of the coordination octahedra (0° = ReO_3 type, 30° = hexagonal closest packing) for FeF_3 at different pressures [211, 212].

p/GPa	a/pm	c/pm	$c/(a\sqrt{3})$	x	Angle of rotation/°
10^{-4}	520.5	1332.1	1.48	0.412	17.0
1.54	503.6	1340.7	1.54	0.385	21.7
4.01	484.7	1348.3	1.61	0.357	26.4
6.42	476	1348	1.64	0.345	28.2
9.0	469.5	1349	1.66	0.335	29.8

the ideal values of 1.633 and 0.333 (Table 14.2). The same behaviour is known for TiF_3 and CrF_3 under pressure [211, 217].

In α-Al_2O_3 the Al positions are symmetry equivalent. By reduction of the symmetry from $R\overline{3}2/c$ to $R\overline{3}$ the corresponding position c splits into two independent positions that can be occupied by atoms of two different elements (c_1 and c_2 of $R\overline{3}$ in Fig. 14.1). Ilmenite, $FeTiO_3$, has this structure. It has edge-sharing octahedra (little boxes side by side) occupied by atoms of the same element. Occupation of the edge-sharing octahedra with two different elements is possible in the space group $R3c$ (positions a_2 and a_3 of $R3c$ in Fig. 14.1); that is the structure of $LiNbO_3$. In addition, ilmenite as well as $LiNbO_3$ have pairs of face-sharing octahedra that are occupied by two different atoms.

Structures of further compounds can be included in the tree in Fig. 14.1. In WCl_6 the tungsten atoms occupy one-sixth of the octahedral voids in a hexagonal-closest packing of chlorine atoms. The space group $R\overline{3}$ is the only appropriate one in this tree; the Wyckoff position a is occupied, the remaining ones remain vacant.

The mentioned structure types and some further examples are listed in Table 14.3. Table 14.4 contains the corresponding crystal data.

Fig. 14.5 The kind of linkage of the octahedra in the hypothetical structure of WCl_3, space group $R32$ [206].

In addition to the RhF_3 and BiI_3 types, Fig. 14.1 shows a third possible structure type for the composition AX_3 in the space group $R32$ if the point orbit c_3 is occupied and c_1 and c_2 remain vacant. In this case there are pairs of face-sharing octahedra, the pairs being connected by sharing vertices, as shown in Fig. 14.5. As yet, no representative is known for this structure type. A compound that could adopt this structure is WCl_3. Trivalent tungsten is known to have a tendency towards structures with face-sharing octahedra, for example in the $W_2Cl_9^{3-}$ ion, because the closeness of the W atoms enables a W≡W bond. Similar pairs of octahedra are also known for $ReCl_4$. Maybe it is worth trying to look for a corresponding modification of WCl_3. (Another modification that has W_6Cl_{18} clusters is known, see Exercise 14.2.)

The space group $R32$ also offers a possibility for a variant of ilmenite, which could occur with $AlTiO_3$, a compound not studied yet. With Al at c_2 and Ti at c_3 of $R32$ (Fig. 14.1) the linkage of the octahedra would be like in ilmenite. However, the Ti atoms would not be placed in edge-sharing octahedra, but in

Table 14.3 Known structure types with space groups according to Fig. 14.1.

Space group	Structure type	Formula type*	Point orbits of the occupied octahedral voids	Number of known representatives
$P6_3/mmc$	hex.-closest packing	□X	–	> 35
	NiAs	AX	a: Ni	> 70
$P\bar{3}m1$	CdI_2	A□X_2	a: Cd	> 75
$R\bar{3}c$	RhF_3 (VF_3)	A□$_2X_3$	b: Rh	c. 20
	α-Al_2O_3	□B_2X_3	c: Al	c. 15
$R\bar{3}$	BiI_3	A□$_2X_3$	c_1: Bi	11
	$FeTiO_3$ (ilmenite)	AB□X_3	c_1: Fe c_2: Ti	c. 25
	α-WCl_6	A□$_5X_6$	a: W	3
	$LiSbF_6$	AB□$_4X_6$	a : Li b: Sb	> 50
	Na_2UI_6	AB_2□$_3X_6$	a : U c_1: Na	5
	$Na_2Sn(NH_2)_6$	AB_2□$_3X_6$	a : Sn c_2: Na	2
	$NiTi_3S_6$	A□$_2C_3X_6$	a : Ti b: Ni c_1: Ti	4
$R3c$	$LiNbO_3$	AB□X_3	a_2: Li a_3: Nb	> 10
$R3$	Ni_3TeO_6	A□$_2C_3X_6$	a_1: Te a_2: Ni a_4: Ni a_6: Ni	1
	Li_2TeZrO_6	ABC_2□$_2X_6$	a_1: Zr a_2: Li a_3: Li a_6: Te	6

* A, B, C = atoms in octahedral voids; □ = unoccupied octahedral void

Table 14.4 Crystal data of some of the name-giving rhombohedral structure types of Table 14.3.

Space group	Compound	$a^{(hex)}$ /pm	$c^{(hex)}$ /pm	Wyckoff position	Element	x	y	z	Ideal coordinates x	y	z	References
$R\bar{3}c$	RhF_3	487.3	1355.0	6b	Rh	0	0	0	0	0	0	[221]
				18e	F	0.652	0	$\frac{1}{4}$	0.667	0	$\frac{1}{4}$	
$R\bar{3}c$	α-Al_2O_3	476.0	1300.0	12c	Al	$\frac{2}{3}$	$\frac{1}{3}$	–0.019	$\frac{2}{3}$	$\frac{1}{3}$	0.0	**
				18e	O	0.694	0	$\frac{1}{4}$	0.667	0	$\frac{1}{4}$	
$R\bar{3}$	BiI_3	752.5	2070.3	6c	Bi	$\frac{2}{3}$	$\frac{1}{3}$	0.002	$\frac{2}{3}$	$\frac{1}{3}$	0.0	[224]
				18f	I	0.669	0.000	0.246	0.667	0.0	0.25	
$R\bar{3}$	ilmenite	508.8	1408.5	6c	Ti	$\frac{2}{3}$	$\frac{1}{3}$	–0.020	$\frac{2}{3}$	$\frac{1}{3}$	0.0	[225]
				6c	Fe	0	0	0.145	0	0	0.167	
				18f	O	0.683	–0.023	0.255	0.667	0.0	0.25	
$R\bar{3}$	α-WCl_6	608.8	1668.0	3a	W	0	0	0	0	0	0	[226]
				18f	Cl	0.667	0.038	0.253	0.667	0.0	0.25	
$R\bar{3}$	$LiSbF_6$	518	1360	3a	Li	0	0	0	0	0	0	[227]
				3b	Sb	$\frac{1}{3}$	$\frac{2}{3}$	$\frac{1}{6}$	$\frac{1}{3}$	$\frac{2}{3}$	$\frac{1}{6}$	
				18f	F	0.598	–0.014	0.246	0.667	0.0	0.25	
$R3c$	$LiNbO_3$	514.8	1386.3	6a	Li	$\frac{2}{3}$	$\frac{1}{3}$	–0.053	$\frac{2}{3}$	$\frac{1}{3}$	0.0	[228],
				6a	Nb	$\frac{1}{3}$	$\frac{2}{3}$	0.0	$\frac{1}{3}$	$\frac{2}{3}$	0.0	[229]
				18b	O	0.704	0.048	0.271	0.667	0	0.25	

** Approximately 40 determinations.

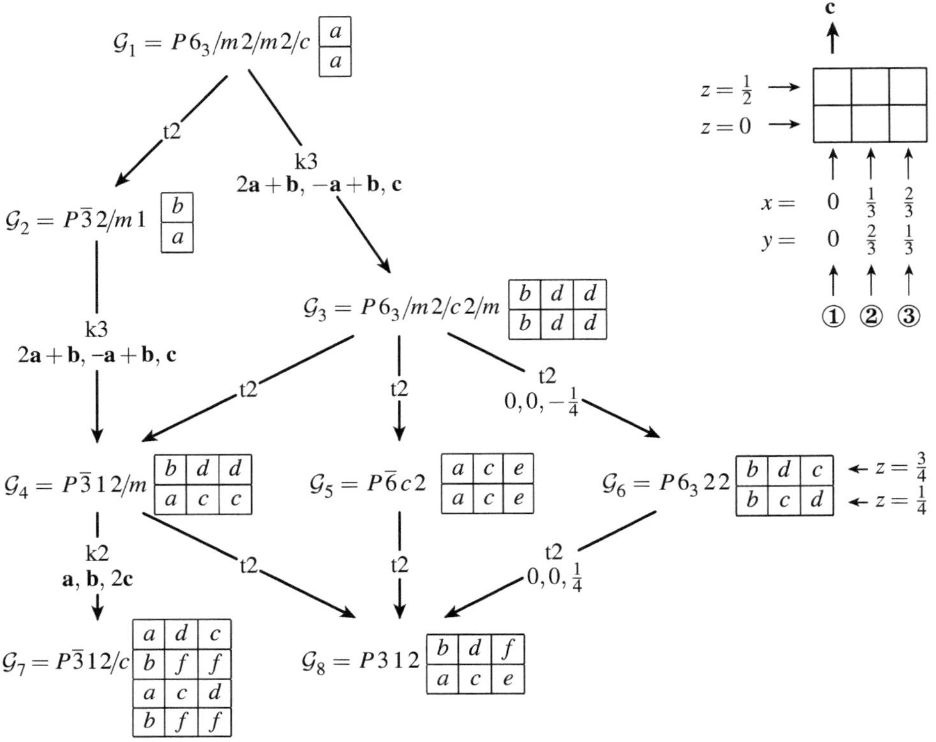

Fig. 14.6 Bärnighausen tree of some hexagonal and trigonal hettotypes of the hexagonal-closest packing of spheres. The little boxes represent the octahedral interstices, the letters are the corresponding Wyckoff letters. The image at the top right shows the corresponding coordinates (cf. Fig. 14.2).

pairs of face-sharing octahedra. Titanium would be trivalent in $AlTiO_3$ and would have an extra valence electron per atom; this could favour the occupation of pairs of face-sharing octahedra by titanium atoms, forming Ti–Ti bonds.

The examples of the postulated possible structures of WCl_3 and $AlTiO_3$ show how conceivable new structure types can be predicted with the aid of Bärnighausen trees. The systematic prediction of structure types is the subject of Chapter 19.

14.2.2 Hexagonal and trigonal hettotypes of the hexagonal-closest packing of spheres

Consider two more possibilities to enlarge the unit cell of the hexagonal-closest packing of spheres. A triplication is possible if the *a*–*b* base plane is increased as for the rhombohedral hettotypes, but the hexagonal basis vector **c** is retained (*H* cell, Fig. 14.2). The resulting Bärnighausen tree is shown in Fig. 14.6. In addition, it contains the group $\mathcal{G}_7$, which is a subgroup with doubled *c* axis ($P\bar{6}c2$ and $P6_322$ have no maximal subgroups with doubled *c* axis). Corresponding structures are mentioned in Tables 14.5 and 14.6.

The compound $AgInP_2S_6$ is mentioned in Table 14.5 (space group $\mathcal{G}_7 = P\bar{3}12/c$). Actually, the octahedron centre at the position *c* of $\mathcal{G}_7$ ($\frac{1}{3}, \frac{2}{3}, \frac{1}{4}$) is not occupied in this case; instead there is a pair of P atoms (coordinates $\frac{1}{3}, \frac{2}{3}$, 0.164

Table 14.5 Known structure types with space groups according to the tree of Fig. 14.6.

Space group	Structure type	Formula type*	Wyckoff positions of occupied octahedral voids	Number of known representatives
$P6_3/mmc$	hex.-closest pack.	□X	–	> 35
	NiAs	AX	$2a$: Ni	> 70
$P\bar{3}m1$	CdI_2	A□X_2	$1a$: Cd	> 75**
$P6_3/mcm$	TiI_3-hex.	A$□_2X_3$	$2b$: Ti	8†
$P\bar{3}1m$	ε-Fe_2N	A□X_2	$1a$: N $2d$: N	c. 7
	OAg_3	A$□_2X_3$	$2c$: O	1
	Li_2ZrF_6	$AB_2□_3X_6$	$1a$: Zr $2d$: Li	c. 17
	$Li_2Pt(OH)_6$	$AB_2□_3X_6$	$1a$: Pt $2c$: Li	3
	Hg_3NbF_6	A$□_2B_3X_6$	$1a$: Nb $1b$: Hg $2d$: Hg	1
$P\bar{6}c2$	$LiScI_3$	AB□X_3	$2a$: Li $2c$: Sc	1
$P6_322$	Ni_3N	A$□_2X_3$	$2c$: N	6
$P\bar{3}1c$	$FeZrCl_6$	AB$□_4X_6$	$2a$: Fe $2c$: Zr	4
	$CaPt(OH)_6$	AB$□_4X_6$	$2b$: Pt $2c$: Ca	2
	Li_2UI_6	$AB_2□_3X_6$	$2a$: Li $2c$: U $2d$: Li	2
	Na_3CrCl_6	A$□_2C_3X_6$	$2a$: Na $2c$: Cr $4f$: Na	2
	$LiCaAlF_6$	ABC$□_3X_6$	$2b$: Ca $2c$: Al $2d$: Li	9
	Cr_2Te_3	□B_2X_3	$2b$: Cr $2c$: Cr $4f$: Cr	3
	Cr_5S_6	□B_5X_6	$2a$: Cr $2b$: Cr $2c$: Cr $4f$: Cr	1
	$AgInP_2S_6$	AB(C_2)$□_3X_6$	$2a$: Ag $2d$: In $2c$: 2P‡	4
$P312$	$KNiIO_6$	ABC$□_3X_6$	$1a$: K $1d$: Ni $1f$: I	4

* A, B, C atoms in octahedral voids; □ vacant octahedral voids.
** Including hydroxides.
† 14 if distorted orthorhombic variants are included, see text.
‡ See text.

and $\frac{1}{3}, \frac{2}{3}, 0.336$). The octahedral void, whose middle is the position c, therefore is occupied by a P_2 dumbbell; it forms a $P_2S_6^{4-}$ ion together with the six S atoms at the octahedron vertices.

The Bärnighausen trees of Figs 14.1 and 14.6 do not cover structural variants that result from distortions with an additional symmetry reduction. Consider the space group $\mathcal{G}_3 = P6_3/m2/c2/m$ in Fig. 14.6; if the Wyckoff position b ($0,0,0$ and $0,0,\frac{1}{2}$) is occupied and d is left vacant, the result is a structure type that consists of strands of face-sharing octahedra parallel to **c**. This structure type (hexagonal TiI_3 type) has been observed among high-temperature modifications of a few trihalides. It occurs only with odd numbers of d electrons (d^1, d^3, d^5). However, most of these trihalides are not hexagonal, but orthorhombic at room temperature ($RuBr_3$ type; space group $Pmnm$). Atoms in the centres of face-sharing octahedra are rather close to each other, so that this arrangement is unfavourable for electrostatic reasons.[1] However, for atoms with odd d electron numbers there can be a preference for face-sharing because this

[1] This is in accordance with Pauling's third rule for polar compounds: edge-sharing and especially face-sharing of coordination polyhedra is unfavourable [11].

Table 14.6 Crystal data of some of the structures according to Table 14.5.

Space group	Compound	a /pm	c /pm	Wyckoff position	Ele-ment	x	y	z	Ideal coordinates x	y	z	Refe-rences
$P6_3/mcm$	TiI_3 hex.	714.2	651.0	2b	Ti	0	0	0	0	0	0	[231]
				6g	I	0.683	0	$\frac{1}{4}$	0.667	0	$\frac{1}{4}$	
$P\bar{3}1m$	Ag_3O	531.8	495.1	2c	O	$\frac{1}{3}$	$\frac{2}{3}$	0	$\frac{1}{3}$	$\frac{2}{3}$	0	[232]
				6k	Ag	0.699	0	0.277	0.667	0	0.25	
	Li_2ZrF_6	497.3	465.8	1a	Zr	0	0	0	0	0	0	[233]
				2d	Li	$\frac{1}{3}$	$\frac{2}{3}$	$\frac{1}{2}$	$\frac{1}{3}$	$\frac{2}{3}$	$\frac{1}{2}$	
				6k	F	0.672	0	0.255	0.667	0	0.25	
$P\bar{6}c2$	$LiScI_3$	728.6	676.8	2a	Sc	0	0	0	0	0	0	[234]
				2c	Li	$\frac{1}{3}$	$\frac{2}{3}$	0	$\frac{1}{3}$	$\frac{2}{3}$	0	
				6k	I	0.673	−0.004	$\frac{1}{4}$	0.667	0	$\frac{1}{4}$	
$P6_322$	Ni_3N	462.2	430.6	2c	N	$\frac{1}{3}$	$\frac{2}{3}$	$\frac{1}{4}$	$\frac{1}{3}$	$\frac{2}{3}$	$\frac{1}{4}$	[235]
				6g	Ni	0.328	0	0	0.333	0	0	
$P\bar{3}1c$	$FeZrCl_6$	628.4	1178.8	2a	Fe	0	0	$\frac{1}{4}$	0	0	$\frac{1}{4}$	[236]
				2c	Zr	$\frac{1}{3}$	$\frac{2}{3}$	$\frac{1}{4}$	$\frac{1}{3}$	$\frac{2}{3}$	$\frac{1}{4}$	
				12i	Cl	0.667	−0.023	0.131	0.667	0.0	0.125	
	$LiCaAlF_6$	500.8	964,3	2b	Ca	0	0	0	0	0	0	[237]
				2c	Al	$\frac{1}{3}$	$\frac{2}{3}$	$\frac{1}{4}$	$\frac{1}{3}$	$\frac{2}{3}$	$\frac{1}{4}$	
				2d	Li	$\frac{2}{3}$	$\frac{1}{3}$	$\frac{1}{4}$	$\frac{2}{3}$	$\frac{1}{3}$	$\frac{1}{4}$	
				12i	F	0.623	−0.031	0.143	0.667	0.0	0.125	
	$AgInP_2S_6$	618.2	1295.7	2a	In	0	0	$\frac{1}{4}$	0	0	$\frac{1}{4}$	[238]
				2d	Ag	$\frac{2}{3}$	$\frac{1}{3}$	$\frac{1}{4}$	$\frac{2}{3}$	$\frac{1}{3}$	$\frac{1}{4}$	
				4f	P	$\frac{1}{3}$	$\frac{2}{3}$	0.164	see text			
				12i	S	0.657	−0.028	0.120	0.667	0.0	0.125	

enables the formation of metal–metal bonds through the shared face of neighbouring polyhedra. In that case, the atoms shift towards each other in pairs, that is, they shift away from the polyhedron centres towards the common polyhedron face (Fig. 14.7). The consequence for the $RuBr_3$ type is that the inversion centres at the occupied octahedron centres and the 6_3 axes cannot be retained, while the mirror planes perpendicular to **c** can be retained. The space group is a subgroup of $P6_3/mcm$ (Fig. 14.8). The symmetry reduction involves a *translationengleiche* subgroup of index 3, and therefore the $RuBr_3$ type is predestinated to form triplet twins (twinned crystals with three kinds of domains; see Chapter 16). In fact, the hexagonal → orthorhombic phase transition entails the formation of such twinned crystals that cause hexagonal symmetry to be feigned for the orthorhombic modification. Ba_3N crystallizes as antitype in the space group $P6_3/mcm$ [230].

Lone electron pairs can be another reason for a shift of atoms out of the octahedron centres, as observed at the P atom of phosphorus triiodide. PI_3

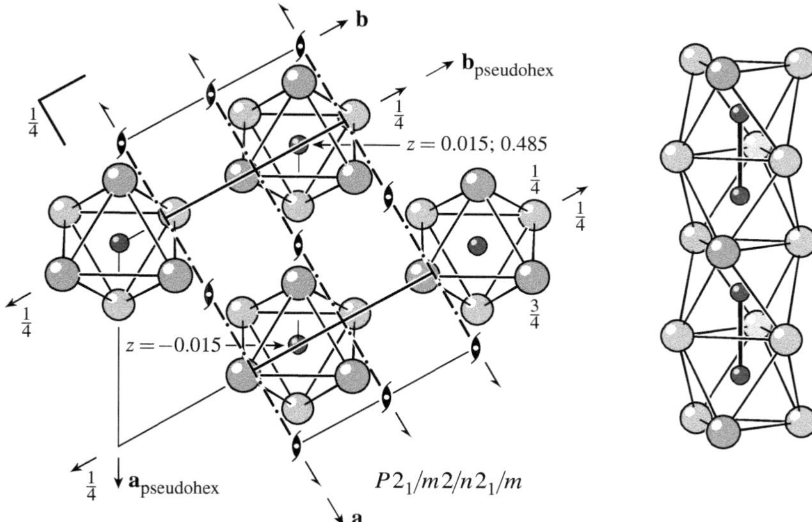

Fig. 14.7 $RuBr_3$: view along a strand of face-sharing octahedra and side view of a strand.

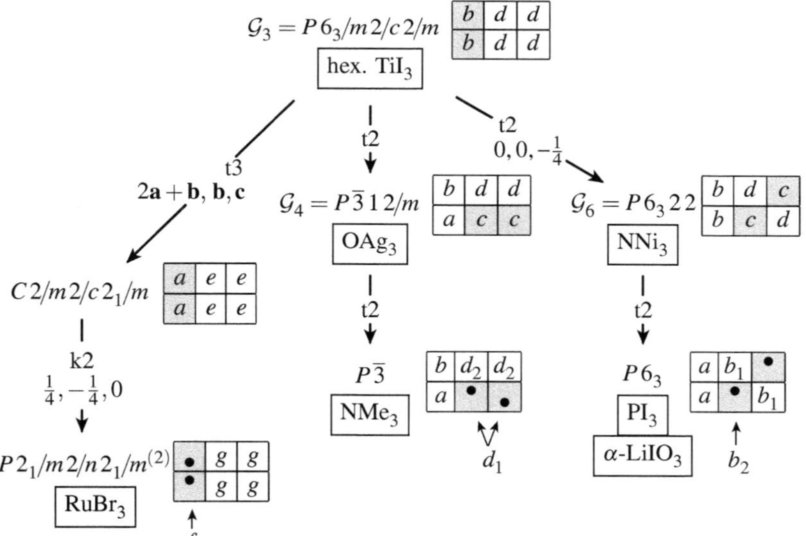

Fig. 14.8 Distorted variants of structures from Fig. 14.6 with subgroups of the space groups $\mathcal{G}_3$, $\mathcal{G}_4$, and $\mathcal{G}_6$. The dots • mark how the Ru, N, P, and iodate-I atoms are shifted from the octahedron centres parallel to **c**. Occupied octahedral voids are displayed in grey; α-$LiIO_3$ additionally has Li^+ ions at the octahedral voids $2a$.

crystallizes with a hexagonal-closest packing of iodine atoms. The distribution of the phosphorus atoms corresponds to the Wyckoff position $2c$ of the space group $P6_3 22$ ($\mathcal{G}_6$ in Fig. 14.6). However, they have been shifted parallel to **c** towards an octahedron face each, which results in three short P–I bonds and three longer P···I contacts. The space-group symmetry has been reduced to $P6_3$ (Fig. 14.8). The same situation is observed for the iodate ions in α-$LiIO_3$; it additionally contains lithium ions in the octahedral voids at the position $2a$. A similar situation arises for crystalline trimethylamine, whose methyl groups form a hexagonal-closest packing. The N atoms are alternately shifted from the Wyckoff position $2c$ of the space group $P\bar{3}1m$ ($\mathcal{G}_4$ in Fig. 14.6) in the directions +**c** and –**c**, reducing the symmetry to $P\bar{3}$.

Table 14.7 Crystal data of $RuBr_3$, PI_3, α-$LiIO_3$, and NMe_3.

	Space group	a /pm	c /pm	Wyckoff position	Element	x	y	z	Ideal coordinates x	y	z	References
$RuBr_3$	$Pmnm^{(2)}$	1125.6	587.3	$4f$	Ru	$\frac{1}{4}$	0.746	0.015	$\frac{1}{4}$	0.75	0.0	[239]
		$b = 649.9$		$2a$	Br	$\frac{1}{4}$	0.431	$\frac{1}{4}$	$\frac{1}{4}$	0.417	$\frac{1}{4}$	
				$2b$	Br	$\frac{1}{4}$	0.052	$\frac{3}{4}$	$\frac{1}{4}$	0.083	$\frac{3}{4}$	
				$4e$	Br	0.597	0.407	$\frac{1}{4}$	0.583	0.417	$\frac{1}{4}$	
				$4e$	Br	0.408	0.903	$\frac{1}{4}$	0.417	0.917	$\frac{1}{4}$	
PI_3	$P6_3$	713.3	741.4	$2b$	P	$\frac{1}{3}$	$\frac{2}{3}$	0.146	$\frac{1}{3}$	$\frac{2}{3}$	0.25	[240]
				$6c$	I	0.686	0.034	0	0.667	0.0	0	
α-$LiIO_3$	$P6_3$	581.3	517.2	$2a$	Li	0	0	0.234	0	0	0.25	[241]
				$6g$	I	$\frac{1}{3}$	$\frac{2}{3}$	0.162	$\frac{1}{3}$	$\frac{2}{3}$	0.25	
				$6c$	O	0.658	−0.095	0	0.667	0.0	0	
NMe_3	$P\bar{3}$	613.7	685.2	$2d$	N	$\frac{1}{3}$	$\frac{2}{3}$	0.160	$\frac{1}{3}$	$\frac{2}{3}$	0.0	[242]
				$6g$	C	0.575	−0.132	0.227	0.667	0.0	0.25	

The crystal data are presented in Table 14.7 in comparison to the expected values without distortions. Not surprisingly, the deviations from the ideal values are largest for the molecular compounds PI_3 and NMe_3. In the case of trimethylamine the deviations are also due to the fact that the C atoms are not exactly in the centres of the methyl ‘spheres’.

14.3 Occupation of octahedral and tetrahedral interstices in the cubic-closest packing of spheres

14.3.1 Hettotypes of the NaCl type with doubled unit cell

If all octahedral voids of a cubic-closest packing of spheres are occupied with one species of atoms, the result is the NaCl type. This is also the structure of the high-temperature modification of $LiFeO_2$ (>670 °C), with random distribution of Li and Fe atoms [244]. In the low-temperature form the Li and Fe atoms are ordered and the doubled unit cell has a c/a ratio of 2.16. If one of the metal atom positions is vacant, this corresponds to the structure of anatase, $Ti\square O_2$, with $c/a = 2.51$ (Fig. 14.9). The corresponding group–subgroup relations are shown in the left part of Fig. 14.10 [247].

The crystal structure of SnF_4 can be described as a cubic-closest packing of fluorine atoms with a doubled unit cell, with one-quarter of the octahedral voids occupied, in such a way that there are layers of vertex-sharing octahedra (Fig. 14.9 and Fig. 14.10 right). The distortion of the sphere packing is small; the c/a ratio is 1.96 instead of 2.

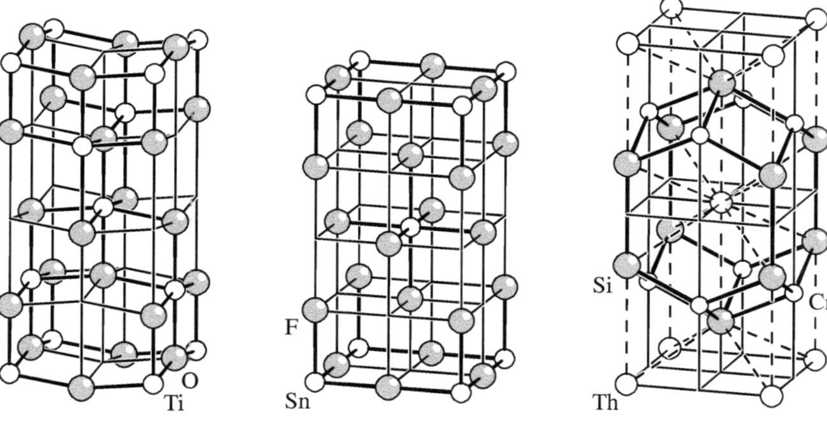

Fig. 14.9 Unit cells of anatase (TiO_2, $I4_1/amd$), SnF_4, and $ThCr_2Si_2$ (both $I4/mmm$).

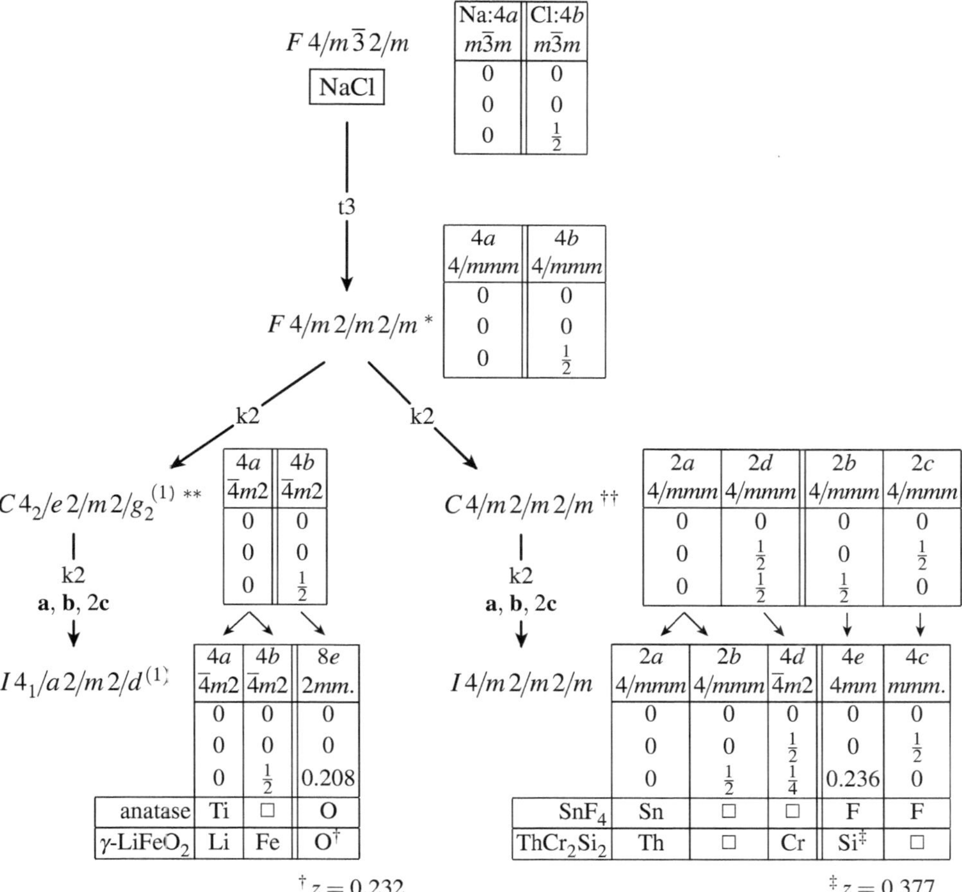

Conventional settings, cell $\frac{1}{2}(\mathbf{a}-\mathbf{b}), \frac{1}{2}(\mathbf{a}+\mathbf{b}), \mathbf{c}$: * $I4/m2/m2/m$ ** $P4_2/n2/n2/m$ †† $P4/m2/m2/m$

Fig. 14.10 Hettotypes of the NaCl type with doubled unit cell. Opinion can differ whether it is sensible to include $ThCr_2Si_2$ here; see text. □ = vacant site. References: anatase [243]; γ-$LiFeO_2$ [244]; SnF_4 [245]; $ThCr_2Si_2$ [246].

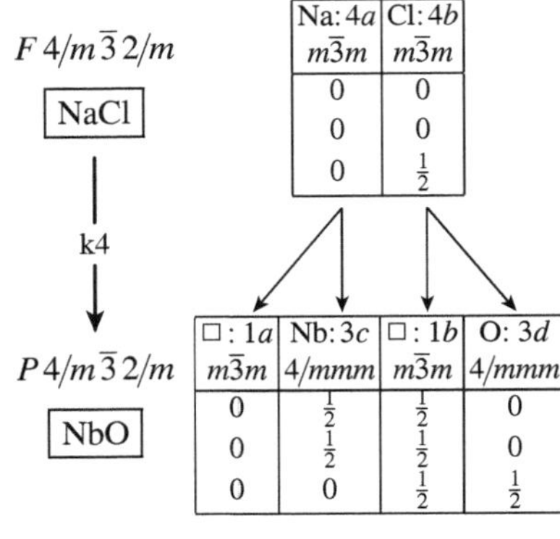

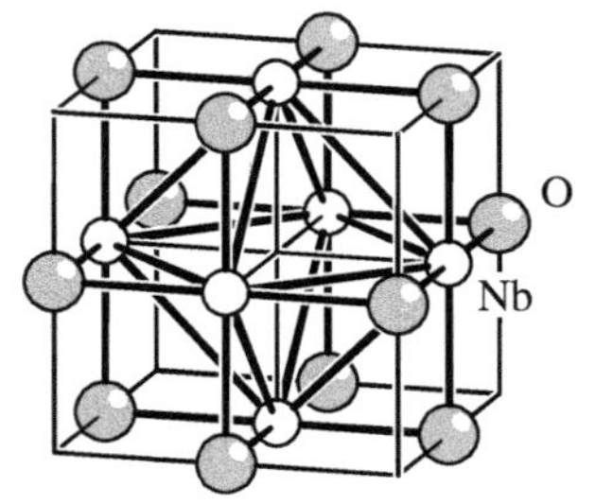

Fig. 14.11 Description of NbO as a defective variant of the NaCl type [248].

On the right side of Fig. 14.10 the $ThCr_2Si_2$ type has been included as a derivative of the NaCl type. It has three-quarters of the metal atom positions occupied, and one-half of the anion positions are vacant. The distortion is more marked, with $c/a = 2.62$. In particular, the free z parameter of the Si atom deviates considerably from the ideal value ($z = 0.377$ instead of $z = 0.25$). Therefore, two Si atoms are shifted away from the Th atom while eight others are closer, resulting in a coordination number of 10 for the Th atom (marked by dotted lines in Fig. 14.9). The Cr atoms have a nearly tetrahedral coordination. The Si atoms are close to one another in pairs, resulting in a direct bond.

We mention $ThCr_2Si_2$ as an example *where the application of group theory comes to the limit of becoming meaningless.* The geometric and chemical deviations are so severe that it is not really justified to regard $ThCr_2Si_2$ as a defective derivative of the NaCl type. In the relation NaCl–$ThCr_2Si_2$ according to Fig. 14.10, group theory serves as nothing more than a formal tool, without chemical or physical justification. However, the relation is acceptable as a mnemonic aid to memorizing the structure of $ThCr_2Si_2$. About 90 representatives are known of the $ThCr_2Si_2$ type.

On the other hand, the mental derivation of the structure of NbO as a defective variant of the NaCl type seems to be natural, having one-quarter of vacant cation and anion sites (Fig. 14.11). The special aspect of NbO are Nb–Nb bonds by formation of octahedral clusters of Nb atoms that are connected via common vertices to form a network. Every Nb atom has a square coordination by four O atoms.

14.3.2 Rhombohedral hettotype of the NaCl type

The hexagonal sheets of Cl^- ions in the NaCl type are oriented perpendicular to the body diagonals of the face-centred cubic unit cell. If only every other sheet of the octahedral interstices perpendicular to one of the diagonals is occupied and the other ones remain vacant, the result is the layered crystal structure of $CdCl_2$. The layers are of the same kind as in CdI_2, but the stacking sequence of the halogen atoms is *ABCABC*. The symmetry is reduced in two steps to the rhombohedral space group $R\bar{3}m$ (Fig. 14.12).

As mentioned in Fig. 14.12, there is an intermediate space group in the symmetry reduction having the same space group type $R\bar{3}m$; it is observed for antiferromagnetic nickel oxide below of the Néel temperature of 523 K. The NaCl cell is only slightly compressed along of one body diagonal (magnetostriction; the lattice parameter $\alpha = 90°$ of the face-centred unit cell is increased to $\alpha = 90.03°$ at 295 K). In the hexagonal setting of the rhombohedral unit cell the ratio c/a is decreased from $\sqrt{6} = 2.449$ to 2.447. However, that is only valid for the unit cell as determined by X-ray diffraction, where the different spin orientations of the unpaired electrons remain invisible. For neutron diffraction the symmetry is a magnetic space group [249, 250].

When all cation sheets are occupied with two alternating kinds of cations, we have the α-$NaFeO_2$ type. It has two kinds of hexagonal sheets with octahedrally coordinated Na and Fe atoms that alternate in the direction of $\mathbf{c}_{hex}$ (Fig. 14.12). For α-$NaFeO_2$ the lattice is somewhat elongated in this stacking direction, having $c/a = 5.32$ (instead of $2\sqrt{6} = 4.90$). Among the numerous

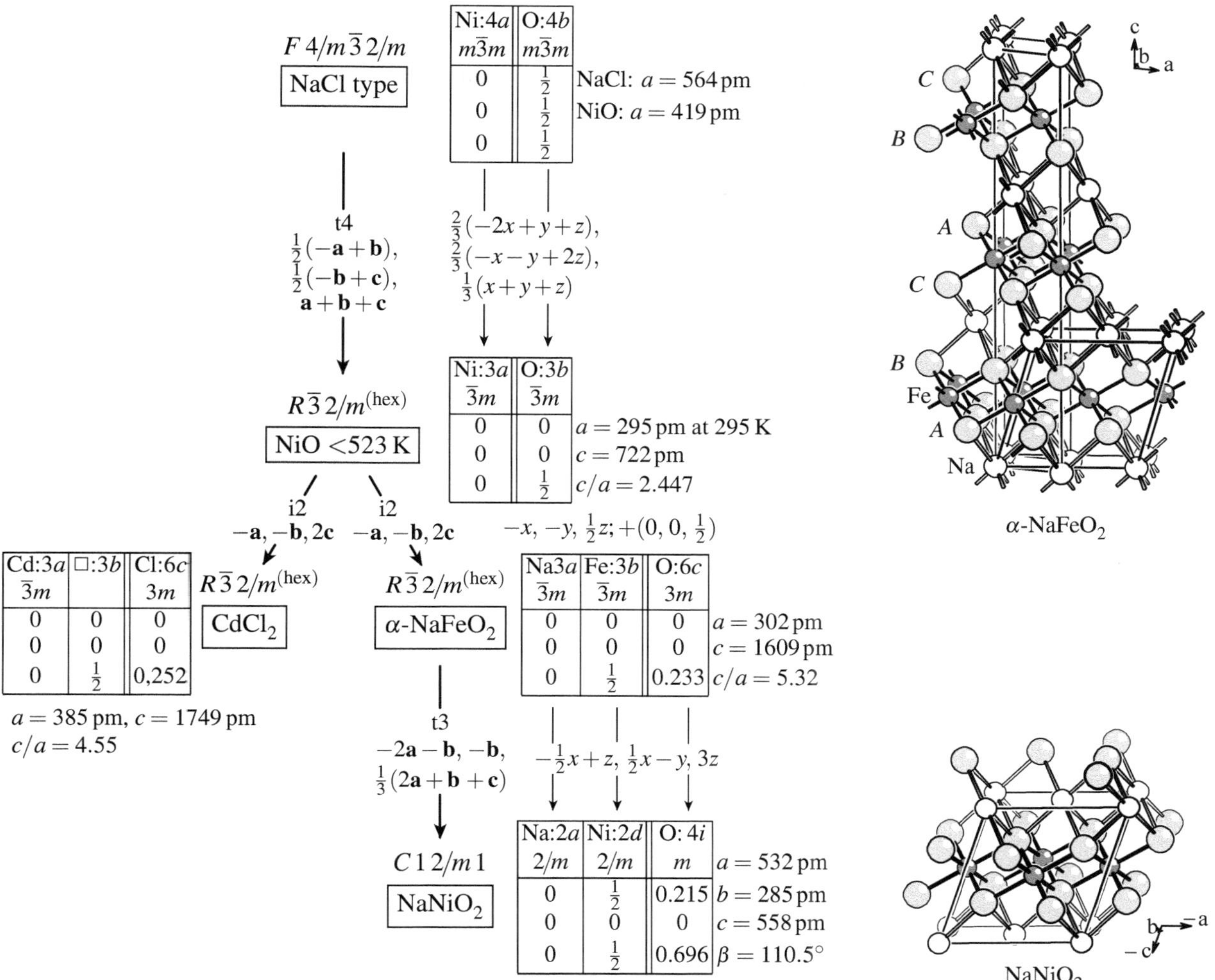

Fig. 14.12 Symmetry reduction from the NaCl type to structures with sheets perpendicular to one of the body diagonals of the cubic unit cell. If every other cation sheet is occupied and the other ones are vacant, that is the $CdCl_2$ type. If the cation sheets are alternately occupied by two kinds of cations, it is the α-$NaFeO_2$ type. $NaNiO_2$ is a variant of the α-$NaFeO_2$ type where the octahedra occupied by Ni atoms are elongated due to the Jahn–Teller effect. References: NiO [249]; $CdCl_2$ [251]; α-$NaFeO_2$ [252]; $NaNiO_2$ [253].

representatives of this structure type, most oxides have c/a values between 4.9 and 5.0.

In $NaNiO_2$ the Ni(III) atoms have the electron configuration d^7; in an octahedral environment that causes a Jahn–Teller distortion with elongated octahedra. As a consequence, the symmetry is reduced from $R\bar{3}\,2/m$ to its subgroup $C\,1\,2/m\,1$ (Fig. 14.12).

If the electron configuration is increased to d^8, the two more distant ligands of the expanded coordination octahedron are moved even further away resulting in a square coordination, such as for Cu(III) in $NaCuO_2$. The hexagonal sheets with the Na^+ ions become somewhat displaced in the direction of **a** so that the stacking sequence of the oxygen atoms is no longer exactly *ABC*.

14.3.3 Hettotypes of the CaF_2 type with doubled unit cell

CaF_2 (fluorite type) can be regarded as a cubic-closest packing of Ca^{2+} ions in which all tetrahedral interstices are occupied by F^- ions. A selection of structure types that result from partial vacation of the tetrahedral interstices with a unit cell unchanged in size and also with a doubled unit cell are mentioned in Fig. 14.13. Views of the unit cells are depicted in Fig. 14.14.

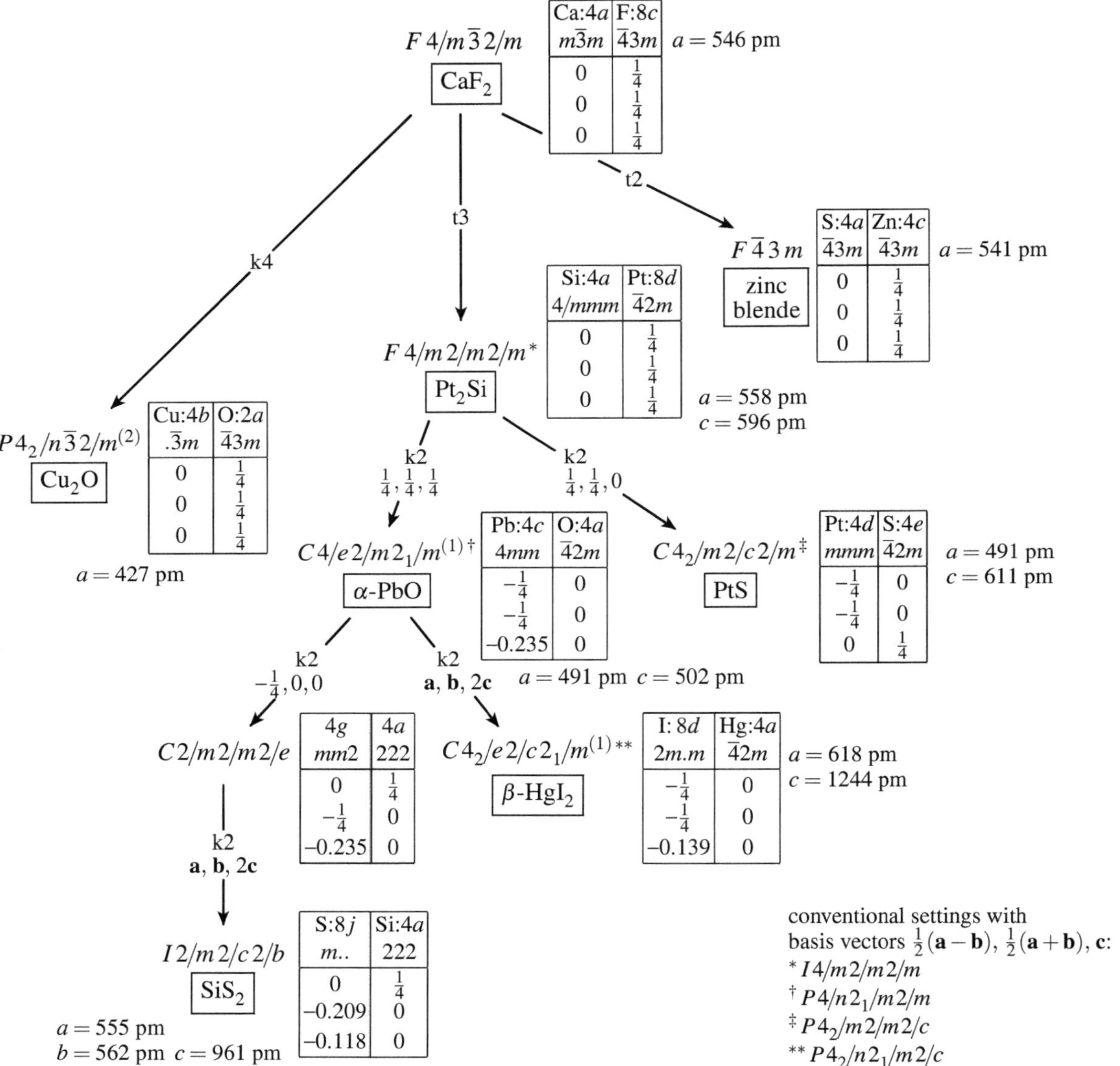

Fig. 14.13 Group–subgroup relations for some structures that are derived from the CaF_2 type by successive removal of atoms from the tetrahedral interstices. The atoms mentioned to the left in the boxes form the packing of spheres. Note the multiplicities of the atomic positions as compared to the chemical composition. If, at a group–subgroup step, the multiplicity of the position mentioned to the right side is halved or the one mentioned to the left side is doubled, then the position mentioned to the right side splits into two positions, keeping their site symmetries; one of them becomes vacant (not mentioned). Cu_2O [254]; Pt_2Si [255]; α-PbO [256]; PtS [257]; HgI_2 [258, 259]; SiS_2 [260].

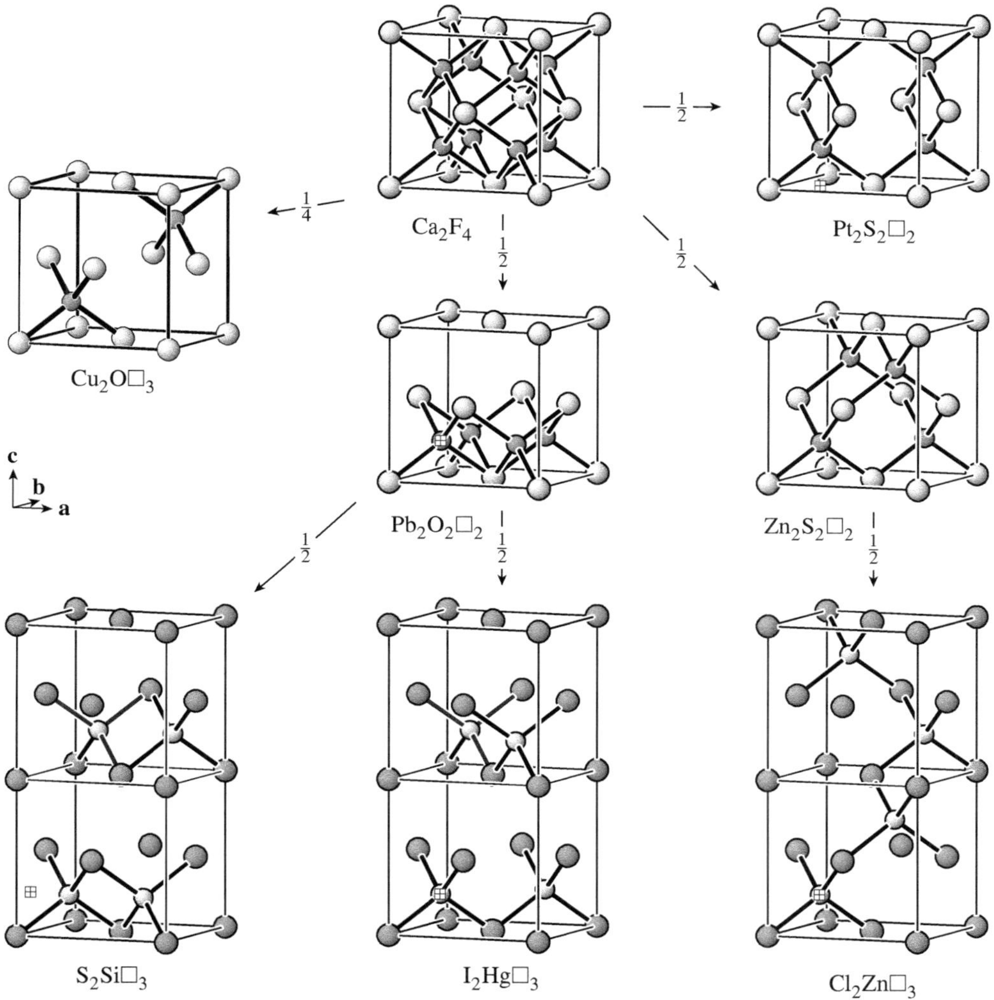

Fig. 14.14 Unit cells of the structures of CaF_2, Cu_2O, α-PbO, zinc blende, PtS, SiS_2, β-HgI_2, and α-$ZnCl_2$. CaF_2 has all tetrahedral voids occupied (= centres of the octants of the cube). Each arrow marks the reduction of the number of occupied tetrahedral voids by the factor given in the arrow. Metal atoms have been drawn light grey, non-metal atoms dark grey. The atoms mentioned first in the formulae form the cubic-closest packing. Schottky symbols □ designate vacant tetrahedral interstices (α-$ZnCl_2$ is not mentioned in Fig. 14.13).
The conventional origin positions of the cells of α-PbO, SiS_2, β-HgI_2, α-$ZnCl_2$, and PtS are not as drawn, but at the sites marked by ⊞. The coordinates mentioned in Fig. 14.13 in each case in the first column then correspond to the atoms drawn at the lower-left corners of the cells. At high pressures, SiS_2 consists of polymorphic forms that have tetrahedra joined via common edges and vertices [261].

If variants with vacancies among the spheres of the packing are included in the considerations, rather different structure types are obtained, with different coordination polyhedra and kinds of linkage of the polyhedra (Table 14.8). In all cases without vacancies in the sphere packing, the atoms in the tetrahedral interstices keep the tetrahedral coordination. Partial occupation of the tetrahedral interstices results in different coordination polyhedra for the atoms of the sphere packing. These coordination polyhedra are derived from the coordination cube of the Ca^{2+} ion when its vertices are being removed (Fig. 14.15). Of

Table 14.8 Coordination polyhedra and polyhedron linkage for the atoms of the sphere packing for the structure types mentioned in Fig. 14.13 and Fig. 14.16. The atoms mentioned first in the formulae form the sphere packing.

Structure type	Coordination polyhedron	Linkage by	
CaF_2	cube	all edges	framework
zinc blende	tetrahedron	all vertices	framework
α-PbO	square pyramid	all basal edges	layers
PtS	rectangle	all edges	framework
I_2Hg	tetrahedron	4 edges	layers
S_2Si	tetrahedron	2 edges	chains
α-Cl_2Zn	angle, c.n. 2	all vertices	framework
Cu_2O	linear, c.n. 2	all vertices	2 interwoven frameworks
β-CuI	cube, 2 missing vertices	3 edges + vertices	cf. Fig. 14.16
$CuFeO_2$	octahedron and linear	3 edges + vertices	cf. Fig. 14.16

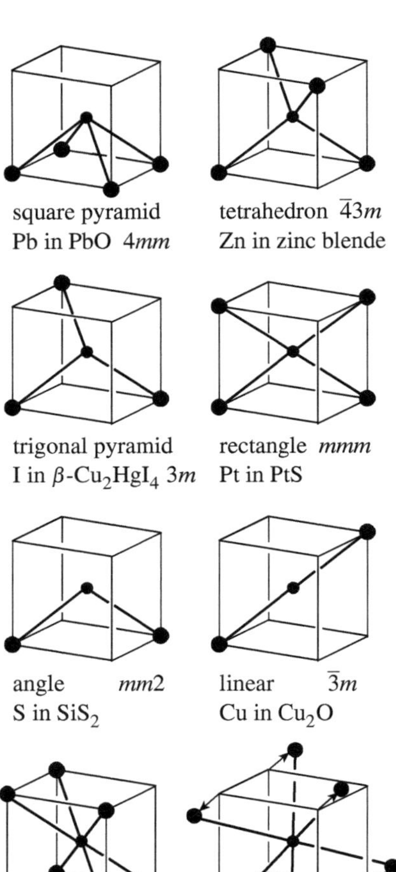

Fig. 14.15 Coordination polyhedra derived from a cube by removal of vertices.

course, there is a group-theoretical connection: The point groups of the coordination polyhedra are subgroups of the point group $m\overline{3}m$ of the cube (for the linear arrangement the point group within the crystal is maximally $\overline{3}m$ and not ∞/mm).

We mention another structure type, the delafossite type, named after a mineral with the composition $CuFeO_2$. It can be derived as a rhombohedral hettotype ($R\overline{3}m$) of the CaF_2 type with one-half of the tetrahedral interstices being vacant (Fig. 14.16). In this case, we obtain two kinds of metal atoms: Linearly coordinated Cu(I) atoms and Fe(III) atoms surrounded by six oxygen atoms. These six O atoms are the remains of the cube of eight F^- ions (substituted by O atoms) that surround a Ca^{2+} ion in CaF_2 after two atoms at two opposite vertices have been removed (Fig. 14.15, bottom). By elongation along the body diagonal running through the vacant vertices (parallel to $\mathbf{c}^{(\text{hex})}$), the remains of the cube becomes an octahedron that hosts an Fe(III) atom; that causes a slight increase of the ideal c/a ratio from 4.90 to 5.45 (hexagonal setting). Without this elongation, the coordination polyhedron is a flattened octahedron; this can be found in the uncommon structure type of β-CuI (at temperatures around 610 K, Fig. 14.16, right). [263]

The results are hexagonal sheets of anions perpendicular to the $\mathbf{c}^{(\text{hex})}$ axis that have the stacking sequence *AABBCC*. The octahedra occupied by the Fe(III) atoms are between the sheet pairs *AB*, *BC*, and *CA*. Linearly coordinated Cu(I) atoms are placed in between of the sheet pairs *AA*, *BB*, and *CC* [262]. Compared to α-$NaFeO_2$ (Fig. 14.12), the atoms in $CuFeO_2$ are arranged and linked in a fundamentally different way. And yet, the space groups $R\overline{3}m$ and the occupied Wyckoff positions 2*a*, 2*b*, and 6*c* are the same in both structure types. Compared with α-$NaFeO_2$ the c/a ratio is slightly increased to 5.65 (instead of 5.32), and the only further difference is the z coordinate of the Wyckoff position 6*c* $(0, 0, \pm z)$ of the oxygen atoms: it is $z = 0.2334 \approx \frac{1}{4}$ for α-$NaFeO_2$ and $z = 0.1066 \approx \frac{1}{8}$ for $CuFeO_2$; together with the symmetry-

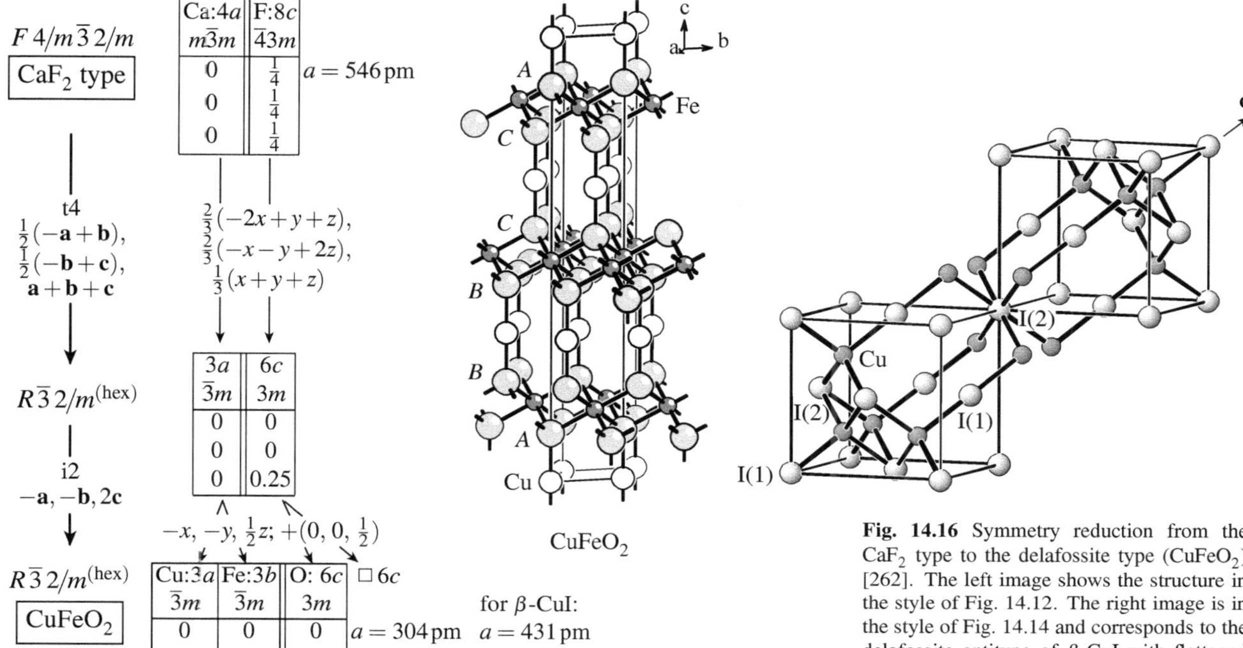

Fig. 14.16 Symmetry reduction from the CaF_2 type to the delafossite type ($CuFeO_2$) [262]. The left image shows the structure in the style of Fig. 14.12. The right image is in the style of Fig. 14.14 and corresponds to the delafossite antitype of β-CuI with flattened coordination octahedra around I(2) and linear coordination of I(1).

equivalent atom at $z = -0.1066$ the latter yields the stacking sequence AA. The cations in both structures nearly correspond to a cubic closest arrangement of spheres. The example shows that the same space group type and the same sequence of occupied Wyckoff positions are not sufficient criteria for isotypism if free parameters among the coordinates differ (see Section 9.5 and Table 9.3).

If we evaluate the parameter of deviation Δ between α-$NaFeO_2$ and $CuFeO_2$ according to eqn (9.3), page 124, we obtain $\Delta = 0.166$. The example shows that the value of Δ gives an indication, but no reliable clue, whether two compounds are isotypic or not. If we describe $CuFeO_2$ as being built up from Fe^{3+} and linear CuO_2^{3-} ions, the structural analogy to isotypic compounds such as $MgCN_2$ and β-NaN_3 becomes apparent. For a comparison of more representatives if the α-$NaFeO_2$ and delafossite types, see Example 24.4, page 315.

Exercises

Solutions on page 332.

14.1. Boron triiodide crystallizes with a hexagonal-closest packing of iodine atoms; the boron atoms occupy triangular interstices whose centres take the position $0, 0, \frac{1}{4}$ of the sphere packing. $a = 699.1$ pm, $c = 736.4$ pm; $P6_3/m$; B: $2c$, $\frac{1}{3}, \frac{2}{3}, \frac{1}{4}$; I: $6h$ 0.318, 0.357, $\frac{1}{4}$ [264]. The van der Waals radius of an iodine atom is 198 pm. Derive the structure of BI_3 from the hexagonal-closest packing of spheres.

14.2. The C and I atoms of the cluster compound CZr_6I_{12} commonly form a cubic-closest packing. W_6Cl_{18} has a cubic-closest packing of Cl atoms with vacancies at the centres of the clusters. The metal atoms occupy part of the octahedral interstices. Derive the structures from the cubic-closest packing of spheres. The van der Waals radii are: Cl 175 pm; I 198 pm.

CZr_6I_{12} [265] $R\bar{3}^{(hex)}$
$a = 1450.8$ pm, $c = 1000.7$ pm

		x	y	z
Zr	$18f$	0.1430	0.0407	0.1301
C	$3a$	0	0	0
I 1	$18f$	0.3114	0.2308	0.0008
I 2	$18f$	0.6155	0.4598	0.0085

W_6Cl_{18} [266] $R\bar{3}^{(hex)}$
$a = 1493.5$ pm, $c = 845.5$ pm

		x	y	z
W	$18f$	0.1028	0.1182	0.1383
Cl 1	$18f$	0.2129	0.5296	−0.0116
Cl 2	$18f$	0.1032	0.2586	0.0024
Cl 3	$18f$	0.4397	0.0777	0.0281

14.3. The crystal structure of Sn_2OF_2 can be regarded as a hettotype of the cassiterite structure (SnO_2, rutile type). There are distortions due to the lone electron pairs at the tin atoms; in addition, one-quarter of the O atom positions are vacant [267]. Derive the symmetry relations between cassiterite and Sn_2OF_2.

SnO_2 $P4_2/mnm$
$a = 473.6$ pm, $c = 318.5$ pm

		x	y	z
Sn	$2a$	0	0	0
O	$4g$	0.305	−0.305	0

Sn_2OF_2 $A112/m$
$a = 507$ pm, $b = 930$ pm, $c = 808$ pm, $\gamma = 97.9°$

		x	y	z
Sn1	$4i$	0.486	0.283	$\frac{1}{2}$
Sn2	$4g$	0	0	0.296
O	$4i$	0.803	0.392	0
F	$8j$	0.301	−0.175	0.321

14.4. The unit cell of α-$ZnCl_2$ is shown in Fig. 14.14. The crystal data are [268]:

space group $I\bar{4}2d$; $a = 539.8$ pm, $c = 1033$ pm; Zn $4a$, 0, 0, 0; Cl $8d$, 0,25, $\frac{1}{4}$, $\frac{1}{8}$.

Derive the relation with the zinc blende type (Fig. 14.13).

14.5. The crystal structure of OsO_4 can be described as a cubic-closest packing of O atoms ($a_{cub} \approx 435$ pm) in which one-eighth of the tetrahedral interstices are occupied [269, 270].

$I12/a1$, $a = 863.2$ pm, $b = 451.5$ pm, $c = 948.8$ pm, $\beta = 117.9°$

		x	y	z	
Os	$4e$	$\frac{1}{4}$	0.241	0	transformed from [270],
O1	$8f$	0.699	0.517	0.618	setting $C12/c$
O2	$8f$	0.420	0.463	0.113	

Make a drawing of the structure in a projection along **b** (one unit cell) and derive the metric relations to the unit cell of the CaF_2 type. Supplement the tree of Fig. 14.13, starting from the space group $C2/m2/m2/e$. (The task is exacting due to several necessary basis transformations and origin shifts and because the correct subgroup among several subgroups on a par has to be chosen.)

14.6. The chlorine and caesium atoms of $CsTi_2Cl_7$-II commonly form a double-hexagonal-closest packing of spheres (stacking sequence *ABAC*). The titanium atoms occupy one-quarter of the octahedral interstices in pairs of face-sharing octahedra [271]. Derive the structure from the double-hexagonal-closest packing of spheres. Remark: Consider first what the metric relations between the unit cells are; the distance between adjacent spheres of the packing is approximately 360 pm. □ = centre of an octahedral interstice.

double-hexagonal-closest packing
$P6_3/mmc$ $c/a = 3.266$

		x	y	z
X1	$2a$	0	0	0
X2	$2d$	$\frac{2}{3}$	$\frac{1}{3}$	$\frac{1}{4}$
□	$4f$	$\frac{1}{3}$	$\frac{2}{3}$	0.125

$CsTi_2Cl_7$-II $P112_1/m$
$a = 728.0$ pm, $b = 635.4$ pm, $c = 1163.0$ pm, $\gamma = 91.5°$

		x	y	z
Cs	$2e$	0.633	0.946	$\frac{1}{4}$
Ti	$4f$	0.139	0.592	0.111
Cl1	$2e$	0.907	0.444	$\frac{1}{4}$
Cl2	$2e$	0.353	0.435	$\frac{1}{4}$
Cl3	$2e$	0.127	0.888	$\frac{1}{4}$
Cl4	$4f$	0.112	0.268	−0.001
Cl5	$4f$	0.634	0.253	0.007

Crystal structures of molecular compounds

15

The treatment of inorganic compounds in the preceding chapters could produce the impression that the application of group–subgroup relations is not appropriate to be applied to crystal structures of more complicated molecular compounds. In fact, the vast number of such compounds crystallize in space groups of lower symmetry, and the molecules frequently take positions with the site symmetries 1 or $\bar{1}$. The most frequent space group among organic compounds is $P2_1/c$ (Table 15.1). For racemates consisting of chiral molecules the statistics is almost the same. Enantiopure molecules can adopt only one of the 65 Sohncke space-group types (cf. Section 9.3, page 119); for them the space groups $P2_12_12_1$ (54 %) and $P2_1$ (34 %) are the most frequent [272]. For this reason, it has even been argued that a 'principle of symmetry avoidance' holds for molecular compounds. However, this is not really true, as is shown by the examples of this chapter. The symmetry principle is also effective for crystal structures of molecular compounds. However, aspect number 2 of the symmetry principle as stated in Section 1.1 is of great importance. Frequently, the low symmetries of a large number of molecules do not allow for packings with molecules occupying special positions with high site symmetries.

Table 15.1 Frequency of space groups among crystal structures of 636756 molecular compounds (2022) at the Cambridge Structural Database [6,7], site symmetries adopted by 96000 molecules in these space groups (2003) [285], and frequency of point groups among 456683 molecules (2010), extracted with CSDSymmetry [286, 287]. Only organic compounds with no more than one kind of molecule in the crystal.

Space group	Frequency %	Site symmetries of the molecules in this space group in %					
$P2_1/c$	37.5	1	86	$\bar{1}$	14		
$P\bar{1}$	21.8	1	81	$\bar{1}$	19		
$P2_12_12_1$ *	9.2	1	100				
$C2/c$	7.1	1	48	$\bar{1}$	10	2	42
$P2_1$ *	6.3	1	100				
$Pbca$	3.9	1	88	$\bar{1}$	12		
$Pna2_1$	1.6	1	100				
Cc	1.0	1	100				
$Pnma$	0.9	1	2	$\bar{1}$	1	m	97
$P1$	0.9	1	100				
146 others	9.8	1	55				

Point group	Frequency among molecules in %
1	82.65
$\bar{1}$	10.11
2	4.26
m	1.53
3	0.54
$\bar{4}$	0.23
$2/m$	0.20
$mm2$	0.07
222	0.06
4	0.04
mmm	0.01

* Sohncke space group.

Molecules of arbitrary shape arrange themselves in a crystal as close as possible [273–275]. This has been shown not only by experience, but also by numerous energetic computations using interatomic potentials [276, 277]. However, even with powerful computer programs, the prediction of how given kinds of molecules will pack themselves in a crystal is only possible in a restricted way. The energy differences between several feasible polymorphic forms are often much too small to establish reliable conclusions, although considerable progress has been attained [278–283]. KITAIGORODSKII dealt intensively with the question of how molecules with irregular shape can be packed so as to occupy space in a most parsimonious way, and he investigated which symmetry elements are compatible with packings of the kind 'proturberance clicks into place with a re-entrant angle' [273, 284]. Mainly, inversion centres, glide planes, and 2_1 axes are appropriate.

Statistics also show that centrosymmetric molecules crystallize in more than 99% of all cases into centrosymmetric space groups and predominantly occupy centrosymmetric positions [285, 288]. If a molecule has a twofold rotation axis, then this is kept in 45% of all cases on a corresponding rotation axis of the crystal. A threefold molecular axis is maintained with a frequency of 47% and a mirror plane with 24%. Space groups that have reflection planes usually occur only if at least one kind of molecule is placed on the planes [274, 285].

Statistics look different for inorganic compounds; higher-symmetry space groups are much more frequent. Among the 100 444 entries of the year 2006 in the *Inorganic Crystal Structure Database* [3], $Pnma$ is the most frequent space group (7.4%), followed by $P2_1/c$ (7.2%), $Fm\overline{3}m$ (5.6%), $Fd\overline{3}m$ (5.1%), $P\overline{1}$ (4.0%), $I4/mmm$ (4.0%), $C2/c$ (3.8%), and $P6_3/mmc$ (3.4%) [289]. However, these statistical data are misleading in that non-molecular compounds having higher-symmetry crystal structures were frequently studied repeatedly, at different temperatures, pressures, and foreign-atom dopings.

The structures of molecules in crystals usually differ only marginally from their structures in the gaseous state or in solution. The intermolecular forces in the crystal are usually too weak to influence the molecular structure significantly. Apart from slight distortions, only conformation angles experience changes [290]. However, there are exceptions, especially if a structural change allows for a better packing density or if the association of molecules upon crystallization is exothermic. For example, PCl_5 consists of trigonal–bipyramidal molecules in the gas phase; upon crystallization a rearrangement to PCl_4^+ and PCl_6^- ions takes place. Dimeric Al_2Cl_6 molecules associate to polymeric layers; N_2O_5 molecules crystallize as $NO_2^+NO_3^-$; Cl_2O_6 molecules crystallize as $ClO_2^+ClO_4^-$.

15.1 Symmetry reduction due to reduced point symmetry of building blocks

Many crystal structures consisting of complicated molecules or ions can be traced back to simple structure types if a molecule or molecular ion as a whole is considered to be only one building block. For example, if a $PtCl_6^{2-}$ ion in K_2PtCl_6 is considered to be one particle, the packing of $PtCl_6^{2-}$ and K^+ ions

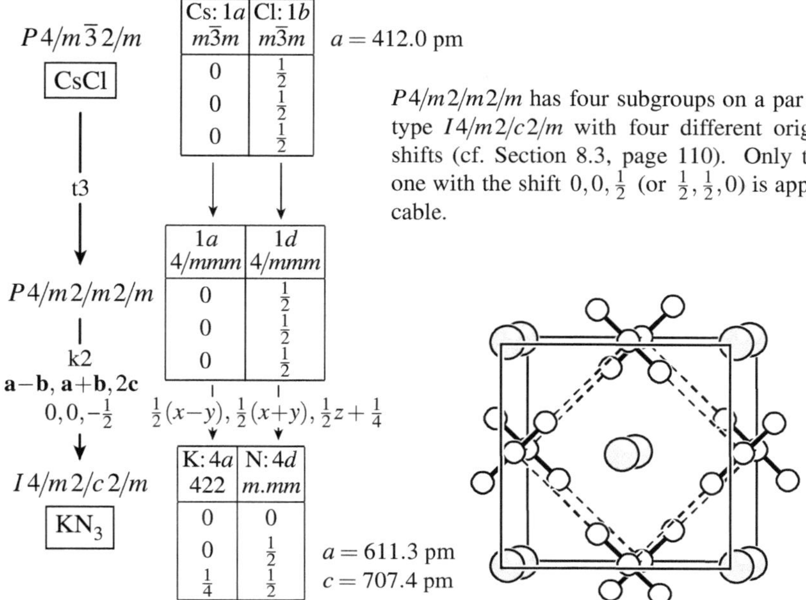

Fig. 15.1 The relation between CsCl and KN_3. Only coordinates of the particle centres are stated. Dotted lines: unit cell corresponding to the CsCl type [291].

corresponds to the CaF_2 type. In this case even the space group $Fm\bar{3}m$ is the same, since the point group $m\bar{3}m$ of the $PtCl_6^{2-}$ ions is the same as the site symmetry of the Ca^{2+} ions in CaF_2. Crystal structures related this way are also considered to be homeotypic (see text after Definition 9.3, page 125).

In most cases the point group of a molecule (or molecular ion) is lower than the site symmetry of the idealized structure type, which we consider to be the aristotype. In that case the symmetry of the space group has to be reduced. In the actual hettotype, the site symmetry of the centre of gravity of the molecule has to be a common subgroup of the site symmetry in the aristotype and the molecular symmetry.

In potassium azide, KN_3, the K^+ and N_3^- ions are arranged like in caesium chloride. However, the N_3^- ions are not spherical, but rod-shaped, point group ∞/mm. In order to achieve a favourable packing, the azide ions are oriented in two mutually perpendicular directions perpendicular to **c**, with a slight expansion of the lattice along **a** and **b**. The symmetry of the space group is reduced to $I4/mcm$ (Fig. 15.1). The site symmetry mmm of the N_3^- ions is a common subgroup of ∞/mm and the site symmetry $m\bar{3}m$ of the Cl^- ions in CsCl. The axis ratio of $c/a = 1.16 < \sqrt{2}$ is an expression of the expansion of the lattice.

15.2 Molecular packings after the pattern of sphere packings

Some molecules and molecular ions are more or less 'round' (i.e. they resemble spheres). This is obvious for the fullerene molecule C_{60}. Such molecules tend to arrange themselves as in a packing of spheres. In addition, they often exhibit dynamical behaviour in the temperature range immediately below the melting point, with molecules that rotate in the crystal or perform strong rotational vibrations. This is called a plastic phase. Molecules like MoF_6 and ions like PF_6^- and BF_4^- are notorious for this kind of behaviour. Upon cooling, a phase transition takes place, the molecules adopt a definite orientation, and a symmetry reduction takes place.

MoF_6 is body-centred cubic above −9.8 °C up to the melting point (17.4 °C), with rotating molecules. It thus corresponds to a body-centred cubic packing of spheres. The molecules become oriented below −9.8 °C and a phase transition to an orthorhombic phase takes place; this is a modification that could only be regarded as a packing of MoF_6 'spheres' with some intellectual acrobatics. However, it fits a description as a double-hexagonal packing of fluorine atoms (stacking sequence *ABAC*) in which one-sixth of the octahedral voids are occupied by Mo atoms [292–294].

In other cases the positions of the molecules do indeed correspond to a packing of spheres. For C_{60} this is not surprising. Its molecules are packed in a cubic-closest packing of spheres in the space group $F\,4/m\bar{3}\,2/m$ [295]. The molecules rotate in the crystal, albeit with a preferential orientation that corresponds to the site symmetry $2/m\bar{3}$. $2/m\bar{3}$ is the common subgroup of the molecular symmetry $2/m\bar{3}\,\bar{5}$ (icosahedron symmetry) and the site symmetry $4/m\bar{3}\,2/m$ in $F\,4/m\bar{3}\,2/m$. If all molecules were ordered, the space group would be $F\,2/m\bar{3}$, a maximal subgroup of $F\,4/m\bar{3}\,2/m$ [296]. Below 249 K the molecules become ordered in the subgroup $P\,2/a\bar{3}$ and the site symmetry is reduced to $\bar{3}$ [297].

The octahedral interstices in a packing of C_{60} molecules can be occupied, with some distortion and expansion of the packing being tolerated. The compound $C_{60}{\cdot}Se_8{\cdot}CS_2$ offers an example (Fig. 15.2). The C_{60} molecules are ordered like in a distorted hexagonal-closest packing of spheres, with Se_8 and

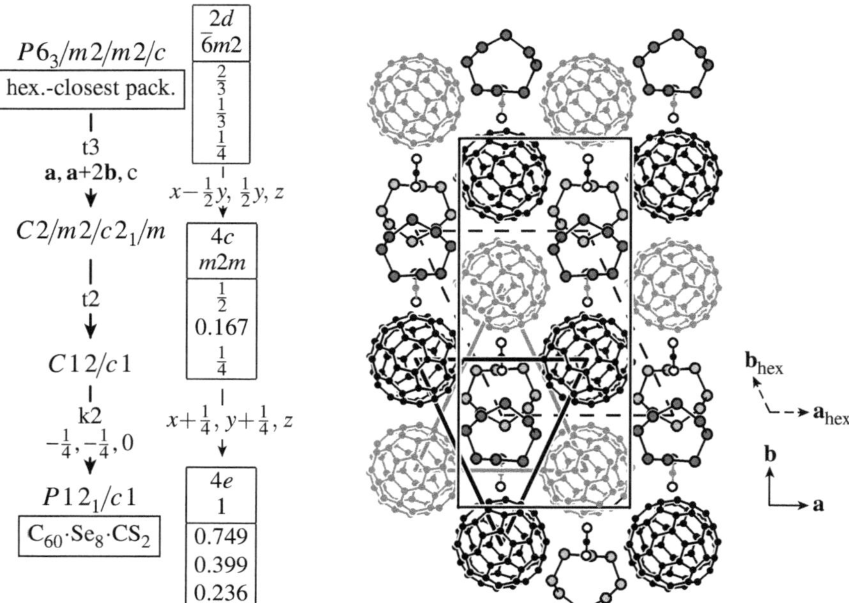

Fig. 15.2 The relation between the hexagonal-closest packing of spheres and the structure of $C_{60}{\cdot}Se_8{\cdot}CS_2$. The coordinates designate the centres of the C_{60} molecules. The unit cell of the sphere packing is marked by dotted lines at the right side. The two triangles are two faces of an octahedron spanned by the centres of six C_{60} molecules [298].

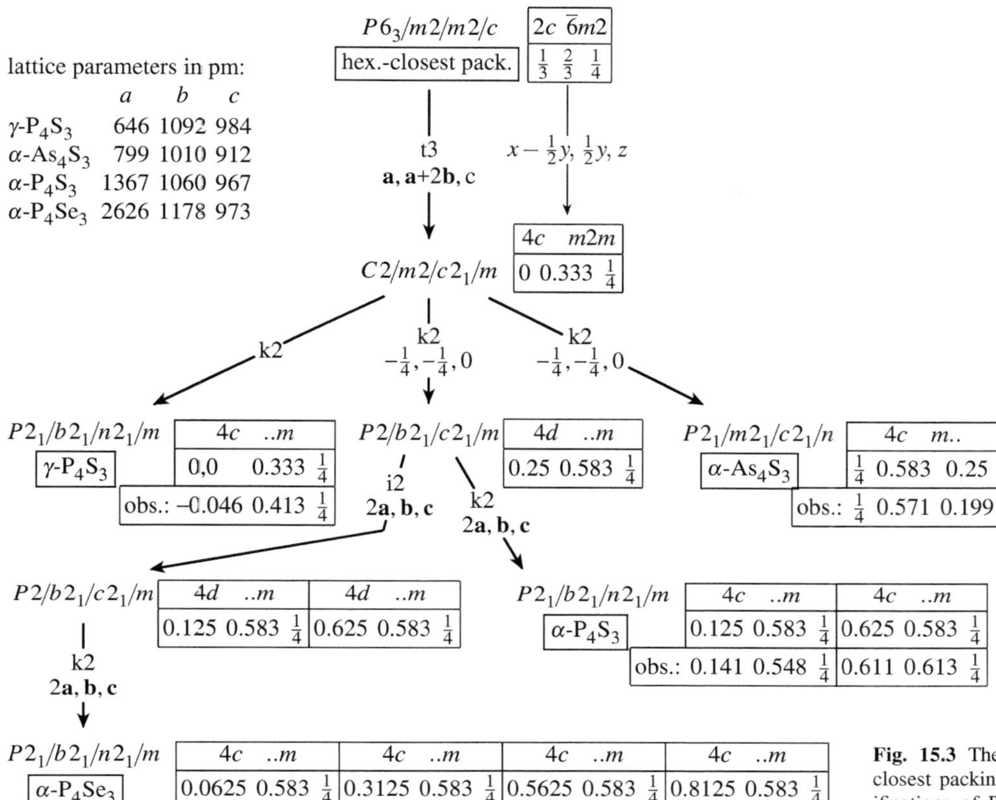

Fig. 15.3 The relations of the hexagonal-closest packing of spheres with some modifications of P_4S_3 and As_4S_3 [299]. Only coordinates of the centres of the molecules are mentioned. *Pmcn* and *Pbnm* are non-conventional settings of *Pnma*.

CS_2 molecules placed in the 'octahedral' voids. The octahedral voids of the sphere packing are located on the 6_3 axes of the space group $P6_3/m2/m2/c$. These screw axes have been lost, just as the mirror planes and the twofold rotation axes have, resulting in the actual subgroup $P12_1/c1$.

For the packing of the nearly ellipsoid-shaped C_{70} molecules see Section 17.1, page 234, and Fig. 17.2.

The less spherical cage molecules of P_4S_3 and As_4S_3 crystallize like in a hexagonal-closest packing of spheres. Several modifications are observed, which is an expression that none of them permits packing free of stress. The four modifications mentioned in Fig. 15.3 have four different space groups, but all of them belong to the space-group type *Pnma*. In all cases the site symmetry of the centres of gravity of the molecules is *m*, and the deviation of the molecular shape from being spherical is absorbed by different kinds of distortion: For γ-P_4S_3 it is a slight shift from the ideal positions (cf. *y* coordinate in Fig. 15.3; Fig. 15.4). For α-P_4S_3 and α-As_4S_3 the lattice has experienced an expansion in the direction of **a**, including two different orientations of the molecules. For α-P_4Se_3 the deviations from the ideal packing are the least, but this entails four symmetry-independent molecules with differently rotated molecules [299].

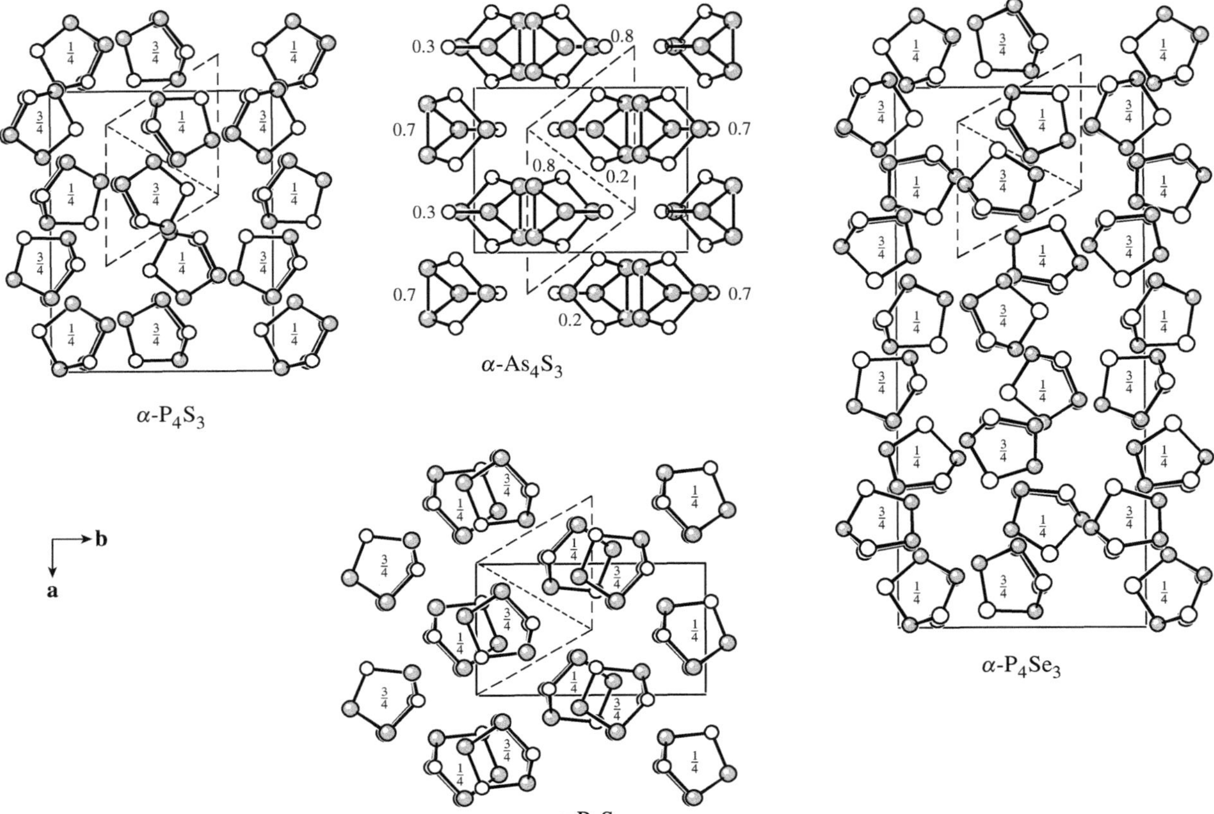

Fig. 15.4 The packing in four modifications of the cage-like molecules E_4X_3. Dotted lines: pseudohexagonal unit cells. Numbers: *z* coordinates of the molecule centres.

Even for more complicated particles the packing can often be derived from a simple structure type. Consider as an example the compound (Na-15-crown-5)$_2$[ReCl$_6$]·4CH$_2$Cl$_2$ [300]. It contains octahedral $ReCl_6^{2-}$ ions; two Na^+ ions are coordinated to two opposite octahedron faces of the $ReCl_6^{2-}$ ion, and each of them has attached a crown-ether molecule; in addition, there are intercalated dichloromethane molecules. The building block [15-crown-5–Na^+–$ReCl_6^{2-}$–Na^+–15-crown-5] is far from being spherical. And yet these blocks are arranged like in a hexagonal-closest packing of spheres (Fig. 15.5 right). If each of these blocks is considered as one unit, the relation is as shown in Fig. 15.5. The symmetry reduction consists of the minimum of three steps that is needed to reduce the site symmetry $\overline{6}m2$ of a sphere in the packing of spheres down to the site symmetry 1. The coordinates of the Re atoms correspond nearly to the ideal values and the axis ratio $b/a = 1.756 \approx \sqrt{3}$ is in accordance with the pseudohexagonal symmetry. The voids in the strands of face-sharing 'octahedra' running along *c* offer more space than in the packing of spheres, due to the size of the building blocks. They form channels that contain the CH_2Cl_2 molecules.

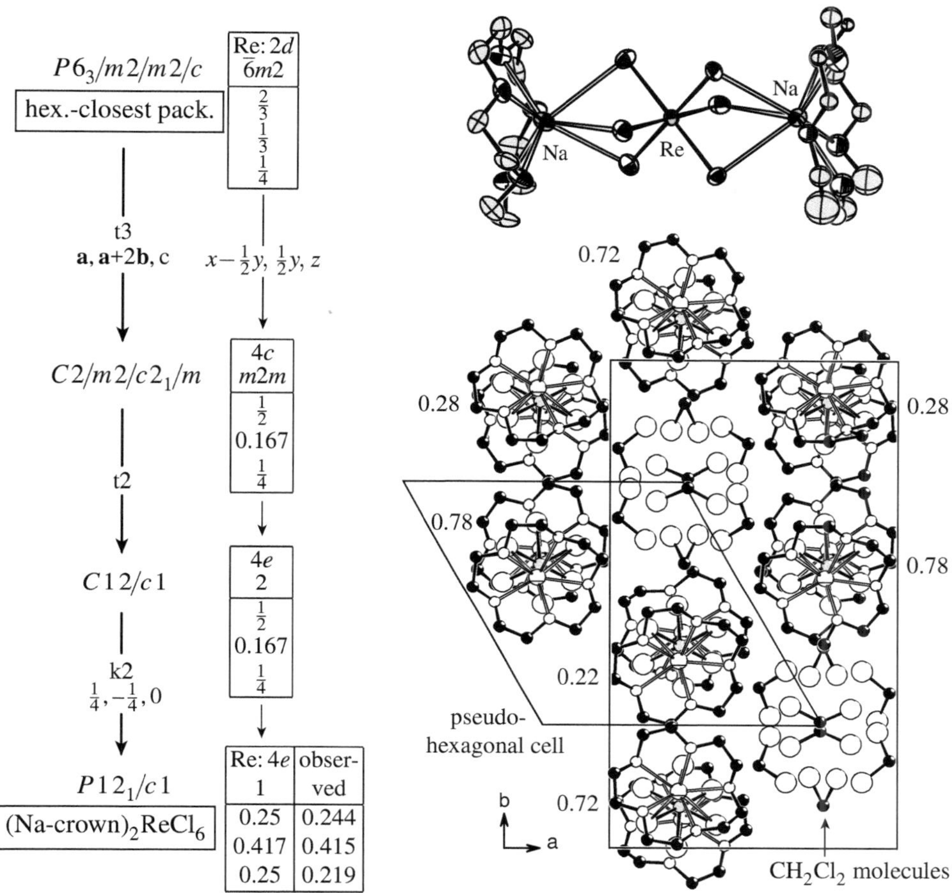

Fig. 15.5 The packing in (Na-15-crown-5)$_2$[ReCl$_6$]·4CH$_2$Cl$_2$ and its relation with the hexagonal-closest packing of spheres. Only coordinates of the Re atoms are mentioned. Top right: One building block (15-crown-5–Na)[ReCl$_6$](Na–15-crown-5); these building blocks are aligned approximately along *c*, in the direction of view of the lower image; numerical values are *z* coordinates of the Re atoms [300].

15.3 The packing in tetraphenylphosphonium salts

Tetraphenylphosphonium ions, $P(C_6H_5)_4^+$, and similar species like $As(C_6H_5)_4^+$ and $Li(NC_5H_5)_4^+$, are popular cations among chemists; they are used to stabilize unstable anions. Generally, the compounds are soluble in weakly polar solvents like CH_2Cl_2 and, as a rule, they crystallize easily.

The space requirement of the phenyl groups in a $P(C_6H_5)_4^+$ ion entails a preferred conformation, such that the ion adopts the point symmetry $\bar{4}$. In crystalline tetraphenylphosphonium salts the ions are frequently stacked into columns, two phenyl groups of one ion being rotated by 90° relative to two phenyl groups of the next ion (Fig. 15.6) [178]. The columns are arranged in parallel in the crystal, parallel to the direction we designate by **c**. The distance between two neighbouring cations in the column amounts to 740–800 pm.

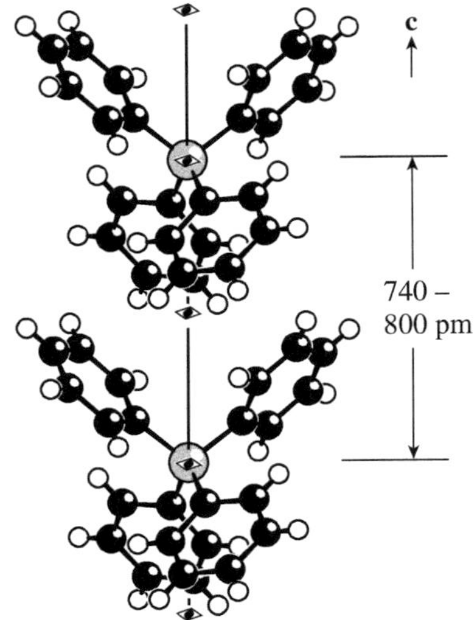

Fig. 15.6 Stacking of $P(C_6H_5)_4^+$ ions into columns.

Most frequent is a tetragonal packing in the space group $P4/n$, where all phosphorus atoms are at the same height at $z = 0$. The cations take positions with the site symmetry $\bar{4}$, whereas the anions are placed on the fourfold rotation axes of the space group $P4/n$ (site symmetry 4). This kind of packing is observed if the anions have (at least) one fourfold axis of rotation, namely with square anions like $AuCl_4^-$, swastika-like anions like $Au(SCN)_4^-$, tetragonal-pyramidal anions like $VOCl_4^-$, and octahedral anions like $SbCl_6^-$ (Fig. 15.8). After a compound whose structure was determined at an early time, this is called the $As(C_6H_5)_4[RuNCl_4]$ type; strictly speaking, that only corresponds to compounds with tetragonal-pyramidal anions [301]. What happens if the symmetry of the anions is not compatible with the site symmetry 4? The consequence is a symmetry reduction, that is, the space group is then a subgroup of $P4/n$ with a correspondingly reduced site symmetry for the anions. The following are examples:

$SnCl_5^-$ ions have a trigonal-bipyramidal structure, point group $\bar{6}2m$. This symmetry is only compatible with the packing of the $As(C_6H_5)_4[RuNCl_4]$ type if the space-group symmetry is reduced from $P4/n$ to $P112/n$, the $SnCl_5^-$ ions being aligned along the former fourfold axis with one of their twofold axes [302]. This requires only one step of symmetry reduction (Fig. 15.7).

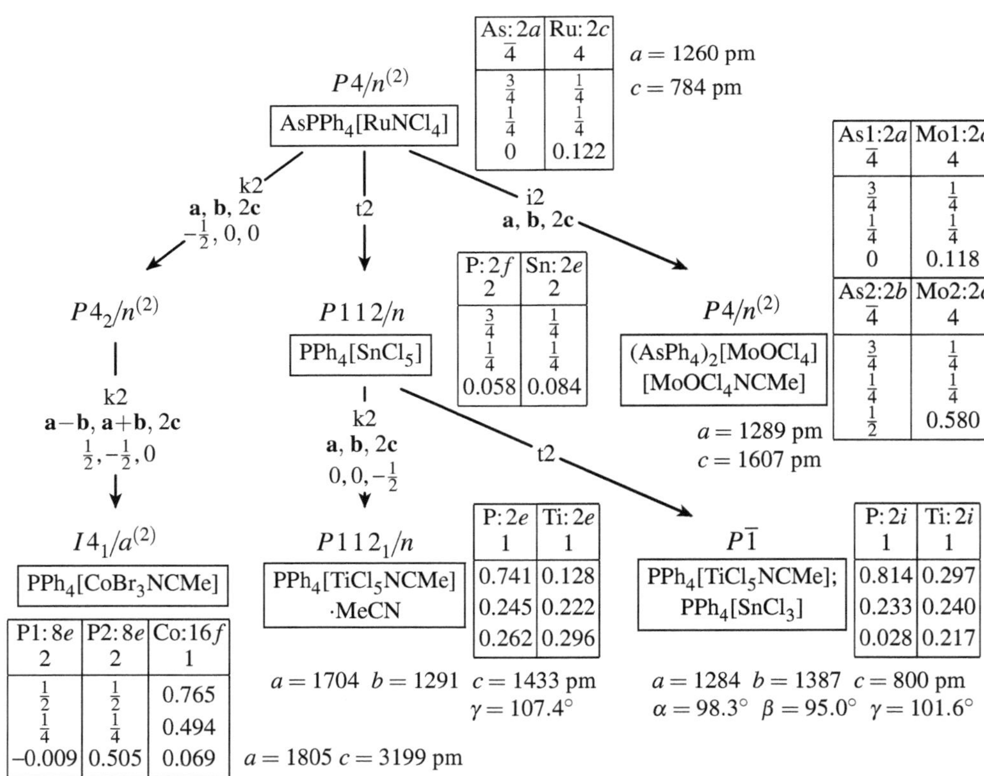

Fig. 15.7 Tree of group–subgroup relations for several tetraphenylphosphonium salts [178, 179].

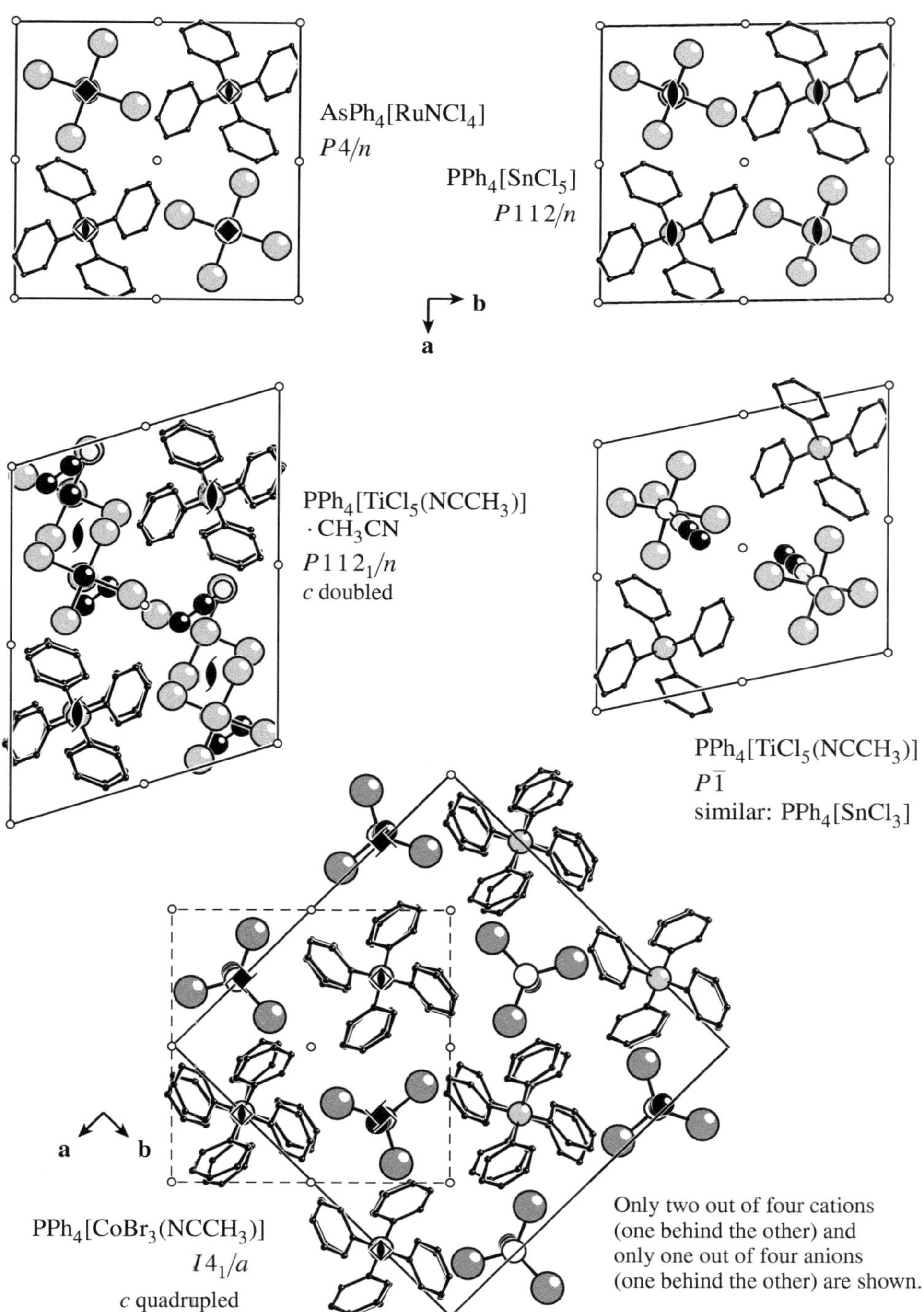

Fig. 15.8 Unit cells of several tetraphenylphosphonium salts.

If the anions have no twofold axes, the symmetry has to be reduced even more. The packing, in principle, is retained with $SnCl_3^-$ ions, but the space-group symmetry is reduced to $P\bar{1}$ [303].

A symmetry reduction can also be necessary if the anions still have a fourfold rotation axis, but their size does not permit them to be aligned along the fourfold rotation axis of the crystal within the predetermined packing of PPh_4^+ ions. For example, $[TiCl_5(NCCH_3)]^-$ ions have point symmetry $4mm$ (not taking into account the H atoms, or if the methyl group is free to rotate). However, along the molecular axis the ions have a length of approximately 1060 pm, whereas the period of translation in the columns of PPh_4^+ ions (c axis) is no longer than 800 pm. Nevertheless, the packing is retained; the $[TiCl_5(NCCH_3)]^-$ ions are tilted relative to the c axis. As a consequence, there can no longer be a rotation axis in this direction, the space-group symmetry is reduced to $P\bar{1}$ (Figs 15.7 and 15.8) [304].

If one $[MoOCl_4]^-$ ion and one $[MoOCl_4(NCCH_3)]^-$ ion are aligned one behind the other on a fourfold axis of rotation, together they have a length of approximately 1600 pm. That is twice the length of 800 pm between two neighbouring $AsPh_4^+$ ions. The packing of $(AsPh_4)_2[MoOCl_4][MoOCl_4(NCCH_3)]$ remains tetragonal and also the space-group type $P4/n$ is retained, albeit with doubled c axis. $P4/n$ has an isomorphic subgroup $P4/n$ with doubled c axis [305].

Even $PPh_4[TiCl_5(NCCH_3)]{\cdot}CH_3CN$ follows the same packing principle. To obtain the space required for the additional acetonitrile molecule, the columns of the PPh_4^+ ions are separated a little and the $[TiCl_5(NCCH_3)]^-$ ions move away from the former 4 axes. After the symmetry reduction $P4/n \longrightarrow P112/n$ $\text{—2c}\rightarrow$ $P112_1/n$ the $[TiCl_5(NCCH_3)]^-$ ions are placed alternately on two sides of a 2_1 axis, each one opposite to one of the additional acetonitrile molecules [304].

Still another packing variant is observed with $[CoBr_3(NCCH_3)]^-$ ions. These are tilted relative to the fourfold axis in four different directions, such that the 4 axis is converted to a 4_1 axis, with fourfold length of c. That requires two steps of symmetry reduction from $P4/n$ to $I4_1/a$. The packing of the PPh_4^+ ions remains almost unaffected, although there are two symmetry-independent PPh_4^+ ions and their site symmetry has been reduced from $\bar{4}$ to 2 [306].

$P\bar{1}$ and $P112_1/n$ ($P2_1/c$) are the most common space-group types of molecular compounds (Table 15.1). Neither in $PPh_4[TiCl_5(NCCH_3)]$ nor in $PPh_4[TiCl_5(NCCH_3)]{\cdot}CH_3CN$ do any particles occupy a special position, and the metrics of their unit cells are far from being tetragonal. Nevertheless, in both cases the close relationship to the tetragonal $As(C_6H_5)_4[RuNCl_4]$ type is evident; the molecular packing is still pseudotetragonal. A 'principle of symmetry avoidance' is out of the question. However, the situation requires a symmetry reduction from the ideal symmetry in the space group $P4/n$, quite in accordance with aspect 2 of the symmetry principle as stated in Section 1.1.

Exercises

Solutions on page 336.

15.1. In urea inclusion compounds the urea molecules form a honeycomb-like host structure via hydrogen bonds in the space group $P6_1 22$. Hydrocarbon molecules or their derivatives are included in the channels of the comb. If the length of the guest molecules or a multiple thereof is compatible with the c lattice parameter of the host structure, the guest molecules can be ordered or disordered, depending on temperature. If the molecular length is not compatible with c, the result is an incommensurate composite crystal. Set up the tree of group–subgroup relations between the space groups of the host structures with the following crystal data [307]:

guest molecule	compound	space group	a/pm	b/pm	c/pm	references
n-hexadecane	$C_{16}H_{34}\cdot[OC(NH_2)_2]_{12}$-I	$P6_1 22$	822	822	1101	[308]
n-hexadecane	$C_{16}H_{34}\cdot[OC(NH_2)_2]_{12}$-II	$P3_2 12$	820	820	2200	[309]
n-hexadecane	$C_{16}H_{34}\cdot[OC(NH_2)_2]_{12}$-III	$P2_1 2_1 2_1$	825	1389	1098	[308]
2,9-decandione	$C_{10}H_{18}O_2\cdot[OC(NH_2)_2]_8$	$P3_1 12$	823	823	4416	[310]
2,7-octandione	$C_8H_{14}O_2\cdot[OC(NH_2)_2]_7$	$P6_1 22$	821	821	7691	[311]

15.2. The $(BN)_3$ ring in the hexachloroborazine molecule, $(BN)_3Cl_6$, occupies approximately as much space as a chlorine atom. The crystal structure can be described as a cubic-closest packing of Cl atoms in which one-seventh of the spheres has been replaced by $(BN)_3$ rings. The Cl···Cl van der Waals distance is approximately 350 pm. Derive the structure from the cubic-closest packing. Z in the table refers to the centre of the molecule.

$R3^{(hex)}$ $a = 883.5$ pm, $c = 1031.3$ pm [312]

	x	y	z
Z	0	0	0.002
Cl 1	0.140	0.404	−0.005
Cl 2	0.408	0.265	0

Symmetry relations at phase transitions

16

16.1 Phase transitions in the solid state

Definition 16.1 A phase transition is an event which entails a discontinuous (sudden) change of at least one property of a material.

Properties that can change discontinuously include volume, density, elasticity, electric, magnetic, optical, or chemical properties. A phase transition in the solid state is accompanied by a structural change, which means a change of crystallographic data for crystalline solids (space group, lattice parameters, occupied positions, atomic coordinates). In the literature, 'structural phase transitions' are indeed distinguished from 'magnetic', 'electronic', and other kinds of phase transitions, but all of these transitions are always accompanied by (sometimes very small) structural changes. For example, a transition from a paramagnetic to a ferromagnetic phase entails spontaneous magnetostriction, i.e. the structure experiences a slight deformation.

Even the so-called isostructural or isosymmetric phase transitions involve structural changes. These are transformations that exhibit no changes of the occupation of Wyckoff positions, the number of atoms in the unit cell, and the space group (within the scope of the parent clamping approximation; Section 12.1, page 147). For example, samarium sulfide, SmS, crystallizes in the NaCl type. At ambient pressure it is a black semiconductor. Under pressure it experiences a phase transition to a metallic, golden lustrous modification, combined with a sudden volume change by -14% and a delocalization of electrons from the $4f$ subshell of the samarium atoms into a metallic $5d6s$ band ('electronic phase transition'; $Sm^{2+}S^{2-} \rightarrow Sm^{3+}S^{2-}e^{-}$). The golden SmS also has the NaCl structure; nevertheless, there is a structural change because the lattice parameter experiences a sudden decrease from $a = 591$ pm to $a = 562$ pm (values at 58 K and 1.13 GPa [313, 314]). For more examples see [315]. Such isosymmetric phase transitions do not involve a group–subgroup relation between the phases; they simply are isotypic phases.

As a rule, the chemical composition does not change at a phase transition.

There exist numerous kinds of phase transitions, and the corresponding theories and experimental findings are the subject of an extensive field of research in physics. This chapter gives only a small insight, group-theoretical aspects being in the foreground. A few additional considerations are dealt with in Chapter 22). First, we introduce a few common distinctions.

16.1.1 First- and second-order phase transitions

A thermodynamically stable phase can become unstable relative to another phase by a change of the external conditions (temperature, pressure, electric field, magnetic field, mechanical forces); this causes a stress that can induce a transformation. The transformation is enantiotropic, i.e. reversible; it can be reversed by returning to the original conditions.[1] If only one variable of state is changed, for example the temperature T or the pressure p, then there is a point of transition T_c or p_c at which the phases are at equilibrium with one another and where the Gibbs free energy change is $\Delta G = 0$.

[1] Monotropic phase transitions are irreversible, i.e. they start from a phase that is only kinetically stable, but thermodynamically unstable at any condition.

Let: G = Gibbs free energy (free enthalpy), H = enthalpy, U = internal energy, S = entropy, and V = volume. They are functions of T and p and further variables of state like electric or magnetic fields; we restrict our considerations to the variables T and p.

For reversible processes, according to the laws of thermodynamics, (negative) entropy and volume are the first derivatives of the Gibbs free energy $G = H - TS = U + pV - TS$ with respect to temperature and pressure. The partial derivatives express that this is valid if the other variable(s) of state are held constant. The second derivatives with respect to T and p express the specific heat at constant pressure, $C_p = \partial H/\partial T$; the compressibility of the volume V at constant temperature, $\kappa V = -\partial V/\partial p$; and the thermal expansion of the volume, $\alpha V = \partial V/\partial T$, α being the coefficient of thermal expansion.

$$\frac{\partial G}{\partial T} = -S \qquad \frac{\partial G}{\partial p} = V \qquad (16.1)$$

$$\frac{\partial^2 G}{\partial T^2} = -\frac{\partial S}{\partial T} = -\frac{1}{T}\frac{\partial H}{\partial T} = -\frac{C_p}{T}$$

$$\frac{\partial^2 G}{\partial p^2} = \frac{\partial V}{\partial p} = -\kappa V$$

$$\frac{\partial^2 G}{\partial p \partial T} = \alpha V$$

Definition 16.2 after EHRENFEST (1933). At a **first-order phase transition** at least one of the first derivatives of the Gibbs free energy G experiences a discontinuous change, i.e. $\Delta S \neq 0$ or $\Delta V \neq 0$. It is accompanied by the exchange of conversion energy (latent heat) $\Delta H = T\Delta S$ with the surroundings. At a **second-order phase transition** volume and entropy experience a continuous variation, but at least one of the second derivatives of G exhibits a discontinuity. At a phase transition of n-th order, a discontinuity appears for the first time at the n-th derivative (however, third and higher order transitions do not really occur).

The mentioned distinction of the order of phase transitions after EHRENFEST is based on purely thermodynamic arguments and macroscopic measured variables, without taking into account the interatomic interactions and structures of the substances.

Since the times of EHRENFEST, theory and experimental measurement techniques have advanced considerably. The mentioned classification has turned out to be neither sufficiently ample nor sufficiently precise. In phase-transition physics, the distinction of discontinuous and continuous phase transitions has replaced the classification after EHRENFEST:

Definition 16.3 At a **discontinuous phase transition** entropy as well as an order parameter change discontinuously. At a **continuous phase transition** the entropy and order parameter change continuously (in infinitesimal small steps).

Section 16.2.2 explains what an order parameter is.

Often there is hardly any difference between the old and the new classification. The terms 'first' and 'second order' continue to be used. 'First-order transition' is often used as a synonym for 'discontinuous phase transition' since every discontinuous transition is also a first-order transition according to EHRENFEST. 'Second-order transition' usually means 'continuous phase transition'. In any case, all phase transitions, including the continuous transitions, exhibit a discontinuous behaviour of certain thermodynamic functions.

In the theory of continuous phase transitions, critical points are of essential importance, and the corresponding physical laws are called *critical phenomena*. Continuous phase transitions exhibit a number of properties in common ('universality'), in contrast to the discontinuous transitions. In particular, they do not depend on the kind of interaction between the atoms, but on their range of action and on the number of space dimensions in which they are active. For more details see Chapter 22.

A discontinuous phase transition is always accompanied by an exchange of latent heat $\Delta H = T\Delta S$ with the surroundings. This exchange cannot occur instantly, i.e. with an infinitely fast transfer of heat. Accordingly, discontinuous phase transitions exhibit *hysteresis*: The transition lags behind the causing temperature or pressure change. When the point of transition is reached, nothing happens; under equilibrium conditions the transition does not get started. Only after the point of transition has been crossed, i.e. under non-equilibrium conditions, does the transition get started, provided that nuclei of the new phase are being formed which then grow at the expense of the old phase. Simultaneously, a temperature gradient is being built up with the surroundings, thus enabling the flow of latent heat. Predominantly, nuclei are formed at defect sites in the crystal.

In contrast, continuous phase transitions show no latent heat, no hysteresis, and no occurrence of metastable phases.

16.1.2 Structural classification of phase transitions

A classification of phase transitions frequently encountered in the literature is due to BUERGER [37]:

1. **Reconstructive phase transitions**: Chemical bonds are broken and rejoined; the reconstruction involves considerable atomic motions. Such conversions are always first-order transitions.
2. **Displacive phase transitions**: Atoms experience small shifts.
3. **Order–misorder transitions**:[2] Different kinds of atoms that statistically occupy the same crystallographic point orbit in a crystal become ordered in different orbits or vice versa. Or molecules that statistically take several orientations become ordered in one orientation.

[2] The common term is term 'order–disorder transition'. However, we take the liberty of replacing the unfortunate term 'disorder' with the more precise term 'misorder' because there is still some order in the 'disordered' state; it is merely an order with faults.

The mentioned classification is often made according to the known structures before and after the phase transition, without experimental evidence of the actually occurring processes and atomic motions during the transition. Nevertheless, frequently there are no doubts as to what category a phase transition is to be assigned. However, the qualitative kind of the classification does not always permit a clear assignment. There exists no unanimous opinion in the literature, of what has to be considered to be a displacive transition and how it should be delimited, on one hand from reconstructive, on the other from order–misorder transitions. For example, some authors call transitions displacive only if the atomic shifts are not only small, but also take place continuously. Some other authors call them displacive if there is a group–subgroup relation between the involved space groups, and reconstructive if not.

Having said that, BUERGER's classification sometimes permits subtle distinctions. For example, $PbTiO_3$ and $BaTiO_3$ both crystallize at high temperatures in the cubic perovskite type ($Pm\overline{3}m$). $PbTiO_3$ has a Ti atom at the centre of its coordination octahedron. In $BaTiO_3$, however, it does not seem to be placed at the octahedron centre, but at eight positions slightly sideways from the centre, with occupancy probabilities of $\frac{1}{8}$ each. At the cubic → tetragonal transition the Ti atoms of $PbTiO_3$ slightly shift away from the centres of the octahedra, parallel to +**c**; that is a displacive transition. However, upon cooling, $BaTiO_3$ becomes tetragonal in that a Ti atom becomes more ordered by occupying only four out of the eight positions with occupancy probabilities of $\frac{1}{4}$ each. Upon further cooling it restricts itself to two positions (orthorhombic) and finally to one position (rhombohedral). These are order–misorder transitions [316–318].

16.2 On the theory of phase transitions

16.2.1 Vibrational modes

Vibrations play a major role at phase transitions in the crystalline state. In physics, vibrations in crystals are treated as quasiparticles, as *phonons*, by analogy to the photons of light. A vibration with a certain frequency corresponds to phonons of a certain energy. We do not deal here with the theory of vibrations, but mention only some terms that are important for the understanding of the following [319–323].

In solid-state physics a *mode* (vibrational mode) is a collective and correlated vibrational motion of atoms having certain symmetry properties.

The symmetry of point and space groups does not deal with time. The symmetry of time-dependent functions like vibrations is not covered. *Representation theory* offers the necessary tools to deal with the symmetry properties of vibrating atoms in molecules and crystals; it is not a topic of this book, but there exists plenty of literature (e.g. [17–24]). *Irreducible representations*[3] (symmetry species) are used to designate the symmetry of vibrations. The symbols introduced by PLACZEK are commonly used to designate the symmetry of molecular vibrations [324, 325] (called Mulliken symbols in American literature because MULLIKEN adopted them to designate the symmetry of wave functions [326]). A list with the most important symbols can be found in Chapter 23. The symbols are also suited for the designation of the symmetry of vibrations in crystals at the Γ point of the Brillouin zone.

In the language of solid-state physics, the Brillouin zone is a polyhedron around the origin of a coordinate system in 'k space'.[4] The origin of the Brillouin zone is called the Γ point. Every vibration in the crystal is represented by a point with coordinates k_1, k_2, k_3 in the Brillouin zone.

[3] Often called irrep, for short.

[4] k space corresponds to the reciprocal space common in crystallography, but is expanded by a factor of 2π.

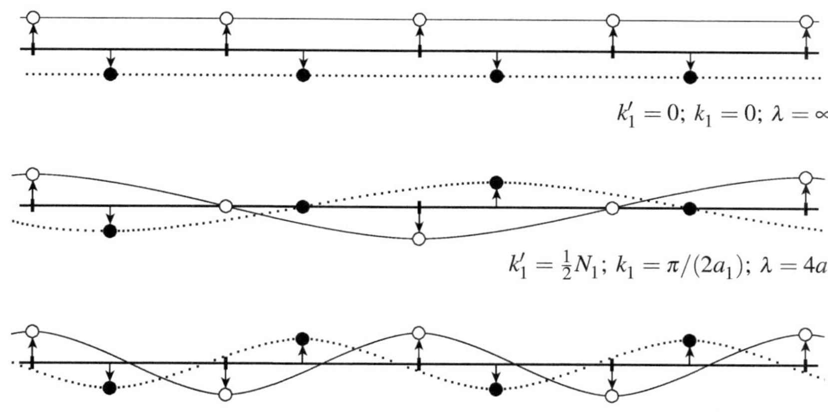

Fig. 16.1 Atomic displacements for three vibrational modes of a chain consisting of two kinds of atoms. Other lattice modes with $0 < k_1 < \pi/a_1$ have wavelengths λ between $2a_1$ and ∞. The number of nodes is $k'_1 = k_1 N_1 a_1/\pi$ with N_1 = number of unit cells in the chain, a_1 = periodicity length (lattice constant); $0 \leq k'_1 \leq N_1$.

Three vibrations are shown in Fig. 16.1 for the simple case of a chain (= crystal with one-dimensional translation symmetry) consisting of two kinds of atoms. N_1 is the number of unit cells in the chain of finite length, a_1 is the lattice constant, and k'_1 is the number of vibrational nodes of the standing wave. The standing wave with $k'_1 = 0$ has no nodes; all translation-equivalent atoms move in the same way and synchronously in all unit cells (in other words: the particle lattice of one kind of atoms vibrates relative to the particle lattice of the other kind of atoms). The vibration with $k'_1 = \frac{1}{2}N_1$ has half as many nodes as unit cells. The vibration with $k'_1 = N_1$ has exactly one node per unit cell; translation-equivalent atoms in adjacent cells move exactly in opposite directions (antisymmetric; opposite in phase). There can be no more nodes than unit cells. Instead of designating the number of nodes by integral numbers k'_1 that run from zero to N_1, numbers $k_1 = k'_1\pi/(N_1 a_1)$ are used that are independent of the number of cells. The maximal possible number of nodes $k'_1 = N_1$ then corresponds to the value $k_1 = \pi/a_1$. Three numbers $0 \leq k_1 \leq \pi/a_1$, $0 \leq k_2 \leq \pi/a_2$, $0 \leq k_3 \leq \pi/a_3$ are needed for a crystal with translational symmetry in three dimensions, and nodal surfaces replace nodal points.

A point at the boundary of the Brillouin zone has $k_i = \pi/a_i$. Special points at the Brillouin zone boundary are designated with capital latin letters (e.g. K, M, X). Vibrational modes at the Γ point, $k_1, k_2, k_3 = 0$, have no nodal surfaces; they have atoms that perform the same vibrational motions synchronously in all unit cells of the crystal. Only modes that are very close to the Γ point can be measured by optical techniques (infrared and Raman spectroscopy). Modes outside of the Γ point can be measured by inelastic neutron scattering.

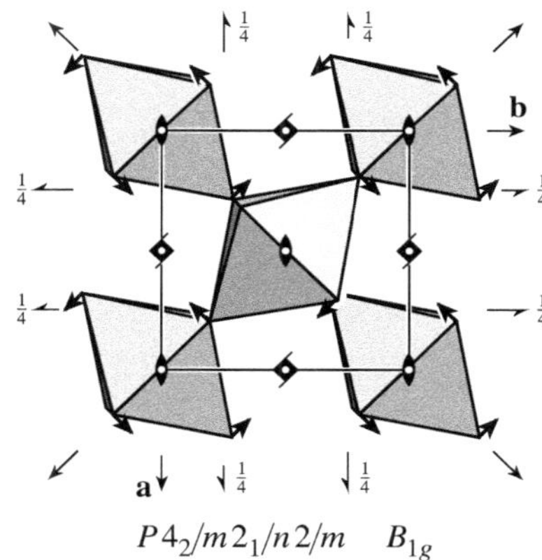

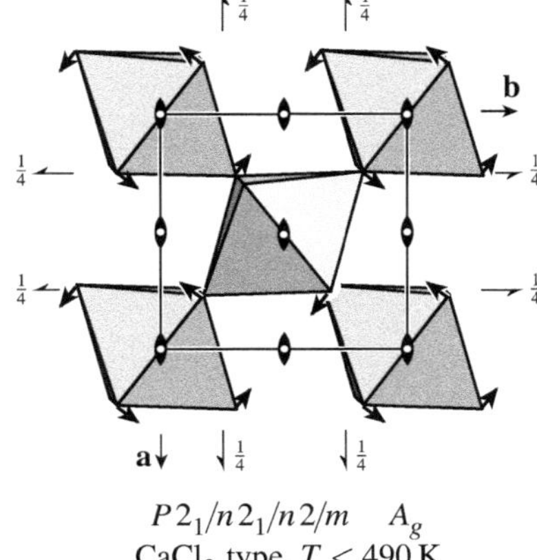

Fig. 16.2 Mutual rotations of the coordination octahedra of the soft mode of $CaCl_2$ in the rutile and $CaCl_2$ types.

For the example of $CaCl_2$, one vibrational mode of the rutile-type modification is marked by arrows in Fig. 16.2 (top). This mode involves a mutual rotational vibration of the strands of edge-sharing coordination octahedra that run along **c**; the motion is the same and synchronous in all unit cells and is thus at the Γ point. In tetragonal $CaCl_2$ ($P4_2/m2_1/n2/m$) this mode is of symmetry species B_{1g}. As explained in Table 23.1 (page 297), in this case B means antisymmetric with respect to the 4_2 axes, that is, after execution of a 4_2 operation the direction of motion of the atoms has to be reversed. The lower-case 1 means symmetric with respect to the 2_1 axes parallel to **a** (and **b**). The index g means symmetric with respect to the points of inversion.[5]

[5] g from the German *gerade*, meaning even.

In the orthorhombic low-temperature modification of $CaCl_2$ ($P2_1/n2_1/n2/m$; Fig. 16.2, bottom), the same kind of mode is symmetric with respect to all symmetry elements, it is *totally symmetric* (also called the *identity* or *unit representation*). Totally symmetric means that the space-group symmetry is completely fulfilled at any time of the vibrational motion. The symmetry species is A_g. The B_{1g} mode of the tetragonal as well as the A_g mode of the orthorhombic form are observable by Raman spectroscopy.

16.2.2 The Landau theory of continuous phase transitions

In 1937, L. D. LANDAU introduced a phenomenological theory[6] to treat continuous phase transitions in a uniform way. The theory has been considerably expanded by E. M. LIFSHITZ and V. L. GINZBURG and more recently by many more authors, and has finally been extended to cover discontinuous phase transitions as well [12, 327–332]

[6] A phenomenological theory relates observations mathematically without tracing them back to a fundamental law.

First, we present the basic concepts of Landau theory for continuous phase transitions by means of the example of the tetragonal $\rightleftharpoons$ orthorhombic phase transition of calcium chloride. An order parameter is defined to trace the course of the phase transition. The *order parameter* is an appropriate, measurable quantity that is apt to account for the essential differences of the phases.

In the first instance, the order parameter is a quantity that can be measured macroscopically. Depending on the kind of phase transition, it can be, for example, the difference of the densities of phases 1 and 2 or the magnetization at a transition from a paramagnetic to a ferromagnetic phase. However, certain changing structural parameters can also be chosen as order parameters. The order parameter must fulfil certain conditions; these include the following:

In the lower-symmetry phase, the order parameter must change continuously with temperature (or pressure or some other variable of state) and finally it must vanish at the critical temperature T_c (or at the critical pressure p_c), that is, it must become zero; in the higher-symmetry phase, after having surpassed T_c, it remains zero. The *critical temperature* corresponds to the point of transition (at a temperature-dependent transition). For transitions in the solid state, the order parameter has to fulfil certain symmetry properties that are explained on the following pages. As explained in Section 22.2, the order parameter η can be used to expand a power series for the Gibbs free energy, $G = G_0 + \frac{1}{2}a_2\eta^2 + \frac{1}{4}a_4\eta^4 + \frac{1}{6}a_6\eta^6 + \cdots$. This power series comprises only even powers of η and thus is invariant to changes of the sign of η.

$CaCl_2$ is tetragonal in the rutile type above the transition temperature $T_c =$ 490 K ($P4_2/mnm$); below it is orthorhombic in the $CaCl_2$ type (Figs 1.2 and 1.3, page 6). As compared to the tetragonal form, the strands of edge-sharing coordination octahedra running parallel to **c** have been mutually rotated in the orthorhombic form. The angle of rotation becomes larger the further the temperature is below T_c. As with all (static) structural descriptions, the stated position of the octahedra refers to their mean positions, that is, the equilibrium positions of the atomic vibrations.

Even with the slightest rotation of the octahedra the symmetry can no longer be tetragonal. The lattice parameters a and b become unequal and the mirror planes running diagonally through the unit cell of the rutile structure cannot be retained (in addition to some more symmetry elements). The symmetry of the $CaCl_2$ type has to be that orthorhombic subgroup of $P4_2/mnm$ in which these mirror planes have been omitted; that is the space group $P2_1/n2_1/n2/m$ ($Pnnm$). Universally it holds:

First criterion of Landau theory: At a continuous phase transition, the space group $\mathcal{H}$ of one phase needs to be a subgroup of the space group $\mathcal{G}$ of the other phase: $\mathcal{H} < \mathcal{G}$ ('the symmetry is broken'). It does not have to be a maximal subgroup.

Starting at a temperature $T > T_c = 490$ K, upon reduction of temperature, the frequency of the B_{1g} mode of the octahedra of tetragonal $CaCl_2$ shifts towards zero. This kind of behaviour is called a *soft mode*. After the temperature has fallen below the critical temperature $T_c = 490$ K, the equilibrium position of the octahedra has switched from that of the tetragonal to that of the orthorhombic form. The strands of octahedra now perform rotational vibration about the new equilibrium position, which now corresponds to the symmetry species A_g of space group $Pnnm$. The frequency shift is reversed, with rising frequencies the more the temperature is below T_c.

Since the motion of atoms during the phase transition is directly connected with the vibrational mode, it is said that 'the phase transition is driven by a soft mode' or 'the mode condenses' or 'the mode freezes'. The symmetry species of the higher-symmetry phase is called the *active representation*.

Second criterion of Landau theory: The symmetry breach is due to only one irreducible representation (symmetry species), the active representation. In the higher-symmetry phase it cannot be the identity representation (totally symmetrical vibration). Upon the phase transition it becomes the identity representation of the lower-symmetry phase.

In Fig. 16.3 the vector $\vec{\eta}$ has been drawn that marks the shift of the equilibrium position of a Cl atom as compared to the position in the high-temperature form. The vector shows how the coordination octahedra have been rotated. At a given temperature $T < T_c$ the octahedra have been rotated by a certain angle, as compared to the rutile type, and the vector $\vec{\eta}$ has some specific length η. Rotation of the octahedra in the opposite direction $-\vec{\eta}$ is completely equivalent energetically. In the tetragonal high-temperature modification the value is $\eta = 0$. The value of η of the shift can be used as an order parameter.

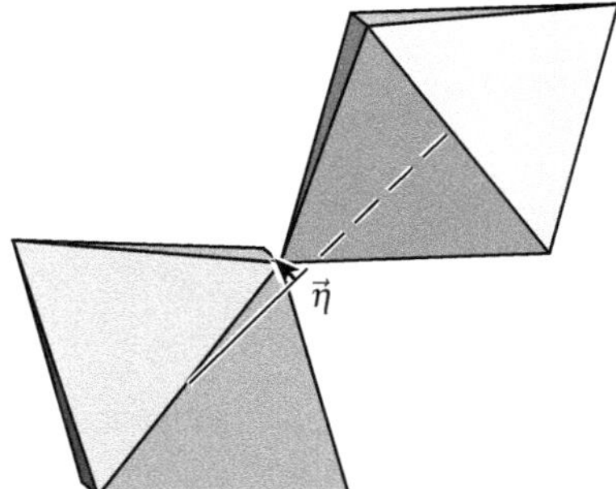

Fig. 16.3 Two connected octahedra in the low-temperature modification of $CaCl_2$ and shift vector $\vec{\eta}$ of the common Cl atom as compared to the position in the high-temperature modification.

According to the theory, the order parameter must behave with respect to symmetry in the same way as the active representation (it must be 'transformed like the active representation'); for $CaCl_2$ this is the symmetry species B_{1g} of $P4_2/mnm$. As can be checked in corresponding tables (in the 'character table' of the point group $4/mmm$, see e.g. [17, 18, 24]), a 'basis function' $x^2 - y^2$ belongs to B_{1g}, x and y being Cartesian coordinates. In our case, the coordinates x and y of the Cl atom can be taken. η being proportional to $x - y$, and with $x^2 - y^2 = (x - y)(x + y)$ and $x + y \approx$ constant, η is proportional to $x^2 - y^2$ to a good approximation and is thus an appropriate order parameter. A couple of further variables can be used as approximate order parameters for $CaCl_2$; due to $\sin\varphi \approx \tan\varphi \approx \varphi$ this is valid for the (small) angle of rotation φ of the octahedra and for the ratio $(b-a)/(b+a)$ of the lattice parameters (the so-called spontaneous deformation).

In the lower-symmetry phase ($T < T_c$) the order parameter follows a power law as mentioned in the margin. A is a constant and β is the *critical exponent*. As explained in Chapter 22, $\beta \approx 0.5$ holds if there are long-range interactions between the particles; this is fulfilled for $CaCl_2$, due to the interconnection of all octahedra. For short ranges of interaction, for example, for magnetic interactions, $\beta \approx 0.33$.

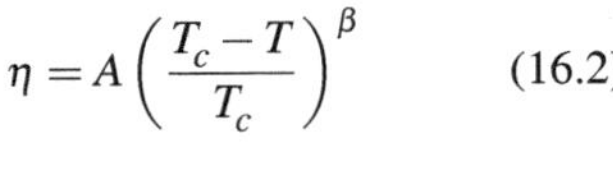

$$\eta = A\left(\frac{T_c - T}{T_c}\right)^{\beta} \qquad (16.2)$$

Landau theory agrees well with the experimental data shown in Fig. 16.4. There is no volume discontinuity at the point of transition. The order parameter $\eta' = (b-a)/(b+a)$ is consistent with the power law (16.2) over a large temperature range, a critical exponent of $\beta = 0.45$ matching the experimental dependence quite well. One lattice mode of symmetry species B_{1g} of $P4_2/mnm$

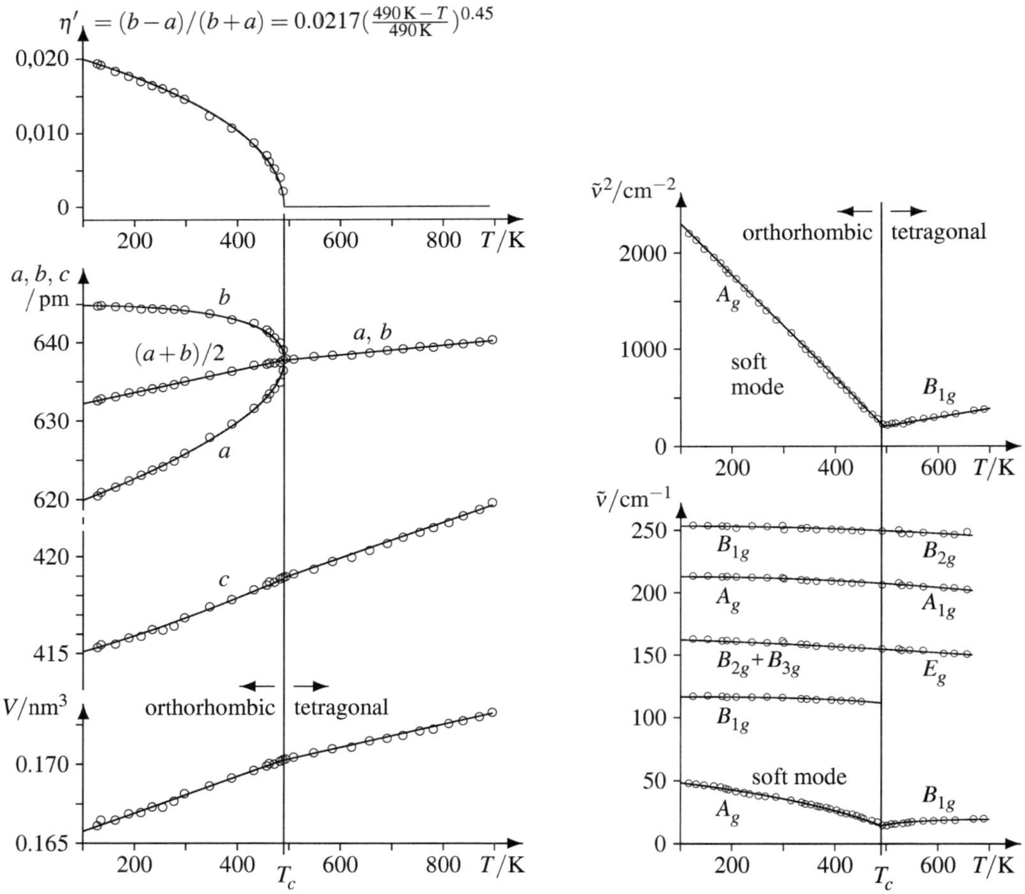

Fig. 16.4 Upper left: Plot of the temperature-dependent course of the order parameter η' of $CaCl_2$. Lower left: Temperature dependence of the lattice parameters and the unit cell volume of $CaCl_2$ [38, 40]. Right: Temperature dependence of the frequencies of Raman-active vibrations of $CaCl_2$ [333]. The B_{1g} mode of the orthorhombic modification at 120 cm^{-1} changes to the symmetry species A_{2g} of the tetragonal modification and becomes Raman-inactive.

exhibits soft mode behaviour and switches to the symmetry species A_g of $Pnnm$, while the frequencies of all other modes remain nearly temperature independent, even across the point of transition. The minimum of the soft mode frequency is exactly at the transition temperature found by crystallography.

The same kind of rutile type $\rightleftharpoons$ $CaCl_2$ type phase transition has been observed with further compounds (e.g. $CaBr_2$, MgF_2, NiF_2, ZnF_2, stishovite-SiO_2, SnO_2). If the phase transition is driven by pressure, the $CaCl_2$ type is the high-pressure modification.

The sketched model of a continuous phase transition, driven by a lattice mode at the Γ point of the Brillouin zone, matches the notion of a continuous change from one structure to the other with the synchronous participation of all atoms of the crystal (or of a large domain of the crystal). A change of space group takes place at the point of transition. In addition, certain physical properties change abruptly. For example, $CaCl_2$ becomes ferroelastic below the point of conversion ('pure and proper ferroelastic' [334]).

If a lattice mode at the Γ point is involved, like in the case of the rutile type $\rightleftharpoons$ $CaCl_2$ type transformation, the space groups of both phases are *translationengleiche* (their primitive unit cells have the same size). If the freezing mode is at a boundary point of the Brillouin zone, atoms in unit cells adjacent in the corresponding direction vibrate with exactly opposite displacements; after the mode has frozen, the volume of the unit cell has been doubled. This means that the symmetry breach must involve a step of symmetry reduction with a *klassengleiche* group–subgroup relation of index 2.

The symbols listed in Chapter 23 are not applicable to vibrational modes outside the Γ point, that is, for modes that have atomic motions that differ from unit cell to unit cell. For this case, other symbols are used [335, 336].

If a non-totally symmetrical lattice mode freezes, the result is a distorted structure whose space group is a subgroup of the original space group. Such a subgroup is called an *isotropy subgroup*. What isotropy subgroups occur depending on symmetry species has been listed for all symmetry species of all space groups [323, 336]. They can also be determined with the computer program ISOTROPY [337]. For the theoretically important interplay between vibrational modes and isotropy subgroups refer to, for example, [327,338,339].

16.3 Domains and twinned crystals

A phase transitions in the solid state often results in a *domain structure* of crystalline phases. The domain structure is the result of nucleation and growth processes. If the arrangement of the building blocks in both phases is rather different, there is no crystallographic relation between the orientations of the initial and the new phase. The nuclei form with orientations at random, depending on the energetic conditions at the defect sites in the crystal. This case is frequently encountered among molecular crystals.

However, if the lattices of the two crystalline phases match to some degree, for a nucleus of the new structure that develops at a defect site of the old structure, it is more favourable to keep the orientation of the old phase. The orientational relations between the phases before and after the transformation, as a rule, are not the result of a homogeneous process involving a simultaneous motion of all atoms in a single crystal. The crystalline matrix of the substrate rather governs the preferred orientation adopted by the nuclei that are formed during the course of the nucleation process. The crystallites that result from the subsequent growth of the nuclei maintain their orientations.

Even at a continuous phase transition that is driven by a soft mode, as a rule, it is not the whole crystal that transforms at once. The ever-existing faults in a crystal and the mosaic structure of real crystals impede lattice modes running uniformly through the whole crystal, and the gradients that inevitably emerge in a crystal during a temperature or pressure change cause different conditions in different regions of the crystal. In addition, fluctuations arise when the point of transition is approached; they differ statistically in different regions of the crystal (cf. Section 22.3 for the fluctuations).

Therefore, the phase transition starts in different regions of the crystal simultaneously or at consecutive time intervals, such that domains result. The domains grow until they meet. If two matching domains meet, they can combine to form a larger domain; otherwise a domain boundary results between them. Once formed domain boundaries can advance through the crystal, one of the domains growing at the expense of the other.

The resulting system of intergrown crystallites is called a *topotactic texture* after W. KLEBER [340].

Under these circumstances, aspect 3 of the symmetry principle, as stated in Section 1.1, is fully effective [13, 121, 342, 343]. A phase transition that is connected with a symmetry reduction will result in new phases that consist of

twin domains
if the formed phase belongs to a crystal class of reduced symmetry, or

antiphase domains
if translational symmetry is lost.

Therefore, twinned crystals are formed if the new phase belongs to a *translationengleiche* subgroup, and antiphase domains if it is a *klassengleiche* subgroup. In the physical literature, phase transitions between *translationengleiche* space groups are called *ferroic transition*s, the lower-symmetry phase being the ferroic phase and the higher-symmetry phase the *para* phase. Non-ferroic transitions take place between *klassengleiche* space groups. For stricter definitions of these terms see [341].

Twin domains are dealt with in Section 16.5, antiphase domains in Section 16.6.

The total number of domains formed, of course, depends on the number of nucleation sites. The number of different *domain kinds*, however, is ruled by the index of the symmetry reduction. At a *translationengleiche* symmetry reduction of index 3 (t3 group–subgroup relation) we can expect twins with three kinds of domains, having three different orientations. An isomorphic subgroup of index 5 (i5 relation), since it is a *klassengleiche* symmetry reduction, will entail five kinds of antiphase domains. If the symmetry reduction includes several steps (in a chain of several maximal subgroups), the domain structure will become more complicated. With two t2 group–subgroup relations, we can expect twins of twins.

The actual number of observable domain kinds may be less than expected if a domain kind is not formed during nucleation. This can be controlled by the experimental conditions. For example, an external electric field can suppress the formation of more than one kind of differently oriented ferroelectric domains at a phase transition from a paraelectric to a ferroelectric modification. Among very small crystals (in the nanometre range), it can happen that only one nucleus is formed per crystal and that its growth is fast enough to fill the whole crystal before a second nucleus is formed, with the result of single-domain crystals [344].

As a rule, among temperature-driven phase transitions, the high-temperature form has the higher symmetry. Among phase transitions induced by electric or magnetic fields, the lower-symmetry phase is the one persisting while the field is being applied. No corresponding rules can be stated for phase transitions induced by pressure; under pressure, the more dense phase is the more stable one.

The symmetry relations are also valid for *topotactic reactions*, when a chemical reaction takes place in the crystalline solid and the orientation of the domains of the product is determined by the original crystal (Section 17.1).

16.4 Can a reconstructive phase transition proceed via a common subgroup?

Some substances exhibit two subsequent *displacive* phase transitions at two different pressures with small atomic displacements, such that the first transition involves a symmetry reduction, the second a symmetry enhancement.

Silicon offers an example. When increasing pressures are exerted upon silicon, it experiences several phase transitions:

Si-I	$\overset{10.3\ \text{GPa}}{\rightleftharpoons}$	Si-II	$\overset{13.2\ \text{GPa}}{\rightleftharpoons}$	Si-XI	$\overset{15.6\ \text{GPa}}{\rightleftharpoons}$	Si-V	$\overset{38\ \text{GPa}}{\rightleftharpoons}$	further
diamond		$I4_1/amd$		$Imma$		$P6/mmm$		modifications
type		β-Sn type						

Silicon-V ($P6/m2/m2/m$) has a simple hexagonal structure [345]. There exists no group–subgroup relation between $I4_1/a2/m2/d$ (β-tin type) and $P6/m2/m2/m$ and they have no common supergroup. However, the structure of Si-XI is intimately related with both that of Si-II and Si-V, and the space group of Si-XI, $I2_1/m2_1/m2_1/a$, is a common subgroup of $I4_1/a2/m2/d$ and $P6/m2/m2/m$ (Fig. 16.5).

Assuming no atomic displacements, the calculated coordinates of a Si atom of Si-XI would be $0, \frac{1}{4}, -0.125$ when derived from Si-II, and $0, \frac{1}{4}, 0.0$ when derived from Si-V. The actual coordinates are halfway between. The metric deviations of the lattices are small. The lattice parameter c of the hexagonal structure is approximately half the value of a of tetragonal Si-II. Taking into account the basis transformations given in Fig. 16.5, the expected lattice parameters for Si-XI, calculated from those of Si-V, would be $a_{\text{XI}} = a_{\text{V}}\sqrt{3} =$ 441.5 pm, $b_{\text{XI}} = 2c_{\text{V}} =$ 476.6 pm, and $c_{\text{XI}} = a_{\text{V}} =$ 254.9 pm. Compare this with the observed values (Fig. 16.5).

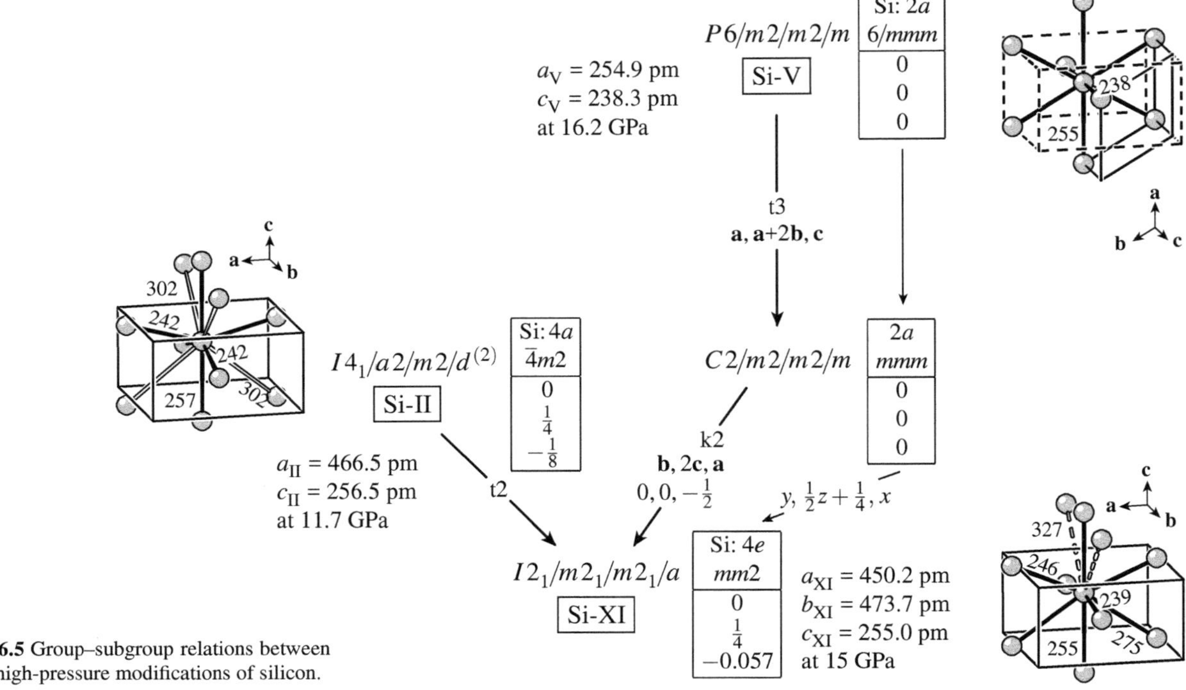

Fig. 16.5 Group–subgroup relations between three high-pressure modifications of silicon.

The coordination of an Si atom is (contact distances in pm up to 340 pm):

Si-II	4 × 242		2 × 257	4 × 302		$I4_1/amd$
	↙	↘	↓	↙	↘	
Si-XI	2 × 239	2 × 246	2 × 255	2 × 275	2 × 327	$Imma$
	↓	↘	↓	↙		
Si-V	2 × 238		6 × 255			$P6/mmm$

At the points of transition, volume discontinuities were observed experimentally, amounting to 0.2% for Si-II ⇌ Si-XI and to 0.5% for Si-XI ⇌ Si-V, and the atomic coordinates shift abruptly [345]. Therefore, the transitions are first-order transitions. However, the abrupt changes are small, and the model of two displacive phase transitions with small atom shifts seems plausible. Most notably, two separate phase transitions are actually observed. In a certain pressure range, the whole crystal actually consists of stable Si-XI. In this case, a group-theoretical relation between Si-II and Si-V via the common subgroup of Si-XI is justified.

A completely different situation is that of *reconstructive* phase transitions when there is no group–subgroup relation between the involved space groups. For such cases, the success of the theory of continuous phase transitions has misled researchers to conceive of mechanisms according to which two subsequent transformations take place via an intermediate phase, similar to the case of the transformations of Si-II to Si-V via Si-XI. The hypothetical intermediate phase is assumed to have a space group that is a common subgroup of the initial and the final space groups.

Reconstructive phase transitions are always first-order transitions and show hysteresis. Therefore, a synchronous motion of atoms in the whole crystal is excluded. The transformation can proceed only by nucleation and growth, with coexistence of both phases during the transformation. Between the growing new and the receding old phase there are phase boundaries. The reconstruction of the structure occurs at and only at these boundaries. Any intermediate state is restricted to this interface between the two phases. The situation is fundamentally different from that of silicon where the existence of Si-XI is not restricted to an interface.

The space groups on either side of an interface are different. They cannot be mapped one onto the other by a symmetry operation; therefore, no symmetry element can exist at the interface. In particular, an intermediate phase whose existence would be restricted to the region of the interface cannot have any space group whatsoever. There is an additional aspect: Space groups describe some static state. During a snapshot of a few femtoseconds duration, no crystal fulfils a space group because the atoms are vibrating and nearly all of them are displaced from their equilibrium positions. Mean atomic positions that fulfil a space group can only be discerned after a somewhat longer time interval has elapsed. During a phase transition, the interface advances through the crystal and the atoms are in motion; there do not exist any mean atomic positions.

Devised reaction mechanisms for reconstructive phase transitions via hypothetical common subgroups are disconnected from reality. Approximately 10 different mechanisms were published merely for the transformation of the NaCl type to the CsCl type, with different common subgroups (see e.g. [346–348] and references therein). The very number is suspect. Formally, arbitrarily many such mechanisms can be devised, as there are always an infinite number of common subgroups.

In such papers, detailed paths of motion of the atoms have been described, depicted in one unit cell. Tacitly, a synchronous motion of the atoms in all unit cells of the crystal is thus suggested, proceeding until they snap into place in the assumed intermediate space group, followed by a synchronous motion from the intermediate phase to the final phase. Synchronous motion, however, is excluded for a first-order transition. But perhaps it has not been meant to be this way, but tacitly only a shift of atoms in one cell has been assumed, followed by further cells, like in a row of falling dominoes. This would be a realistic model with nucleation and subsequent growth. According to theoretical calculations, certain locally confined structural motifs can emerge for short times [349]. However, this does not require and is not in accordance with the occurrence of a hypothetical intermediate phase. It may well be that, for a few picoseconds, the atoms are arranged in a couple of cells as in one cell of the assumed intermediate phase. It is, however, impossible to assign a space group to this state.

In addition, in the case of phase transitions driven by temperature changes, a succession via a primary symmetry reduction followed by a symmetry enhancement is most improbable because a change of temperature should entail a sequence of only symmetry enhancements or only symmetry reductions.

16.5 Growth and transformation twins

Definition 16.4 An intergrowth of two or more macroscopic, congruent, or enantiomorphic individuals of the same crystal species is a twin, if there is a crystal-symmetric relative orientation between the individuals [351].

The individuals are the twin components or twin domains. A *twin operation* specifies their mutual orientation. A twin operation is a symmetry operation that maps one domain onto the other. The twin element is the corresponding symmetry element. This symmetry operation does not belong to the point group of the crystal.

The domain boundary or twin interface is a planar or irregular surface at which the twin components are in contact. If the structures of the adjacent domains have to be harmonized by certain distortions on either side of the boundary, the domain boundary is in the midst of a domain wall that can have a thickness of up to a few unit cells. If the domain boundary is planar, on the atomic scale, its symmetry is a layer group.

The occurrence of twins is a widespread phenomenon. It has to be distinguished whether the twins have developed during the growth of the crystals or by a phase transition in the solid state.

Among **growth twins**, the formation of nuclei of crystallization decides how the individuals are intergrown. In this case, group–subgroup relations between space groups are irrelevant.

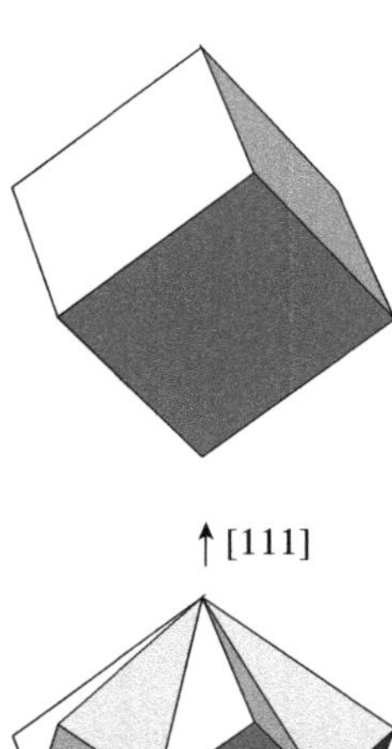

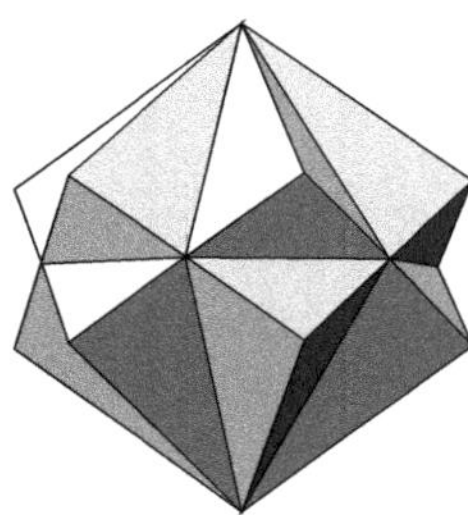

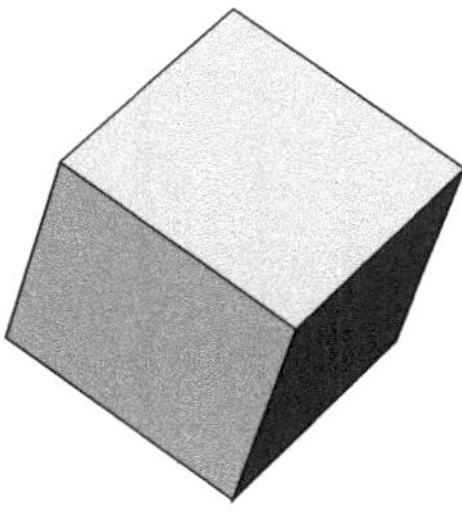

Fig. 16.6 Centre: View of a twin of the mineral fluorite having the twin axis [111]. The two cubes are mutually rotated about the space diagonal [111] by exactly 180°.

The shape of a kind of the frequently occurring twins of the cubic mineral fluorite (CaF_2) is shown in Fig. 16.6. This is an example of a *penetration twin*; these occur only among growth twins. Starting from a nucleus, the crystal grows with two different orientations that are mutually rotated by 180° about [111]. Actually, the twin consists of twelve pyramid-shaped domains, six of each orientation, originating from a common point at the centre of the crystal. Contact twins are simpler; the twin components have a common a plane. A polysynthetic (or lamellar) twin is formed by repetition of contact twins and consists of a sequence of domains with two alternating orientation states.

Transformation twins are formed during a phase transition in the solid state. Frequently, there exists a group–subgroup relation between the involved space groups and the kind of twinning depends on this relation. The group–subgroup relation between the two modifications of calcium chloride is shown in Fig. 1.2 (page 6). $Pnnm$ is a *translationengleiche* subgroup of index 2 of $P4_2/mnm$, so we can expect the formation of twins with two orientational states when $CaCl_2$ is transformed from the tetragonal rutile type to the orthorhombic $CaCl_2$ type. The lattice parameters a and b, being equal in the tetragonal structure, become unequal during the phase transition, either $a < b$ or $a > b$. If we retain the directions of the basis vectors **a** and **b**, then one of the orientational states of the twin components is that with $a < b$, the other one that with $a > b$ (Fig. 1.3, page 6). The two components can be mapped one onto the other by a symmetry operation that has disappeared at the symmetry reduction, for example by reflection through a plane that runs diagonally through the unit cell of the rutile type. This (former) symmetry element becomes the twin element.

During the phase transition, the distinction of the lattice parameters taking place within the matrix of the tetragonal crystal causes mechanical stress. The occurrence of differently oriented twin domains partly alleviates the stress. Therefore, a crystal consisting of many twin domains will be energetically favoured as compared to a single-domain crystal (Fig. 16.7). A uniformly synchronous motion of all atoms in the whole crystal resulting in a single-domain crystal is less probable.

Potassium sulfate crystallizes from aqueous solution in an orthorhombic modification with a pseudohexagonal structure [352, 353]:

$$Pmcn,\ a = 576.3\ \text{pm},\ b = 1007.1\ \text{pm}\ (\approx a\sqrt{3} = 998.2\ \text{pm}),\ c = 747.6\ \text{pm}$$

Growth twins are formed (depending on crystallization conditions) adopting the shape of pseudo-hexagonal plates that are subdivided in three pairs of sector domains with 60° angles (Fig. 16.7). Upon heating, a first-order phase transformation takes place at 587 °C, at which the domains coalesce to one domain, the crystal becoming hexagonal ($P6_3/mmc$). In the hexagonal modification the SO_4^{2-} ions are misordered in at least two orientations (Fig. 16.8).

When allowed to cool again, the SO_4^{2-} ions regain their order, and again orthorhombic twins with three kinds of domains are formed, but this time as transformation twins with lamellar domains.

Isotypic ammonium sulfate also forms the same kind of growth twins. By slight compression, the domain boundaries can be caused to advance through the crystal until a single-domain single crystal is obtained [357]. The expected phase transition to a hexagonal high-temperature form does not occur because ammonium sulfate melts 'prematurely'. Therefore, in this case, the hexagonal aristotype is a hypothetical modification ('paraphase', 'prototype phase') [358]. In contrast to potassium sulfate, ammonium sulfate exhibits a phase transition to a ferroelectric modification below −50 °C in which the SO_4^{2-} tetrahedrons are tilted. The corresponding group–subgroup relations are shown in Fig. 16.8.

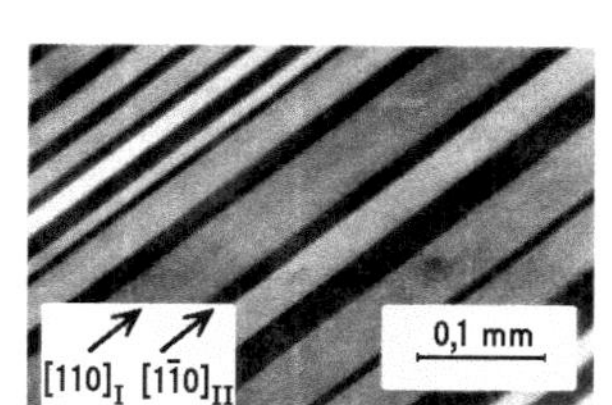

CaBr$_2$

K_2SO_4

Fig. 16.7 Top: Transformation twin with domains I and II of a crystal of $CaBr_2$ ($CaCl_2$ type); direction of view along **c**, photograph taken with crossed polarization filters [350]. Bottom: Growth twin of K_2SO_4 having six domains, precipitated from aqueous solution. View onto (001) of a plate, about 1 mm thick and 5 mm in diameter (taken between polarization filters to enhance the contrast [351]; M. Moret, Milano).

16.6 Antiphase domains

The phase transition of Cu_3Au from its misordered high-temperature phase in the space group $Fm\bar{3}m$ to the ordered phase involves a *klassengleiche* symmetry reduction of index 4 to the space group $Pm\bar{3}m$ (Fig. 1.4, page 7). Therefore, the occurrence of antiphase domains can be expected. The crystal class is retained, and the size of the primitive unit cell is enlarged by a factor of 4 (because the face-centring disappears).

Consider a nucleus of crystallization of the ordered phase and its growth, and let the gold atoms occupy the vertices of the initial unit cell. Let another nucleus be formed at some other site in the crystal, but with a shifted origin, that us, with a gold atom at one of the face centres of the initial unit cell. Somewhere the growing domains will meet. Even though the unit cells have the same size and the same orientation, they do not match with one another because their unit cells are displaced by half of a face diagonal. A domain boundary emerges at which the 'wrong' atoms come to be next to each other.

$P6_3/m2/m2/c$

HT-K_2SO_4

K:2a	K:2c	S:2d	
$\bar{3}m.$	$\bar{6}m2$	$\bar{6}m2$	847 K [354]:
0	$\frac{2}{3}$	$\frac{2}{3}$	$a = 588.6$ pm
0	$\frac{1}{3}$	$\frac{1}{3}$	$c = 811.8$ pm
0	$\frac{3}{4}$	$\frac{1}{4}$	

t3
a, **a**+2**b**, **c**

$x-\frac{1}{2}y, \frac{1}{2}y, z$

$C2/m2/c2_1/m$

4a	4c	4c
$2/m..$	$m2m$	$m2m$
0	$\frac{1}{2}$	$\frac{1}{2}$
0	0.167	0.167
0	$\frac{3}{4}$	$\frac{1}{4}$

k2
$-\frac{1}{4}, -\frac{1}{4}, 0$

$x+\frac{1}{4}, y+\frac{1}{4}, z$

$P2_1/m2_1/c2_1/n$

LT-K_2SO_4
HT-$(NH_4)_2SO_4$

K: 4c	K: 4c	S: 4c	O: 4c	O: 4c	O: 8d	K_2SO_4,	$(NH_4)_2SO_4$,
$m..$	$m..$	$m..$	$m..$	$m..$	1	293 K [352, 353]:	233 K [355]:
$\frac{1}{4}$	$\frac{3}{4}$	$\frac{3}{4}$	$\frac{3}{4}$	$\frac{3}{4}$	0.959	$a = 576.3$ pm	$a = 597.9$ pm
0.295	0.411	0.420	0.416	0.558	0.352	$b = 1007.1$ pm	$b = 1061.2$ pm
0.011	0.674	0.233	0.038	0.296	0.301	$c = 747.6$ pm	$c = 776.2$ pm

t2

$P2_1cn$

LT-$(NH_4)_2SO_4$

N: 4a	N: 4a	S: 4a	O: 4a	O: 4a	O: 4a	O: 4a	
1	1	1	1	1	1	1	219.5 K [356]:
0.258	0.753	$\frac{3}{4}$	0.773	0.777	0.923	0.525	$a = 595.6$ pm
0.299	0.397	0.421	0.399	0.557	0.349	0.380	$b = 1056.9$ pm
0.023	0.682	0.243	0.060	0.281	0.334	0.302	$c = 783.0$ pm

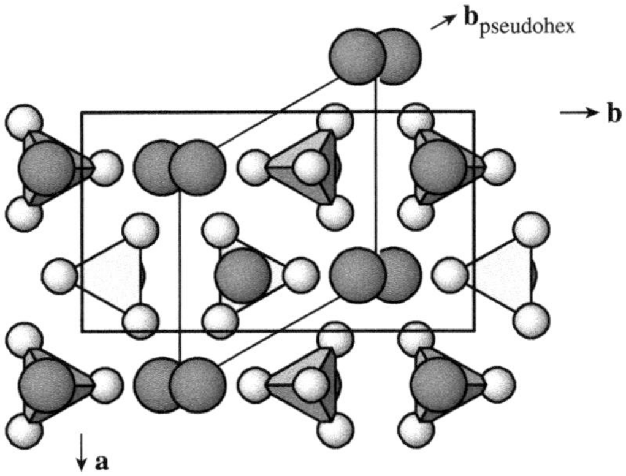

$\mathbf{b}_{pseudohex}$
b
a

Fig. 16.8 Crystal structure of the low-temperature form of potassium sulfate and group–subgroup relations from the hexagonal high-temperature form, in which the sulfate ions are misordered (for this reason no O atom coordinates are given). Ammonium sulfate forms a non-centrosymmetric, ferroelectric low-temperature form.

This is an antiphase boundary between antiphase domains (they have shifted phases). Since the gold atoms have the choice to occupy one out of four positions of the initial cell ($0,0,0,\ \frac{1}{2},\frac{1}{2},0,\ \frac{1}{2},0,\frac{1}{2},\ 0,\frac{1}{2},\frac{1}{2}$), four kinds of domains result. Antiphase domains are also called translation domains.

Antiphase domains do not show up in X-ray diffraction; the structure determination is not hampered (except when the domains are very small and part of the Bragg reflections spreads to diffuse scattering). Antiphase boundaries can be made visible by electron microscopy.

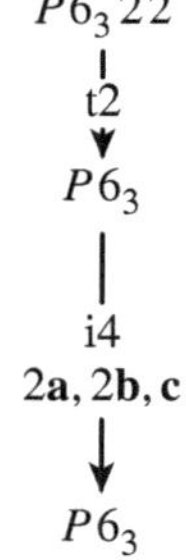

Consider the example of the structures of the high- and low-temperature modifications of $BaAl_2O_4$ shown in Fig. 16.9. These are stuffed tridymite structures, with a framework of vertex-sharing AlO_4 tetrahedra and Ba^{2+} ions in hexagonal channels. At high temperatures $BaAl_2O_4$ is paraelectric in the space group $P6_322$. Upon cooling, a phase transition takes place between 670 and 400 K, resulting in a ferroelectric modification in the space group $P6_3$, with the AlO_4 tetrahedra being mutually tilted and with doubled lattice parameters a and b. This requires the two steps of symmetry reduction shown in the margin.

The *translationengleiche* step causes the occurrence of twin domains that are distinct by opposite orientations (opposite **c** directions) and that alternate

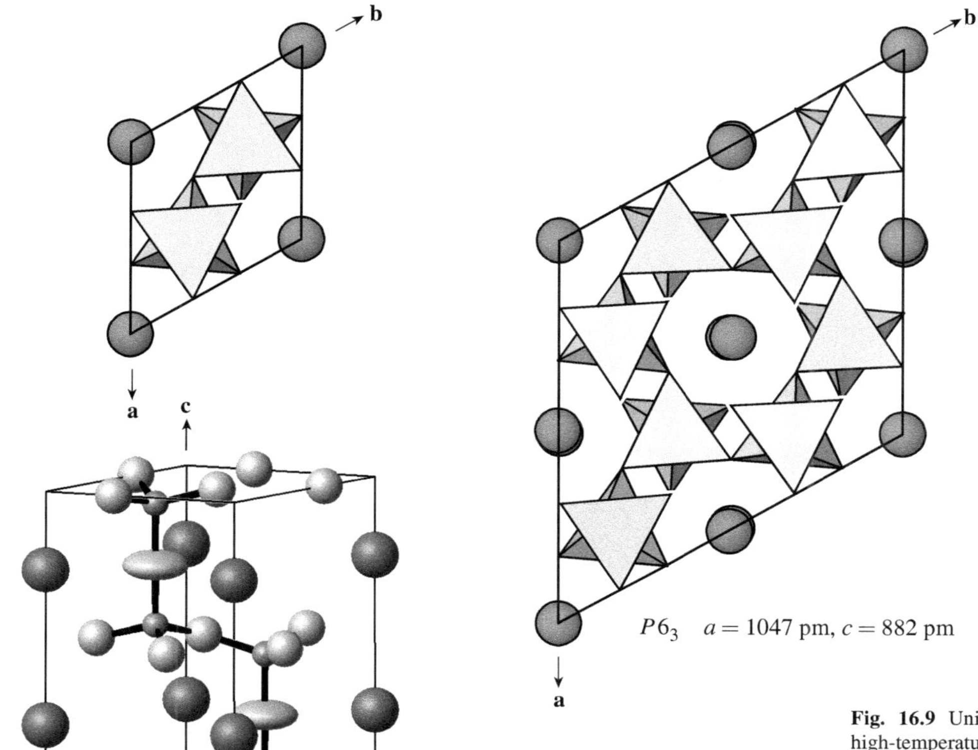

Fig. 16.9 Unit cells of $BaAl_2O_4$. Left: high-temperature form [359]. Right: low-temperature form with unit cell enlarged by a factor of 4 and tilted coordination tetrahedra [360].

in lamellae along c (Fig. 16.10 left). In addition, due to the isomorphic subgroup of index 4, four kinds of antiphase domains are observed. Figure 16.11 shows schematically how the unit cells of the four kinds of domains are mutually displaced. The domain boundaries can be discerned with an electron microscope (Fig. 16.10 centre and right).

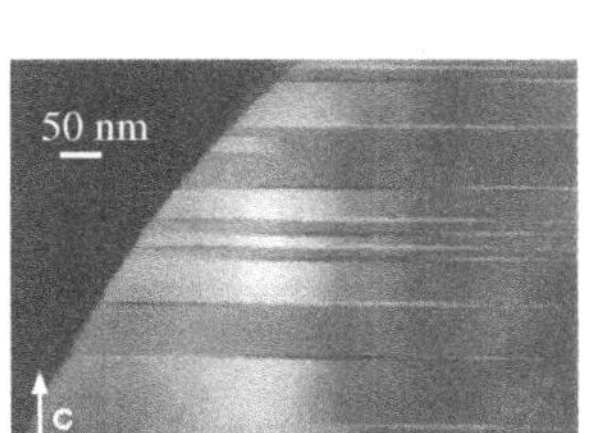

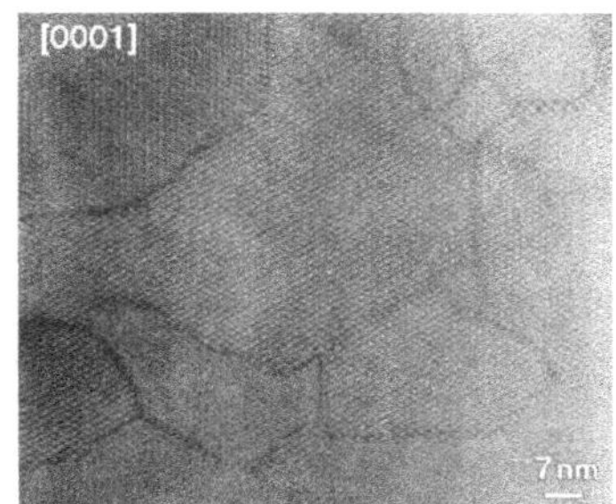

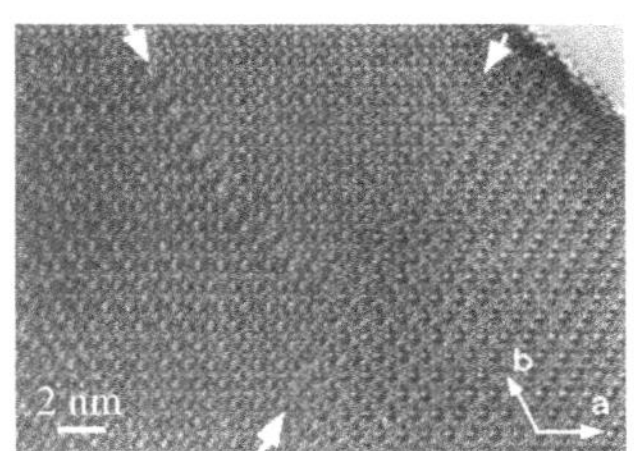

Fig. 16.10 Transmission electron-microscopic frames of $BaAl_2O_4$ [361]. Left: View along [100] showing twin domains. Centre and right: View along [001]; arrows mark antiphase boundaries.

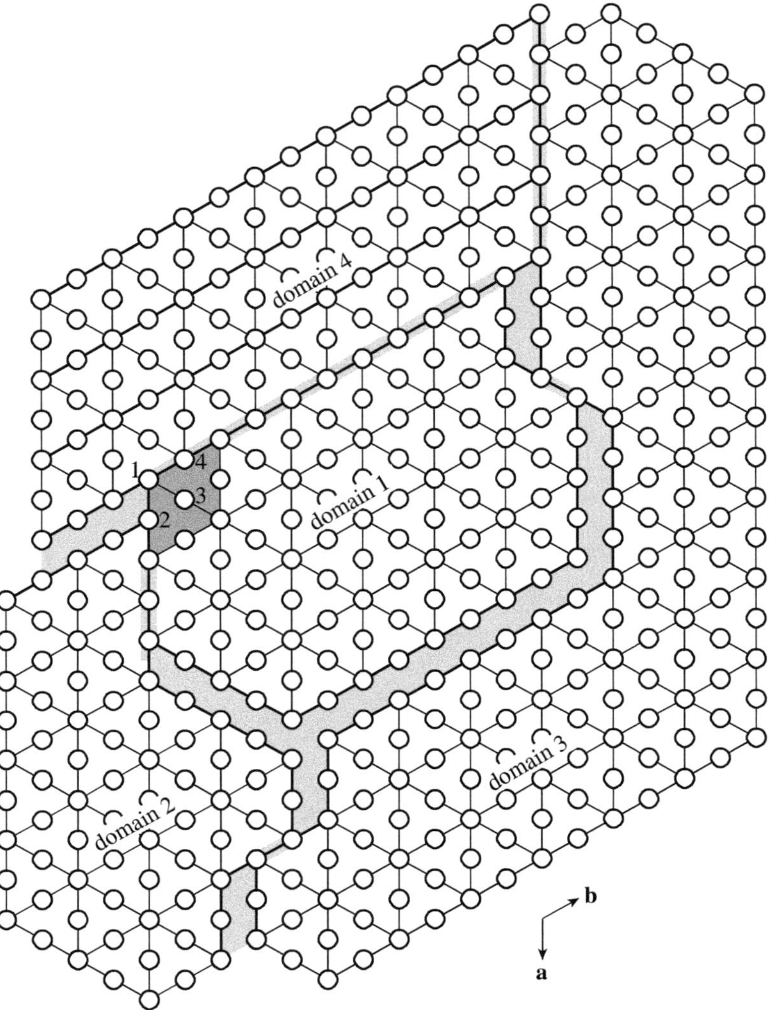

Fig. 16.11 Schematic drawing with the four kinds of antiphase domains of $BaAl_2O_4$ [361]. The circles represent the Ba atoms. Domain boundaries are enhanced in grey. One unit cell is marked in domain 1. The numbers at the Ba atoms designate the four different origin positions of the four domains.

16.7 The role of non-realizable intermediate groups in a Bärnighausen tree

In Bärnighausen trees it often occurs that between the aristotype and the hettotype one or more intermediate groups have to be considered, although they seemingly cannot be realized because the occupation of the atomic sites already fulfills the higher symmetry of the aristotype. Take the example of the compound K_3NiO_2 [362]. It contains strands consisting of K^+ ions and NiO_2^{3-} dumbbells (Fig. 16.12). At temperatures above of 150 °C (423 K) the strands run in straight lines diagonally across the unit cell. Upon cooling a phase transition takes place at 423 K with the formation of twinned crystals. The centrosymmetric (achiral) space group $P4_2/mnm$ is converted to the pair of enantiomorphic (chiral) space groups $P4_1 2_1 2$ and $P4_3 2_1 2$, including a doubling

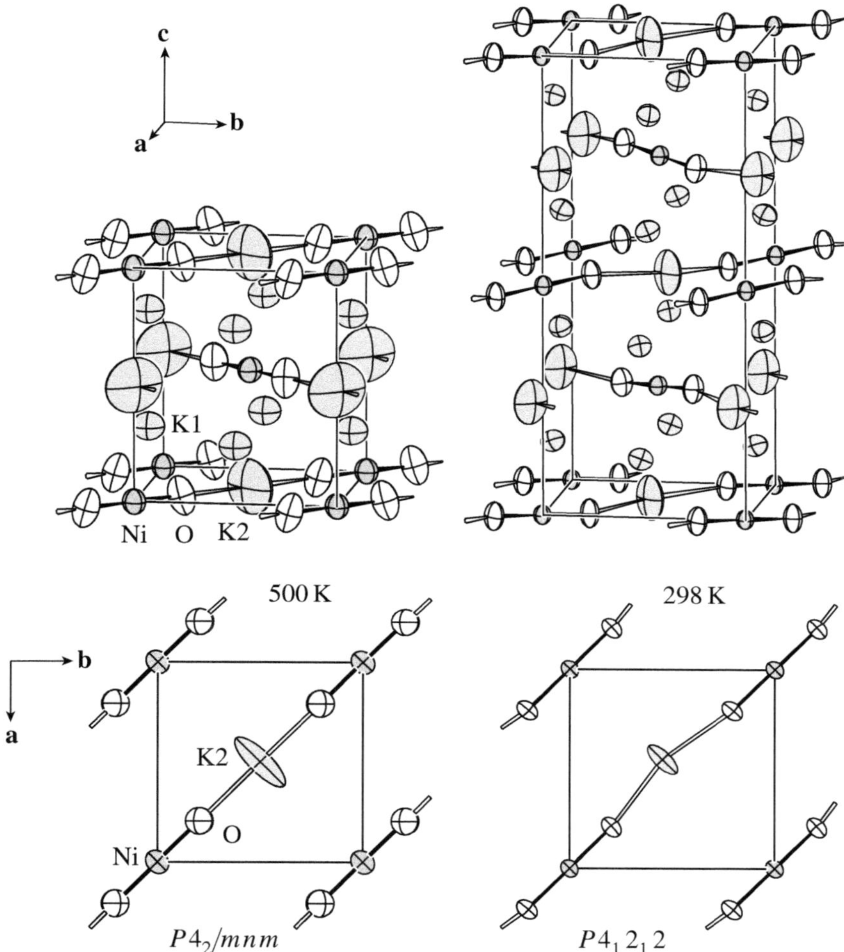

Fig. 16.12 Crystal structure of K_3NiO_2 at 500 K and at 298 K. Seeming ellipsoids of thermal displacement at a probability level of 75 %.

of c. The strands undergo a slight bending in that the potassium ions $K2^+$ slip away from the special position $\frac{1}{2},\frac{1}{2},0$ to one side of the diagonal.

The group–subgroup relation is shown in Fig. 16.13. In the relation of a non-Sohncke space group (here $P4_2/mnm$) and a pair of enantiomorphic (chiral) subgroups (here $P4_1 2_1 2$ and $P4_3 2_1 2$) there is always an intermediate non-chiral Sohncke space group (here $P4_2 2_1 2$). The first step of the symmetry reduction is *translationengleiche* and entails the formation of inversion twins. Chiral right and left twin domains should appear here. However, the equilibrium positions of the atoms in the space group $P4_2 2_1 2$ still fulfil all of the symmetry of the achiral space group $P4_2/mnm$. The chirality can only be due to the fact that the occupied positions in $P4_2 2_1 2$ have chiral site symmetries and that the symmetry of the thermal motion of the atoms corresponds to their site symmetries. Usually, the thermal displacement of an atom is depicted by a centrosymmetric ellipsoid, based on the assumption of harmonic vibrations of the atoms. If an atom is placed on a chiral position, its vibrations cannot be harmonic and its whereabouts are chiral; therefore, it cannot properly be depicted

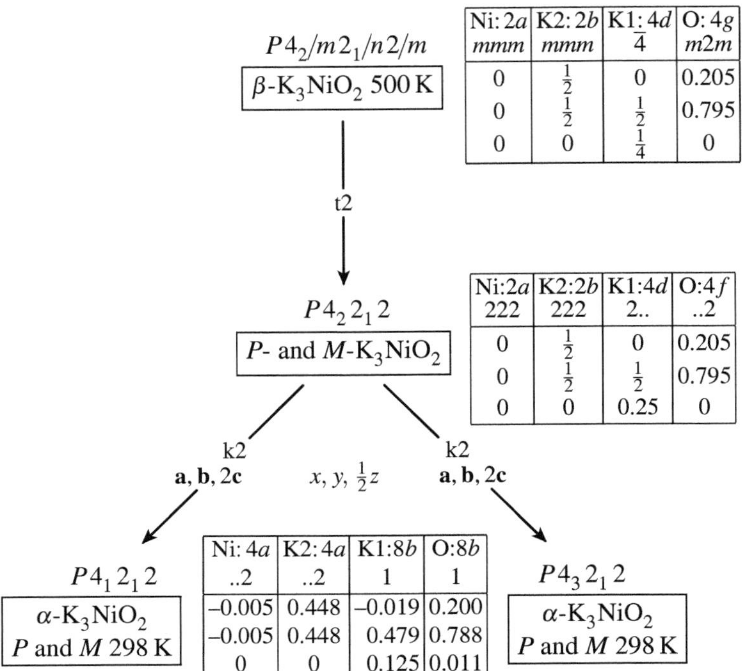

Fig. 16.13 Group–subgroup relations among the modifications of K_3NiO_2. [362]

by an ellipsoid. The ellipsoids shown in Fig. 16.12, as usual, were calculated under the assumption of harmonic vibrations. The large ellipsoid for K2 is conspicuous; at 500 K it probably is due to a dynamic misorder; however, at 298 K, the thermal motion should be depicted by some other figure than an ellipsoid. At 50 K the structure resembles the one 298 K, but the ellipsoids are small and not conspicuos [362].

It is possible that at a temperature a little above 243 K, a modification of K_3NiO_2 in the space group $P4_22_12$ could exist. Should it be possible to get hold of it, it would be absolutely necessary to calculate the parameters of anharmonic vibrations. However, it is likely that such a modification would prove to be elusive. It would be twinned due to the *translationengleiche* group–subgroup relation. The vibrational motion in every one of the twin domains would either be right-handed (*P*) or left-handed (*M*). The X-ray diagrams of both domain kinds would be exactly superposed. With the experimental techniques available in 2023 it is hopeless to distinguish such twins in the space group $P4_22_12$, that differ only in the handedness of their thermal motion.

The second step of the symmetry reduction is *klassengleiche* and thus twinning is precluded in this case. Instead, the two enantiomorphic subgroups $P4_12_12$ and $P4_32_12$ emerge, so that antiphase domains are to be found. In summary, four kinds of domains are to be expected:

$$P4_12_12\text{-}P \qquad P4_12_12\text{-}M \qquad P4_32_12\text{-}M \qquad P4_32_12\text{-}P$$

$P4_12_12$-*P* and $P4_32_12$-*M* constitute an enantiomorphic pair, and so do $P4_12_12$-*M* and $P4_32_12$-*P*. These two pairs are diastereomeric.

Even when the aristotype and the hettotype are centrosymmetric and the occupation of the positions in the intermediate space group fulfils the symmetry of the aristotype, the site symmetries in the intermediate group are lower than in the aristotype.

Exercises

Solutions on page 336.

16.1. What are experimentally observable properties that can be used to decide whether a phase transition is of second order (after EHRENFEST) or is continuous?

16.2. May or must an 'isosymmetric' phase transition be a continuous transition?

16.3. In Fig. 16.4 (left) the curve for the cell volume V has a kink at T_c. May this be the case for a second-order phase transition?

16.4. The packing of several modifications of P_4S_3 and similar cage-like molecules is shown in Fig. 15.4 (page 204), and the corresponding symmetry relations are given in Fig. 15.3. All modifications have the same packing pattern, but with differently rotated molecules. Could these modifications be transformed one to the other by continuous phase transitions?

16.5. At high temperatures, $BaTiO_3$ has the cubic perovskite structure, space group $Pm\bar{3}m$. Upon cooling it is distorted, adopting the space group $P4mm$. Will the crystals of the low-symmetry structure be twinned? If so, with how many kinds of domains?

16.6. $SrTiO_3$ has the cubic perovskite structure ($Pm\bar{3}m$). Upon cooling below 105 K, the coordination octahedra are mutually rotated and the symmetry is reduced to $I4/mcm$, including an enlargement of the unit cell by a factor of 4 ($a' = b' = a\sqrt{2}$, $c' = 2c$). Will the crystals of the low-symmetry structure be twinned? If so, with how many kinds of domains?

16.7. $SrCu_2(BO_3)_2$ exhibits a second-order phase transition at 395 K. Above 395 K its space group is $I4/mcm$, below it is $I\bar{4}2m$ with the same lattice parameters [363]. Is the formation of twins to be expected upon cooling beyond the point of transition? If so, will these cause problems for X-ray crystal structure determination?

16.8. The structure of $ErCo_2$ experiences a distortion during the phase transition from the paramagnetic to the ferrimagnetic modification at 32 K. At $T > 32$ K it is cubic, space group $Fd\bar{3}m$, $a = 713.3$ pm; below 32 K it is rhombohedral, space group $F\bar{3}m$, $a = 714.5$ pm, $\alpha = 89.91°$ (non-conventional face-centred setting for $R\bar{3}m$ with rhombohedral axes) [364]. Why do the X-ray reflections of the ferrimagnetic phase show a broadening or splitting?

16.9. According to quantum-mechanical calculations, sillimanite, Al_2SiO_5, should exhibit a phase transition if pressure is exerted to more than 30 GPa. The atoms are expected to move together, with an increase of the coordination number of the Si atoms from 4 to 5. The calculated change of the corresponding Si–O distances and of the lattice parameter a as a function of pressure is shown below (arrows mark the directions of pressure increase and decrease) [365]. The remaining lattice parameters and the structure in general are changed only marginally. The space group is $Pnma$ before and after the transition. Is this a continuous phase transition? Is the transformation reconstructive or displacive?

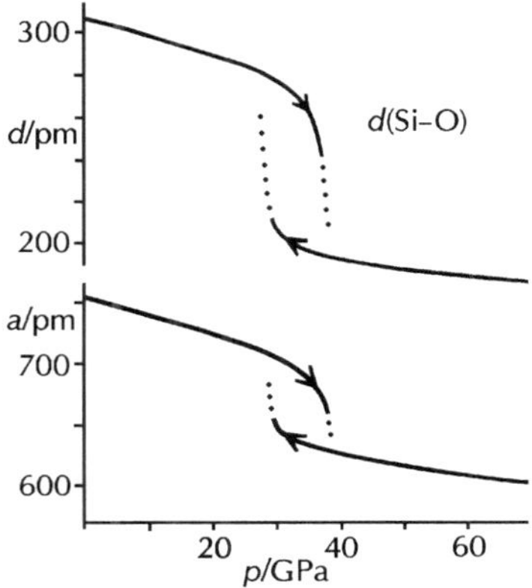

Topotactic reactions

A topotactic reaction is a chemical reaction that takes place in the solid state such that the orientation of the product crystal is determined by the orientation of the initial material. As an example, Fig. 17.1 shows three images taken by electron diffraction. The first one was made from a trigonal single crystal of $ZnFe_2O_2(OH)_4$, the last one from cubic spinel $ZnFe_2O_4$ that was obtained from this after heating. The central image shows an intermediate stage at which the reaction had proceeded halfway; it shows reflections of the initial and the product crystal. The direction of view is along the trigonal axis [001] which becomes the cubic axis [111].

In this case, like with many other topotactic reactions, there is *no* crystallographic group–subgroup relation between the initial and the final product. The formation of the topotactic texture is rather to be understood by analogy with the formation of growth twins. In $ZnFe_2O_2(OH)_4$, space group $P\bar{3}m1$, the O atoms are arranged in a hexagonal-closest packing with metal atoms in octahedral interstices. In spinel, space group $Fd\bar{3}m$, it is a cubic-closest packing with metal atoms in octahedral and tetrahedral interstices. The orientational states result from the orientation of nuclei of crystallization which is predetermined by the matrix of the initial crystal.

This is in accordance with a similar, much investigated example. $Mg(OH)_2$, brucite, is homeotypic to CdI_2; it has a hexagonal-closest packing of O atoms in the space group $P\bar{3}m1$. By thermal dehydration, a single crystal of $Mg(OH)_2$ is converted to MgO (periclase), NaCl type, space group $Fm\bar{3}m$, with cubic-closest packing of O atoms [367, 368]. Thereby a topotactic texture is formed that consists of a large number of MgO crystallites that are oriented like and only like the cubic twins with a twin axis [111] (Fig. 16.6). These are orientations associated with the terms 'obverse' and 'reverse' in crystallography. The analogy between the macroscopically well-shaped twins according to Fig. 16.6 and the fine texture of periclase crystallites points to a common kind of nascent state. The growth twin originates from only two nuclei of the obverse and reverse orientations, whereas the topotactic texture originates from a large number of nuclei that have emerged at energetically equivalent sites in the interior of the initial crystal, either obverse or reverse with statistical probability.

Nucleation generally is the controlling factor of topotactic reactions. The nuclei are formed in the matrix of the original crystal; nuclei that have the appropriate orientation have a higher probability of being formed and of continuing to grow. Frequently, particles have to diffuse through the crystal, for example the water eliminated in the preceding examples. This causes the formation of defects in the crystal; these are new nucleation sites. In addition, the

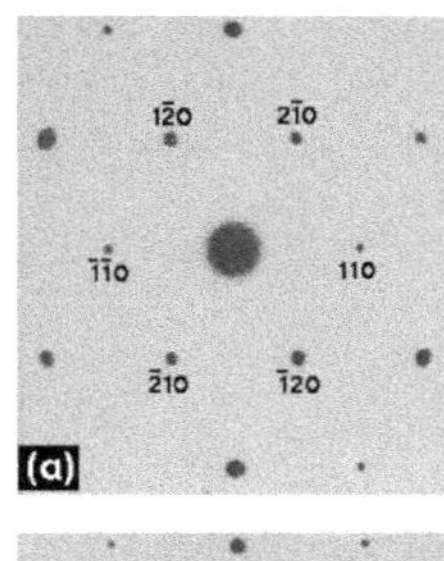

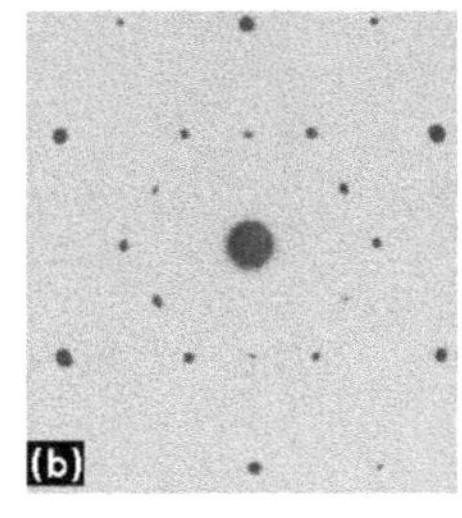

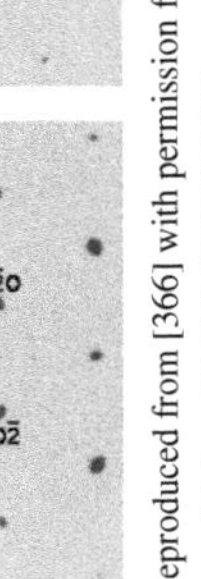

Fig. 17.1 Single-crystal electron diffraction images of $ZnFe_2O_2(OH)_4$. (a) δ-FeOOH type, (b) after five days at 210 °C, (c) spinel structure $ZnFe_2O_4$ after five days at 260 °C [366].

topotactic texture becomes porous, but the morphology of the initial crystal does not collapse.

This phenomenon has been known for a long time in mineralogy under the term *pseudomorphism*. A pseudomorph is a mineral that has been altered after its formation but has kept its appearance. An example is malachite, $Cu_2CO_3(OH)_2$, formed from azurite, $Cu_3(CO_3)_2(OH)_2$, in that it has taken up OH^- ions from surrounding water and released CO_3^{2-} ions.

17.1 Symmetry relations among topotactic reactions

Aside from reactions as dealt with in the preceding section, there also exist topotactic reactions with retention of symmetry or with a group–subgroup relation. For example, some molecules dimerize or polymerize within the crystal in such a way that the reacting molecules move together and join with each other, but on the whole the molecules keep their places. The space group may be retained or it may change to a subgroup or supergroup. In organic chemistry, such reactions are called topochemical.

Crystals of the fullerene C_{70} that were obtained by sublimation consist of a hexagonal-closest packing of molecules ($P6_3/mmc$); in the mean, the rotating molecules are spherical and the ratio $c/a = 1.63$ corresponds to the ideal value of a packing of spheres [369, 370]. Upon cooling, a first-order phase transition is observed at approximately 50 °C, but the structure retains the hexagonal space group, albeit with $c/a = 1.82$. There is no group–subgroup relation. The molecules, which have an elongated shape keep rotating, but henceforth they are oriented parallel to the hexagonal axis and rotate about this direction. At about 20 °C, this rotation also freezes. Due to the molecular symmetry $\overline{10}2m$ (D_{5h}), the crystal symmetry then cannot continue to be hexagonal; it becomes orthorhombic, space group $Pbnm$ (Fig. 17.2). The symmetry reduction includes a *translationengleiche* relation of index 3 (Fig. 17.3), and accordingly transformation twins are formed with the three orientations as shown in Fig. 8.1 (page 105). The intermediate group $Cmcm$ is not observed because the molecules are turned along their fivefold axes in such a way that their twofold axes do not coincide with those of $Cmcm$ (Fig. 17.2).

Under pressure (2 GPa) C_{70} polymerizes in the crystalline state at 300 °C [371]. The starting point at this temperature is the hexagonal high-temperature form. The joining in the polymer now forces the molecules into an orientation that fulfils the space group $Cmcm$ (Fig. 17.3). The crystals of the polymer are triplets (twins with three orientational states of the domains). Upon polymerization, the C_{70} molecules approach each other, causing a decrease of the lattice parameter c from 1853 pm to 1792 pm; the polymeric molecules are aligned along c.

Polymerization of C_{70} is an example of a topotactic reaction with a group–subgroup relation between the space groups of the reactant and the product. Such reactions proceed by nucleation and growth in a similar way to the formation of transformation twins. For the formation of domains, the same rules apply as mentioned in Section 16.3 (page 220) for phase transitions.

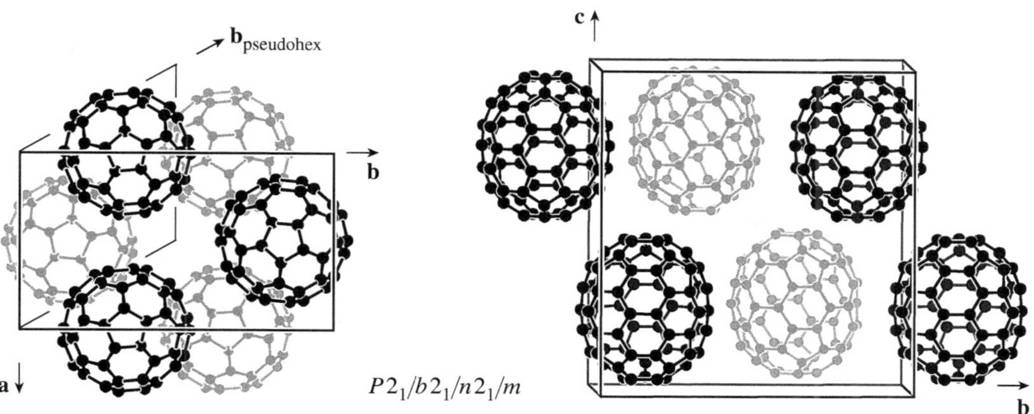

Fig. 17.2 The low-temperature modification of C_{70}. $a = 1002$ pm, $b = 1735$ pm $= a\sqrt{3}$, $c = 1853$ pm. Left: view along the pseudohexagonal axis.

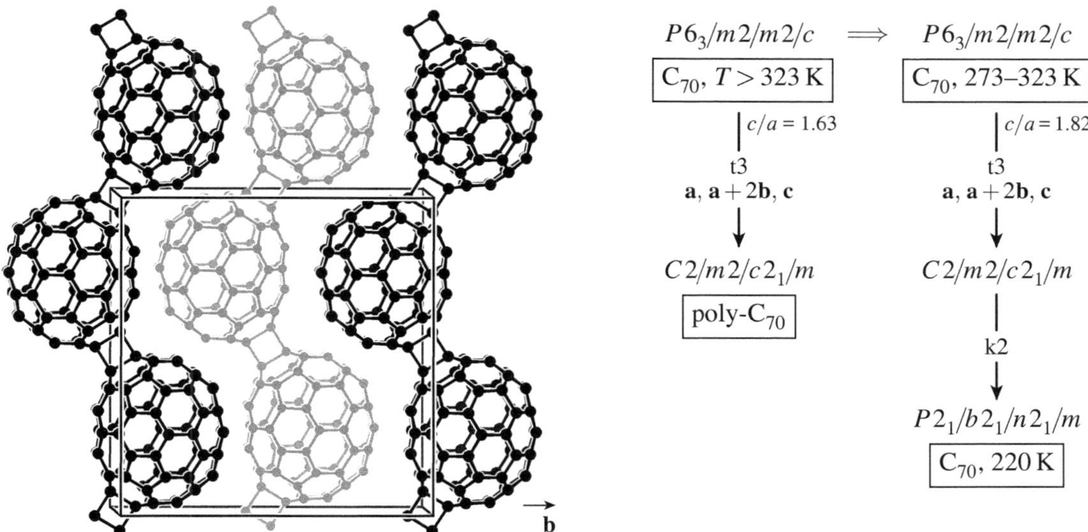

Fig. 17.3 Crystal structure of polymeric C_{70}, space group $C2/m2/c2_1/m$ ($Cmcm$), $a = 999$ pm, $b = 1730$ pm, $c = 1792$ pm. Right: symmetry relations between the C_{70} modifications and with poly-C_{70}.

17.2 Topotactic reactions among lanthanoid halides

The reduction of rare-earth trihalides with the corresponding metals at high temperatures yields subhalides of the general formula Ln_nX_{2n+1}. Their crystal structures are so-called vernier structures, in which the halogen atoms alternate in closer and less closer rows like in a vernier scale. The arrangement of the metal atoms is roughly face-centred cubic (Fig. 17.4) [372–381].

The vernier structures can be derived from the CaF_2 type, albeit with a slight excess of halogen atoms. The structures of the dihalides $DyCl_2$ ($SrBr_2$ type, $P4/n$) [382–384] and $TmCl_2$ (SrI_2 type, $Pcab$) [385, 386] can also be derived from the CaF_2 type (Fig. 17.5). In both cases, the metal atoms have distorted, pseudo-face-centred cubic arrangements. In the case of the $SrBr_2$ type, the halogen atoms are arranged in layers of squares as in CaF_2, but half of the layers are turned by 45°. In the SrI_2 type, the squares are distorted to rhombuses and the layers are corrugated (Fig. 17.5).

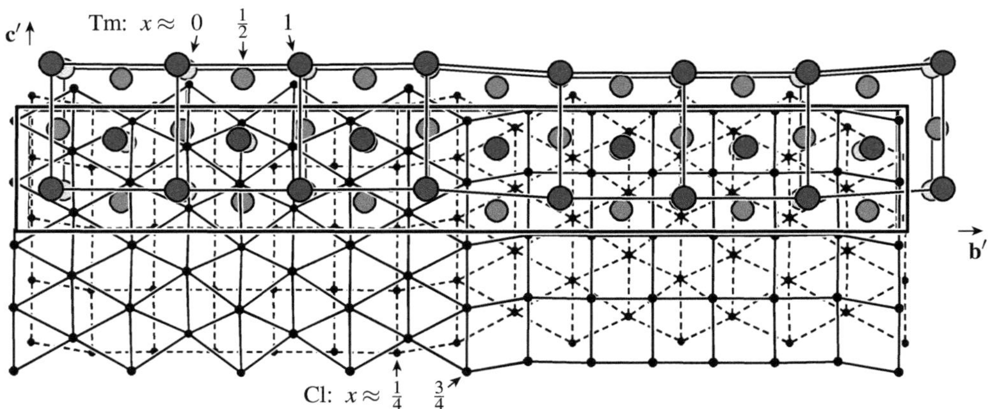

Fig. 17.4 The vernier structure of Tm_7Cl_{15}. Seven pseudo-face-centred 'cells' of Tm atoms are shown in the upper part. The Cl atoms are located at the junctions of the drawn nets; only the nets are shown in the lower part. Seven rows of Cl atoms with the pseudosquare pattern are located on top of eight rows with the pseudohexagonal pattern.

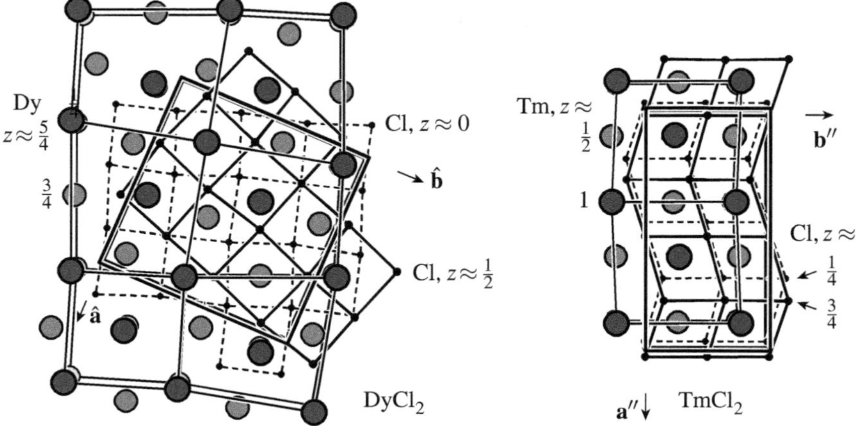

Fig. 17.5 The crystal structures of $DyCl_2$ ($SrBr_2$ type) and $TmCl_2$ (SrI_2 type). Six pseudo-face-centred 'cells' of metal atoms are shown for $DyCl_2$, two for $TmCl_2$. The Cl atoms are located at the junctions of the drawn nets; the nets of $TmCl_2$ are corrugated.

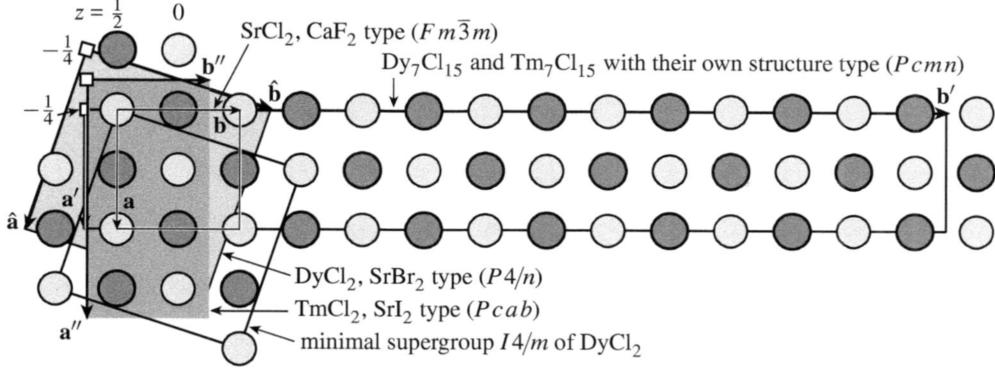

Fig. 17.6 Structural relations between the strontium halides and the vernier structure Ln_7Cl_{15}. Only the metal atom positions of the CaF_2 type are shown. The axes in the direction of view are $\hat{\mathbf{c}} \approx \mathbf{c}'' \approx \mathbf{c}' \approx \mathbf{c}$.

How the unit cells are related with that of the CaF_2 type is shown in Fig. 17.6. From the figure, we calculate the lattice parameters listed in the margin (in pm), assuming $a = b = c = 682$ pm (Dy compounds) and $a = b = c = 678$ pm (Tm compounds) for the CaF_2 type.

$DyCl_2$ $P4/n$ [381] calculated	observed
$\hat{a} = \frac{1}{2}\sqrt{10}a = 1078$	1077.5
$\hat{b} = \frac{1}{2}\sqrt{10}a = 1078$	1077.5
$\hat{c} = c = 682$	664.3
Dy_7Cl_{15} $Pcmn$ [375] calculated	**observed**
$a' = a = 682$	667.4
$b' = 7b = 4774$	4818
$c' = c = 682$	709.7
Tm_7Cl_{15} $Pcmn$ [375, 387] calculated	**observed**
$a' = a = 678$	657.1
$b' = 7b = 4746$	4767.7
$c' = c = 678$	700.1
$TmCl_2$ $Pcab$ [386, 387] calculated	**observed**
$a'' = 2a = 1356$	1318.1
$b'' = b = 678$	671.4
$c'' = c = 678$	697.7

When these compounds are heated, their anion partial structure melts before the actual fusion takes place; that is, there is a phase transition at which the cations, due to the mutual repulsion of their higher charges, retain the order as in the CaF_2 type, whereas the intervening anions start floating. This melting of the anion partial structure is well known among the strontium halides and other compounds having the CaF_2 structure; this includes a high ionic conductivity of the high-temperature form [388] and a high transformation entropy, which can be even higher than the entropy of fusion [389]. The quasiliquid state of the anions at the high temperatures of preparation of the vernier compounds is compatible with a non-stoichiometric composition. Upon cooling, the cations determine a matrix in which the anions organize themselves. Depending on composition, several compounds crystallize in common, their crystals being intergrown in a regular manner. If, for example, the high-temperature phase has the composition $DyCl_{2.08}$, $DyCl_2$ and Dy_7Cl_{15} (= $DyCl_{2.14}$) crystallize in common.

The X-ray diffraction pattern of a crystal obtained in this way, at first glance, is confusing (Fig. 17.7 left). Taking into account the orientational relations according to Fig. 17.6 and the group–subgroup relations (Fig. 17.8), the pattern can be interpreted as an intergrowth of $DyCl_2$ and Dy_7Cl_{15} (Fig. 17.7 right). The tetragonal c^* axis of $DyCl_2$ is exactly coincident with a reciprocal axis of Dy_7Cl_{15}; for this reason, it was chosen to be the c^* axis, resulting in the non-conventional space-group setting $Pcmn$ ($Pnma$ would be conventional).

The strong reflections in Fig. 17.7 that result from superposition of reflections from both compounds correspond to reflections of the CaF_2 type, which, however, is not present. Since the lattice parameters do not match exactly, the reflections are not superimposed exactly, but appear broadened.

Atomic coordinates calculated from the CaF_2 type and observed coordinates of the Tm atoms of Tm_7Cl_{15} are compared in Table 17.1. As can be seen in

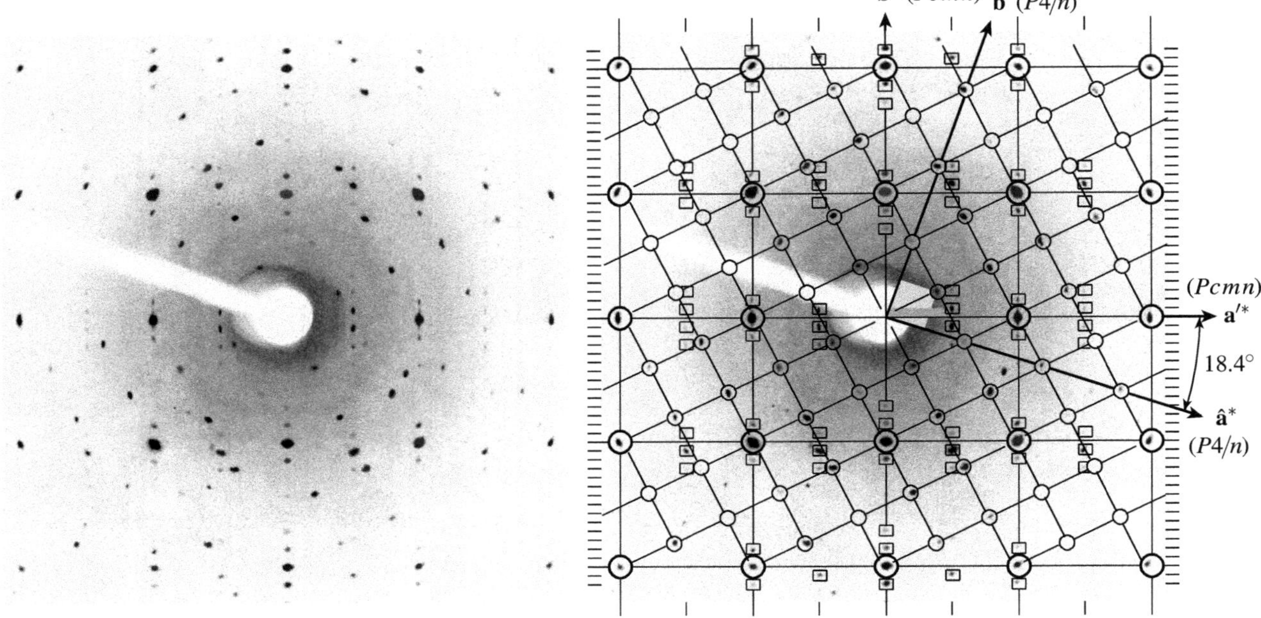

Fig. 17.7 Buerger precession photograph of a 'single crystal' of $DyCl_x$ with $x \approx 2.08$ [381] (Mo$K\alpha$ radiation; $K\beta$ reflections and streaks originating from *bremsstrahlung* have been removed).
Right: Interpretation as topotactic texture of $DyCl_2$ and Dy_7Cl_{15} by superposition of *hk*0 reflections of both compounds. Small circles: $DyCl_2$ in the $SrBr_2$ type ($P4/n$); rectangles: Dy_7Cl_{15} in the Tm_7Cl_{15} type ($Pcmn$); large circles: superimposed reflections from both substances. Due to the *n* glide planes, all reflections with $h+k=2n+1$ are absent. 18.4° = arc tan $\frac{1}{3}$.

Table 17.1 Atomic coordinates for the cations of Tm_7Cl_{15} calculated from the CaF_2 type according to $x, \frac{1}{7}y+\frac{1}{28}, z+\frac{1}{4}$; $\pm(0,\frac{1}{7},0)$; $-(0,\frac{2}{7},0)$ and comparison with the observed atomic coordinates [375]. The projection onto the CaF_2 type refers to observed coordinates that have been retransformed according to $x, 7y-\frac{1}{4}, z-\frac{1}{4}$.

	Calculated			Observed			Obs. projected on CaF_2			Ideal values		
Atom	x	y	z	x	y	z	x	y	z	x	y	z
Tm(1)	0.0	0.0357	0.25	−0.0242	0.03946	0.3346	−0.0242	0.0262	0.0846	0	0	0
Tm(2)	0.0	−0.1071	0.25	0.0224	−0.11122	0.2624	0.0224	−1.0285	0.0124	0	−1	0
Tm(3)	0.0	0.1786	0.25	−0.0200	0.18094	0.3258	−0.0200	1.0166	0.0758	0	1	0
Tm(4)	0.0	$-\frac{1}{4}$	0.25	0.0178	$-\frac{1}{4}$	0.2626	0.0178	−2	0.0126	0	−2	0

the figure of the structure (Fig. 17.4), the arrangement of the rare-earth atoms matches quite well a cubic-closest packing, which is ideally fulfilled in $SrCl_2$. The calculated coordinates deviate only slightly from the observed ones. A good way to recognize the coincidence is to project the observed coordinates onto the CaF_2 type (i.e. to 'recalculate' them back to the aristotype).

The topotactic intergrowth observed between $TmCl_2$ and Tm_7Cl_{15} is similar to that of $DyCl_2$ and Dy_7Cl_{15}. The relative location of the unit cells shown in Fig. 17.6 can be discerned clearly in X-ray diffraction diagrams (Fig. 17.9). The basis vectors of both compounds have exactly the same directions, the long

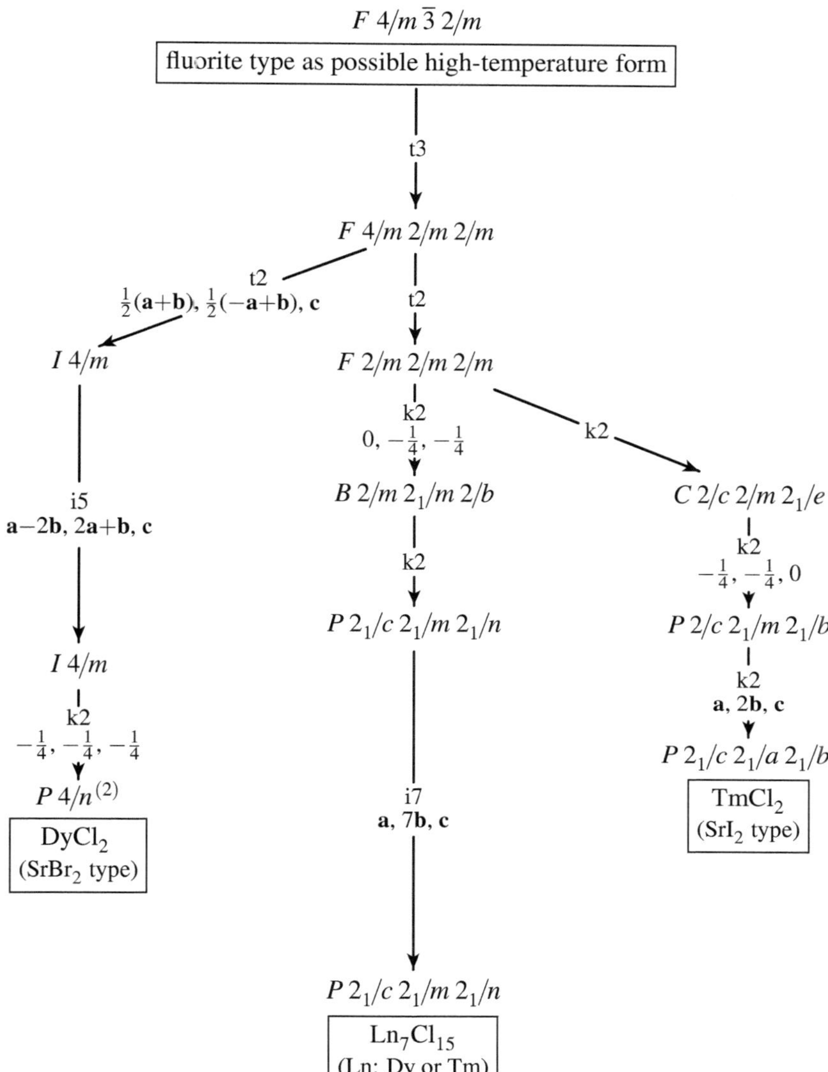

Fig. 17.8 Derivation of the crystal structures of some rare-earth chlorides from the aristotype fluorite by group–subgroup relations.

vector **a**″ of $TmCl_2$ being perpendicular to the long vector **b**′ of the vernier compound. Note the reflections 0 4 0 and 4 4 0 of $TmCl_2$ in Fig. 17.9 next to 0 28 0 and 2 28 0 of Tm_7Cl_{15}. They are not superposed exactly because $b\,(TmCl_2) = 671.4$ pm $< \frac{1}{7}b\,(Tm_7Cl_{15}) = 681.1$ pm. $\frac{1}{2}a\,(TmCl_2) = 659.05$ pm is scarcely larger than $a\,(Tm_7Cl_{15}) = 657.1$ pm; however, the difference can be perceived by the mutual shift of the reflection 8 4 0 of $TmCl_2$ and 4 28 0 of Tm_7Cl_{15} in the horizontal direction (shorter distances in the reciprocal lattice of the X-ray diagram correspond to longer distances in real space). Since the reflections $hk0$ with $h+k = 2n+1$ are absent for Tm_7Cl_{15}, there is no superposition with the reflections 2 0 0, 6 0 0 (and 10 0 0, 14 0 0, not shown in Fig. 17.9) of $TmCl_2$, so that these can be measured accurately and permit an accurate determination of a''.

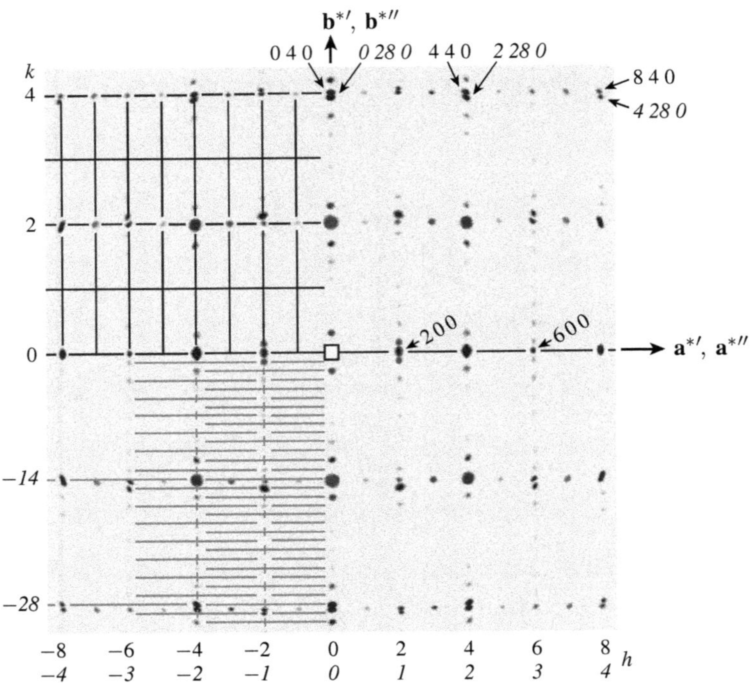

Fig. 17.9 Simulated precession photograph of the reflections $hk0$ of a topotactic texture of $TmCl_2$ and Tm_7Cl_{15}. In the upper-left quadrant, the reciprocal lattice lines of $TmCl_2$ have been included, with blanks at the sites of reflections. In the lower-left quadrant the reciprocal lattice lines of Tm_7Cl_{15} have been drawn. The simulation is based on a Weissenberg photograph and on measured X-ray diffraction data from a 'single crystal' [381]. Due to the glide planes, the reflections $h+k=2n+1$ of Tm_7Cl_{15} ($Pcmn$) are absent; $TmCl_2$ ($Pcab$) has absent reflections $k=2n+1$ and $h00$ if $h=2n+1$. Miller indices in italics refer to Tm_7Cl_{15}.

Exercises

Solutions on page 336.

17.1. (Demanding exercise). The oxides Ln_2O_3 of the rare earths occur in a trigonal A form and a monoclinic B form. When a melt of a mixture of La_2O_3 and Sm_2O_3 solidifies (24–28% amount-of-substance fraction of La_2O_3), a segregation takes place; two kinds of intergrown crystals are obtained, those of the A and the B form, which show up by X-ray diffraction by a superposition of the interference patterns of both forms. There is a characteristic orientational relation (see figure). Pure A-Sm_2O_3 is not known; however, it can be stabilized by doping. The following lattice parameters were taken from A-Sm_2O_3 stabilized by Zr [390]; the atomic coordinates of the table are those of A-Ce_2O_3 [391].

A-Sm_2O_3: $P\bar{3}2/m1$, $a_A = 377.8$ pm, $c_A = 594.0$ pm, $Z=1$;

B-Sm_2O_3: $C12/m1$, $a_B = 1420$ pm, $b_B = 362.7$ pm, $c_B = 885.6$ pm, $\beta_B = 99.99°$, $Z=6$.

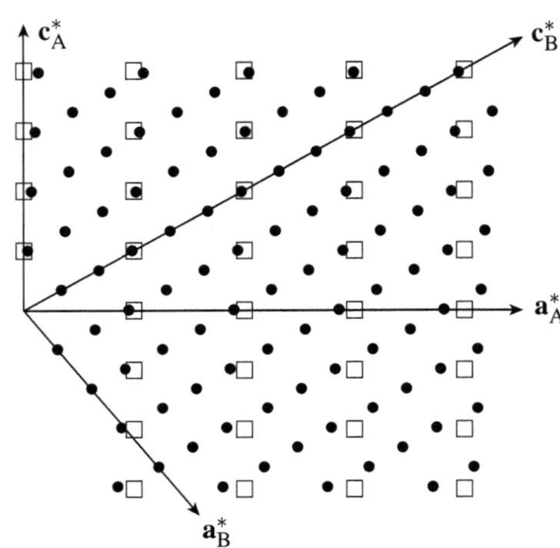

Section $h0l$ of the reciprocal lattice of a topotactic intergrowth of A-Ln_2O_3 (squares) and B-Ln_2O_3 (dots). The absent reflections $h=2n+1$ of B-Ln_2O_3 have not been included [392].

		x	y	z
A-Sm_2O_3, $P\bar{3}2/m1$				
Sm	$2d$	$\frac{1}{3}$	$\frac{2}{3}$	0.246
O(1)	$2d$	$\frac{1}{3}$	$\frac{2}{3}$	0.647
O(2)	$1a$	0	0	0
B-Sm_2O_3, $C12/m1$ [393]				
Sm(1)	$4i$	0.135	0	0.490
Sm(2)	$4i$	0.190	0	0.138
Sm(3)	$4i$	0.467	0	0.188
O(1)	$4i$	0.129	$\frac{1}{2}$	0.286
O(2)	$4i$	0.175	$\frac{1}{2}$	−0.027
O(3)	$4i$	−0.202	$\frac{1}{2}$	0.374
O(4)	$4i$	0.026	0	0.657
O(5)	$2a$	0	0	0

(a) Enter the indices of the reflections of the A and the B form in the figure close to the origin of the reciprocal lattice. Mind the extinctions. Express the basis vectors of the reciprocal cell $\mathbf{a}^*_A, \mathbf{b}^*_A, \mathbf{c}^*_A$ of the A form by linear combinations of the basis vectors $\mathbf{a}^*_B, \mathbf{b}^*_B, \mathbf{c}^*_B$ and write down the result in matrix form. The lattices do not match accurately; be satisfied with an obvious approximation. Consider that $\mathbf{b}^*_A$ forms a 60° angle with $\mathbf{a}^*_A$ and that the reciprocal basis vectors are transformed like the atomic coordinates.

(b) What is the transformation matrix A $\rightarrow$ B for the basis vectors of real space? Check the metric relations between the unit cells. Draw a sketch that shows the relations of the basis vectors.

(c) Set up the group–subgroup relations from the A to the B form. For the relation $C12/m1$ —i3→ $C12/m1$ you will find a different basis transformation in *International Tables A1* to the one needed here (different setting of the monoclinic cell).

(d) Multiply the transformation matrices for the consecutive steps of symmetry reductions. Do you notice something familiar?

(e) Map the atoms of the B form onto the trigonal unit cell of the A form by transformation of the coordinate triplets x_B, y_B, z_B; compare them with those of the A form.

(f) The B form tends to form twins. Under a microscope the twin operation is easy to recognize, but it seems to be curious [394]: it is the twofold rotation about [1 3 2]. By transformation from $[1\,3\,2]_B$ to $[u, v, w]_A$, show that the twin operation is sensible.

(g) Crystals of the B modification form thin plates with $\{\bar{2}01\}$ as the main face. The lateral faces $\{1\,0\,1\}$, $\{1\,1\,1\}$, and $\{1\,\bar{1}\,1\}$ are narrow. Transform the mentioned face indices to $\{hkl\}$ triplets of the trigonal cell. Note that Miller indices are transformed like the basis vectors of the unit cell.

Group–subgroup relations as an aid for structure determination

One of the most frequent errors in crystal structure determinations is the choice of the wrong space group. As a consequence, the atomic coordinates and all derived values are unreliable. A procedure to detect wrong space groups is described in Section 9.4. Among the reasons for wrong structure determinations are deficient knowledge of crystallography and the uncritical acceptance of computer results that have often been produced with computer programs' automated routines. But even specialists can become victims of pitfalls, in spite of meticulous work. In addition, there exist certain problems that can complicate structure determination.

There exist many crystals with a seemingly normal X-ray diffraction pattern,[1] the analysis of which yields a structural model that appears reasonable. And yet, the structure determination can be wrong. There exist certain alarm signals that should always be taken seriously:

- systematically absent reflections that do not match any space group;
- split reflections that are increasingly separate the higher the angle of diffraction θ;
- seeming misorder, expressed by split atomic positions (atoms occupy several positions randomly with occupancy probabilities of less than 1);
- suspicious parameters of 'thermal' displacement of certain atoms (i.e. conspicuously large, elongated, or flat ellipsoids of 'thermal motion');
- high correlations between certain parameters during refinement by the method of least squares (do not skip the corresponding table of the computer output!);
- unreasonable bond lengths and angles.

In the following, using examples, we point out how certain problems can be solved and how certain errors can be avoided with the aid of group–subgroup relations.

[1] Normal means: without satellite reflections, without diffuse scattering, no quasicrystals.

18.1 What space group should be chosen?

At the beginning of any crystal structure determination by diffraction techniques, the space group has to be established. First of all, three observations are used for its determination:

1. the metrics of the unit cell;
2. the Laue symmetry, which can be deduced from the equality or non-equality of the intensities of certain reflections;
3. extinct reflections, which point to the presence of centrings, glide planes, or screw axes.

Frequently, it is not possible to establish unambiguously the space group from these data and there remain several possibilities among which to choose.

If the structures of related compounds are already known, group–subgroup relations can help to choose the correct space group.

Numerous salts of tetraphenylphosphonium and -arsonium ions crystallize with tetragonal symmetry in one of the space groups $P4/n$ (if the anion has a fourfold rotation axis) or $I\bar{4}$ (if the anion is tetrahedral). In both cases, the lattice parameters are close to $a = b \approx 1300$ pm and $c \approx 780$ pm. The packing in the crystal is governed by the space requirements of the cations, which are stacked along c into columns; see Section 15.3. The packing in the space group $P4/n$ has phosphorus (or As) atoms at $z = 0$; the packing in the space group $I\bar{4}$ has mutually shifted columns, with P (As) atoms at $z = 0$ and $z = \frac{1}{2}$.

$P(C_6H_5)_4[I(N_3)_2]$
$a = 1499$ pm, $b = 1006$ pm, $c = 800$ pm,
$\alpha = \beta = 90°$, $\gamma = 91.7°$
all reflections $h+k+l = 2n+1$ absent

Crystals of tetraphenylphosphonium diazidoiodate(1–), $P(C_6H_5)_4[I(N_3)_2]$, are monoclinic with the lattice parameters and the reflection condition mentioned in the margin. This leaves three space groups to choose from: $I\,1\,1\,2/m$, $I\,1\,1\,2$, or $I\,1\,1\,m$. According to the cell metrics, a packing similar to that of one of the mentioned tetragonal structures can be expected. In that case, the correct space group should be a subgroup of either $P4/n$ or $I\bar{4}$. This is true only for $I\,1\,1\,2$, which is a subgroup of $I\bar{4}$. Therefore, $I\,1\,1\,2$ is probable, and the packing of the ions should correspond to that of the structures with tetrahedral anions. Actually, this is the case. It was possible to start with the refinement of the structure (at first without the N atoms) right away, without a preceding phase determination, starting from the atomic coordinates of a substance that crystallizes in the space group $I\bar{4}$ [395].

	Di-Co-DF1-L13A (modif. 1)	Di-Mn-DF1-L13G
a/nm	8.978	8.930
b/nm	14.772	14.640
c/nm	3.760	3.820
space group	$C222_1$	$P2_12_12_1$

$C222_1$
k2
$-\frac{1}{4},0,0$
$P2_12_12_1$

18.2 Solving the phase problem of protein structures

The structure solution mentioned in the preceding section, using only a group–subgroup relation, can help to solve the structures of proteins. The solution of the phase problem is one of the main obstacles for the crystal structure determination of proteins and other very large structures. If the crystal structure of a protein with a similar unit cell is already known, it is worthwhile to search for a group–subgroup relation.

The crystal data of two protein–metal complexes are given in the margin. The crystal structure of the manganese complex (the one mentioned second) had been known. The similarity of the lattice parameters adumbrates a group–supergroup relation with the cobalt complex. The asymmetric unit of $C222_1$ contains half as many atoms as that of $P2_1\,2_1\,2_1$. If there is a group–supergroup relation, the structure of the manganese complex should have pseudorotation axes of order 2, which are real rotation axes in the supergroup. If these axes were present, there should exist pairs of atoms whose coordinates are equivalent by the relation $\frac{1}{2}-x, y, \frac{1}{2}-z$ (in the coordinate system of $P2_1\,2_1\,2_1$). The table in the margin lists the calculated coordinates of manganese atoms according to this relation (marked by primes ′) in comparison to the observed coordinates of other atoms (the coordinates were converted from data of the protein database [8]).

The pseudosymmetry seems to be fulfilled. Using the atomic coordinates of the manganese protein (after the necessary origin shift), it was possible to determine the crystal structure of the cobalt protein [396].

	x	y	z
Mn1′	0.268	0.125	0.278
Mn2	0.273	0.124	0.221
Mn3′	0.102	0.297	0.585
Mn5	0.108	0.298	0.627
Mn4′	0.063	0.300	0.589
Mn6	0.068	0.301	0.637
Mn7′	0.450	0.016	0.034
Mn8	0.450	0.012	−0.026
Mn10′	0.167	0.040	−0.008
Mn12	0.167	0.042	−0.072

18.3 Superstructure reflections, suspicious structural features

Superstructure reflections are rather weak X-ray diffraction reflections that appear in between the strong main reflections. If they are missed, the derived unit cell is too small and the space group is wrong. Four-circle diffractometers that begin with the measurement procedure by searching for a number of reflections in order to determine the unit cell can be the first source of error. The search routine sometimes does not register any of the weak reflections. Diffractometers with an area detector are, in principle, less prone to errors of this kind, but even then the weak reflections can be missed or rejected by the computer software. In powder diffractograms, the superstructure reflections can sometimes hardly be detected on top of the background radiation.

Superstructure reflections not considered at first require an enlargement of the unit cell, and the real space group is a *klassengleiche* subgroup of the space group assumed initially. Don't get irritated by the terms; the space group of a 'superstructure' is a subgroup.

Zr_2Co_2In crystallizes in the Mo_2FeB_2 type. The Zr atoms span cubes and pairs of trigonal prisms that have common faces (Fig. 18.1). The indium atoms are in the midst of the cubes. The cobalt atoms inside the prisms form Co_2 dumbbells. If the dumbbells consist of Si atoms and uranium atoms take the cube vertices and centres, this is the U_3Si_2 type. In both cases, the space group is $P4/m2_1/b2/m$.

At first glance, the X-ray diffraction pattern of Hf_2Ni_2Sn corresponds to the Mo_2FeB_2 type. Not until after having looked thoroughly, can a few superstructure reflections be discerned that require a doubling of the c axis (Fig. 18.2). Patterns from single crystals are more convenient to reveal the weak superstructure reflections. Refinement of the structure without the superstructure reflections resulted in an excellent residual index of only $R_1 = 0.0216$, and yet the structure was not correct. The position of the Ni atom had to be refined with a split position (two sites with occupancy of one half at $z = \pm\, 0.068$). In addition, the parameter $U_{11} = U_{22}$ of the 'thermal' displacement of the Hf

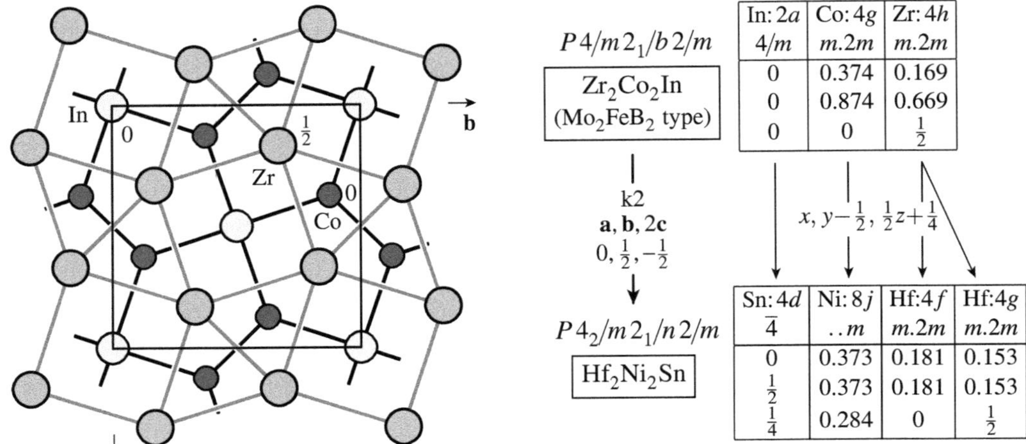

Fig. 18.1 Unit cell of Zr_2Co_2In (Mo_2FeB_2 type; U_3Si_2 type if the Zr and In positions are occupied by atoms of the same element) and group–subgroup relations to Hf_2Ni_2Sn [397].

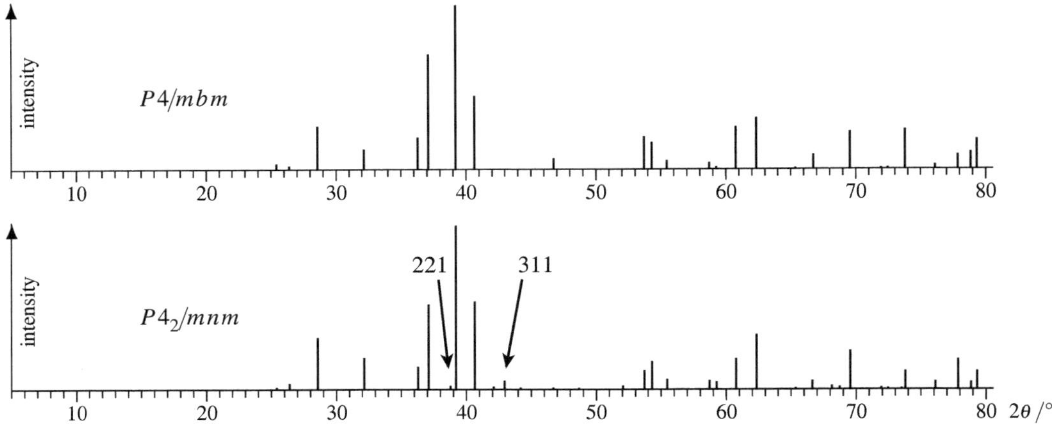

Fig. 18.2 Simulated (calculated) X-ray powder diffraction diagrams of Hf_2Ni_2Sn for the space groups $P4/m\,2_1/b\,2/m$ and $P4_2/m\,2_1/n\,2/m$. Arrows mark the two most prominent superstructure reflections.

atom was suspiciously large (Table 18.1). The correct space group, taking into account the superstructure reflections, is $P4_2/mnm$. No split positions are then needed for the nickel atoms. Instead of one, there are two independent hafnium positions with reasonable displacement parameters.

Another example are the two modifications of iodine azide, IN_3. The substance is difficult to handle; it is very explosive and it requires handling at low temperatures. It consists of chain-like polymeric molecules $(\text{–I–N}_3\text{–})_\infty$ with the shape of double combs, whose teeth (azide groups) point to opposite sides, and that are interleaved in the crystal forming layers:

Table 18.1 Refinement results for Hf_2Ni_2Sn [397].

Without superstructure reflections,
$P4/m2_1/b2/m$, $a = 703.1$ pm, $c = 338.1$ pm, $R_1 = 0.0216$, 199 reflections

atom	x	y	z	$U_{11} = U_{22}/\text{pm}^2$	U_{33}/pm^2
Sn	0	0	0	44	74
Hf	0.16738	$x+\frac{1}{2}$	0	176	42
Ni	0.3733	$x+\frac{1}{2}$	0.0679	half occupancy	

With superstructure reflections,
$P4_2/m2_1/n2/m$, $a = 703.1$ pm, $c = 676.1$ pm, $R_1 = 0.0219$, 370 reflections

atom	x	y	z	$U_{11} = U_{22}/\text{pm}^2$	U_{33}/pm^2
Sn	0	$\frac{1}{2}$	$\frac{1}{4}$	36	70
Hf1	0.18146	x	0	48	40
Hf2	0.15310	x	$\frac{1}{2}$	51	49
Ni	0.3733	x	0.2838	48	62

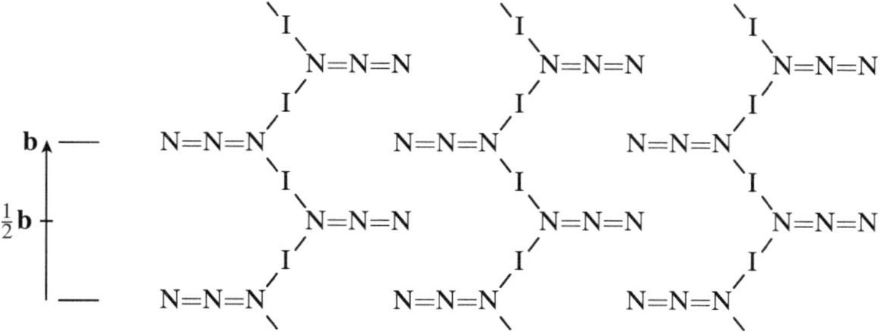

In α-IN_3 the layers are slightly corrugated, in β-IN_3 they are strongly corrugated.

The structure of α-IN_3 was determined twice by two independent research groups [398, 399]; β-IN_3 was discovered by the second research group. Both modifications crystallize in the space group *Pnma*, and in both structures all N atoms are situated on the mirror planes of the space group at $y = 0$ and $y = \frac{1}{2}$, with the iodine atoms halfway in between at $y = \frac{1}{4}$ and $y = \frac{3}{4}$. As a consequence, all X-ray reflections *hkl* with odd *k* values are very weak. They were sorted out by the computer software at all three (automated) structure determinations. Because of the missed reflections, the lattice parameters *b* of the initially determined unit cells had only one-half of their true values, $\mathbf{b}' = \frac{1}{2}\mathbf{b}$, and the derived space groups were incorrect (*Pcma* instead of *Pnma*). And yet, the structures could be refined with the incorrect unit cells and space groups with excellent matches of observed and calculated structure factors ($R_1 = 0.022$ for 502 and $R_1 = 0.025$ for 543 reflections, respectively).

The structural models of both modifications exhibited a seeming misorder with half-occupied positions for the nitrogen atoms, with coincidence of the

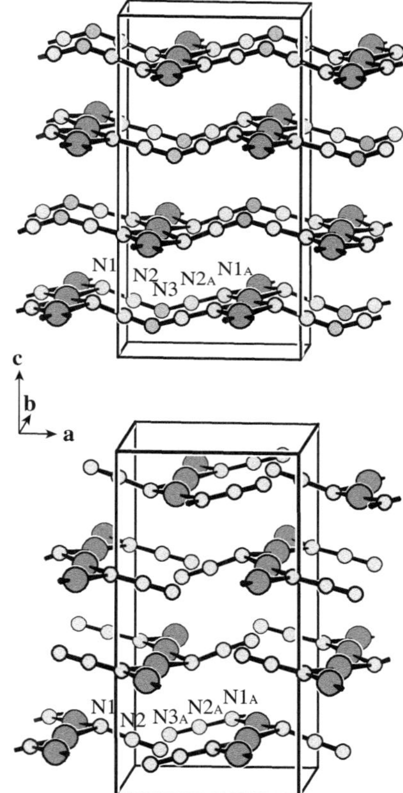

Fig. 18.3 Top: Structure of α-iodine azide as obtained without the weak reflections with odd k indices, entailing too small a unit cell $\frac{1}{2}\mathbf{b}$ and a wrong space group. The N atoms drawn in light grey have half occupancy and the terminal N atoms of the azide groups collide. The iodine atoms have a chemically unreasonable coordination. $a = 655.5$ pm, $b = 398.4$ pm, $c = 1277.6$ pm.
Bottom: Correct structure of α-iodine azide as obtained with the weak reflections, correct unit cell, and correct space group. The polymeric molecules run parallel to **b**. $a = 655.6$ pm, $b = 796.9$ pm, $c = 1277.9$ pm.

positions of N3 and N3A; in addition, the iodine atoms had fourfold planar coordination. That is not a sensible model (Fig. 18.3 top). That caused a human crystallographer to inspect the original area-detector files, and that disclosed the actual presence of the weak reflections.

A redetermination of the structures, taking into account the weak reflections, revealed the mentioned structure models of polymeric $(\text{–I–N}_3\text{–})_\infty$ molecules with linear coordination of the iodine atoms (as to be expected in the presence of three lone electron pairs). The correct structures could be derived directly from the misordered models because the real space group *Pnma* is a subgroup of *Pcma* with doubled cell parameter *b*. The correct model of α-IN_3 is shown at the bottom of Fig. 18.3. The group–subgroup relation is depicted in Fig. 18.4 [400]. The situation is similar for β-IN_3. Refinements with the old X-ray data yielded indices of coincidence of $R_1 = 0.025$ for 704 reflections for α-IN_3 and of $R_1 = 0.019$ for 778 reflections for β-IN_3.

Conclusion: It always pays off to inspect the primary X-ray diffraction data with crystallographic expertise before relying on an automated structural determination by computer.

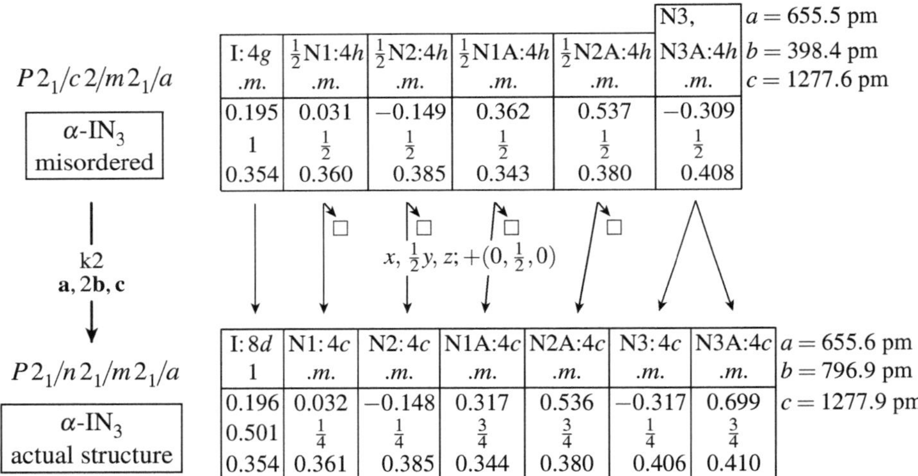

I:4*g*	½N1:4*h*	½N2:4*h*	½N1A:4*h*	½N2A:4*h*	N3, N3A:4*h*
.*m*.	.*m*.	.*m*.	.*m*.	.*m*.	.*m*.
0.195	0.031	−0.149	0.362	0.537	−0.309
1	½	½	½	½	½
0.354	0.360	0.385	0.343	0.380	0.408

I:8*d*	N1:4*c*	N2:4*c*	N1A:4*c*	N2A:4*c*	N3:4*c*	N3A:4*c*
1	.*m*.	.*m*.	.*m*.	.*m*.	.*m*.	.*m*.
0.196	0.032	−0.148	0.317	0.536	−0.317	0.699
0.501	¼	¼	¾	¾	¼	¾
0.354	0.361	0.385	0.344	0.380	0.406	0.410

Fig. 18.4 Group–subgroup relation between the misordered (incorrect) structural model of α-IN_3 and the correct structure. The half-occupied positions of the N atoms split each into a fully occupied and an unoccupied position (with $y + \frac{1}{2}$, marked by □).

18.4 Detection of twinned crystals

The X-ray diffraction data of $Er_{2.30}Ni_{1.84}In_{0.70}$ indicate a structure with the Mo_2FeB_2 type. There are no superstructure reflections. The diffraction pattern corresponds to the Laue symmetry $4/mmm$ (equal intensities of reflections *hkl* and *khl*). Refinement of the structure, assuming the Mo_2FeB_2 type, led to suspiciously large parameters of thermal motion of the erbium atoms in the *a*–*b* plane. This suggested a reduced symmetry, for example a lower Laue class. In fact, the Laue symmetry $4/m$ is correct. In this case, the reason is merohedral twins with exact superposition of the reflections.[2] The reflections *hkl* of one

[2] The components of merohedral twins (twins by merohedry) have coincident lattices [351].

kind of domain are superposed exactly on the reflections *khl* of the other kind. If both kinds of domains are present with about the same volume fractions, the diffraction pattern of the higher Laue symmetry is feigned. The true space group is $P4/m$, a maximal subgroup of $P4/m2_1/b2/m$. Since it is a *translationengleiche* subgroup, there are no superstructure reflections. Refinement in the space group $P4/m$, taking into account the twins, yielded a satisfactory result. The cube about the origin of the unit cell (Fig. 18.1) is slightly larger and that about $\frac{1}{2}, \frac{1}{2}, 0$ is smaller [401].

Solid-state reactions are usually performed at high temperatures. If a substance forms several polymorphic forms, at first a high-temperature form will be formed. During the subsequent cooling, unnoticed phase transitions with symmetry reductions can take place. If such a transition involves a *translationengleiche* group–subgroup relation, the formation of twins can be expected (cf. Section 16.3). The twin operation corresponds to a symmetry operation of the high-temperature form; therefore, the X-ray reflections from the twin domains will be nearly or exactly superposed in a systematic manner. This can feign a wrong space group, which, in many cases, is a supergroup of the real space group, like in the example of $Er_{2.30}Ni_{1.84}In_{0.70}$. However, the symmetry of a subgroup can also be feigned, as shown with the following example.

The space groups and lattice parameters from the literature of three substances are listed in the margin. All three have the same structure (Fig. 18.5) and nearly the same lattice parameters, and therefore, the occurrence of three different space groups is out of the question. It should be easily possible to distinguish the three space groups $C2/c$, $P2/c$, and $P12_1/n1$ from the reflection conditions in the X-ray patterns. Therefore, there must be some fundamental error. Twins are the cause, and group–subgroup relations can help to trace the error. The actual space group is $C2/c$ [405].

	$CaCrF_5$	$CaMnF_5$	$CdMnF_5$
space group	$C2/c$	$P2/c$	$P2_1/n$
a/pm	900.5	893.8	884.8
b/pm	647.2	636.9	629.3
c/pm	753.3	783.0	780.2
β/°	115.9	116.2	116.6
reference	[402]	[403]	[404]

The structure contains zigzag-shaped MF_5^{2-} chains of vertex-sharing octahedra along **c**. If the chains are linear, the symmetry is enhanced to $I2/m2/m2/m$ with half the size of the unit cell. At the conditions of preparation at high temperature, this enhanced symmetry seems to be true. Upon cooling, a phase transition takes place, with symmetry reduction to $C2/c$ (Fig. 18.5). This involves a *translationengleiche* subgroup of index 2, and thus the formation of twins with two orientational states. Corresponding to the orthorhombic metric of the supergroup, the metrical relations of the monoclinic cells are such that there is a superposition of the X-ray reflections from one twin component with those of the other. Part of the reflections of one component appear right there where the other one has extinct reflections. As a consequence, the C centring of the space group $C2/c$ cannot be recognized by the absent reflections $h+k=2n+1$ (for all hkl). However, the reflections with $h+k=2n+1$ are still absent if l is even. This is a reflection condition that does not match with any space group; hence it was disregarded, although this should have been an alarm signal.

The refinement of the structure of $CaMnF_5$ with the too-low space-group symmetry $P2/c$, a subgroup of $C2/c$, yielded unreliable atomic coordinates and bond lengths and the wrong conclusion, 'unexpectedly, this is the first case of Mn^{3+} exhibiting both possible distortions due to the Jahn–Teller effect, elongated and compressed coordination octahedra' [403].

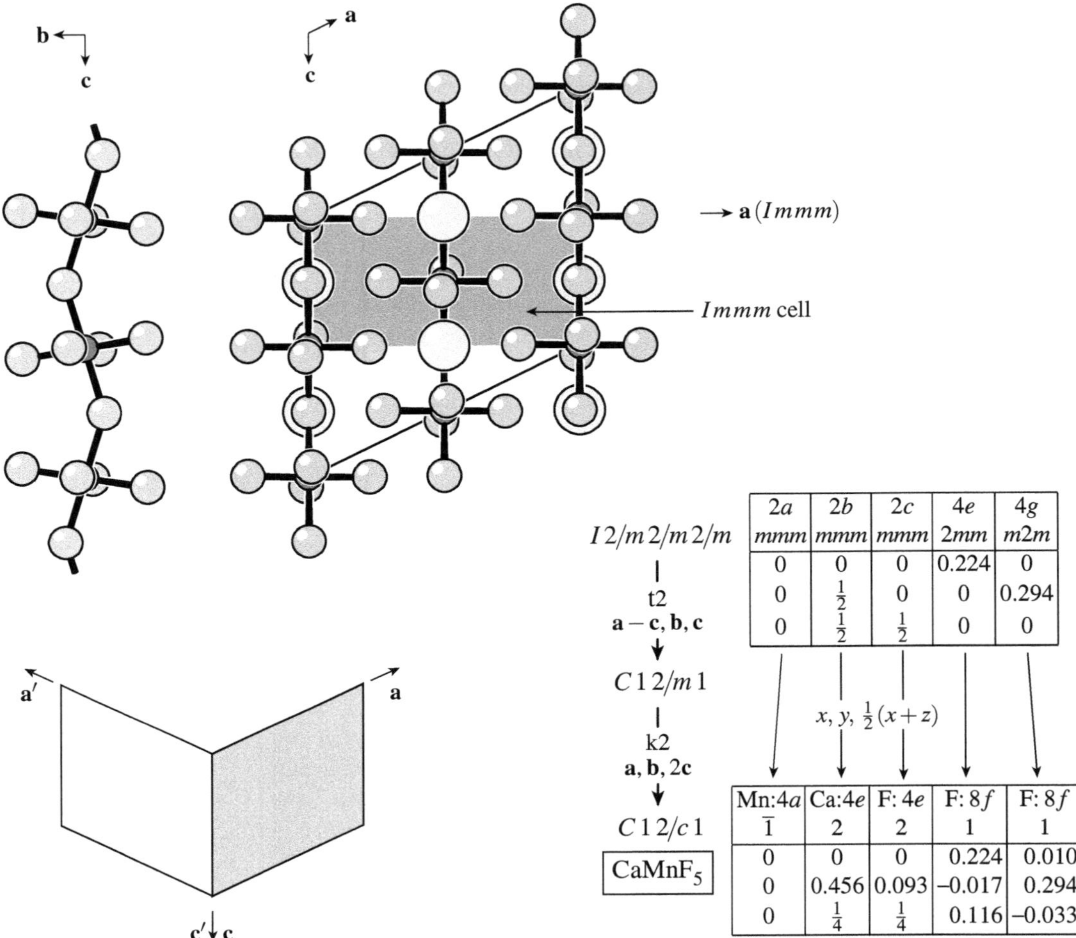

Fig. 18.5 Structure of the compounds $CaCrF_5$, $CaMnF_5$, and $CdMnF_5$ and group–subgroup relations from the probable high-temperature aristotype. Upper left: Section from a chain of the vertex-sharing octahedra; in the high-temperature form the chain is linear (or misordered in two orientations). Lower left: Mutual orientation of the unit cells of the twin domains.

Further information concerning twins and the solution of twinned crystal structures is available at:

`www.cryst.chem.uu.nl/lutz/twin/gen_twin.html`

Exercises

Solutions on page 338.

18.1. HoRhIn (= $Ho_3Rh_3In_3$) crystallizes in the ZrNiAl type, space group $P\bar{6}2m$, Laue class $6/mmm$, with the following coordinates:

		x	y	z
Ho	$3g$	0.4060	0	$\frac{1}{2}$
Rh1	$1b$	0	0	$\frac{1}{2}$
Rh2	$2c$	$\frac{1}{3}$	$\frac{2}{3}$	0
In	$3f$	0.7427	0	0

The structure of $Gd_3Rh_2In_4$ is similar, but the position of the Rh2 atom is taken by Rh and In atoms with half occupancy each [406]. According to the X-ray reflection intensities, the Laue symmetry is still $6/mmm$. There are no superstructure reflections. Refinement yields a suspiciously large displacement parameter $U_{11} = U_{22}$ for the Gd atom at the position $3g$. What is the probable problem of the structure determination? What is the actual space group?
Hint: The positions of the Rh and In atoms are occupied in an ordered manner.

18.2. An X-ray powder-diffraction pattern of Eu_2PdSi_3 can initially be indexed with a small hexagonal unit cell with $a = 416$ pm and $c = 436$ pm. The pattern of the reflection intensities agrees with an AlB_2-like structure ($P6/mmm$, Al: $1a$ 0, 0, 0; B: $2d$ $\frac{1}{3}$, $\frac{2}{3}$, $\frac{1}{2}$; Fig. 12.4). An ordered distribution of the palladium and silicon atoms within the layers of hexagonal rings is not compatible with this symmetry and cell. A detailed inspection of the powder diffractogram reveals the presence of weak reflections that require a doubling of the a and b axes [407]. Use group–subgroup relations to search for a model with ordered distribution of the atoms of the silicide Eu_2PdSi_3.

18.3. The structure of $CsMnF_4$ contains distorted coordination octahedra that share vertices, forming corrugated, tetragonal layers. Initially, the space group $P4/nmm$ was assumed [408]. However, several aspects remained unsatisfactory: The ellipsoids of 'thermal' motion of the bridging fluorine atoms were suspiciously large; these ellipsoids had their semi-major axes approximately parallel to the covalent bonds (normally they are approximately perpendicular); each Mn atom had four long and two short Mn–F bonds (4× 201 pm in the layer, 2× 181 pm terminal), but the distortion at Mn^{3+} ions due to the Jahn–Teller effect should result in four short and two long bonds; the observed ferromagnetism (at low temperatures) is only compatible with elongated and not compressed coordination octahedra [409]. These deficiencies were eliminated by a new determination of the structure [410]. How was it possible to solve the problem?
Hints: Compare the image of the structure with the figure of the symmetry elements in *International Tables A*, space group $P4/nmm$ (No. 129), origin choice 2; the correct structure is tetragonal and the Mn atoms occupy a centrosymmetric position.

Original values:

$P4/n2_1/m2/m^{(2)}$, $a = 794.4$ pm, $c = 633.8$ pm		x	y	z
Cs1	$2b$	$\frac{3}{4}$	$\frac{1}{4}$	$\frac{1}{2}$
Cs2	$2c$	$\frac{1}{4}$	$\frac{1}{4}$	0.449
Mn	$4d$	0	0	0
F1	$8i$	$\frac{1}{4}$	−0.003	0.048
F2	$8j$	−0.028	x	0.281

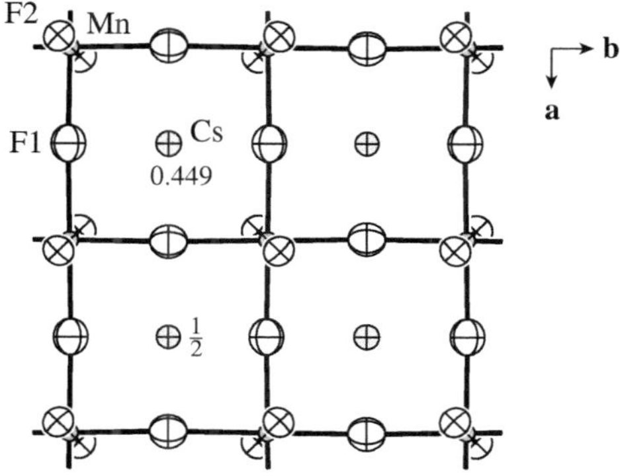

$CsMnF_4$ ($P4/nmm^{(2)}$), projection along c; ellipsoids of 'thermal' motion at 70 % probability

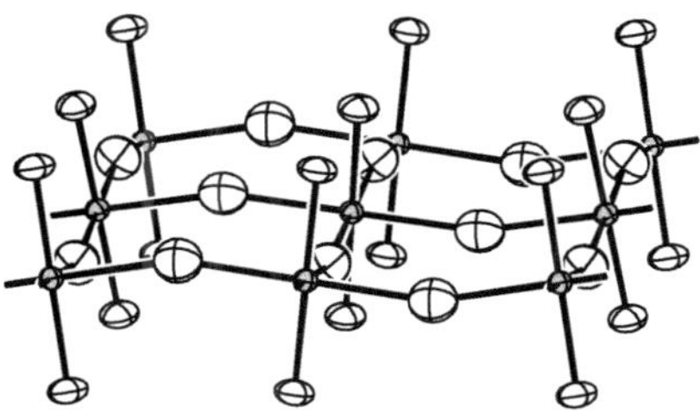

MnF_4^- layer

Prediction of possible structure types

19

19.1 Derivation of hypothetical structure types with the aid of group–subgroup relations

Previous chapters deal with relationships between *known* structure types. The starting point is always a high-symmetry structure type, the aristotype, from which all other structures can be derived. As a rule, additional subgroups of the space group of the aristotype can be found that belong to as yet unknown structure types. We have presented two examples: the conceivable structure of an unknown modification of WCl_3 and a variant of the ilmenite type which could be the structure of $AlTiO_3$ (Section 14.2.1, page 184; Fig. 14.5).

A systematic search for new possible structures can be performed with the aid of group–subgroup relations (i.e. possible structure types can be predicted). What sense does it make to look for hypothetical structure types, the number of known structures being already immense? There are several reasons:

1. It is still possible to reveal relatively simple structure types for which no representatives are as yet known. It would be worthwhile to specifically search for them or to answer the question, why they do not exist.
2. If a microcrystalline compound was prepared, the powder diffraction diagram of which cannot be interpreted for the moment, model computations with feasible structure types could be of help.
3. Hypothetical models can be useful for the structure determination of crystals exhibiting misorder with diffuse X-ray scattering, although at first glance this does not seem to be an adequate approach. For an example see the end of Section 19.4, page 270.

The starting point is always an aristotype and a structural principle. For example, the aristotype can be the hexagonal-closest packing of spheres, and the structural principle can be the partial occupation of the octahedral interstices of this packing. Of course, only such structure types will be found that meet these initial conditions. To put it in another way: Among the infinite set of conceivable structures, we restrict ourselves to a subset that causes the infinite problem to become more manageable. However, in order to reduce the infinite problem to a finite one, some more restrictions are necessary. Such restrictions can be: The chemical composition; a given molecular structure; a maximum number of symmetry-independent atoms for every atom species; a maximum size of the unit cell. Finally, energy computations, such as are also being performed without group-theoretical considerations [412, 413], can help to settle what are the most probable structures.

To explain how to proceed, we examine what structures are possible for compounds $A_aB_b...X_x$ if the X atoms form a hexagonal-closest packing of spheres and the remaining atoms occupy its octahedral interstices. We thus extend the considerations of Sections 14.2.1 and 14.2.2 to include as yet unknown structure types of this family of structures.

In this case, the aristotype has the space group $P6_3/m2/m2/c$, and the space groups of the sought structure types are subgroups thereof. The unit cell of the hexagonal-closest packing of spheres contains two X atoms and two octahedral interstices (cf. Fig. 14.2, page 181). If the unit cell is enlarged by a factor of Ξ, the cell contains 2Ξ octahedral voids and 2Ξ X atoms. We treat unoccupied octahedral voids in the same way as occupied octahedral voids; in a way, they are being occupied by Schottky defects (symbol $\square$). For the chemical composition $A_aB_b...\square_sX_x$ we then have: $a+b+\cdots+s=x$.

All numbers $2a\Xi/x$, $2b\Xi/x$, ..., $2s\Xi/x$ have to be integral. That means that, depending on chemical composition, the unit cell of the packing of spheres has to be enlarged by an appropriate factor of Ξ. For example, for a trihalide $A_1\square_2X_3$, the cell must be enlarged by a factor of $\Xi = 3$ or a multiple thereof; otherwise it is not possible to occupy one-third of the octahedral voids by A atoms in an ordered manner.

The first consideration to be made is how to enlarge the cell of the aristotype in an appropriate way. Figure 19.1 shows the possibilities of enlargement of the hexagonal cell by the factors 2, 3, and 4. The enlargement factor Ξ always refers to the primitive cells. If all possible kinds of cell enlargement for a given factor Ξ are being sought, some caution is necessary with the larger values of Ξ; perhaps some possibility might be missed, or seemingly, though not really different possibilities could be considered several times.

Enlargement of the (primitive) unit cell means loss of translational symmetry. Therefore, there must be at least one *klassengleiche* group–subgroup relation. We set up a tree of group–subgroup relations and look for subgroups that correspond to the selected cell enlargement. In so doing, we have to monitor how the point orbits that are to be occupied with atoms develop from each group to its subgroups. The necessary information can be taken from *International Tables* A1. Most important are those group–subgroup relations that involve a splitting of the considered orbit into independent orbits. For our example, we have to monitor what and how many independent orbits result from the orbit of the octahedron centres. The symmetry reduction has to be tracked only until all octahedral voids have become symmetry independent in the enlarged cell, provided that distortion variants are not considered. Subgroups that do not involve splittings of the orbits do not have to be considered unless they are needed as intermediate groups.

If the partial substitution of the X atoms is also considered, the development of the orbits of the X atoms also has to be monitored. However, they should also be tracked if only one kind of X atom is present to be sure of how many non-equivalent positions the X atoms will be distributed into. Space groups with many symmetry-independent X atoms are less probable according to the symmetry principle.

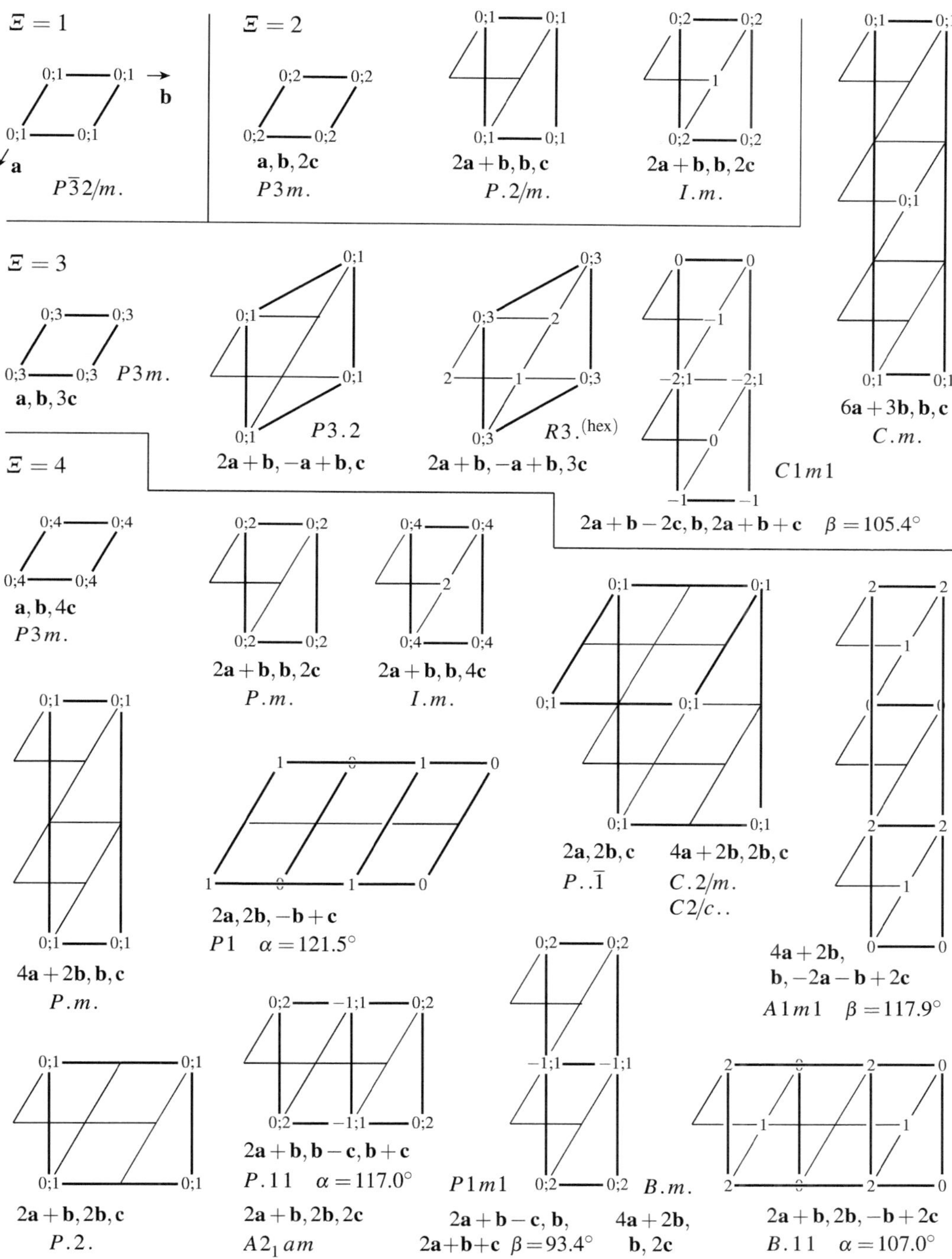

Fig. 19.1 Possibilities for the enlargement of a hexagonal unit cell by the factors $\Xi = 2$, 3, and 4. Numbers indicate the heights of the unit cell vertices and of translation-equivalent points in the direction of view as multiples of the hexagonal lattice constant c. Symmetry symbols state the minimum symmetry retained if the packing of spheres stays undistorted and atoms are being inserted into the centres of the octahedra (1 means a direction that remains without symmetry).

Some subgroups of the tree permit certain distortions if the involved coordinates are no longer fixed by symmetry. Additional distortions require further symmetry reductions with *translationengleiche* subgroups. They can be regarded in retrospect; we disregard them here.

Consider as an example the tripled, rhombohedral unit cell that is shown in Fig. 19.1 under $\Xi = 3$, $2\mathbf{a}+\mathbf{b}, -\mathbf{a}+\mathbf{b}, 3\mathbf{c}$. In this case, there is a ninefold enlargement of the cell in the hexagonal axes setting, but due to the rhombohedral centring the primitive cell is only tripled. The corresponding tree of group–subgroup relations is shown in Fig. 14.1, page 181. The lowest-symmetry space group in this tree is $R3$. This is the most symmetrical subgroup of $P6_3/mmc$ that has all octahedral voids symmetrically independent. Additional space groups that could have been included as intermediate groups between $P6_3/mmc$ and $R3$ are not mentioned because they have no more symmetry-independent octahedral voids than preceding space groups. Space groups not mentioned in Fig. 14.1 cannot occur with the considered kind of cell enlargement, as long as the packing of spheres consists of only one kind of atoms and provided that no distortions require additional symmetry reductions.

Known structure types corresponding to the tree of Fig. 14.1 are mentioned in Section 14.2.1. In addition, the unknown structure is mentioned that could exist for WCl_3 in the space group $\mathcal{G}_7 = R32$ if its point orbit c_3 is occupied by W atoms (Fig. 14.5, page 184). For the space groups mentioned in Fig. 14.1, some more structure types are feasible, depending on what orbits are occupied by atoms of different elements. For example, the $LiSbF_6$ type is known for the composition $AB\square_4X_6$ ($R\bar{3}$, occupied orbits a and b; Fig. 14.1). There are two more possibilities for this composition, namely in the space group $R3$:

1. Atom A at orbit a_2, B at a_3 of $R3$; that is a BiI_3 derivative with substitution of the Bi positions by alternating A and B atoms within a layer.
2. Atom A at orbit a_3, B at a_6 of $R3$; that is a derivative of the hypothetical WCl_3 with one A and B atom each in the pairs of face-sharing octahedra.

This way, we have revealed two new possible structure types for compounds ABX_6. There are no additional possible structures for this composition with one of the space groups of Fig. 14.1. For example, in $R3$, if A were to occupy a_1 and B were to occupy a_3, this would be the same structure type as with occupation of a_2 and a_3. Occupation of a_1 and a_5 would result in the $LiSbF_6$ type which, however, does not belong to space group $R3$ because this distribution of atoms can already be realized in space group $R\bar{3}$. Therefore, when predicting possible structure types by systematic occupation of all point orbits to be considered, caution should be exercised with respect to two aspects:

1. Several distributions of atoms that at first seem to be different may refer to the same structure type; only one of them is to be considered.
2. Some atom distributions are realizable with higher-symmetry space groups and have to be assigned to them.

Errors can be avoided if, for a given chemical composition, it has been calculated in how many ways different kinds of atoms can be allocated to the orbits of a space group; see the next section.

Derivations of possible crystal structure types by this method have been described for the following structural principles: Occupation of octahedral voids in the hexagonal-closest packing of spheres [206, 416], occupation of octahedral voids in the cubic-closest packing of spheres [247,417,418], occupation of octahedral and tetrahedral voids in the cubic-closest packing of spheres [419].

19.2 Enumeration of possible structure types

In this section we present mathematical procedures to enumerate the different structures that are possible for every subgroup, given an aristotype and structural principle. Even when we try to determine completely all different substitution variants of the NaCl type with doubled unit cell (cf. Section 14.3.1), without mathematical support, we run into difficulties .

19.2.1 The total number of possible structures

Given a certain number of kinds of atom and a certain number of point orbits, how many non-equivalent ways do there exist to distribute the atoms? The total number of possibilities can be calculated with Pólya's enumeration theorem [420,421]. The theorem can be used, for example, to enumerate the isomers of organic molecules [422, 423]. Consider the example of a trigonal prism; how many different ways are there to mark the six vertices with at most six different colours (or to occupy them with at most six kinds of atoms)?

The six vertices of the prism are numbered in Fig. 19.2. The notation (123)(456) is a way to express one possible way to permute the symmetry-equivalent vertices. (123) means cyclic exchange of the vertices ① → ② → ③ → ①. (13) means exchange of the vertices ① and ③. (5) means vertex ⑤ keeps its position. If there are n numbers between the parentheses, we write s_n for short. We multiply all values s_n belonging to a permutation; this product is called a cycle structure term. The group of permutations of the trigonal prism is the group of all possibilities of permuting its vertices in accordance with its symmetry, wich is the point group $\overline{6}m2$. The permutation group has an order of 12; it comprises the 12 permutations listed in Fig. 19.2.

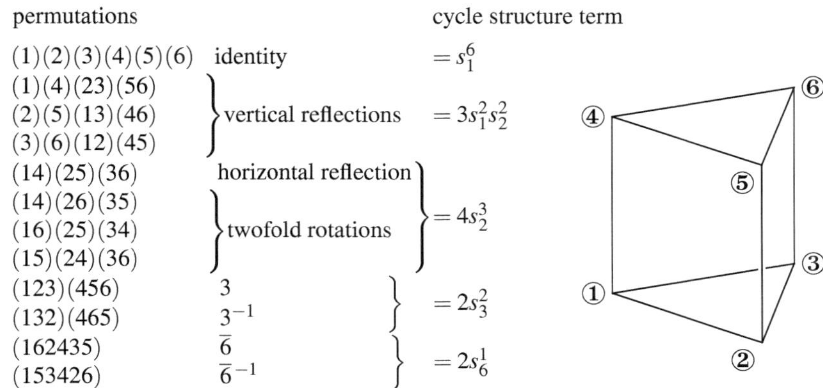

Fig. 19.2 Permutation group of the trigonal prism.

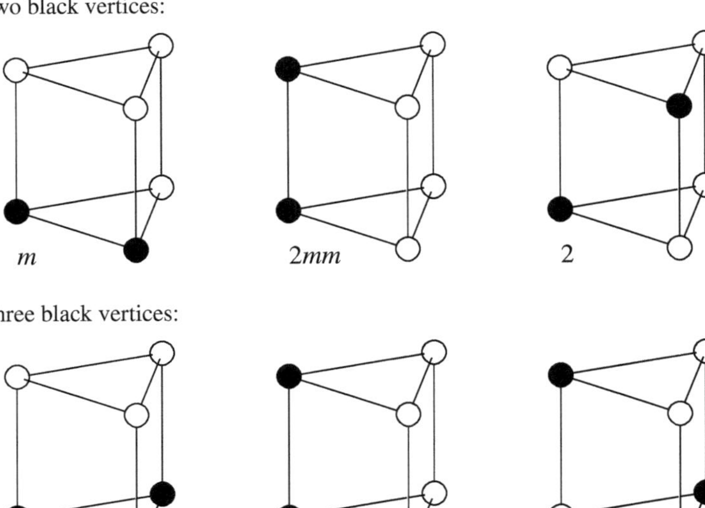

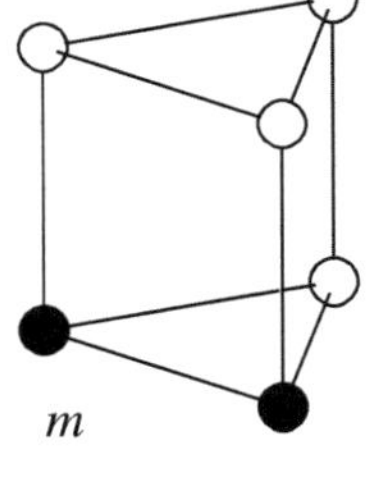

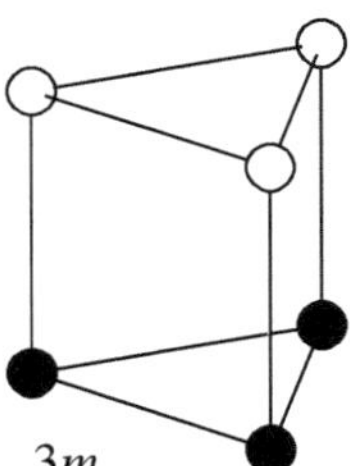

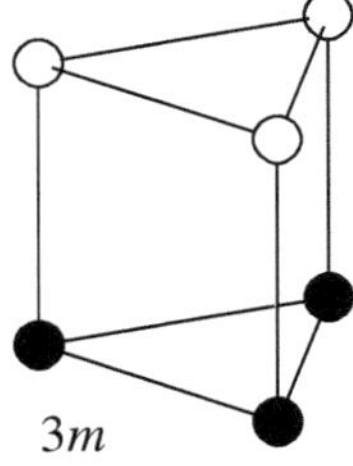

Fig. 19.3 The three possible colourings of the vertices of a trigonal prism with two colours if one of the colours is used to mark two and three vertices, respectively.

Cycle index of the trigonal prism:
$Z = \frac{1}{12}(s_1^6 + 3s_1^2 s_2^2 + 4s_2^3 + 2s_3^2 + 2s_6^1)$

$s_n = x_1^n + x_2^n + \cdots + x_k^n$

The sum of all cycle structure terms, divided by the order of the permutation group, is the *cycle index Z*. Every one of the $k \leq 6$ colours (or atom species) is designated by one variable x_i, for example x_1 = white, x_2 = black. Now we substitute sums of powers x_i^n for the quantities s_n. Applying this substitution in the cycle index yields C, the *generating function*. With two colours x_1 and x_2 we have $s_n = x_1^n + x_2^n$; the generating function is then:

$$\begin{aligned} C &= \tfrac{1}{12}[(x_1+x_2)^6 + 3(x_1+x_2)^2(x_1^2+x_2^2)^2 + 4(x_1^2+x_2^2)^3 + 2(x_1^3+x_2^3)^2 + 2(x_1^6+x_2^6)] \\ &= x_1^6 + x_1^5 x_2 + 3x_1^4 x_2^2 + 3x_1^3 x_2^3 + 3x_1^2 x_2^4 + x_1 x_2^5 + x_2^6 \end{aligned} \quad (19.1)$$

The powers correspond to the possible colour markings. For example, $3x_1^4 x_2^2$ means that there are three possible ways to mark four vertices with the colour x_1 and two vertices with x_2. These three markings are shown in Fig. 19.3; in addition, the three colour markings for $3x_1^3 x_2^3$ are shown.

For vertices that are not coloured (or that are left vacant of atoms), one variable x_i can be replaced by $x^0 = 1$. For example, if some of the vertices are left colourless and the remaining ones are marked with the two colours x_1 and x_2, the generating function is:

$$\begin{aligned} C = {} & \tfrac{1}{12}[(1+x_1+x_2)^6 + 3(1+x_1+x_2)^2(1+x_1^2+x_2^2)^2 + 4(1+x_1^2+x_2^2)^3 \\ & +2(1+x_1^3+x_2^3)^2 + 2(1+x_1^6+x_2^6)] \\ = {} & x_1^6 + x_1^5 x_2 + 3x_1^4 x_2^2 + 3x_1^3 x_2^3 + 3x_1^2 x_2^4 + x_1 x_2^5 + x_2^6 \\ & +x_1^5 + 3x_1^4 x_2 + 6x_1^3 x_2^2 + 6x_1^2 x_2^3 + 3x_1 x_2^4 + x_2^5 \\ & +3x_1^4 + 6x_1^3 x_2 + 11x_1^2 x_2^2 + 6x_1 x_2^3 + 3x_2^4 \\ & +3x_1^3 + 6x_1^2 x_2 + 6x_1 x_2^2 + 3x_2^3 \\ & +3x_1^2 + 3x_1 x_2 + 3x_2^2 \\ & +x_1 + x_2 \\ & +1 \end{aligned} \quad (19.2)$$

In the preceding result, all variants that have the same number of colourless vertices have been collected in one line each (beginning with the second = sign; the sum of the powers of each product is the same for all products in one line). The first line (sum of powers equal to 6 each, i.e. no colourless vertices), of course, corresponds to eqn (19.1). Since colourless vertices could also be marked with an additional (third) colour, the result is the same as for a marking with three colours and no colourless vertices; therefore, the coefficient $11x_1^2x_2^2 = 11x_1^2x_2^2 \times 1^2$ has the same meaning as $11x_1^2x_2^2x_3^2$ for three colours and no colourless vertices.

If the permutation group is that of an achiral point group, the presented arithmetic technique does not distinguish pairs of enantiomers in the case of chiral colour markings. A pair of enantiomeric molecules is counted as one isomer. To find out how many enantiomeric pairs exist, the calculation is repeated, but with the permutation group that consists only of rotations. In the case of the trigonal prism, these are the identity, three twofold rotations, and two threefold rotations. The cycle index is then Z'. With two colours x_1 (white) and x_2 (black) the resulting generating function is:

$Z' = \frac{1}{6}(s_1^6 + 3s_2^3 + 2s_3^2)$

$$\begin{aligned} C' &= \tfrac{1}{6}[(x_1+x_2)^6 + 3(x_1^2+x_2^2)^3 + 2(x_1^3+x_2^3)^2] \\ &= x_1^6 + x_1^5x_2 + 4x_1^4x_2^2 + 4x_1^3x_2^3 + 4x_1^2x_2^4 + x_1x_2^5 + x_2^6 \end{aligned} \qquad (19.3)$$

By subtraction of the generating functions C' minus C from eqn (19.1), we obtain the number of enantiomeric pairs:

$$C' - C = x_1^4x_2^2 + x_1^3x_2^3 + x_1^2x_2^4$$

That is one enantiomeric pair with two, one with three, and one with four black vertices. They are those shown in Fig. 19.3 for the point groups 2 and 1; for each of them there exists another enantiomer in addition to the ones shown.

If the permutation group corresponds to a chiral point group, all colour markings are chiral and are counted once each.

The number of possible colour markings has been calculated and listed for many convex polyhedra [424].

If we are interested in the occupation of points in a crystal, the procedure is the same. Starting from a set of symmetry-equivalent points, the possible permutations within one unit cell are determined. Consider the six octahedral interstices shown in Fig. 14.2 (page 181), placed at $z = 0$ and $z = \frac{1}{2}$ in the positions ①, ②, and ③ within the unit cell. Referred to the small unit cell marked by the grey background, all of these points are symmetry equivalent. In the tripled unit cell, they can be permuted and marked in the same way as the vertices of a trigonal prism. The point in the centre of the prism ①–②–③ in Fig. 14.2 has the site symmetry $\bar{6}m2$ in the space group of the aristotype $P6_3/m2/m2/c$; that is the point group of the prism.

19.2.2 The number of possible structures depending on symmetry

The result of a calculation with Pólya's theorem is the total number of possible colour markings; there is no statement as to their symmetries. However, their point groups must be subgroups of the point group of the most symmetrical colouring. The point group of the trigonal prism is $\overline{6}m2$ if all vertices are equal. The point groups of the prisms with differently coloured vertices are subgroups thereof. The point groups mentioned in Fig. 19.3 are such subgroups.

With reference to the tree of group–subgroup relations, WHITE has expanded Pólya's theorem, so that the colour markings can be enumerated depending on symmetry [425].[1] The application to crystallographic problems was explained by MCLARNAN [426]; see also [411]. MCLARNAN used the procedure to enumerate the stacking sequences of closest packings of spheres if the stacking sequence of the hexagonal layers with positionings *A*, *B*, and *C* is repeated after $N = 2, 3, 4, \ldots, 50$ layers [427]. He also enumerated the stacking variants of CdI_2, ZnS, and SiC polytypes [427, 428] and of sheet silicates [429]. Another technique to enumerate stacking variants in simpler cases has been described by IGLESIAS [430].

[1] The paper can hardly be understood by non-mathematicians.

Prerequisites for the application of WHITE's method are the tree of group–subgroup relations and a survey of the orbits of the concerned atomic positions. Starting from the aristotype, all subgroups have to be considered that entail splittings of orbits into non-equivalent orbits. The tree needs to be traced only down to the point where all atomic positions of interest have become symmetrically independent (not taking into account distortion variants) or until a predetermined limit of symmetry reduction has been reached.

We designate the space groups (for molecules: the point groups) of the tree by $\mathcal{G}_1, \mathcal{G}_2, \ldots$, the numbering following the hierarchy of decreasing symmetry; for two groups $\mathcal{G}_i$ and $\mathcal{G}_j$, $i < j$, the index of $\mathcal{G}_i$ in $\mathcal{G}_1$ must be smaller than or equal to the index of $\mathcal{G}_j$ in $\mathcal{G}_1$. $\mathcal{G}_1$ is the aristotype. A matrix **M** is calculated having the matrix elements:

$$m_{ij} = \frac{1}{|\mathcal{G}_j|} \sum_{g \in \mathcal{G}_1} \chi(g\mathcal{G}_i g^{-1} \subseteq \mathcal{G}_j) \quad (19.4)$$

$$= I_j \frac{[\mathcal{G}_i \subseteq \mathcal{G}_j]}{[\mathcal{G}_i]} \quad (19.5)$$

$\lvert\mathcal{G}_j\rvert$	= order of the group $\mathcal{G}_j$
g	= symmetry operation of $\mathcal{G}_1$
I_j	$= \frac{\lvert\mathcal{G}_1\rvert}{\lvert\mathcal{G}_j\rvert}$ = index of $\mathcal{G}_j$ in $\mathcal{G}_1$
$[\mathcal{G}_i]$	= number of subgroups conjugate to $\mathcal{G}_i$ in $\mathcal{G}_1$
$[\mathcal{G}_i \subseteq \mathcal{G}_j]$	= number of these conjugates that are also subgroups of $\mathcal{G}_j$
$\chi(\text{COND})$	= 1 if the condition COND is fulfilled
$\chi(\text{COND})$	= 0 if the condition COND is not fulfilled

$\chi(\text{COND})$ is called the 'characteristic function'. In our case the condition is COND = $g\mathcal{G}_i g^{-1} \subseteq \mathcal{G}_j$, that is, it has to be established whether application of the symmetry operations g of the aristotype to the group $\mathcal{G}_i$ generates the group $\mathcal{G}_j$. To compute the sum of eqn (19.4), all symmetry operations of the aristotype have to be processed. This includes the translations of the aristotype that correspond to the largest unit cell of the considered subgroups; that is, of the unit cell of the aristotype as many adjacent unit cells have to be considered as it corresponds to the largest unit cell among the subgroups. For tetragonal, trigonal, hexagonal, and cubic aristotypes the number of considered cells has to be in accordance with the crystal system; for example, considering the hettotypes of a tetragonal aristotype having a doubled, two cells of the aristotype not only in the direction of **a** have to be considered, but also in the direction of **b**; otherwise the group properties of the tetragonal aristotype would be violated. For larger trees of subgroups the computational expenditure is considerable and can be handled in reasonable time only by a fast computer. For the computation, the symmetry operations are represented by 4×4 matrices [416,418,419].

To calculate by hand, eqn (19.5) is more convenient. For that purpose it is required to have an overview of how many conjugates in $\mathcal{G}_1$ exist for every subgroup. As explained in Section 8.3, this can be evaluated with the aid of the normalizers. For example, using the table of the Euclidean normalizers of *International Tables A*, we find the relations mentioned in the margin for $\mathcal{G}_6$ (hexagonal axes setting) of the tree of Fig. 14.1, page 181.

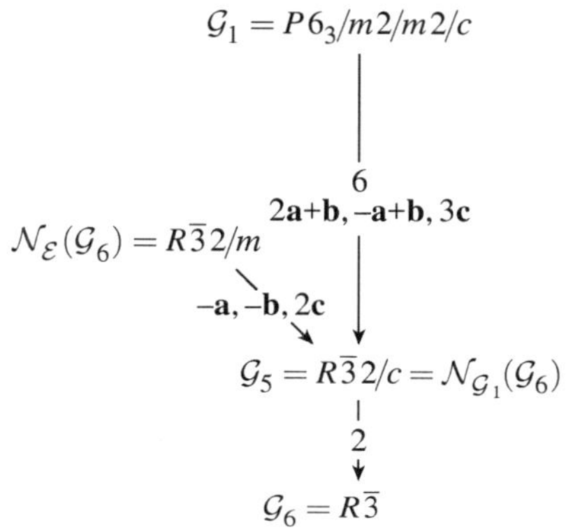

$\mathcal{G}_1 = P6_3/m2/m2/c$

6

2a+b, –a+b, 3c

$\mathcal{N}_{\mathcal{E}}(\mathcal{G}_6) = R\bar{3}2/m$

–a, –b, 2c

$\mathcal{G}_5 = R\bar{3}2/c = \mathcal{N}_{\mathcal{G}_1}(\mathcal{G}_6)$

2

$\mathcal{G}_6 = R\bar{3}$

The index of $\mathcal{N}_{\mathcal{G}1}(\mathcal{G}_6)$ in $\mathcal{G}_1$ is 6; accordingly, $\mathcal{G}_6$ belongs to a conjugacy class consisting of six subgroups that are conjugate in $\mathcal{G}_1$. They differ by three positions of their origins (which are translationally equivalent in $\mathcal{G}_1$), with two orientations each, rhombohedral obverse and reverse.

The matrix **M** is triangular, that is, all elements above the main diagonal are $m_{ij} = 0$. All components are zeros or positive integers, and the first column contains only ones.

Consider the tree of Fig. 14.1 as an example. The space groups are numbered from $\mathcal{G}_1$ to $\mathcal{G}_9$. The corresponding matrix is:

$$\mathbf{M} = \begin{pmatrix} 1 &&&&&&&& \\ 1 & 2 &&&&&&& \\ 1 & 0 & 2 &&&&&& \\ 1 & 2 & 2 & 4 &&&&& \\ 1 & 0 & 2 & 0 & 1 &&&& \\ 1 & 2 & 2 & 4 & 1 & 2 &&& \\ 1 & 0 & 2 & 0 & 1 & 0 & 2 && \\ 1 & 0 & 2 & 0 & 3 & 0 & 0 & 6 & \\ 1 & 2 & 2 & 4 & 3 & 6 & 6 & 6 & 12 \end{pmatrix} \qquad (19.6)$$

For example, the matrix elements m_{64} and m_{65} of this matrix can be calculated with eqn (19.5) as follows:

$I_4 = 2 \times 2 = 4; \quad I_5 = 2 \times 3 = 6;$

$[\mathcal{G}_6] = 6$: there are six conjugates to $\mathcal{G}_6$ in $\mathcal{G}_1$ (see above);

$[\mathcal{G}_6 \subseteq \mathcal{G}_4] = 6$: all six conjugates to $\mathcal{G}_6$ (in $\mathcal{G}_1$) are also subgroups of $\mathcal{G}_4$;

$[\mathcal{G}_6 \subseteq \mathcal{G}_5] = 1$: only one of the conjugates to $\mathcal{G}_6$ is also a subgroup of $\mathcal{G}_5$, namely the one with the same origin position and the same orientation (obverse or reverse).

$$m_{64} = I_4 \frac{[\mathcal{G}_6 \subseteq \mathcal{G}_4]}{[\mathcal{G}_6]} = 4\frac{6}{6} = 4; \qquad m_{65} = I_5 \frac{[\mathcal{G}_6 \subseteq \mathcal{G}_5]}{[\mathcal{G}_6]} = 6\frac{1}{6} = 1$$

An element m_{ij} is always and only equal to zero if the space group $\mathcal{G}_i$ is not a subgroup of $\mathcal{G}_j$. If the matrix **M** is calculated by a computer with the aid of eqn (19.4), the results automatically show if there is a group–subgroup relation between two groups, even for non-maximal subgroups. This ensures that errors are avoided with complicated trees of subgroups. For the calculation of **M** there is not even the need to plot the tree; a complete, hierarchically ordered list of the subgroups is sufficient. *However, the list may not contain subgroups conjugate in* $\mathcal{G}_1$; only one representative of every conjugacy class is permitted.

This is because two subgroups $\mathcal{G}_i$ and $\mathcal{G}_j$ are conjugate by definition if the condition $g\mathcal{G}_i g^{-1} \subseteq \mathcal{G}_j$ is fulfilled for at least one g. According to eqn (19.4) the result is then $m_{ij} > 0$. Since conjugate subgroups have the same order, there can be no group–subgroup relation between them. A computed value $m_{ij} > 0$ for two subgroups of the same order indicates that they are conjugate, and one of them has to be eliminated from the list. On the other hand, non-conjugate subgroups on a par (Section 8.3, page 111) may not be excluded from the list.

For the subsequent computation, the inverse matrix of $\mathbf{M}$ is needed, $\mathbf{B} = \mathbf{M}^{-1}$. Since it is a triangular matrix, the matrix inversion is easy to calculate. With some practice, it is easily performed by mental arithmetic. All diagonal elements of $\mathbf{B}$ are the reciprocals of the diagonal elements of $\mathbf{M}$, $b_{ii} = 1/m_{ii}$. All other elements result from the formula in the margin. When the matrix $\mathbf{B}$ is computed row by row from top to bottom, the b_{kj} needed to calculate the b_{ij} are known.

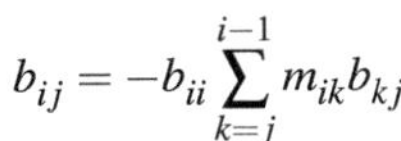

$$b_{ij} = -b_{ii} \sum_{k=j}^{i-1} m_{ik} b_{kj}$$

The inverse of the above-mentioned matrix (19.6) is:

$$\mathbf{B} = \mathbf{M}^{-1} = \begin{pmatrix} 1 \\ -\frac{1}{2} & \frac{1}{2} \\ -\frac{1}{2} & 0 & \frac{1}{2} \\ \frac{1}{4} & -\frac{1}{4} & -\frac{1}{4} & \frac{1}{4} \\ 0 & 0 & -1 & 0 & 1 \\ 0 & 0 & \frac{1}{2} & -\frac{1}{2} & -\frac{1}{2} & \frac{1}{2} \\ 0 & 0 & 0 & 0 & -\frac{1}{2} & 0 & \frac{1}{2} \\ 0 & 0 & \frac{1}{3} & 0 & -\frac{1}{2} & 0 & 0 & \frac{1}{6} \\ 0 & 0 & -\frac{1}{6} & \frac{1}{6} & \frac{1}{2} & -\frac{1}{4} & -\frac{1}{4} & -\frac{1}{12} & \frac{1}{12} \end{pmatrix}$$

If an element is $m_{ij} = 0$, then $b_{ij} = 0$ also holds. All b_{ij} of one row of $\mathbf{B}$ always add up to 0 (the first row excepted).

$\mathcal{G}_6 = R\overline{3}$

c_2	b	c_2
a	c_1	c_1

c_2	b	c_2
a	c_1	c_1

c_2	b	c_2
a	c_1	c_1

In addition to the matrix $\mathbf{B}$, for a given chemical composition, we also have to calculate in how many ways the different kinds of atoms can be allocated to the considered orbits for every space group of the tree. This is done according to the rules of combinatorics. For example, in the space group $\mathcal{G}_9$ of Fig. 14.1 all six octahedral interstices are symmetry independent; if two of them are to be occupied by two equal atoms and four remain unoccupied, there are $v_9 = \binom{6}{2} = 15$ combinatorial distributions. In the space group $\mathcal{G}_6 = R\overline{3}$ there are only three possibilities to occupy two octahedral interstices by two equal atoms, as shown in the margin (boxes in grey represent occupied octahedral interstices). A general formula for the computation of the combinatorial possibilities is given in the next section.

In this way, one number v_1 to v_9 is computed for each space group $\mathcal{G}_1$ to $\mathcal{G}_9$. These numbers are combined into a column $\mathbf{v} = (v_1, \ldots, v_9)^{\mathrm{T}}$. They specify the number of possible combinatorial distributions of atoms among the orbits of every space group for a given chemical composition. Among the lower-symmetry space groups, they include those distributions that in reality have a higher symmetry, and equal structures may have been counted repeatedly. For example, among the three possibilities shown in the margin, the first one (a and b occupied) does not correspond to the space group $\mathcal{G}_6 = R\overline{3}$ because it fulfils

Table 19.1 Numbers of non-equivalent structural variants (= components of **z**), depending on space group and chemical composition for the occupation of octahedral interstices in the hexagonal-closest packing with tripled, rhombohedral unit cell (Fig. 14.1) [206]. X = atoms of the sphere packing, A, B, C = atoms in octahedral interstices, □ = unoccupied octahedral interstices.

	AX □X		ABX_2 $A\square X_2$		AB_2X_3 $A\square_2X_3$ $\square B_2X_3$		$ABCX_3$ $AB\square X_3$		AB_5X_6 $A\square_5X_6$ $\square B_5X_6$		ABC_4X_6 $AB\square_4X_6$ $A\square C_4X_6$		$AB_2C_3X_6$ $AB_2\square_3X_6$ $A\square_2C_3X_6$ $\square B_2C_3X_6$	
	v	**z**	**v**	**z**	**v**	**z**	**v**	**z**	**v**	**z**	**v**	**z**	**v**	**z**
$\mathcal{G}_1 = P6_3/m2/m2/c$	1	1	0	0	0	0	0	0	0	0	0	0	0	0
$\mathcal{G}_2 = P\bar{3}2/m1$	1	0	2	1	0	0	0	0	0	0	0	0	0	0
$\mathcal{G}_3 = P\bar{3}12/c$	1	0	0	0	0	0	0	0	0	0	0	0	0	0
$\mathcal{G}_4 = P\bar{3}$	1	0	2	0	0	0	0	0	0	0	0	0	0	0
$\mathcal{G}_5 = R\bar{3}2/c$	1	0	0	0	1	1	0	0	0	0	0	0	0	0
$\mathcal{G}_6 = R\bar{3}$	1	0	4	1	3	1	6	3	2	1	2	1	4	2
$\mathcal{G}_7 = R32$	1	0	0	0	3	1	6	3	0	0	0	0	0	0
$\mathcal{G}_8 = R3c$	1	0	0	0	3	0	6	1	0	0	0	0	0	0
$\mathcal{G}_9 = R3$	1	0	20	1	15	0	90	4	6	0	30	2	60	4
sum (≙ Pólya)		1		3		3		11		1		3		6

the higher symmetry of the space group $\mathcal{G}_5 = R\bar{3}c$; the other two (occupation of c_1 and c_2) represent the same structure type; only one of them should be considered. The numbers $z_1, \ldots, z_9$ of *non-equivalent* structure types, assigned to the correct space groups, result from the product of **B** and **v**.

Number of non-equivalent structure types:

$\mathbf{z} = \mathbf{Bv}$ with $\mathbf{z} = (z_1, \ldots, z_9)$

These numbers are listed in Table 19.1 for different chemical compositions, referred to the tree of subgroups of Fig. 14.1. The total numbers for each composition result from eqn (19.2), page 258; in accordance with the chemical formula $A_aB_b\square_sX_6$, take x_1 = A and x_2 = B and look at the coefficient of the product of powers $x_1^a x_2^b$. For the composition $A_2B_2\square_2X_6$, the coefficient $11x_1^2x_2^2$ indicates eleven possible structures; for $AB_2\square_3X_6$ there are six ($6x_1x_2^2$).

In order to compute the combinatorial distributions contained in the column **v**, for every space group $\mathcal{G}_i$ it must be known how many different symmetry-equivalent atomic positions have to be considered and which multiplicities these positions have. Therefore, starting from the aristotype, we have to monitor how the positions develop from a group to its subgroups, and particularly how they split into non-equivalent positions. The necessary information is contained in *International Tables* A1.

19.3 Combinatorial computation of distributions of atoms among given positions

Given:

1. The unit cell of a space group with a (finite) number of points that can be occupied by atoms. The points are subdivided into $Z_1 \times n_1$ points of multiplicity Z_1, $Z_2 \times n_2$ points of multiplicity Z_2, etc.; the multiplicity in each case is the number of symmetry-equivalent points in the *primitive* unit

cell. $n_1, n_2, \ldots$ indicate how many points of the multiplicities $Z_1, Z_2, \ldots$ are present in the primitive cell.

2. A set of a A atoms, b B atoms, ..., that are to be distributed among the given points of the unit cell in accordance with symmetry. The total number of atoms is no more than the total number of available points. Some points may remain unoccupied.

The number P of combinations to distribute the atoms among the points can be calculated with the following recurrence formula [431]:

$$P = \sum_{j=0}^{N_1} [\binom{n_1}{j} \sum_{k=0}^{N_2} [\binom{n_2}{k} \sum_{l=0}^{N_3} \cdots \sum_{y=0}^{N_{z-1}} [\binom{n_{z-1}}{y}) [\binom{n_z}{N_z} P_{j,k,\ldots,y}]] \ldots]] \qquad (19.7)$$

with:

$$\left.\begin{array}{l} \binom{n}{0} = 1 \quad \binom{0}{0} = 1 \quad \binom{n}{k} = \dfrac{n!}{k!\,(n-k)!} \\ N_1 = \min(n_1, \operatorname{int}(a/Z_1)) \\ N_2 = \min(n_2, \operatorname{int}(\frac{a-jZ_1}{Z_2})) \\ N_3 = \min(n_3, \operatorname{int}(\frac{a-jZ_1-kZ_2}{Z_3})) \\ \text{etc.} \end{array}\right\} \qquad (19.8)$$

$\min(u, v)$ = the lesser of the two numbers u and v

$\operatorname{int}(\frac{x}{y})$ = integer result of the division x/y, with or without remainder

The sequence $Z_1 \leq Z_2 \leq Z_3 \ldots$ is to be satisfied. The number of addends (sigma signs) in eqn (19.7) is such that the last but one number is $N_{z-1} > 0$, and z is less than or equal to the number of occurring multiplicities.

If the integer division int(...) leaves a remainder when computing the last number N_z or if the result of this division yields $Z_z > N_z > 0$ or $\operatorname{int}(\ldots) > n_z$, then $P_{j,k,\ldots,y} = 0$. Otherwise $P_{j,k,\ldots,y} = 1$ if there are no B atoms; if there are B atoms, $P_{j,k,\ldots,y}$ again has to be computed according to eqn (19.7), but with the values $n_1' = n_1 - j$, $n_2' = n_2 - k, \ldots$ instead of $n_1, n_2, \ldots$ and with the number b of B atoms instead of a in eqns (19.8). If there are c C atoms, the whole procedure is repeated with the number c in eqns (19.8), etc.

Example 19.1

Consider the distribution of $a = 2$ A atoms and $b = 3$ B atoms among points of the Wyckoff positions $1a$, $1b$, $2c$, $2d$, $2e$, and $3f$. $1a$ and $1b$ are $n_1 = 2$ positions of multiplicity $Z_1 = 1$; $2c$, $2d$, and $2e$ are $n_2 = 3$ positions of multiplicity $Z_2 = 2$; $3f$ are $n_3 = 1$ positions of multiplicity $Z_3 = 3$. There exist 16 combinations, shown in Fig. 19.4.

Formulated to detail, the recurrence formula (19.7) yields:

Initial values:

$$\begin{array}{ll} a = 2 & b = 3 \\ n_1 = 2 & Z_1 = 1 \\ n_2 = 3 & Z_2 = 2 \\ n_3 = 1 & Z_3 = 3 \end{array}$$

Because there are $z = 3$ different multiplicities, a maximum of three numbers N_1, N_2, and N_3 appear in eqn (19.7), and a maximum of two ($= z - 1$) sigma signs with the running indices j and k.

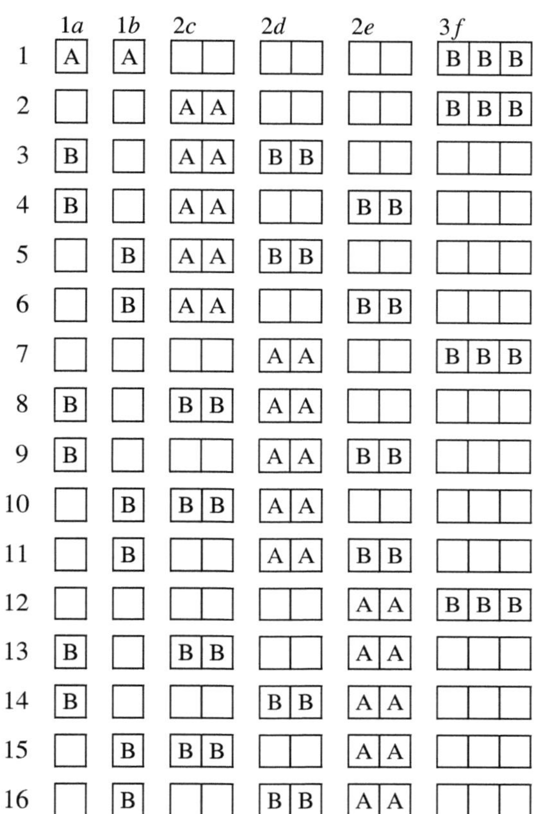

Fig. 19.4 The 16 combinations for the distribution of two A and three B atoms among points of the positions $1a$, $1b$ $2c$, $2d$, $2e$, and $3f$.

$N_1 = \min(2, \text{int}(\frac{2}{1})) = 2$; the index j runs from 0 to 2
for $j = 0$ we have $N_2 = \min(3, \text{int}(\frac{2-0}{2})) = 1$; the index k runs from 0 to 1
for $j = 0$ and $k = 0$ we have $N_3 = \min(1, \text{int}(\frac{2-0-0}{3})) = 0$ with remainder on division; due to the remainder we have $P_{0,0} = 0$
for $j = 0$ and $k = 1$ we have $N_3 = \min(1, \text{int}(\frac{2-0-1\times2}{3})) = 0$, no remainder
for $j = 1$ we have $N_2 = \min(3, \text{int}(\frac{2-1\times1}{2})) = 0$ with remainder;
because $N_2 = 0$ there is no running index k;
due to the remainder we have $P_1 = 0$
for $j = 2$ we have $N_2 = \min(3, \text{int}(\frac{2-2\times1}{2})) = 0$ without remainder;
because $N_2 = 0$ there is no running index k
Equation (19.7) therefore yields:

$$
\begin{aligned}
P &= \underbrace{\binom{n_1}{0}\binom{n_2}{0}\binom{n_3}{N_3}P_{0,0}}_{j=0,\,k=0} + \underbrace{\binom{n_1}{0}\binom{n_2}{1}\binom{n_3}{N_3}P_{0,1}}_{j=0,\,k=1} + \underbrace{\binom{n_1}{1}\binom{n_2}{N_2}P_1}_{j=1,\text{ no }k} + \underbrace{\binom{n_1}{2}\binom{n_2}{N_2}P_2}_{j=2,\text{ no }k} \\
&= \binom{2}{0}\binom{3}{0}\binom{1}{0}\times 0 + \binom{2}{0}\binom{3}{1}\binom{1}{0}P_{0,1} + \binom{2}{1}\binom{3}{0}\times 0 + \binom{2}{2}\binom{3}{0}P_2 \\
&= 3P_{0,1} + 1P_2 \qquad (19.9)
\end{aligned}
$$

The computation of $P_{0,1}$ is performed again with eqn (19.7). In the following, j', k', N'_1, N'_2, N'_3, and $P'_{j',k'}$ replace the variables j, k, N_1, etc.

$n'_2 = n_2 - k = 3 - 1 = 2$ replaces n_2; $n'_1 = n_1 = 2$, $n'_3 = n_3 = 1$. In eqns (19.8), $b = 3$ replaces a. $P'_{j',k'} = 1$ holds because there are no C atoms, unless $P'_{j',k'} = 0$ due to a remainder from one of the integer divisions. We obtain:

$N'_1 = \min(2, \text{int}(\frac{3}{1})) = 2$
for $j' = 0$: $N'_2 = \min(2, \text{int}(\frac{3-0}{2})) = 1$
for $j' = 0$ and $k' = 0$: $N'_3 = \min(1, \text{int}(\frac{3-0-0}{3})) = 1$ without remainder,
$\rightarrow P'_{0,0} = 1$
for $j' = 0$ and $k' = 1$: $N'_3 = \min(1, \text{int}(\frac{3-0-1\times2}{3})) = 0$ with remainder,
$\rightarrow P'_{0,1} = 0$
for $j' = 1$: $N'_2 = \min(2, \text{int}(\frac{3-1\times1}{2})) = 1$
for $j' = 1$ and $k' = 0$: $N'_3 = \min(1, \text{int}(\frac{3-1\times1-0}{3})) = 0$ with remainder,
$\rightarrow P'_{1,0} = 0$
for $j' = 1$ and $k' = 1$: $N'_3 = \min(1, \text{int}(\frac{3-1\times1-1\times2}{3})) = 0$ without remainder,
$\rightarrow P'_{1,1} = 1$
for $j' = 2$: $N'_2 = \min(1, \text{int}(\frac{3-2\times1}{3})) = 0$ with remainder,
$\rightarrow P'_2 = 0$, no running index k'

This results in:

$$\begin{aligned}
P_{0,1} &= \underbrace{\binom{n'_1}{0}\binom{n'_2}{0}\binom{n'_3}{N'_3}P'_{0,0}}_{j'=0,\,k'=0} + \underbrace{\binom{n'_1}{0}\binom{n'_2}{1}\binom{n'_3}{N'_3}P'_{0,1}}_{j'=0,\,k'=1} + \underbrace{\binom{n'_1}{1}\binom{n'_2}{0}\binom{n'_3}{N'_3}P'_{1,0}}_{j'=1,\,k'=0} \\
&\quad + \underbrace{\binom{n'_1}{1}\binom{n'_2}{1}\binom{n'_3}{N'_3}P'_{1,1}}_{j'=1,\,k'=1} + \underbrace{\binom{n'_1}{2}\binom{n'_2}{N'_2}P'_2}_{j'=2,\text{ no }k'} \\
&= \binom{2}{0}\binom{2}{0}\binom{1}{1}\times 1 + \binom{2}{0}\binom{2}{1}\binom{1}{0}\times 0 + \binom{2}{1}\binom{2}{0}\binom{1}{0}\times 0 \\
&\quad + \binom{2}{1}\binom{2}{1}\binom{1}{0}\times 1 + \binom{2}{2}\binom{2}{0}\times 0 \\
&= 5
\end{aligned} \tag{19.10}$$

The value of 5 corresponds to each of the five combinations for the B atoms shown in Fig. 19.4 under the numbers 2–6, 7–11, and 12–16, when the A atoms occupy one of the twofold positions 2*c*, 2*d*, and 2*e*, respectively.

P_2 is computed in the same way, with the new values $n'_1 = n_1 - j = 2 - 2 = 0$ instead of n_1, $n'_2 = n_2 = 3$, $n'_3 = n_3 = 1$ and $b = 3$ instead of $a = 2$. Because of $n'_1 = 0$, $N'_1 = 0$ and j' can adopt only the value of 0:

$N'_2 = \min(3, \text{int}(\frac{3-0}{2})) = 1$
for $k' = 0$: $N'_3 = \min(1, \text{int}(\frac{3-0-0}{3})) = 1$ without remainder, $\rightarrow P'_{0,0} = 1$
for $k' = 1$: $N'_3 = \min(1, \text{int}(\frac{3-0-1\times2}{3})) = 0$ with remainder, $\rightarrow P'_{0,1} = 0$

$$\begin{aligned}
P_2 &= \underbrace{\binom{n'_1}{0}\binom{n'_2}{0}\binom{n'_3}{N'_3}P'_{0,0}}_{j'=0,\,k'=0} + \underbrace{\binom{n'_1}{0}\binom{n'_2}{1}\binom{n'_3}{N'_3}P'_{0,1}}_{j'=0,\,k'=1} \\
&= \binom{0}{0}\binom{3}{0}\binom{1}{1}\times 1 + \binom{0}{0}\binom{3}{0}\binom{1}{0}\times 0 \\
&= 1
\end{aligned} \tag{19.11}$$

This value of 1 corresponds to the one possible way to distribute the B atoms when the A atoms occupy the positions $1a$ and $1b$ (Fig. 19.4, combination 1).
The total number of combinations is obtained by substitution of eqns (19.10) and (19.11) into eqn (19.9):

$$P = 3 \times 5 + 1 \times 1 = 16$$

To compute eqn (19.7), a computer program is convenient that activates a procedure that computes the expression after one of the sigma signs. If another sigma sign appears, the procedure activates itself once more and continues doing so until all sums have been calculated.

19.4 Derivation of possible crystal structure types for a given molecular structure

In Section 19.1 we considered how atoms can be distributed among given positions (e.g. octahedral voids). In this section we choose a different approach; we start from a given molecular structure and consider how the molecules can be packed and what space groups may occur. For molecules with irregular shape, like most organic molecules, geometrical considerations are of little help. This is different for inorganic molecules, which can often be regarded as sections from a packing of spheres. We choose the example of the dimeric pentahalides to explain the procedure.

Pentachlorides, bromides, and iodides like $(NbCl_5)_2$, $(UCl_5)_2$, $(WBr_5)_2$, and $(TaI_5)_2$ consist of molecules having the point symmetry $2/m2/m2/m$ (Fig. 19.5). The halogen atoms span two edge-sharing octahedra. Edge-sharing octahedra also exist in closest packings of spheres (Fig. 14.2, page 181). In fact, $(MX_5)_2$ molecules pack themselves in such a way that the halogen atoms form a closest packing of spheres, with one-fifth of the octahedral voids being occupied. This entails three important restrictions: (1) pairs of adjacent octahedral voids have to be occupied; (2) all octahedral voids next to a molecule must remain unoccupied; (3) the molecules can only adopt certain orientations in the packing of spheres. In addition, we take for granted that no position of a sphere is vacant (this actually is always fulfilled).

Figure 19.5 shows how an $(MX_5)_2$ molecule fits itself into a hexagonal-closest packing of X atoms. Only one of the twofold rotation axes of the molecule conforms with a twofold rotation axis of the sphere packing (parallel to $\mathbf{b}_{hex}$), namely the one that runs through the two M atoms. The remaining rotation axes of the molecule are inclined by 35.3° and −54.7° relative to the c axis of the sphere packing. There are two opposed inclinations, one if the occupied octahedral voids are situated between layers A and B, the other if they are between layers B and A of the stacking sequence $ABAB\ldots$.

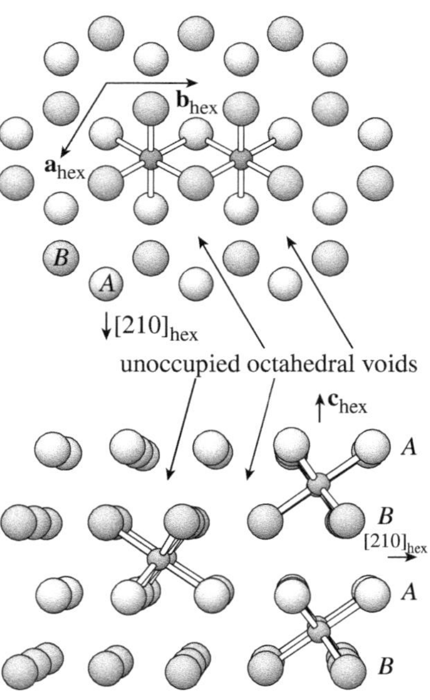

Fig. 19.5 Embedding of an $(MX_5)_2$ molecule into the hexagonal-closest packing of spheres.

Because coincidence between symmetry axes of the molecule and the packing of spheres is only present in one direction, the full symmetry neither of the molecules nor of the packing of spheres can be retained. The most symmetrical point group for the molecules is $2/m$, a subgroup of $2/m2/m2/m$. Among the

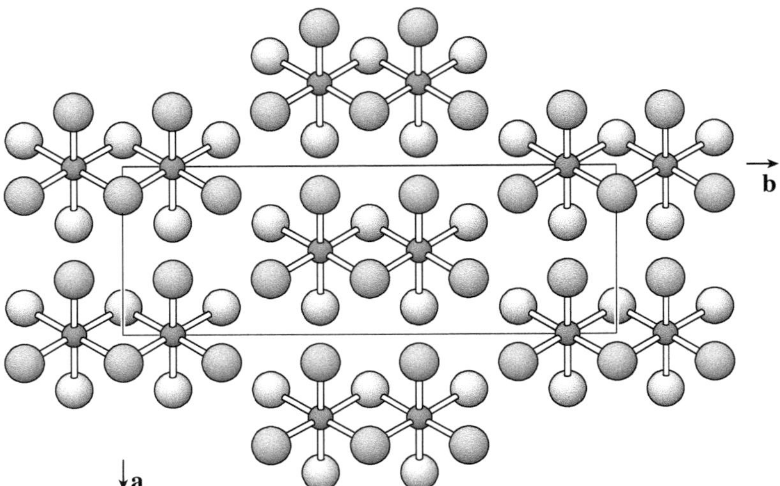

Fig. 19.6 The only way for a close packing of $(MX_5)_2$ molecules in one layer; layer group $c\,1\,2/m\,1$.

symmetry axes of the space group of the packing of spheres, $P6_3/m2/m2/c$, the 6_3, $\overline{6}$, $\overline{3}$, and 3 axes cannot be retained. Due to the condition that the octahedral voids next to a molecule must remain unoccupied, the reflection planes perpendicular to **c** and the 2 axes parallel to [210] (last 2 in the Hermann–Mauguin symbol) cannot be retained because they run through the X atoms.

There are additional geometric restrictions. If $\frac{2}{5}$ of the octahedral voids are being occupied only between layers *A* and *B* in the stacking sequence *ABAB*... of X atoms, and those between *B* and *A* remain vacant (stacking sequence $A\gamma_{2/5}B\square$; $\gamma =$ atoms in octahedral voids), the result is layers of molecules in which there exists only one possible arrangement for the molecules (Fig. 19.6). The layer of molecules has the layer group $c\,1\,2/m\,1$. The layers can be stacked with different mutual displacements, and they can be mutually turned by 120°.

If octahedral voids are being occupied between *A* and *B* as well as between *B* and *A*, stacking sequence $A\gamma_{2/5-n}B\gamma'_n$, the closest packing can be realized only if molecules of equal inclination are stacked to columns parallel to $\mathbf{c}_{\text{hex}}$, like the two molecules shown to the right of the lower image in Fig. 19.5. Twofold rotation axes and reflection planes are then only possible in the symmetry direction **b** (parallel to the connecting M–M line in a molecule). In other symmetry directions, 2_1 axes and glide planes are possible, provided that $n = \frac{1}{5}$ (stacking sequence $A\gamma_{1/5}B\gamma'_{1/5}$); in this case the numbers of columns of molecules with the two inclinations are the same. *c* glide planes are possible perpendicular to **b**. For more details see [432].

Keeping in mind that centrosymmetric molecules crystallize in centrosymmetric space groups, almost without exceptions (page 200), we do not consider non-centrosymmetric space groups. In addition, the number of M atoms not equivalent by symmetry should be kept low. Taking into account all of the mentioned restrictions, only a few subgroups of $P6_3/m2/m2/c$ can occur. The corresponding Bärnighausen tree reveals which space groups are possible when $(MX_5)_2$ molecules crystallize with a hexagonal-closest packing of X atoms (Fig. 19.7). One-fifth of the octahedral voids being occupied, in any case there must be a step with an isomorphic subgroup of index 5. The space

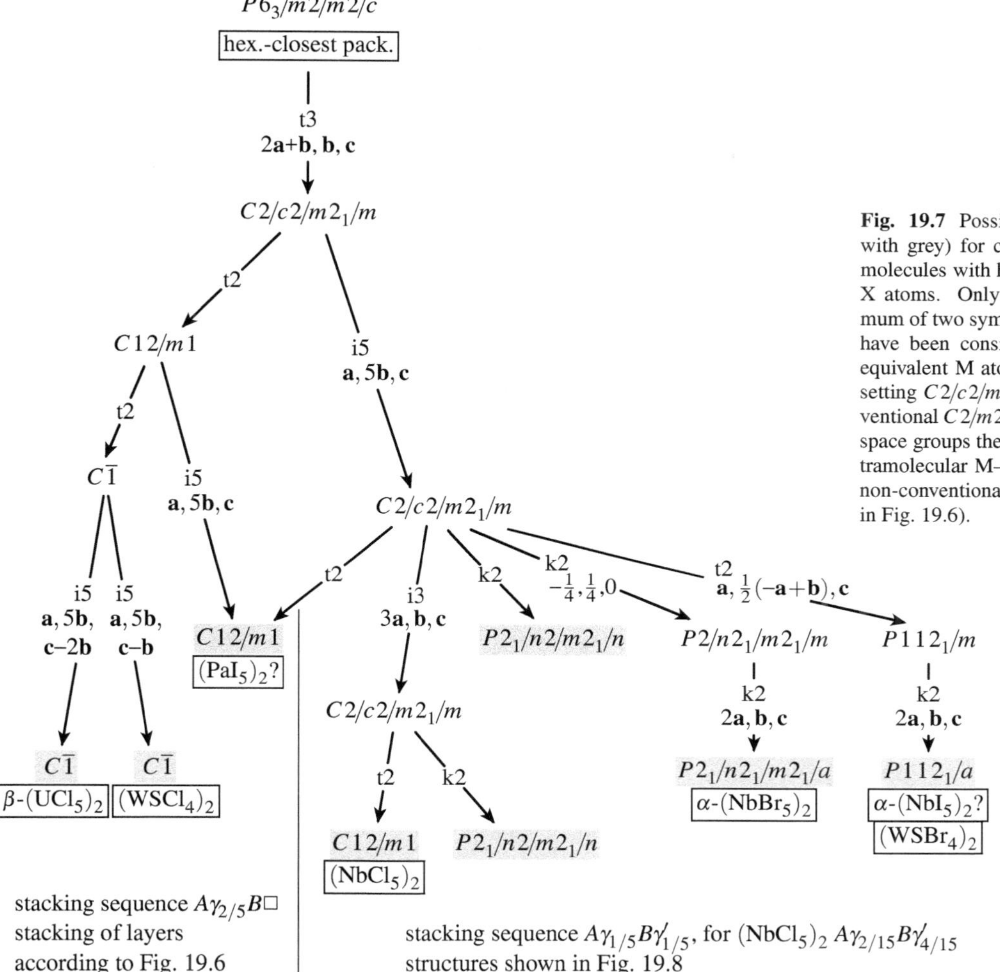

Fig. 19.7 Possible space groups (marked with grey) for crystal structures of $(MX_5)_2$ molecules with hexagonal-closest packing of X atoms. Only space groups with a maximum of two symmetry-independent M atoms have been considered, but with symmetry-equivalent M atoms within a molecule. The setting $C2/c2/m2_1/m$ has been chosen (conventional $C2/m2/c2_1/m$) to ensure that for all space groups the **b** axes are parallel to the intramolecular M–M connecting line. $C\bar{1}$ is a non-conventional setting of $P\bar{1}$ (centring as in Fig. 19.6).

groups considered in Fig. 19.7 refer only to structures in which there are no more than two symmetrically independent metal atoms. The corresponding molecular packings are shown in Fig. 19.8. For further possibilities, including other packings of spheres and the corresponding Bärnighausen trees, see [432]. No representatives are as yet known for some of the possible space groups.

Similar considerations were applied to derive the possible space groups and crystal structures for other halides, for example MX_6 molecules [433], MX_4 chains of edge-sharing octahedra [434], and MX_5 chains of vertex-sharing octahedra [435].

The intermolecular forces between the molecules differ only marginally for all of the conceivable crystal structures. Correspondingly, different polymorphic forms are observed for these pentahalides; for example, four modifications are known for $(MoCl_5)_2$, all of which correspond to one of the predicted packings [436].

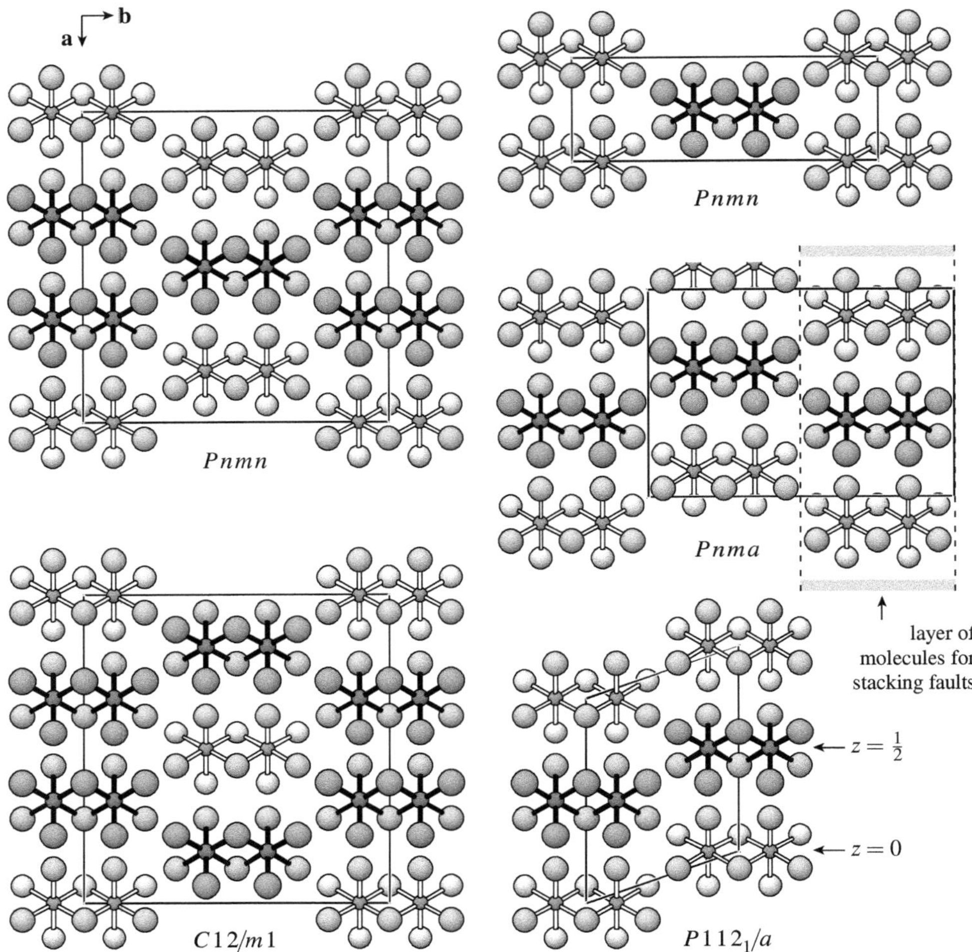

Fig. 19.8 Possible ways to pack $(MX_5)_2$ molecules with M atoms at $z = 0$ and $z = \frac{1}{2}$ $(A\gamma_{2/5-n}B\gamma'_n)$.

[2] The term 'misorder' is explained in margin note no. 2 in Section 16.1.2 (page 213).

In addition, pentahalides $(MX_5)_2$ show a tendency to crystallize with some misorder.[2] The X-ray diffraction patterns then exhibit diffuse streaks besides sharp Bragg reflections. Diffuse streaks imply the presence of layers that consist of ordered molecules, but the sequence of the layers has no periodic order. In the direction of **b** the number of molecular contacts is least, and the intermolecular forces are the weakest in this direction. The layers of ordered molecules are oriented perpendicular to **b**; they are mutually shifted at random parallel to +**a** and −**a** (one such layer is marked in the image of the space group *Pnma* in Fig. 19.8). The misordered structure ultimately is a randomly mixed form between the structures with the space groups *Pnma* and $P112_1/a$.

An 'average structure' is obtained if only the Bragg reflections are evaluated. The average structure results when all layers are projected into one layer. This average structure has the space group $C2/c2/m2_1/m$, which is the common supergroup of all possible space groups (the triclinic ones excepted); in Fig. 19.7 it is the one mentioned after the first i5 step. In the average structure the octahedral voids show a seeming partial occupancy. The real structure can be determined only by evaluation of the intensities along the diffuse streaks (which is cumbersome). In this case, the usual methods of crystal structure solution (Patterson synthesis, direct methods) fail. However, taking into account the above-mentioned structural possibilities, it still is possible to obtain a suitable structure model. The misordered structure variants of $(NbBr_5)_2$, $(TaI_5)_2$, and, similarly, $(MoCl_4)_6$ were determined this way [437–439].

Exercises

Solutions on page 338.

19.1. Take the prism shown in Fig. 19.2, but assume that the edges ②–③ and ⑤–⑥ are shorter than the remaining edges, so that the point group is not $\overline{6}m2$ but $2mm$. Compute in how many non-equivalent ways two and three vertices can be marked with one colour.

19.2. Compute the number of possible isomers of square-pyramidal molecules of compositions MX_3YZ and MX_2Y_2Z (M = central atom; X, Y, Z = ligands). How many are chiral?

19.3. Take the right branch of the tree of Fig. 14.10, page 191, down to the subgroup $I4/mmm$. Compute the number of possible non-equivalent structure types for the compositions ABX_2 and ABX_4 in space group $I4/mmm$.

Historical remarks

20

For centuries, the objective of crystallography was little more than the description of crystals that were found as minerals or that grew from solutions or from melts. They attracted attention because of their regular shapes, planar faces, and cleavability parallel to these faces. Early findings were:

> The law of constancy of interfacial angles between crystal faces, stated by NIELS STENSEN in 1669; he came from Copenhagen and his geological work in Tuscany is considered to be the origin of modern geology.
> The law of symmetry by RENÉ JUST HAÜY (1815): if the shape of a crystal is altered, corresponding parts (faces, edges, angles) of the crystal are simultaneously and similarly modified.
> The law of rational indices by HAÜY (1784), according to which any crystal face can be specified by a set of three, usually small integral numbers.

Cleavage of calcite crystals resulting in smaller crystals of the same shape led HAÜY to the assumption of a smallest parallelepiped ('molécule intégrante') as a building block of a crystal. In 1824, LUDWIG SEEBER explained certain physical properties of crystals by placing chemical molecules at the vertices of the parallelepipeds. The concepts of unit cell and of translational symmetry were thus introduced.

In the nineteenth century interest turned to the mathematical treatment of symmetry. Based on the law of rational indices, LUDWIG FRANKENHEIM (1826), JOHANN F. C. HESSEL (1830), and AXEL GADOLIN (1867) derived the 32 crystal classes, some of which had never been observed among crystals.

The classification of the 14 lattices of translations by AUGUSTE BRAVAIS (1850) was followed by the classification of infinite regular systems of points.

CHRISTIAN WIENER, in 1863, stated in *Foundations of World Order* [440]:

> Regular arrangement of equal atoms takes place when every atom has placed the others around itself in a coincident manner.

In LEONHARD SOHNCKE's *Development of a Theory of Crystal Structure* (1879) [441], a crystal is mentally replaced by a system of mass points which always have a minimum mutual distance:

> Around every point the arrangement of the others is the same.

In SOHNCKE's book, extensive reference is made to earlier historical developments, such as work by R. HOOKE (1667), CHR. HUYGENS (1690), and W. H. WOLLASTON (1813).

SOHNCKE, EVGRAF STEPANOVICH FEDOROV (1891), ARTHUR SCHOENFLIES (1891), and WILLIAM BARLOW (1894) then turned to the investigation of the underlying plane groups and space groups. F. HAAG (1887) postulated that crystal structures should be regular arrangements of atoms and that crystal symmetry should be a space group. However, these conjectures were speculations at that time, as were the atom packings described by SOHNCKE and BARLOW, such as a model of the NaCl structure (which they did not assign to any substance).

It was not until 1912 that this was proven experimentally by the first X-ray diffraction experiment by WALTHER FRIEDRICH and PAUL KNIPPING, suggested by MAX VON LAUE. The first crystal structure determinations then followed with simple inorganic materials (such as NaCl, KCl, and diamond) by WILLIAM HENRY BRAGG and WILLIAM LAWRENCE BRAGG.

The presentations by SCHOENFLIES and FEDOROV of the space groups were not yet appropriate for use in structure determinations with X-rays. The breakthrough came with the geometric description of the space groups by symmetry elements and point positions by PAUL NIGGLI in his book *Geometrische Kristallographie des Diskontinuums* (1919) [442]. Thereupon, RALPH W. G. WYCKOFF prepared tables and diagrams of the unit cells with the symmetry elements and special positions [443]. Together with tables by W. T. ASTBURY and KATHLEEN YARDLEY [444] they were the basis for the trilingual *International Tables for the Determination of Crystal Structures*, edited by CARL HERMANN and published in 1935 [26]. These tables included the Hermann–Mauguin symbols that had been introduced by HERMANN (1928) [56] and supplemented by CHARLES MAUGUIN [57].

In his work *Das Krystallreich* (1920), FEDOROV compiled an extensive collection of crystal-morphological data [445]. In a 159-page elaboration of 1904, he formulated two laws [446]. In the first, which he called a 'deductive law' that is based on 'general principles of hard science', he states that there is:

> A distribution of all crystals in two categories, the cubic and the hypohexagonal ones.

The second law, which is of the 'inductive kind', 'the trueness of which is based solely on countless facts', is:

> All crystals are either pseudotetragonal or pseudohexagonal in a broad sense, i.e. if even such deviations are accepted as extreme cases like 20°. The main value of this law is that deviations are the less frequent the larger they are.

Compare this with aspects 1 and 2 of the symmetry principle as stated in Section 1.1.

In his textbook on mineralogy [447], PAUL NIGGLI expressly supported FEDOROV's opinion. However, due to its unprovability, he scored FEDOROV's symmetry law only as a fruitful working hypothesis or philosophical doctrine. Furthermore, he states:

> Cubic and hexagonal crystal species are the typical representatives of crystalline material, and any deviation from the two highest symmetries has its special causes, attributable to the complexity of the building blocks of the crystal.

Finally NIGGLI emphasizes that there are often 'manifold reminiscences of higher symmetry', but that this 'rather is a confirmation than a proof to the contrary for the tendency towards the highest possible symmetry'.

FRITZ LAVES was the first and, for a long time, the only one who used the symmetry principle as a guideline [448–450]. In his work *Phase Stability in Metals and Alloy Phases* (1967) he states:

> In crystal structures there exists a strong tendency towards formation of high-symmetry arrangements.

That is a restriction to aspect 1 of the symmetry principle in our formulation. For LAVES it was one out of three fundamental principles. The other two were the principle of closest packing and the principle of maximal connectivity of the building blocks of a crystal.

Systematic utilization of group–subgroup relations begins with a paper by HERMANN on subgroups of space groups (1929) [58]. In this paper, the distinction between *zellengleiche* (i.e. *translationengleiche*) and *klassengleiche* subgroups is introduced. *Translationengleiche* subgroups were included in the 1935 edition of *International Tables for the Determination of Crystal Structures*, contrary to the *klassengleiche* subgroups, but they were omitted in the later edition of 1952. It was not until 1983 that the subgroups were again listed in *International Tables* [14], now including the *klassengleiche* subgroups, albeit not in a complete form.

'Atoms of the same type tend to be in equivalent positions' is the essential statement in a fundamental paper by G. O. BRUNNER of 1971 [30]. According to this principle, from the infinity of closest packings of spheres, the cubic-closest and the hexagonal-closest packing are selected. In the formalism of stacking sequences, these are the sequences *ABC* and *AB*. The next possible kind of stacking, *ABAC*, already requires atoms at two crystallographically different (i.e. symmetry-nonequivalent) positions.

An extensive consideration of the interplay between symmetry and crystal packing of organic molecules was published by A. I. KITAIGORODSKII (1955) [273]. However, symmetry is not used as an ordering principle, but the compatibility of crystal symmetry with the packing of organic molecules is studied depending on their symmetries and shapes; the principle of closest packing has priority. A table lists what space groups should be preferred, depending on molecular symmetry.

The complete compilation of all subgroups of the space groups began in 1966 by JOACHIM NEUBÜSER and HANS WONDRATSCHEK [28]. An important part thereof was published in 1983 in the new Volume *A* of *International Tables* [14], but this ample work was not completed before 2004 in cooperation with MOIS AROYO and YVES BILLIET and then published in *International Tables A*1 [16]. HARTMUT BÄRNIGHAUSEN witnessed this from the very beginning during WONDRATSCHEK's and BÄRNIGHAUSEN's common time at the University of Freiburg, Germany. They maintained scientific contact after both had become professors at the University of Karlsruhe. This was the fertile soil that led BÄRNIGHAUSEN to develop the concept of trees of subgroups which are the subject of this book.

With the emergence of the internet, the *Bilbao Crystallographic Server* was launched in 1998 by JUAN MANUEL PÉREZ-MATO and MOIS AROYO at the University of the Basque Country. Its databases and computer programs also form part of the *Symmetry Database* of the International Union of Crystallography.

Part III

Widening Excurses to Special Topics

Isomorphic subgroups

21

As stated by Theorem 7.6 (page 93) and in Section 12.3, indices of maximal isomorphic subgroups may only be prime numbers $p \neq 1$, squares of prime numbers p^2, or cubes of prime numbers p^3. Often only certain prime numbers are permitted. The reasons are explained in this chapter, together with a short digression on number theory.

An isomorphic subgroup always has a unit cell that is enlarged by a factor i, i being the index of the symmetry reduction.

If $i \geq 3$, as a rule, there are i conjugate subgroups that differ by the positions of their origins (translational conjugation, cf. Section 8.1, page 106). This does not apply to cell enlargements in directions in which the origin 'floats' (i.e. to space groups whose origins are not fixed by symmetry). Isomorphic subgroups on a par (different conjugacy classes of the same kind; Definition 8.2, page 111) occur if $i = 2$; in addition, they occur among some trigonal and hexagonal space groups if $i = 3$, and among certain tetragonal, trigonal, and hexagonal space groups if $i \geq 5$ (see the text under Fig. 21.4).

For **triclinic**, **monoclinic**, and **orthorhombic space groups** the index of a maximal isomorphic subgroup is an arbitrary prime number p. However, the prime number $p = 2$ is always excluded if the unit cell is enlarged parallel to any 2_1 axis or in the direction of the glide component of a glide plane. The reason for this can be seen in Fig. 21.1.

In addition, an index of 2 is excluded for base-centred monoclinic and orthorhombic space groups if the unit cell is enlarged in the plane of the centring.

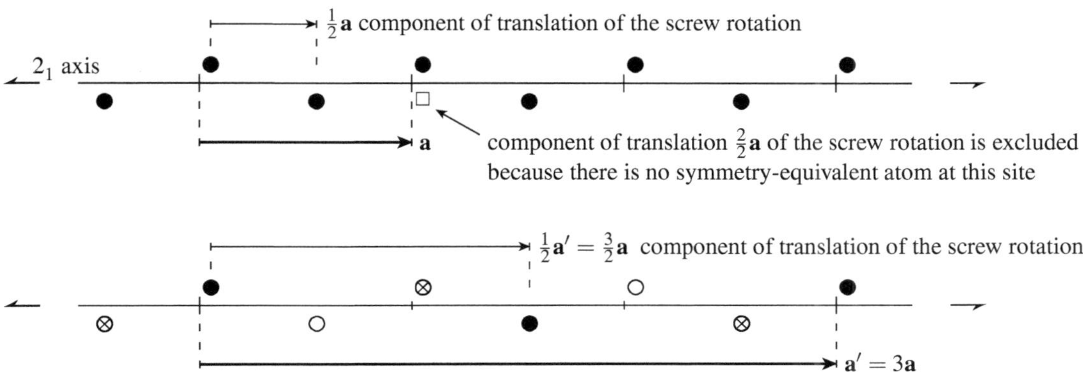

Fig. 21.1 A translation vector parallel to a 2_1 axis cannot be doubled. An elongation by an odd factor is possible.

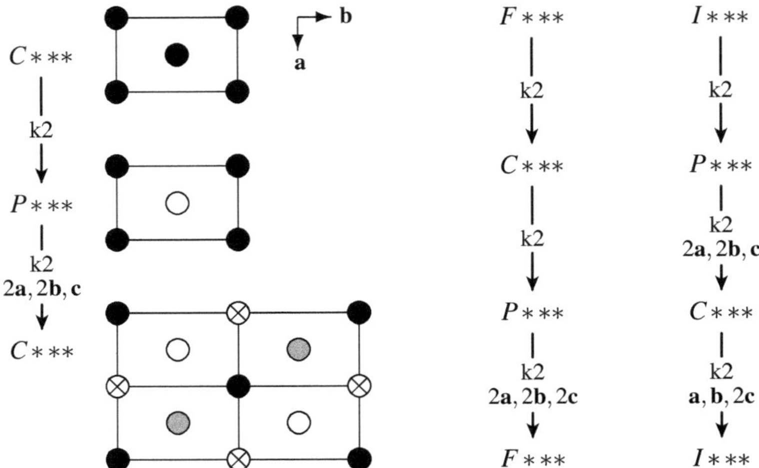

Fig. 21.2 Left: A *C*-centred monoclinic or orthorhombic space group has no isomorphic subgroups of index 2 if it is enlarged in the *a*–*b* plane, but it can have isomorphic subgroups of index 4; ∗∗∗ stands for 121, 1*m*1, 1*c*1, 12/*m*1, 12/*c*1, 222, *mm*2, *m*2*m* 2*mm*, *cc*2, *mmm*, *mcm*, *cmm*, or *ccm*.
Right: Orthorhombic space groups *F*∗∗∗ and *I*∗∗∗, ∗∗∗ = 222, *mm*2, *m*2*m*, 2*mm*, or *mmm*, have no isomorphic subgroups of index 2 and 4, but of index 8.

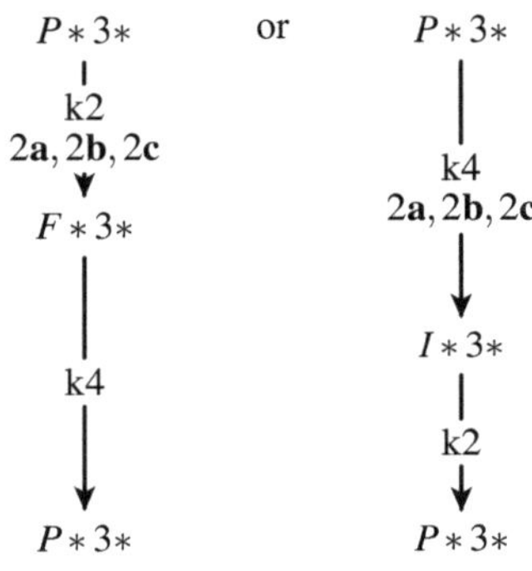

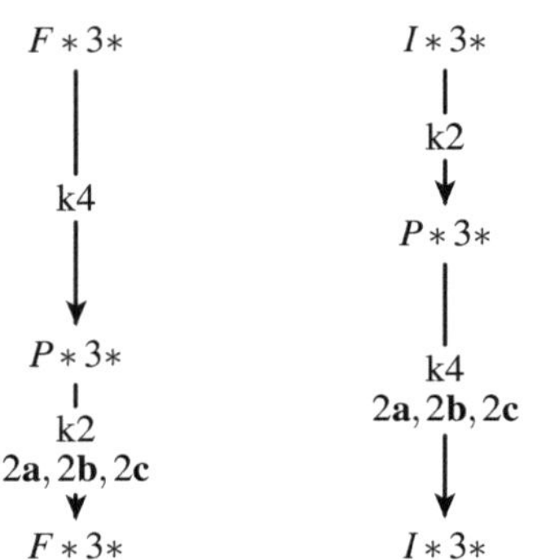

However, some base-centred space groups have non-maximal isomorphic subgroups of index $4n$ (n = integer) that can be reached via a primitive intermediate group (Fig. 21.2 left). An index of 2 is also excluded for face-centred and body-centred orthorhombic space groups; however, in some cases there are non-maximal isomorphic subgroups of index 8 (Fig. 21.2 right).

Cubic space groups have only maximal isomorphic subgroups of index p^3, $p \geq 3$. The reason is the condition that $a' = b' = c'$ must also hold for the subgroups and that the directions of the axes must be retained. Isomorphic subgroups of index 8 ($= 2^3$; $2\mathbf{a}, 2\mathbf{b}, 2\mathbf{c}$) are possible, provided there are no screw axes or glide planes that impede a doubling of lattice parameters. However, the isomorphic subgroups of index 8 are not maximal, but are reached via an intermediate group with another kind of centring. See the relations in the margin; ∗ stands for 2, 4, $\overline{4}$, or *m*; it may be missing in the last symmetry direction; 3 may also be $\overline{3}$.

Tetragonal, trigonal, and hexagonal space groups, cell enlargement parallel to c. By analogy to orthorhombic space groups, the indices of maximal isomorphic subgroups obtained by increasing **c** can be arbitrary prime numbers; however, $p = 2$ is excluded for body- and face-centred cells and if there are 4_2 or 6_3 axes or glide planes with glide components parallel to **c** (glide planes *c*, *d*, or *n* perpendicular to **a** or $\mathbf{a} - \mathbf{b}$). Special restrictions apply to rhombohedral space groups and if there are screw axes 3_1, 3_2, 4_1, 4_3, 6_1, 6_2, 6_4, or 6_5. The corresponding permitted prime numbers are listed in Table 21.1.

Tetragonal space groups, cell enlargement in the *a*–*b* plane. The condition $a' = b'$ and the right angle between $\mathbf{a}'$ and $\mathbf{b}'$ must be preserved for isomorphic subgroups of tetragonal space groups whose cell is enlarged in the *a*–*b* plane.

An index of $p = 2$ requires a basis transformation $\mathbf{a} - \mathbf{b}, \mathbf{a} + \mathbf{b}, \mathbf{c}$ (or $\mathbf{a} + \mathbf{b}, -\mathbf{a} + \mathbf{b}, \mathbf{c}$) and occurs if the cell is primitive and if there are no 2_1 axes and no glide components parallel to **a** and $\mathbf{a} - \mathbf{b}$.

The crystal classes 422, 4*mm*, $\overline{4}2m$, $\overline{4}m2$, and 4/*mmm* have symmetry elements in the symmetry directions **a** and $\mathbf{a} - \mathbf{b}$. The basis transformation

Table 21.1 Permitted indices of symmetry reduction for maximal isomorphic subgroups of tetragonal, trigonal, and hexagonal space groups having screw axes if **c** is multiplied.

Screw axis*	Permitted indices $p^\dagger$	Comments
4_1	$4n-1$	substitution $4_1 \to 4_3$; irrelevant for I cells because they have 4_1 and 4_3 axes
4_1	$4n+1$	
4_2	≥ 3	
4_3	$4n-1$	substitution $4_3 \to 4_1$; irrelevant for I cells because they have 4_1 and 4_3 axes
4_3	$4n+1$	
R	2 or $6n-1$	transformation $\mathbf{a}' = -\mathbf{a}$, $\mathbf{b}' = -\mathbf{b}$ or interchange obverse $\rightleftharpoons$ reverse
R	$6n+1$	
3_1	2 or $6n-1$	substitution $3_1 \to 3_2$
3_1	$6n+1$	
3_2	2 or $6n-1$	substitution $3_2 \to 3_1$
3_2	$6n+1$	
6_1	$6n-1$	substitution $6_1 \to 6_5$
6_1	$6n+1$	
6_2	2 or $6n-1$	substitution $6_2 \to 6_4$
6_2	$6n+1$	
6_3	≥ 3	
6_4	2 or $6n-1$	substitution $6_4 \to 6_2$
6_4	$6n+1$	
6_5	$6n-1$	substitution $6_5 \to 6_1$
6_5	$6n+1$	

* R = rhombohedral space group, hexagonal setting
† p = prime number; n = arbitrary positive integer

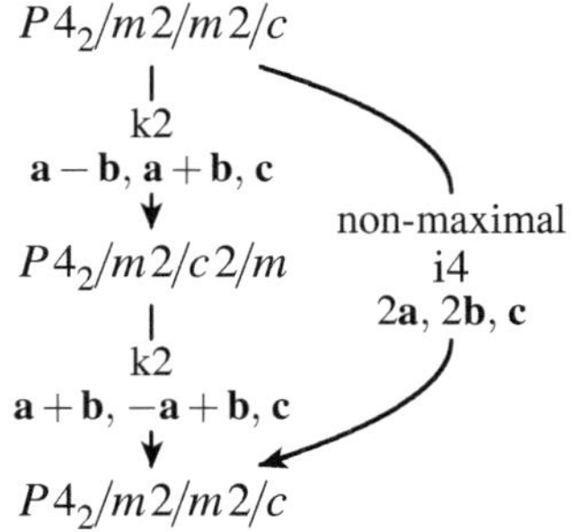

$\mathbf{a}-\mathbf{b}, \mathbf{a}+\mathbf{b}, \mathbf{c}$ ($p = 2$) then only permits maximal isomorphic subgroups if the symmetry elements in the symmetry directions **a** and $\mathbf{a}-\mathbf{b}$ are of the same kind, for example at $P422$, $P4mm$, and $P4/mcc$; otherwise there is an interchange, for example $P4_2mc$ —k2→ $P4_2cm$ or $P\bar{4}2m$ —k2→ $P\bar{4}m2$, and the maximal subgroup is *klassengleiche*, but not isomorphic. If the cell is doubled once more, one returns to the initial directions in the now quadrupled cell. In this case there are non-maximal isomorphic subgroups with index values that are divisible by 4, as shown in the example in the margin.

In a similar way, the space groups $I422$, $I4mm$, and $I4/mmm$ have non-maximal isomorphic subgroups of index 4 that are reached via an intermediate group (basis $\mathbf{a}-\mathbf{b}, \mathbf{a}+\mathbf{b}, 2\mathbf{c}$); $I\bar{4}m2$ and $I\bar{4}2m$ have non-maximal isomorphic subgroups of index 8, basis $2\mathbf{a}, 2\mathbf{b}, 2\mathbf{c}$.

In the crystal classes 422, $4mm$, $\bar{4}m2$, $\bar{4}2m$, and $4/mmm$, maximal subgroups of index $i > 3$ exist only for indices $i = p^2$ (p = arbitrary prime number ≥ 3), basis transformation $p\mathbf{a}$, $p\mathbf{b}$, $\mathbf{c}$.

The crystal classes 4, $\bar{4}$, and $4/m$ have no symmetry elements in the directions **a** and $\mathbf{a}-\mathbf{b}$. The directions of the basis vectors $\mathbf{a}', \mathbf{b}'$ of the subgroup can be different from those of $\mathbf{a}, \mathbf{b}$. There are two possibilities for maximal isomorphic subgroups:

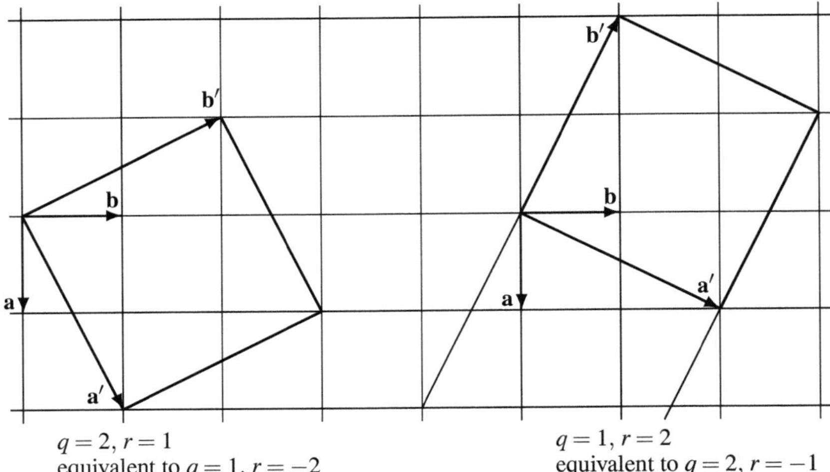

Fig. 21.3 Cell enlargements $\mathbf{a}' = q\mathbf{a} + r\mathbf{b}, \mathbf{b}' = -r\mathbf{a} + q\mathbf{b}, \mathbf{c}$ of tetragonal space groups for the pairs of values $q = 2, r = 1$ and $q = 1, r = 2$ for isomorphic subgroups on a pair of index $p = 5 = 2^2 + 1^2 = 1^2 + 2^2$.

index p^2 with $p = 4n - 1$, basis transformation $p\mathbf{a}$, $p\mathbf{b}$, $\mathbf{c}$; or

index $p = 2$ or $p = q^2 + r^2 = 4n + 1$, integral $q, r, n \neq 0$.

In the case of the index $p = q^2 + r^2 = 4n + 1$ there are exactly two conjugacy classes with p conjugate subgroups each. The conjugacy classes (subgroups on a par) have cells with two orientations, as shown in Fig. 21.3, one with the pair q, r, the other one with interchanged values q and r. As can be seen from the figure, the lattice parameter for both subgroups is $a' = a\sqrt{q^2 + r^2}$ according to Pythagoras, and the basis plane is enlarged by a factor of $p = q^2 + r^2$.

Trigonal and hexagonal space groups, cell enlargement in the *a–b* plane. Isomorphic subgroups of trigonal and hexagonal space groups whose cells are enlarged in the *a–b* plane must meet the condition $a' = b'$ with an angle of 120° between $\mathbf{a}'$ and $\mathbf{b}'$.

The crystal classes 321, 312, 3*m*1, 31*m*, $\bar{3}m1$, $\bar{3}1m$, 622, 6*mm*, $\bar{6}m2$, $\bar{6}2m$, and 6/*mmm* have symmetry elements in the symmetry directions **a** and/or $\mathbf{a} - \mathbf{b}$. There are maximal isomorphic subgroups with basis transformations $p\mathbf{a}$, $p\mathbf{b}$, $\mathbf{c}$, index p^2 for every prime number $p \neq 3$. Maximal isomorphic subgroups of index 3 exist only in the crystal class 622 and for the space groups *P*6*mm*, *P*6*cc*, *P*6/*mmm*, and *P*6/*mcc*, basis transformation $2\mathbf{a} + \mathbf{b}$, $-\mathbf{a} + \mathbf{b}$, $\mathbf{c}$. The remaining space groups of these crystal classes have *klassengleiche* subgroups of index 3 with an interchange *P*321 ⇌ *P*312, *P*3*m*1 ⇌ *P*31*m*, $P6_3mc \rightleftharpoons P6_3cm$, etc. There exist non-maximal isomorphic subgroups of index 3^{2n} with the basis transformations $p\mathbf{a}$, $p\mathbf{b}$, $\mathbf{c}$, $p = 3^n$.

In the crystal classes 3, $\bar{3}$, 6, $\bar{6}$, and 6/*m* there are no symmetry elements in the directions **a** and $\mathbf{a} - \mathbf{b}$. There are two possibilities for maximal isomorphic subgroups:

index p^2 with $p = 2$ or $p = 6n - 1$, basis transformation $p\mathbf{a}$, $p\mathbf{b}$, $\mathbf{c}$; or

index $p = 3$ or $p = q^2 - qr + r^2 = 6n + 1$, $q, r, n =$ positive integers, $q > r > 0$.

In the case of the index $p = q^2 - qr + r^2 = 6n + 1$ there are exactly two conjugacy classes with p conjugate subgroups each. One of them corresponds to the pair of values q, r and the basis vectors $q\mathbf{a} + r\mathbf{b}$, $-r\mathbf{a} + (q - r)\mathbf{b}$, $\mathbf{c}$; at

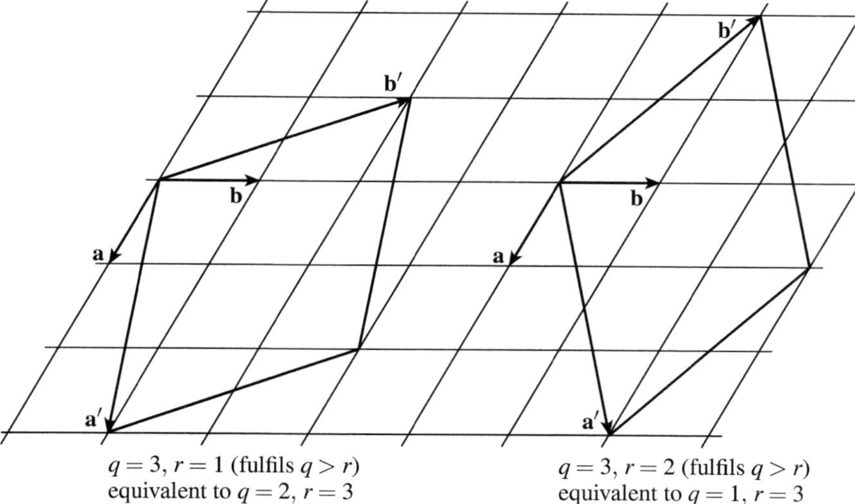

Fig. 21.4 Cell enlargements $\mathbf{a}' = q\mathbf{a} + r\mathbf{b}$, $\mathbf{b}' = -r\mathbf{a} + (q-r)\mathbf{b}$, $\mathbf{c}$ of isomorphic trigonal and hexagonal subgroups for the pairs of values $q = 3, r = 1$ and $q = 3, r = 2$ for the two conjugacy classes of index $p = 7 = 3^2 - 3\cdot 1 + 1^2 = 3^2 - 3\cdot 2 + 2^2$.

the other one, r is replaced by $q - r$ and the basis vectors are $q\mathbf{a} + (q-r)\mathbf{b}$, $(r-q)\mathbf{a} + r\mathbf{b}$, $\mathbf{c}$ (Fig. 21.4). The lattice parameters are $a' = b' = a\sqrt{p}$.

A little number theory: the isomorphic subgroups of the crystal classes 4, $\bar{4}$, 4/*m*, 3, $\bar{3}$, 6, $\bar{6}$, and 6/*m*

Tetragonal space groups. The possible values for the index and the number of isomorphic tetragonal subgroups with the cell enlargement $q\mathbf{a} + r\mathbf{b}$, $-r\mathbf{a} + q\mathbf{b}$, $\mathbf{c}$ (Fig. 21.3) can be examined with number theory [136]. If the sum of two square numbers is a prime number, $q^2 + r^2 = p$, then this prime number is $p = 2$ or $p = 4n + 1$ (Fermat's theorem). Every prime number $p = 4n + 1$ can be expressed as the sum of two square numbers. Therefore, the indices of maximal subgroups can be $i = 2$ and the prime numbers $i = p = 4n + 1$.

If the subgroup is not maximal, there can be several kinds of cell enlargement for the same index, one each for a conjugacy class of subgroups. In order to find out how many conjugacy classes there are, we determine the divisors of the index i and determine how many of them are of the kinds $4n + 1$, $2n$, and $4n - 1$:

$$i \left\{ \begin{array}{lcl} \text{number of divisors } 4n+1 & = & Z_+ \\ \text{number of divisors } 2n & \rightarrow & 0 \\ -\text{number of divisors } 4n-1 & = & \underline{-Z_-} \end{array} \right.$$
$$\text{sum} = \text{number of conjugacy classes}$$

Example 21.1
The number 10 has the divisors 1, 2, 5, and 10:

$$i = 10 \left\{ \begin{array}{ll} \text{divisors } 4n+1:\ 1,\ 5 & Z_+ = 2 \\ \text{divisors } 2n:\ 2,\ 10 & \rightarrow 0 \\ -\text{divisors } 4n-1:\ \text{none} & \underline{-Z_- = 0} \end{array} \right.$$
$$\text{sum} = 2$$

There are two conjugacy classes of isomorphic tetragonal subgroups of index 10. They correspond to the pairs of values $q, r = 1, 3$ and $q, r = 3, 1$; $1^2 + 3^2 = 3^2 + 1^2 = 10$.
They can be reached by two steps of symmetry reduction, one of index 2 and the other of index 5 (or vice versa).

Example 21.2

$$i = 21 \left\{ \begin{array}{ll} \text{divisors } 4n+1: \ 1,\ 21 & Z_+ = \ \ 2 \\ \text{divisors } 2n: \ \text{none} & \quad \to \ \ 0 \\ -\text{divisors } 4n-1: \ 3,\ 7 & -Z_- = -2 \\ \hline & \text{sum} = \ \ 0 \end{array} \right.$$

The number 21 cannot be expressed by a sum of two square numbers. There are no tetragonal isomorphic subgroups of index 21 with cell enlargement in the *a*–*b* plane.

Example 21.3

$$i = 25 \left\{ \begin{array}{ll} \text{divisors } 4n+1: \ 1,\ 5,\ 25 & Z_+ = 3 \\ \text{divisors } 2n: \ \text{none} & \quad \to 0 \\ -\text{divisors } 4n-1: \ \text{none} & -Z_- = 0 \\ \hline & \text{sum} = 3 \end{array} \right.$$

There are three conjugacy classes of index 25. They have the pairs of values $q, r = 4, 3$; $q, r = 3, 4$; $q, r = 5, 0$; $4^2 + 3^2 = 3^2 + 4^2 = 5^2 + 0^2 = 25$.

The procedure is also applicable to prime numbers. All prime numbers $p = 4n + 1$ have the two divisors 1 and p of kind $4n + 1$. For every prime number $p = 4n + 1$ there are exactly two conjugacy classes. The prime numbers $p = 4n - 1$ have the divisor 1 of the kind $4n + 1$ and the divisor $p = 4n - 1$; therefore, $Z_+ = 1$ and $Z_- = 1$, $Z_+ - Z_- = 0$, and none of these prime numbers corresponds to an isomorphic subgroup (this always holds: if $q^2 + r^2 = p$, then $p \neq 4n - 1$).

Trigonal and hexagonal space groups. For the possible values for the index and the number of isomorphic trigonal and hexagonal subgroups with the cell enlargement $q\mathbf{a} + r\mathbf{b}$, $-r\mathbf{a} + (q - r)\mathbf{b}$, $\mathbf{c}$ we have (Fig. 21.4): if the sum $p = q^2 - qr + r^2$ is a prime number, then this prime number is $p = 3$ or $p = 6n + 1$. For every prime number $p = 6n + 1$ two such sums can be found, one with q, r, and the other with $q, (q - r)$; $q > r > 0$.

If the subgroup is not maximal, there can be several kinds of cell enlargement of the hexagonal cell for the same index *i*, each one for one conjugacy class of subgroups. How many conjugacy classes there are follows from the divisors of *i*:

$$i \left\{ \begin{array}{ll} \text{number of divisors } 3n+1 & = \ \ Z_+ \\ \text{number of divisors } 3n & \to \ \ 0 \\ -\text{number of divisors } 3n-1 & = -Z_- \\ \hline \text{sum} & = \text{number of conjugacy classes} \end{array} \right.$$

Example 21.4

$$i=21 \left\{ \begin{array}{ll} \text{divisors } 3n+1: \ 1,\ 7 & Z_+ = 2 \\ \text{divisors } 3n: \ 3,\ 21 & \to 0 \\ -\text{divisors } 3n-1: \ \text{none} & -Z_- = 0 \\ \hline & \text{sum} = 2 \end{array} \right.$$

There are two conjugacy classes of hexagonal isomorphic subgroups of index 21, with the pairs of values $q, r = 5, 1$ and $q, r = 5, 4$; $5^2 - 5\cdot 1 + 1^2 = 5^2 - 5\cdot 4 + 4^2 = 21$.

Example 21.5

$$i=25 \left\{ \begin{array}{ll} \text{divisors } 3n+1: \ 1,\ 25 & Z_+ = 2 \\ \text{divisors } 3n: \ \text{none} & \to 0 \\ -\text{divisors } 3n-1: \ 5 & -Z_- = -1 \\ \hline & \text{sum} = 1 \end{array} \right.$$

There is one conjugacy class of index 25: $q=5, r=0$; $5^2 - 5\cdot 0 + 0^2 = 25$.

Exercises

Solutions on page 340.

21.1. Can the following space groups have isomorphic subgroups of index 2?
$P12/c1$, $P2_12_12$, $P2_1/n2_1/n2/m$, $P4/m2_1/b2/m$, $P6_122$.

21.2. In the text concerning tetragonal space groups it is argued that there are non-maximal isomorphic subgroups of the space groups $I4/m2/m2/m$, index 4, and $I\bar{4}m2$, index 8, although there are no maximal isomorphic subgroups of index 2. Set up the relations.

21.3. Does the space group $P4_2/m$ have isomorphic maximal subgroups of indices 4, 9, 11, and 17 with a cell enlargement in the *a*–*b* plane? If so, what are the necessary transformations of the basis vectors?

21.4. How many isomorphic subgroups of index 65 are there for $P4/n$?

21.5. Crystalline $PtCl_3$ consists of Pt_6Cl_{12} clusters and of chains of edge-sharing octahedra of the composition $PtCl_4$ [451]. To a good approximation, the chlorine atoms, by themselves, form a cubic-closest packing of spheres, but $\frac{1}{37}$ of the positions of the spheres are vacant (in the centres of the Pt_6Cl_{12} clusters). The Pt atoms occupy octahedral voids of the packing of spheres. $PtCl_3$ is rhombohedral, space group $R\bar{3}$, $a = 2121$ pm, $c = 860$ pm. The symmetry of the sphere packing ($a_{\text{cub}} \approx 494$ pm) can be reduced in two initial steps down to the intermediate group $R\bar{3}$:
$Fm\bar{3}m$ —t4; $\frac{1}{2}(-\mathbf{a}+\mathbf{b})$, $\frac{1}{2}(-\mathbf{b}+\mathbf{c})$, $\mathbf{a}+\mathbf{b}+\mathbf{c}$→ $R\bar{3}m^{(\text{hex})}$ —t2→ $R\bar{3}^{(\text{hex})}$
Compute the lattice parameters of this intermediate group. How does the cell of $PtCl_3$ result from this cell? What is the index of the symmetry reduction between the intermediate and the final group?

21.6. Use one of the relations depicted in Fig. 8.7 (page 112) to show why isomorphic subgroups of the space group $P4/m$ of index 5 have exactly two conjugacy classes of subgroups on a par. The Euclidean normalizer of $P4/m$ is $P4/mmm$ with $\frac{1}{2}(\mathbf{a}-\mathbf{b})$, $\frac{1}{2}(\mathbf{a}+\mathbf{b})$, $\frac{1}{2}\mathbf{c}$.

On the theory of phase transitions

22

This chapter is a supplement to Sections 16.1 and 16.2 supposed to enhance the understanding of the physico-chemical background of phase transitions in the solid state. However, this is only a brief glimpse into the theory of phase transitions.

22.1 Thermodynamic aspects concerning phase transitions

In Section 16.1.1, EHRENFEST's definition of first- and second-order phase transitions is presented. The determining factors for the classification are the derivatives of the Gibbs free energy with respect to temperature, pressure, or other variables of state. The role of the derivatives is illustrated in Fig. 22.1. In a schematic way, it is shown how the Gibbs free energies G_1 and G_2 of two phases depend on temperature. A first-order phase transition takes place if the two curves intersect, the intersection point being the point of transition T_c. The more stable phase is the one having the lower G value; the one with the higher G value can exist as a metastable phase. When switching from the curve G_1 to G_2 at the point T_c, the slope of the curve is changed abruptly; the first derivative switches discontinuously from one value to another.

A second-order phase transition entails no metastable phases. The dotted curve G_2 represents the hypothetical (computed or extrapolated) course for the high-temperature modification below T_c. The curve G_1 ends at T_c, where it merges with the curve G_2 with the same slope, but with a different curvature. Mathematically, curvature corresponds to the second derivative; it changes discontinuously at T_c.

During a first-order transition both phases coexist. Properties that refer to the whole specimen, for example the uptake of heat or the magnetization, result from the superposition of the properties from both phases, corresponding to their constituent amounts. Measurement yields a hysteresis curve (Fig. 22.2). Upon heating, phase 1 (stable below T_c) at first lingers on in a metastable state above T_c (dotted line in Fig. 22.2). The more T has surpassed T_c, the more nuclei of phase 2 are formed, and the more they grow, the more their properties dominate until finally only phase 2 is present. Heat uptake follows the right branch of the curve and shows a sluggish progression of the phase transition at $T > T_c$. Upon cooling, the same happens in the opposite direction, with sluggish heat release at $T < T_c$. The width of the hysteresis loop is not a specific

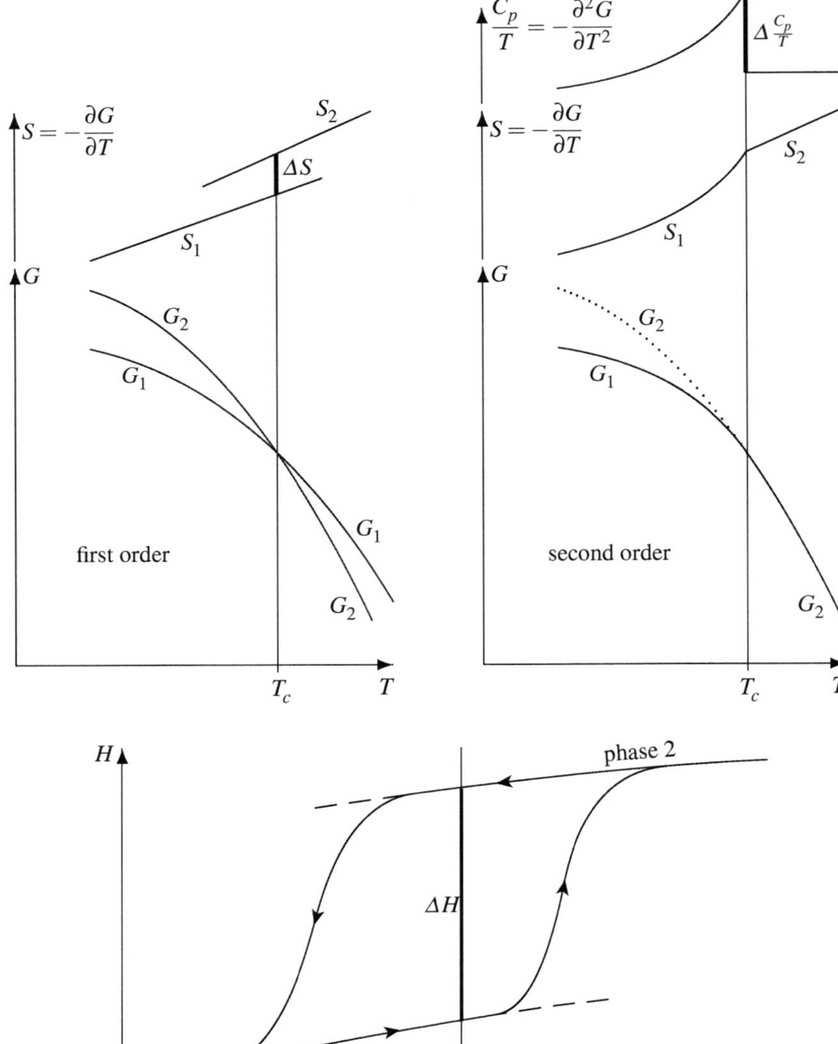

Fig. 22.1 Temperature dependence of the Gibbs free energies G_1 and G_2 and of their derivatives for two phases that are transformed one to the other at the temperature T_c; left, first order; right, second order.

Fig. 22.2 Hysteresis curve of the heat contents (enthalpy) H of a specimen at an enantiotropic, temperature-dependent first-order phase transition. Arrows mark the direction of temperature change.

property of the material. It depends on the kind and the number of defects in the crystal and, therefore, depends on the past history of the specimen.

The order of a phase transition can differ depending on conditions. For example, pressure causes crystalline cerium to undergo a first-order phase transition γ-Ce $\rightarrow$ α-Ce; it is associated with a significant, discontinuous change of volume (-13 % at room temperature) and an electron transition $4f \rightarrow 5d$ [452]. The structures of both modifications, γ-Ce and α-Ce, are cubic-closest packings of spheres ($Fm\overline{3}m$, atoms at position $4a$). The higher the temperature, the lesser is the change of volume, and finally it becomes $\Delta V = 0$ at the critical point [453]. Above the critical temperature (485 K) or the critical pressure (1.8 GPa), there no longer exists a difference between γ-Ce and α-Ce (Fig. 22.3).

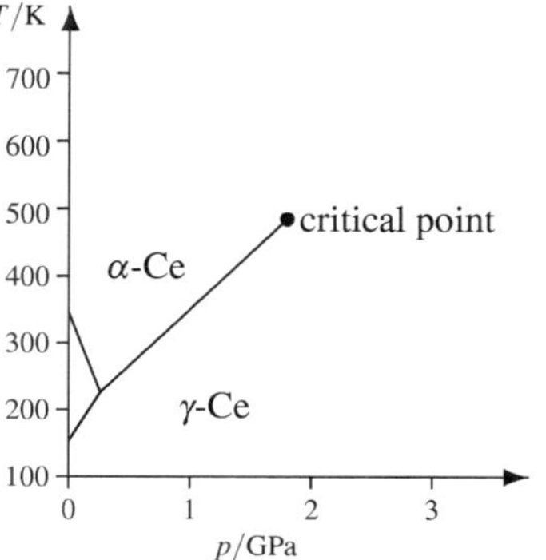

Fig. 22.3 Section from the phase diagram of cerium [452, 454].

Potassium dihydrogenphosphate exhibits a change from a first-order to a second-order transition. At ambient pressure, upon change of temperature, it is converted from a paraelectric to a ferroelectric phase. In the paraelectric modification, space group $I\bar{4}2d$, the H atoms are misordered within hydrogen bonds:

$$\text{O–H}\cdots\text{O} \rightleftharpoons \text{O}\cdots\text{H–O}$$

In the ferroelectric phase, space group $Fdd2$, they are ordered. At ambient pressure, a first-order transition takes place at 122 K with a small volume discontinuity (ΔV approx. -0.5 %). At pressures above 0.28 GPa and temperatures below 108 K (the 'tricritical' point; Fig. 22.4) it is of second order [456, 457].

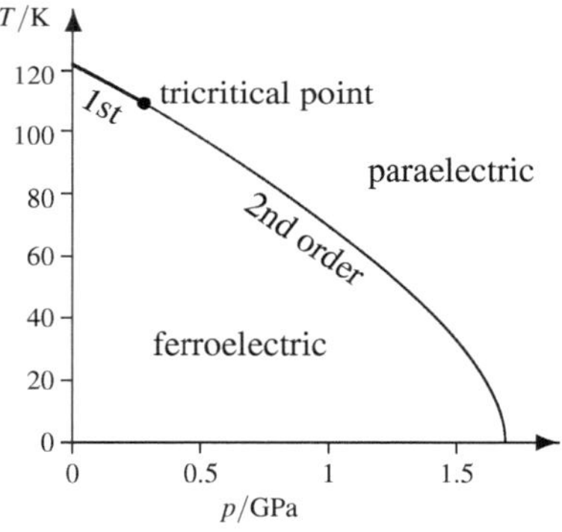

Fig. 22.4 Section from the phase diagram of KH_2PO_4 [455].

In Section 16.1.1, after EHRENFEST's definition, a newer definition is mentioned according to which the continuous change of an order parameter is decisive. At first glance, this seems to be a completely different criterion as compared to EHRENFEST's classification. Actually, there is not so much of a difference. Phase transitions of nth order in terms of EHRENFEST having $n > 2$ are not really observed. The case $n = \infty$ does not describe a phase transition, but a continuous change of properties (e.g. of thermal dilatation; this is not to be confused with a continuous phase transition). Even at a continuous phase transition, invariably some thermodynamic function experiences a discontinuity. That embraces mot just a sudden change of a second derivative of the Gibbs free energy, but any kind of discontinuity or singularity. The change of the order parameter η as a function of some variable of state (e.g. temperature T) exhibits a singularity at the point of transition: on one side ($T < T_c$) an analytical dependence according to a power law holds (eqn (16.2), page 218), but at the point of transition ($T = T_c$) the order parameter vanishes (it becomes $\eta = 0$ for $T \geq T_c$; Fig. 16.4 top left, page 218).

22.2 About Landau theory

In the following explanations, we repeatedly resort to the example of the continuous phase transition of $CaCl_2$ (Section 16.2, page 214). Let the Gibbs free energy G of a crystal have the value G_0 at the point of transition. For small values of the order parameter η, the change of Gibbs free energy relative to G_0 can be expressed by a Taylor series. All terms with odd powers have been omitted from the series, since G has to remain unchanged when the sign of η is changed (it makes no difference whether the octahedra of $CaCl_2$ have been turned to one or the other side; Figs 16.2, 16.3). This proposition is universally valid according to Landau theory: For a continuous phase transition, only even powers may appear in the Taylor series.

Taylor expansion of G with respect to η:

$$G = G_0 + \frac{a_2}{2}\eta^2 + \frac{a_4}{4}\eta^4 + \frac{a_6}{6}\eta^6 + \cdots \qquad (22.1)$$

As long as η is small, the Taylor series can be truncated after a few terms, whereby the term with the highest power has to have a positive coefficient; this is important, because otherwise G would become more and more negative with increasing η, which would be an unstable state. The coefficients $a_2, a_4, \ldots$ depend on the variables of state T and p. Essential is the dependence of a_2 on

T and p. For a temperature-dependent phase transition, $a_2 = 0$ must hold at the point of transition, above it is $a_2 > 0$ and below $a_2 < 0$, while $a_4, a_6, \ldots$ remain approximately constant. If, in accordance with experience, a_2 changes linearly with temperature close to T_c, we have eqn (22.2). The effect of temperature on the Gibbs free energy is shown in Fig. 22.5. At temperatures above T_c, G has a minimum at $\eta = 0$; the structure of $CaCl_2$ is tetragonal. At the point of transition $T = T_c$, there is also a minimum at $\eta = 0$, with a gentle curvature close to $\eta = 0$. Below T_c, the curve has a maximum at $\eta = 0$ and two minima. The more the temperature is below T_c, the deeper are the minima and the more distant they are from $\eta = 0$. At $T < T_c$ the structure is no longer tetragonal and the octahedra have been turned to one side or the other, corresponding to the η value of one of the minima.

$$a_2 = k(T - T_c) \qquad (22.2)$$

with k = positive constant

If $a_4 > 0$, the Taylor series (22.1) can be truncated after the fourth-power term. From the first derivative with respect to η we obtain condition (22.3) for the minima: If $T < T_c$ and thus $a_2 < 0$, the positions of the minima are at $\eta_{1,2}$. Taking $A = \pm\sqrt{kT_c/a_4}$, the order parameter η of the stable structure then follows the power law given in eqn (22.5).

$$\frac{\partial G}{\partial \eta} = a_2\eta + a_4\eta^3 = 0 \qquad (22.3)$$

$$\eta_{1,2} = \pm\sqrt{-\frac{a_2}{a_4}} = \pm\sqrt{\frac{k(T_c - T)}{a_4}} \qquad (22.4)$$

$$\eta = A\left(\frac{T_c - T}{T_c}\right)^{\beta} \qquad (22.5)$$

with $\beta = \frac{1}{2}$ and $T < T_c$

A is a constant and β is the *critical exponent*. The derived value of $\beta = \frac{1}{2}$ is valid under the assumptions made that G can be expressed by a Taylor series neglecting powers higher than 4, that a_2 depends linearly on the temperature difference $T - T_c$, and that $a_4 > 0$ is (nearly) independent of temperature.

In addition, it is assumed that the order parameter does not fluctuate. This, however, cannot be taken for granted close to the point of transition. Just below T_c, the two minima of the curve for G (Fig. 22.5, $T \lesssim T_c$) are shallow and the maximum in between is low. The energy barrier to be surmounted to shift from one minimum to the other is less than the thermal energy, and η can fluctuate easily from one minimum to the other. At the point of transition itself (curve $T = T_c$ in Fig. 22.5) and slightly above, the curvature is so small that fluctuations of η induced by temperature have almost no influence on the Gibbs free energy. The local fluctuations can be different in different regions of the crystal.

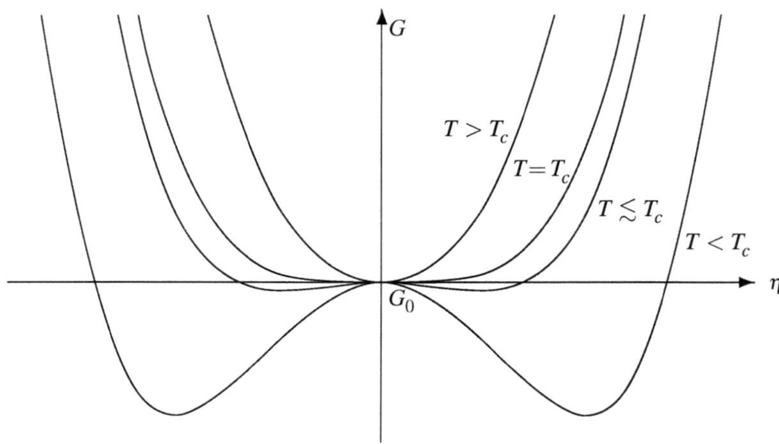

Fig. 22.5 Gibbs free energy G as a function of the order parameter η for different temperatures according to eqn (22.1), truncated after the fourth power.

Actually, there are only minor fluctuations if the interactions in the crystal have a long range. For example, the range of interactions between the tightly connected octahedra of $CaCl_2$ is long; all of the octahedra (or a large number of them) would have to participate in the fluctuations. For long-range interactions, Landau theory is satisfied and the critical exponent amounts to $\beta \approx \frac{1}{2}$.

If $T > T_c$, the minimum for G is at $\eta = 0$, and at this value the first and second derivatives of the Gibbs free energy with respect to temperature are those mentioned in the margin.

At $T > T_c$:

$$\left.\frac{\partial G}{\partial T}\right|_{\eta=0} = \frac{\partial G_0}{\partial T}$$

$$\left.\frac{\partial^2 G}{\partial T^2}\right|_{\eta=0} = \frac{\partial^2 G_0}{\partial T^2}$$

If $T < T_c$, the value of G corresponding to the minima is obtained by inserting eqns (22.2) and (22.4) into eqn (22.1), truncated after the fourth power:

$$\begin{aligned} G|_{\eta_{1,2}} &= G_0 + \frac{k(T-T_c)}{2}\left(\sqrt{\frac{k(T_c-T)}{a_4}}\right)^2 + \frac{a_4}{4}\left(\sqrt{\frac{k(T_c-T)}{a_4}}\right)^4 \\ &= G_0 - \frac{k^2(T_c-T)^2}{4a_4} \qquad \text{at } T < T_c \end{aligned}$$

From this, we obtain the first and second derivatives with respect to temperature for $T < T_c$:

$$\begin{aligned} -S|_{\eta_{1,2}} &= \left.\frac{\partial G}{\partial T}\right|_{\eta_{1,2}} = \frac{\partial G_0}{\partial T} + \frac{k^2}{2a_4}(T_c - T) \\ -\left.\frac{\partial S}{\partial T}\right|_{\eta_{1,2}} &= \left.\frac{\partial^2 G}{\partial T^2}\right|_{\eta_{1,2}} = \frac{\partial^2 G_0}{\partial T^2} - \frac{k^2}{2a_4} \end{aligned}$$

When the point of transition is approached from the side of low temperatures, $(T_c - T)$ and η become 0, and $\partial G/\partial T$ merges at T_c with the value $\partial G_0/\partial T$ that it also adopts above T_c. However, below T_c, $\partial^2 G/\partial T^2$ has a value that is smaller by $\frac{1}{2}k^2/a_4$ than at $T > T_c$. The first derivative of G thus does not experience a sudden change at T_c, but the second derivative does, just as expected for a second-order transition according to EHRENFEST's definition.

Displacing an octahedron of $CaCl_2$ from its equilibrium position requires a force F. Force is the first derivative of energy with respect to displacement; for our purposes that is the first derivative of the Gibbs free energy with respect to η, $F = \partial G/\partial \eta$. The force constant f is the first derivative of F with respect to η; actually, it is only a constant for a harmonic oscillator, then we have $F = f\eta$ holds (Hooke's law; a harmonic oscillator is one for which the Taylor series contains only terms of powers zero and two). For the curves $T > T_c$ and $T < T_c$ in Fig. 22.5, Hooke's law is only an approximation close to the minima. The second derivatives of the Taylor series (22.1), truncated after the fourth power, corresponds to eqn (22.6). If $T > T_c$, with the minimum at $\eta = 0$, the force constant is thus $f \approx a_2$. If $T < T_c$, the minima are at $\eta_{1,2} = \pm\sqrt{-a_2/a_4}$, eqn (22.4). By substitution of these values into eqn (22.6), we obtain the force constant in proximity to these minima.

$$f = \frac{\partial F}{\partial \eta} \approx \frac{\partial^2 G}{\partial \eta^2} = a_2 + 3a_4\eta^2 \quad (22.6)$$

Substitute $\eta_{1,2}$ for η:

$$f|_{\eta_{1,2}} \approx \frac{\partial^2 G}{\partial \eta^2} = -2a_2 \qquad \text{at } T < T_c$$

Together with eqn (22.2), it follows that

$$f \approx k(T - T_c) \ \text{ at } \ T > T_c \qquad (22.7)$$

and

$$f \approx 2k(T_c - T) \ \text{ at } \ T < T_c \qquad (22.8)$$

The square ν^2 of a vibrational frequency is proportional to the force constant. Therefore, the frequency can be chosen as an order parameter according to eqn (22.5) with ν instead of η. When approaching the temperature of transition, $f \to 0$ and thus $\nu \to 0$. The vibration becomes a soft mode.

This is valid for phase transitions in both directions. Starting at a high temperature, from tetragonal $CaCl_2$, the octahedra perform rotational vibrations about the equilibrium position $\eta = 0$. Below T_c, after the phase transition, the octahedra vibrate about the equilibrium positions $\eta_1 > 0$ or $\eta_2 < 0$; these shift increasingly away from $\eta = 0$ the more the temperature difference $T_c - T$ increases. Simultaneously, the square of the vibrational frequency ν^2 again increases proportional to $(T_c - T)$. According to eqn (22.8), the temperature dependence of the squared frequency ν^2 at $T < T_c$ should be twice as large as at $T > T_c$.

Equations (22.6) and (22.8) are not valid in close proximity to the point of transition. Hooke's law is not then valid. In fact, the frequency of the soft mode of $CaCl_2$ does not decrease down to 0 cm^{-1}, but only to 14 cm^{-1}. In addition, the measured temperature dependence of its squared frequency at $T < T_c$ is not twice as large, but 6.45 times larger than at $T > T_c$ (Fig. 16.4, page 218; [333]).

22.3 Renormalization-group theory

Landau theory shows good qualitative agreement with experimental observations. However, there are quantitative discrepancies, and these are quite substantial in the case of short-range interactions, for example magnetic interactions. Especially, this applies in proximity to the point of transition. An improvement has been achieved by K. G. WILSON's *renormalization-group theory* [458, 459].[1]

[1] This is not a branch of normal group theory. The renormalization group is not a group in terms of group theory.

Landau theory is a 'mean-field theory' that assumes equal conditions in the whole crystal (also called molecular field theory in the case of magnetic phase transitions). This is not fulfilled for short-range atomic interactions. In this case, fluctuations cannot be neglected.

Consider a substance like EuO as an example. It exhibits a second-order paramagnetic–ferromagnetic phase transition at $T_c = 69.2$ K. The spins of the Eu atoms tend to line up in the same direction. If the spin of an Eu atom has the orientation ↑, it affects neighbouring Eu atoms to adopt this same ↑ spin. Due to the short range of the magnetic interaction only the nearest-neighbour spins are coupled to each other. Indirectly, however, farther atoms are indeed being influenced. If the spin ↑ of the first atom has induced an ↑ alignment of the second atom, this atom will cause a next (third) atom to adopt an ↑ alignment as well, etc. As a consequence, the spins of all atoms are correlated. However, thermal motion counteracts correlation, such that a spin is occasionally reversed and adopts the 'wrong' orientation ↓. The lower the temperature, the less probable this reorientation is.

Competition between the tendency towards a uniform spin orientation and the thermal introduction of disorder has the consequence that the correlation can only be detected up to a certain distance, the *correlation length*.

At high temperatures, the correlation length is nearly zero. The spins are oriented at random, and they frequently change their orientation. In the mean, the number of ↑ and ↓ spins is balanced, and the material is paramagnetic. As the temperature falls the correlation length increases. Domains appear in which spins mostly point in the same direction. The overall magnetization is still zero because there are domains with either orientation (Fig. 22.6). The

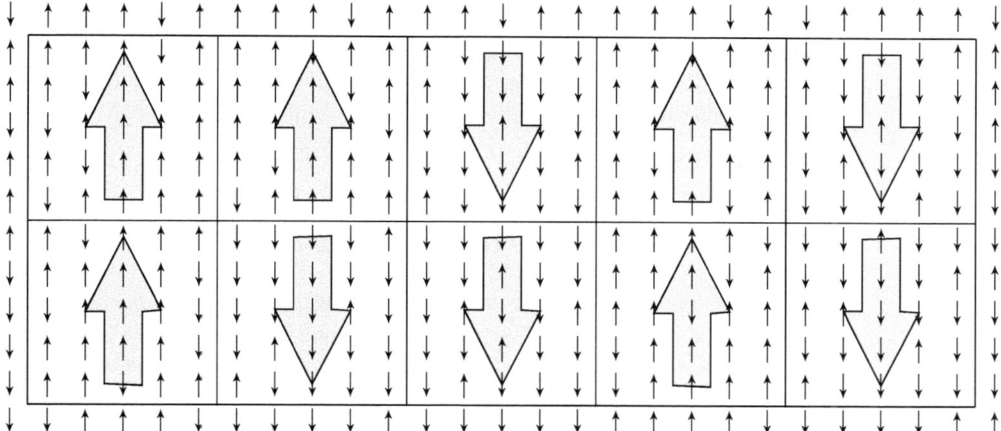

Fig. 22.6 Domains of predominantly parallel spins in a magnetic material.

domains are not rigid; they continuously fluctuate and there are continuous fluctuations of single spins within a domain.

As the temperature approaches the critical temperature T_c, the Curie temperature, the correlation length grows rapidly. The domains become larger. At the Curie temperature the correlation length becomes infinite. Any two spins are correlated, no matter what the distance between them is. Nevertheless, fluctuations persist, and the material remains unmagnetized. However, it is exquisitely sensitive to small perturbations. Holding the spin of a single atom in one direction creates a disturbance that spreads throughout the lattice and gives the entire material a magnetization. Below T_c a ferromagnetic order of the spins arises.

The method of treating the outlined model mathematically is *renormalization*. Let p be the probability of finding another $\uparrow$ spin next to an $\uparrow$ spin. The lattice is divided into blocks of a few spins each. Using the probability p, the number of spins of each orientation in the block is calculated. If the majority of spins in the block is $\uparrow$, an $\uparrow$ spin is assigned to the whole block (thick arrows in Fig. 22.6). The effect is to eliminate all fluctuations in spin direction whose scale of length is smaller than the block size. In the same way, in several steps of iteration, the blocks are collected into larger and even larger blocks. The effects of interactions at small scales of length are thus extended to large scales.

The model is independent of the kind of interaction between the particles (it is not restricted to spin–spin interactions). According to the *universality hypothesis* by GRIFFITHS [460], the laws that control continuous phase transitions, in contrast to discontinuous ones, depend only on the range of interaction and the number of dimensions d and n. d is the number of space dimensions in which the interactions are active; n is the number of 'dimensions' of the order parameter, that is, the number of components needed for its description (e.g. the number of vector components to define the spin vector). A short-range interaction decreases by more than $r^{-(d+2)}$ as a function of distance r; a long-range interaction decreases according to $r^{-(d+\sigma)}$, with $\sigma < d/2$.

From the theory, for short-range interactions, a value of $\beta = \frac{1}{8}$ results for the critical exponent if $d = 2$; this applies, for example, to ferromagnets whose magnetic interactions are restricted to planes and to adsorbed films. If $d = 3$, depending on n, critical exponents of $\beta = 0.302$ ($n = 0$) to $\beta = 0.368$ ($n = 3$) are obtained. Experimental values for EuO ($d = 3$, $n = 3$) yield $\beta = 0.36$. For long-range interactions the results are consistent with Landau theory (i.e. $\beta = 0.5$).

22.4 Discontinuous phase transitions

Finally, let us consider what the consequences are if odd powers appear in the Taylor series, eqn (22.1). A power of one, η^1, is not possible because the order parameter has to be zero at the transition temperature; G as a function of η has to have a minimum at $T > T_c$, the first derivative must be zero at $\eta = 0$, which implies $a_1 = 0$. Including a third power, the series up to the fourth power corresponds to eqn (22.9). If we again assume a linear dependence of a_2 on temperature, $a_2 = k(T - T_0)$, and assume that $a_3 > 0$ and $a_4 > 0$ are nearly independent of temperature, we obtain curves for the Gibbs free energy as shown in Fig. 22.7 (T_0 is not the transition temperature).

$$G = G_0 + \frac{a_2}{2}\eta^2 + \frac{a_3}{3}\eta^3 + \frac{a_4}{4}\eta^4 \tag{22.9}$$

At high temperatures there is only one minimum at $\eta = 0$; there exists one stable high-temperature modification. As temperature falls, a second minimum appears at a negative η value; at first, this minimum is above the minimum at $\eta = 0$ (curve $T > T_c$ in Fig. 22.7). This means that a metastable phase can exist at this value of η. When the temperature is lowered further, such that both minima are at the same height, their G values are equal and $\Delta G = 0$; this is the condition of a state of equilibrium, the corresponding temperature is the transition temperature $T = T_c$. If $T < T_c$, the minimum at $\eta < 0$ is below that at $\eta = 0$. Therefore, there exists a stable phase at $T < T_c$ with a negative value of η and a metastable phase at $\eta = 0$.[2]

[2] Not shown in Fig. 22.7: If $T = T_0 < T_c$ there is only the one minimum at negative η and a saddle point at $\eta = 0$; if $T < T_0$ the curve is similar to that in Fig. 22.5, with a deep minimum at $\eta_1 < 0$, a less deep minimum at $\eta_2 > 0$, and a maximum at $\eta = 0$.

These are the conditions for a discontinuous phase transition. The order parameter η corresponds to the value of η of the lower minimum. Starting from

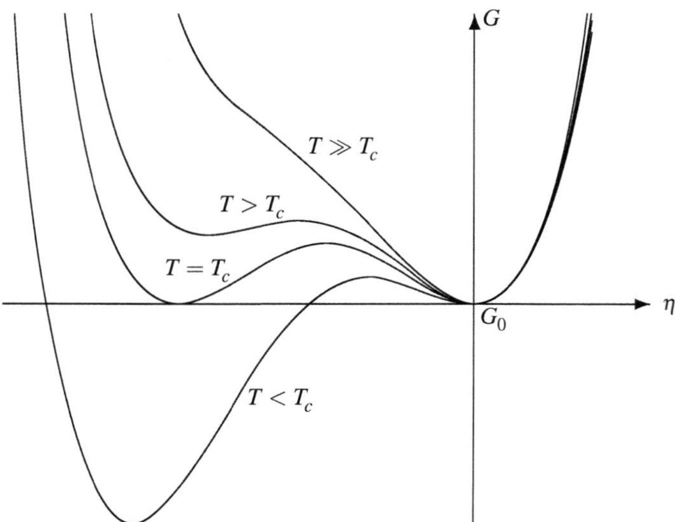

Fig. 22.7 Gibbs free energy G as a function of an order parameter η at different temperatures if a third-power term appears in the Taylor series. If $a_3 < 0$ in eqn (22.9), the curves are mirror-symmetrical with respect to the vertical coordinate axis, with minima at $\eta > 0$.

the high-temperature form with $\eta = 0$, at decreasing temperatures a metastable phase appears at a certain temperature; below T_c this phase becomes stable and it has an order parameter η that is quite different from zero. There is no continuous change of the structure with a continuous change of η, but below T_c the second structure is suddenly the more stable one; the order parameter experiences a sudden change from $\eta = 0$ to $\eta \neq 0$ involving a discontinuous change of structure.

The appearance of a metastable phase entails hysteresis. As the temperature falls, the structure at first persists in being metastable in the high-temperature form with $\eta = 0$ even below T_c; the transformation does not set in before the new, lower minimum has become low enough. The lower the new minimum at $\eta \neq 0$ is, the lower is, viewed from $\eta = 0$, the maximum between the two minima; the energy barrier to be surmounted from the minimum at $\eta = 0$ to the minimum at $\eta \neq 0$ becomes lower. That means decreasing energy of activation and thus a faster transformation the further T is below the point of transition T_c.

Such a model is appropriate only for the description of the thermodynamic conditions of discontinuous phase transitions if both structures are similar, so that an order parameter can be found whose (small, but discontinuous) change results in a straightforward conversion of one structure to the other. The model is applicable to discontinuous displacive phase transitions, but hardly to reconstructive phase transitions. It should also be pointed out that this is a purely thermodynamic point of view by which the Gibbs free energies of two structures are compared with the aid of an order parameter. There is no reference to any mechanism as to how the phase transition actually proceeds. The sudden change of the numerical value of η, which states the shift of the thermodynamic stability from one structure to the other, does not mean that the atoms actually execute a simple displacement in the structure. Mechanism and kinetics rather are a matter of nucleation and growth. The atoms cannot all be displaced simultaneously, otherwise there would be no hysteresis.

Symmetry species

23

Symmetry species are used to describe the symmetry properties when points arranged symmetrically in space are associated with anisotropic properties. A small local coordinate system is attached to each point, which serves to account for the direction-dependent properties at this point. If an atom is situated at the point, such properties can be vibrational motions or the atomic orbitals (wave function = electron as standing wave). If the point is mapped onto another point by a symmetry operation, it is mapped together with its local coordinate system. A symmetry species specifies how a symmetry operation affects the local conditions at the mapped atom.

Table 23.1 Symbols (after PLACZEK [324]) for symmetry species (irreducible representations) for molecules and for crystals at the Γ point of the Brillouin zone, given a point group, referred to a Cartesian coordinate system *xyz* (incomplete list).

A	symmetric to the main axis (z axis; axis of highest order); point groups 222 and *mmm*: symmetric to all twofold axes; cubic point groups: symmetric to the threefold axes
A_1	as A, in addition symmetric to all other rotation axes or (if not present) to reflection planes
A_2	as A, in addition antisymmetric to all other rotation axes or (if not present) to reflection planes
B	antisymmetric to the main axis
B_1	as B and symmetric to the symmetry axis in the direction of x or (if not present) to the reflection plane perpendicular to x; point groups 222 and *mmm*: symmetric to the z axis
B_2	as B and antisymmetric to the symmetry axis in the direction of x or (if not present) to the reflection plane perpendicular to x; point groups 222 and *mmm*: symmetric to the y axis
B_3	point groups 222 and *mmm*: symmetric to the x axis
E	doubly-degenerate ('two-dimensional representation')[1]
T (or F)	triply-degenerate ('three-dimensional representation')[2]
index g	symmetric to the centre of inversion[3,5] (inversion of all three signs of x, y, and z makes no difference)
index u	antisymmetric to the centre of inversion[4,5] (inversion of the signs of x, y, and z is accompanied by an inversion of the sign of the function)
prime $'$	symmetric to the plane of reflection[6]
double prime $''$	antisymmetric to the plane of reflection[6]

Antisymmetric means: after execution of the symmetry operation the value of the function (e.g. the vibrational displacement of an atom) is exactly opposite (sign inversion).

[1] vibration that can be composed by linear combination from two independent vibrations of the same kind; occurs only if there is at least one symmetry axis of order three or higher

[2] only for cubic point groups

[3] g = from German *gerade*, meaning even

[4] u = from German *ungerade*, meaning odd

[5] stated always and only for centrosymmetric structures

[6] mentioned only if the facts do not follow from the other symbols

Mathematically, symmetry operations are represented by matrices, thus the term representation theory. *Irreducible representations* fulfil certain mathematical conditions; in particular, they are independent of each other. Symmetry species are irreducible representations. Representation theory is dealt with in numerous publications [17–24].

Usually, symmetry species are designated with symbols that were developed by PLACZEK for molecular vibrations [324,325], and which then were adopted and further developed for other applications. The most important symbols have been collected in Table 23.1. In the case of atomic wave functions upper case letters as in Table 23.1 are used if they refer to a multi-electron system, and lower case letters if they refer to one electron.

The Bilbao Crystallographic Server

24

The study of symmetry relationships between crystal structures and the setting up of Bärnighausen trees requires frequent consultations of *International Tables A* and *A*1, and involves many calculations. To avoid having to do this by hand, specialized databases and computer programs are helpful, such as the ones found on the *Bilbao Crystallographic Server* [461–464] (cf. Table 24.1), referred to as the Server from here on. The Server can be accessed free of charge by typing its web address[1] in a browser, with no registration required. Similar programs to those described in Sections 24.1 and 24.2 are available at the *Symmetry Database*,[2] operated by the *International Union of Crystallography* and accessible to those who have access to the electronic version of *International Tables*.

Operating since 1998, the Server was founded at the Physics Department of the University of the Basque Country UPV/EHU by JUAN MANUEL PÉREZ-MATO and MOIS I. AROYO in coordination with GOTZON MADARIAGA and HANS WONDRATSCHEK. The databases and programs available on the Server are the result of many people's work.[3] The data and programs are continuously being updated, corrected, and developed. The description in this chapter refers to the state as of June 2023.

The Server is built on a core of databases and its programs are organized into different sections (hereafter referred to as menus for clarity) with common application fields. There are a number of programs facilitating the study of specific problems related to crystallography, crystal chemistry, and solid-state physics involving crystallographic groups and their subgroups (Table 24.1).

Part of the data that can be looked up by inspection of *International Tables A*, *A*1, and *E* can also be retrieved on the Server (i.e. there is access to the crystallographic data on space, subperiodic, plane, and point groups).

Most of the programs on the Server accept data solely in the *standard settings* (or default settings) of the space groups. These are the conventional settings, that is, those settings that are *completely* tabulated in *International Tables A*. In addition, for space groups with more than one description in *International Tables A*, the following settings are chosen as standard:

- unique axis *b* and cell choice 1 for monoclinic space groups;
- hexagonal axes and obverse centring for rhombohedral space groups;
- origin choice 2 (origin at $\bar{1}$) for the centrosymmetric space groups listed with two origin choices.

[1] https://www.cryst.ehu.es

[2] https://symmdb.iucr.org/

[3] Among those who should be highlighted are ELI KROUMOVA, SVET IVANTCHEV, ASEN K. KIROV, CÉSAR CAPILLAS, DANEL OROBENGOA, EMRE TASCI, GEMMA DE LA FLOR, SAMUEL V. GALLEGO, and LUIS ELCORO.

The use of non-standard and non-conventional settings, in many cases, is more convenient to clearly describe the symmetry relationships between crystal structures. Nevertheless, in such cases, the use of the programs of the Server requires a preliminary transformation of the structural data into the corresponding standard settings.

Table 24.1 List of the programs available on the *Bilbao Crystallographic Server* and described in this chapter. The programs `GENPOS`, `WYCKPOS`, `MAXSUB`, and `SERIES` are found in the menu Space-group symmetry; `WYCKSPLIT` and `SUBGROUPGRAPH` in the menu Group–subgroup relations of space groups; and `SETSTRU`, `TRANSTRU`, `EQUIVSTRU`, `COMPSTRU`, and `STRUCTURE RELATIONS` in the menu Structures Utilities.

Program	General description
`GENPOS`[4]	Provides the coordinate triplets and the symmetry operations of the general position (or generators) of a space group in the standard setting as well as in non-conventional settings (cf. Section 24.1.1)
`WYCKPOS`[4]	Lists the Wyckoff positions with their coordinate triplets and site symmetries for a designated space group in the standard setting as well as in non-conventional settings (cf. Section 24.1.2)
`MAXSUB`[4]	Provides the list of maximal subgroups of a chosen space group (cf. Section 24.2.1)
`SERIES`	Lists the maximal isomorphic subgroups of a space group (cf. Section 24.2.1)
`SUBGROUPGRAPH`	Determines the chains of maximal subgroups and presents them in group–subgroup graphs for any group–subgroup pair of space groups (cf. Section 24.2.2)
`WYCKSPLIT`	Calculates the relations between the Wyckoff positions of a space group and its subgroups (cf. Section 24.2.3)
`SETSTRU`	Transforms the crystal-structural data from one space group setting to another, provided that both of them are described in one of the settings tabulated in *International Tables A* (cf. Section 24.3.1)
`TRANSTRU`	Transforms the crystal-structural data from one arbitrary space-group setting to another, including those of a subgroup (cf. Section 24.3.1)
`EQUIVSTRU`	Calculates all equivalent sets of atomic coordinates of a given crystal structure (cf. Section 24.3.2)
`COMPSTRU`	Compares two crystal structures described in the same space-group type and having the same sequence of occupied Wyckoff positions (cf. Section 24.4.1)
`STRUCTURE RELATIONS`	Evaluates the structural relationship between two group–subgroup related crystal structures with the same chemical composition (cf. Section 24.4.2)

[4]These programs are also available in the menu Subperiodic groups: layer, rod and frieze groups. They give access to the databases of the crystallographic subperiodic groups. The input and output are similar to those of the space groups.

24.1 The symmetry database of the space groups

The menu Space-group symmetry of the Server gives access to the crystallographic space-groups database, and programs to calculate the geometric interpretation of symmetry operations or to identify the space group from a set of generators. Among these programs, those retrieving the general position, generators, and Wyckoff positions of space groups are described in detail below.

24.1.1 General position and generators of the space groups

The program `GENPOS` lists the general position and the generators of a space-group type. The space-group number is required as input; if unknown, the space-group type can be selected from a table with the Hermann–Mauguin symbols. The program offers three different options to retrieve the data:

With the option Standard/Default Setting, a table is shown that lists the general position (or the generators) of the selected space group-type in the standard setting, such as shown in Table 24.2 for the space-group type $P2_1/c$. The listed coordinate triplets x, y, z of the general position correspond to the numbered list of the first entry under 'Positions' in *International Tables A* (cf. Section 6.4.2, page 81). The corresponding symmetry operations are also given as 3×4 matrices (cf. Section 3.2, page 20), in the style of *International Tables A* (Section 6.4.3), and as Seitz symbols $\{\mathsf{V} \mid \mathsf{v}\}$.

By selecting the option ITA Settings,[5] a list of conventional and non-conventional settings of the chosen space-group type is provided by the program, such as tabulated in *International Tables A*.[6] By a click, a default matrix–column pair $(\boldsymbol{P}, \boldsymbol{p})$ is stated of how the standard setting can be transformed to the selected setting (cf. Section 3.6). The general position of the selected setting is listed in the same way as in Table 24.2, together with the corresponding listing for the standard setting of the space-group type.

[5] **Abbreviations** used throughout at the Bilbao Crystallographic Server:

BCS	Bilbao Crystallographic Server
ITA	*International Tables A*
HM	Hermann–Mauguin
WP	Wyckoff position
SS	site symmetry
CIF	Crystallographic information file

[6] Table 1.5.4.4 in the sixth edition (2016), Table 4.3.2.1 in previous editions.

Table 24.2 General position of the space-group type $P2_1/c$ (No. 14) in the standard setting (*unique axis b*, *cell choice* 1), as in the output of the program `GENPOS`.

N	(x,y,z) form	Matrix form	ITA	Seitz[7]
1	x, y, z	$\begin{pmatrix} 1 & 0 & 0 & 0 \\ 0 & 1 & 0 & 0 \\ 0 & 0 & 1 & 0 \end{pmatrix}$	1	$\{1 \mid 0\}$
2	$\bar{x}, y+\frac{1}{2}, \bar{z}+\frac{1}{2}$	$\begin{pmatrix} -1 & 0 & 0 & 0 \\ 0 & 1 & 0 & \frac{1}{2} \\ 0 & 0 & -1 & \frac{1}{2} \end{pmatrix}$	$2(0,\frac{1}{2},0)\ 0,y,\frac{1}{4}$	$\{2_{010} \mid 0\frac{1}{2}\frac{1}{2}\}$
3	$\bar{x}, \bar{y}, \bar{z}$	$\begin{pmatrix} -1 & 0 & 0 & 0 \\ 0 & -1 & 0 & 0 \\ 0 & 0 & -1 & 0 \end{pmatrix}$	$\bar{1}\ 0,0,0$	$\{\bar{1} \mid 0\}$
4	$x, \bar{y}+\frac{1}{2}, z+\frac{1}{2}$	$\begin{pmatrix} 1 & 0 & 0 & 0 \\ 0 & -1 & 0 & \frac{1}{2} \\ 0 & 0 & 1 & \frac{1}{2} \end{pmatrix}$	$c\ x,\frac{1}{4},z$	$\{m_{010} \mid 0\frac{1}{2}\frac{1}{2}\}$

[7] The Seitz symbols $\{\mathsf{V} \mid \mathsf{v}\}$ in the last column are shorthand descriptions of the matrix–column representations $(\boldsymbol{W}, \boldsymbol{w})$ of the symmetry operations of the space groups. They consist of two parts: a matrix (or linear) part V and a translation part v. The matrix part V consists of a modified Hermann–Mauguin symbol that specifies the kind of symmetry operation and the orientation of its corresponding symmetry element with respect to the basis. The translation parts v correspond to the last columns of the matrix–column representations (the zero translation column is denoted by a single zero).

With the option User-Defined Setting the program produces the data in any other non-conventional setting. To this end, the matrix–column pair $(\boldsymbol{P},\boldsymbol{p})$ for the transformation from the standard to the new setting must be specified[8] (cf. Sections 3.6 and 10.3). In this case, the program only provides the standard Hermann–Mauguin symbol of the chosen space-group type.

[8] The matrix components of $\boldsymbol{P}$ and $\boldsymbol{p}$ may be given as integer or decimal numbers or as fractional numbers in the style 3/4.

24.1.2 Wyckoff positions of the space groups

The program WYCKPOS provides a list of Wyckoff positions for a designated space-group type in the same way as in the *International Tables A* (cf. Table 24.3 and Section 6.4.2, page 81). It lists the Wyckoff positions in the standard setting (option Standard/Default Setting) as well as in different settings. With the option ITA Settings a list of the conventional and non-conventional settings of the chosen space group-type[9] is shown. An advantage of WYCKPOS is that the Wyckoff symbols and the symmetry-equivalent coordinates of the general position and of all special positions can be retrieved for all of the listed non-conventional space-group settings. Neither the printed nor the online editions of *International Tables A* offer this.

[9] As listed in *International Tables A*, Table 1.5.4.4 in the sixth edition (2016), Table 4.3.2.1 in previous editions.

With the option User-Defined Setting any other setting can be chosen by specifying the matrix–column pair $(\boldsymbol{P},\boldsymbol{p})$ for the transformation. If a transformation involves an increase or decrease of the unit cell, neither the multiplicities nor the additional translation coordinates of the transformed Wyckoff positions are calculated by the program. Only the standard Hermann–Mauguin symbol of the space group type is provided by the program.

An explicit listing of the symmetry operations of the site-symmetry group of a point (see Definition 6.3) can be obtained by directly clicking on its coordinate triplet in the table. Optionally, the symmetry operations of the site-symmetry group of an arbitrary point can be calculated using an auxiliary tool of WYCKPOS. Table 24.4 shows the symmetry operations that constitute the site-symmetry group $m..$ of the point with coordinates $\frac{1}{2}, 0.25, 0.25$ referred to the coordinate system of the space group $Cmm2$.

Table 24.4 List of the symmetry operations of the site-symmetry group $m..$ of a point with coordinates $\frac{1}{2}, 0.25, 0.25$ for the space group $Cmm2$, calculated by the auxiliary tool in WYCKPOS. The point $\frac{1}{2}, 0.25, 0.25$ belongs to the Wyckoff position $4e$.

Coordinate triplet	Matrix form	ITA
x, y, z	$\begin{pmatrix} 1 & 0 & 0 & 0 \\ 0 & 1 & 0 & 0 \\ 0 & 0 & 1 & 0 \end{pmatrix}$	1
$-x+1, y, z$	$\begin{pmatrix} \bar{1} & 0 & 0 & 1 \\ 0 & 1 & 0 & 0 \\ 0 & 0 & 1 & 0 \end{pmatrix}$	$m\ 1/2, y, z$

Table 24.3 Wyckoff positions of the space-group type $Cmm2$ (No. 35) with respect to the standard setting and its non-conventional setting $A2mm$, as displayed by the program WYCKPOS.

Standard/default setting $Cmm2$				ITA setting $A2mm$
Multiplicity	Wyckoff letter	Site symmetry	Coordinates $(0,0,0)+\ (\frac{1}{2},\frac{1}{2},0)+$	Coordinates $(0,0,0)+\ (0,\frac{1}{2},\frac{1}{2})+$
8	f	1	$(x,y,z)\ (-x,-y,z)\ (x,-y,z)\ (-x,y,z)$	$(z,x,y)\ (z,-x,-y)\ (z,x,-y)\ (z,-x,y)$
4	e	$m..$	$(0,y,z)\ (0,-y,z)$	$(z,0,y)\ (z,0,-y)$
4	d	$.m.$	$(x,0,z)\ (-x,0,z)$	$(z,x,0)\ (z,-x,0)$
4	c	$..2$	$(\frac{1}{4},\frac{1}{4},z)\ (\frac{1}{4},\frac{3}{4},z)$	$(z,\frac{1}{4},\frac{1}{4})\ (z,\frac{1}{4},\frac{3}{4})$
2	b	$mm2$	$(0,\frac{1}{2},z)$	$(z,0,\frac{1}{2})$
2	a	$mm2$	$(0,0,z)$	$(z,0,0)$

24.2 Subgroups of the space groups

The programs on the Server to study the group–subgroup relations between space groups can be found in the menus Space-group symmetry and Group–subgroup relations of space groups. The latter contains more programs than those described in this section.[10]

[10] There are programs in the menu Group–subgroup relations of space groups that retrieve the minimal supergroups of space groups (MINSUP) and that calculate the coset decomposition for a group–subgroup pair of space groups (COSETS).

24.2.1 Maximal subgroups of the space groups

In the menu Space-group symmetry the program MAXSUB facilitates the retrieval of the maximal subgroups of a space group. That includes the maximal isomorphic subgroups up to an index of 9.

The style of the output of the program MAXSUB is different from how this information is provided by *International Tables A*1. MAXSUB first provides a list of the maximal subgroup types of the selected space group, not the individual subgroups themselves (Table 24.5). Each subgroup type is specified by its space-group number, Hermann–Mauguin symbol, and index of the subgroup in the group. The complete list of individual subgroups and their distribution in classes of conjugate subgroups is obtained by clicking on 'show..'. For example, the space group *Pbam* has four *klassengleiche* maximal subgroups of the type *Pnma* of index 2 distributed in four conjugacy classes as shown on

Table 24.5 List of the maximal subgroups of the space group *Pbam* (No. 55) up to index 7 as displayed by the program MAXSUB. Subgroups marked with t are *translationengleiche* and those marked with k are *klassengleiche* or *isomorphic.* A click on 'show..' reveals that there are four different subgroups of the type *Pnma* of index 2 distributed in four conjugacy classes, as shown on the right side.

[11] Here the word 'Type' does not mean space-group type.

N	IT number	HM symbol	Index	Type[11]	Transformations
1	10	$P2/m$	2	t	show..
2	14	$P2_1/c$	2	t	show..
3	18	$P2_12_12$	2	t	show..
4	26	$Pmc2_1$	2	t	show..
5	32	$Pba2$	2	t	show..
6	55	$Pbam$	2	k	show..
7	55	$Pbam$	3	k	show..
8	55	$Pbam$	5	k	show..
9	55	$Pbam$	7	k	show..
10	58	$Pnnm$	2	k	show..
11	62	$Pnma$	2	k	show.. →

Subgroups	Transformation matrix
Conjugacy class a	
group No 1	$\begin{pmatrix} 0 & 0 & 1 & 0 \\ 1 & 0 & 0 & 0 \\ 0 & 2 & 0 & 1/2 \end{pmatrix}$
Conjugacy class b	
group No 2	$\begin{pmatrix} 0 & 0 & 1 & 0 \\ 1 & 0 & 0 & 0 \\ 0 & 2 & 0 & 0 \end{pmatrix}$
Conjugacy class c	
group No 3	$\begin{pmatrix} 1 & 0 & 0 & 0 \\ 0 & 0 & 1 & 0 \\ 0 & -2 & 0 & 1/2 \end{pmatrix}$
Conjugacy class d	
group No 4	$\begin{pmatrix} 1 & 0 & 0 & 0 \\ 0 & 0 & 1 & 0 \\ 0 & -2 & 0 & 0 \end{pmatrix}$

Why are there four conjugacy classes of *Pnma*?
Answer an page 341.

the right side of Table 24.5. The matrix–column pairs $(\boldsymbol{P},\boldsymbol{p})$ for the transformation from the standard basis of the space group $\mathcal{G}$ to the standard bases of each of the maximal subgroups of the type $\mathcal{H}$ are provided by the program.

To be sure of obtaining all subgroups, it is necessary to click on 'show..' for *every* subgroup type. For example, in the output list shown first for the space group $Pbam$ (Table 24.5), each *translationengleiche* subgroup type except $P2_1/c$ and $Pmc2_1$ refers to one maximal subgroup, while $P2_1/c$ and $Pmc2_1$ each refer to two differently oriented subgroups in two conjugacy classes. *Pbam* (index 2) and *Pnnm* each refer to two subgroups on a par (Section 13.2).

A second click on the option ChBasis lists the coordinate triplets of the general position of the subgroup $\mathcal{H}$, referred to its standard coordinate system as well as transformed to the coordinate system of the initial group $\mathcal{G}$ (listed under the heading Transformed). The latter allows the identification of the symmetry operations of $\mathcal{G}$ that are being retained in $\mathcal{H}$. The relations between the Wyckoff positions of the space group and those of its maximal subgroups can also be calculated, if the option Show WP Splittings is selected in the input page of the program MAXSUB.

Every space group has an infinity of maximal isomorphic subgroups. Generally, they have an unchanged space group symbol; for the space groups belonging to the 11 pairs of enantiomorphic space-group types, it can also be the symbol of the enantiomorphic space-group type. The index is p, p^2, or p^3, with p being a prime number (not all prime numbers are permitted; for details see Chapter 21). The index is also the factor of enlargement of the unit cell.

In the menu Space-group symmetry the program SERIES allows the retrieval of the series of maximal isomorphic subgroups of space groups. The program provides the parametric descriptions of the series and the individual listings of all maximal isomorphic subgroups of indices up to 27 (125 for cubic space groups). By a click on 'show..' the set of transformation matrices for the conjugate maximal isomorphic subgroups for a selected index are also listed. The output format is similar to that of MAXSUB. As an example, Table 24.6 shows the output of maximal isomorphic subgroups for the space group $P4_122$, which is subdivided into three series.

Table 24.6 Output of the program SERIES for the space group $P4_122$ (No. 91), showing three series of maximal isomorphic subgroups.

	Subgroup	Index	Transformation	Conditions
Series 1	$P4_322$ (95)	p	$\begin{pmatrix} 1 & 0 & 0 & 0 \\ 0 & 1 & 0 & 0 \\ 0 & 0 & p & u \end{pmatrix}$	prime $p > 2$ $p = 4n - 1$ $0 \le u < p$
Series 2	$P4_122$ (91)	p	$\begin{pmatrix} 1 & 0 & 0 & 0 \\ 0 & 1 & 0 & 0 \\ 0 & 0 & p & u \end{pmatrix}$	prime $p > 4$ $p = 4n + 1$ $0 \le u < p$
Series 3	$P4_122$ (91)	p^2	$\begin{pmatrix} p & 0 & 0 & u \\ 0 & p & 0 & v \\ 0 & 0 & 1 & 0 \end{pmatrix}$	prime $p > 2$ $0 \le u < p$ $0 \le v < p$

24.2.2 Group–subgroup chains of maximal subgroups

If a space group $\mathcal{H}$ is a non-maximal subgroup of another space group $\mathcal{G}$, $\mathcal{G} > \mathcal{H}$, their relationship can be represented by a chain of intermediate subgroups $\mathcal{Z}_k$: $\mathcal{G} > \mathcal{Z}_1 > ... > \mathcal{Z}_n = \mathcal{H}$. For a specified index of $\mathcal{H}$ in $\mathcal{G}$ there are, in general, a number of subgroups $\mathcal{H}_j < \mathcal{G}$ isomorphic to $\mathcal{H}$. The menu Group–subgroup relations of space groups gives access to the program `SUBGROUPGRAPH` [466]. It serves to assemble all possible group–subgroup chains of intermediate maximal subgroups $\mathcal{Z}_k$ between the space-group types $\mathcal{G}$ and $\mathcal{H}$ and to present them in group–subgroup graphs.

The input of `SUBGROUPGRAPH` consists of the space-group numbers of $\mathcal{G}$ and $\mathcal{H}$, the index i is optional. There is only the possibility to use the standard settings of space groups, which may be inconvenient in some cases.

Group–subgroup pairs with unspecified indices

If the index is not specified, the program `SUBGROUPGRAPH` lists the possible intermediate space groups $\mathcal{Z}_k$ that relate $\mathcal{G}$ and $\mathcal{H}$. The list is given in the form of a table whose rows correspond to the intermediate space groups $\mathcal{Z}_k$, specified by their Hermann–Mauguin symbols (Table 24.7). The subgroups of each intermediate space group are listed by their space-group numbers and the indices in brackets. Note that $C2$ in Table 24.7 can also be an intermediate group when its isomorphic subgroups and its subgroup $P2$ have to be considered.

By a click on the button Draw the graph the list is presented as a general contracted graph (Fig. 24.1). As in all *contracted* graphs, every node indicates a space-group type and not a space group; all space groups belonging to the same space-group type are represented by just one node in the graph. Every arrow represents one maximal group–subgroup relation having the index given in brackets. A pair of arrows in opposite directions marks relations in both directions. Maximal isomorphic subgroups are represented by loops and can have an infinity of different index values, of which only those up to 9 are given. Therefore, the maximal isomorphic subgroups of cubic space groups, whose lowest index is 27, are not included in the results of `SUBGROUPGRAPH`.

Table 24.7 Output of the program `SUBGROUPGRAPH` for the relations between the space group type $I\bar{4}2d$ (No. 122) and its non-maximal subgroups $C2$ (No. 5) for any arbitrary index.

N	Group	HM symbol	Maximal subgroups
1	**122**	$I\bar{4}2d$	024[2]; 043[2]; 082[2]; 122[3]; 122[5]; 122[7]; 122[9]
2	082	$I\bar{4}$	005[2]; 081[2]; 082[3]; 082[5]; 082[7]; 082[9]
3	081	$P\bar{4}$	003[2]; 081[2]; 081[3]; 081[5]; 081[7]; 081[9]; 082[2]
4	043	$Fdd2$	005[2]; 043[3]; 043[5]; 043[7]
5	024	$I2_12_12_1$	005[2]; 017[2]; 024[3]; 024[5]; 024[7]
6	020	$C222_1$	005[2]; 017[2]; 018[2]; 020[3]; 020[5]; 020[7]
7	018	$P2_12_12$	003[2]; 018[2]; 018[3]; 018[5]; 018[7]
8	017	$P222_1$	003[2]; 017[2]; 017[3]; 017[5]; 017[7]; 018[2]; 020[2]
9	003	$P2$	003[2]; 003[3]; 003[5]; 003[7]; 005[2]
10	**005**	$C2$	003[2]; 005[2]; 005[3]; 005[5]; 005[7]

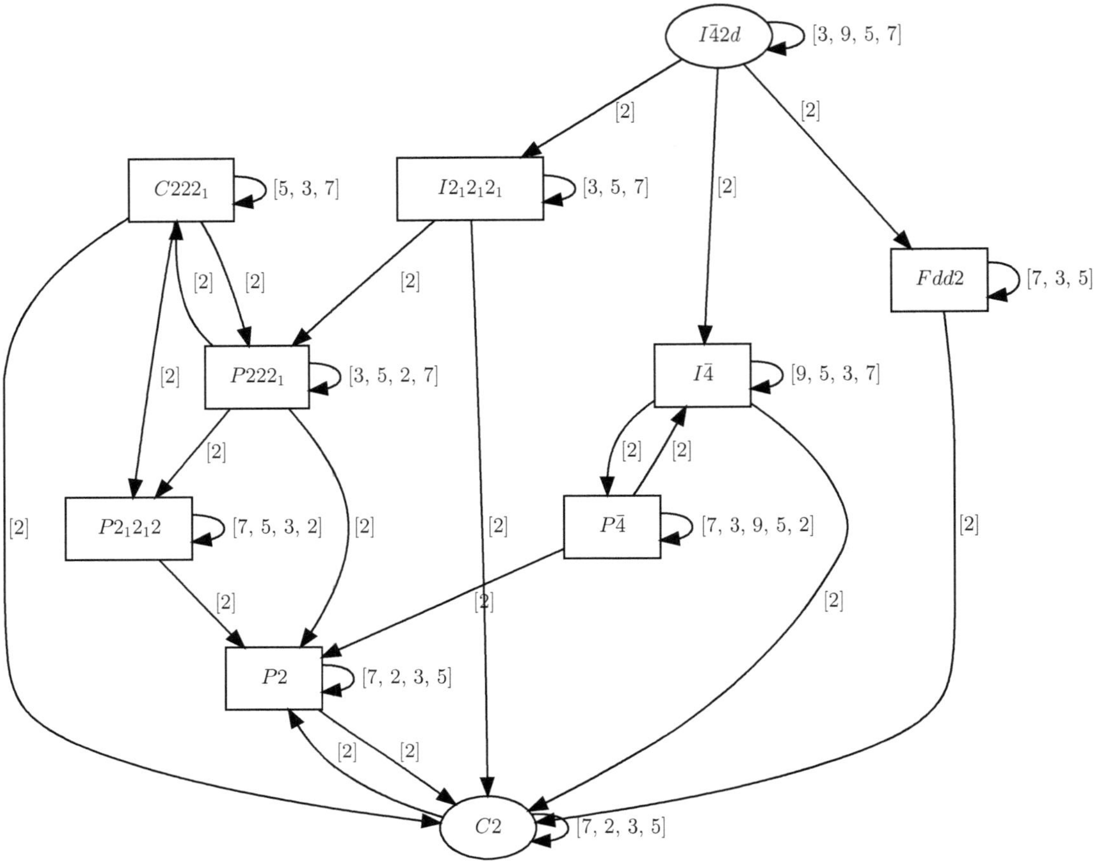

Fig. 24.1 General contracted graph for $I\bar{4}2d$ (No. 122) > $C2$ (No. 5) as given by the program SUBGROUPGRAPH. Each node corresponds to a space-group type that can appear as an intermediate group in the chain of the group–subgroup pair $I\bar{4}2d > C2$. Each arrow of the graph corresponds to a maximal group–subgroup pair of the indicated index [i]. Isomorphic subgroups (with indices up to 9) are shown as loops.

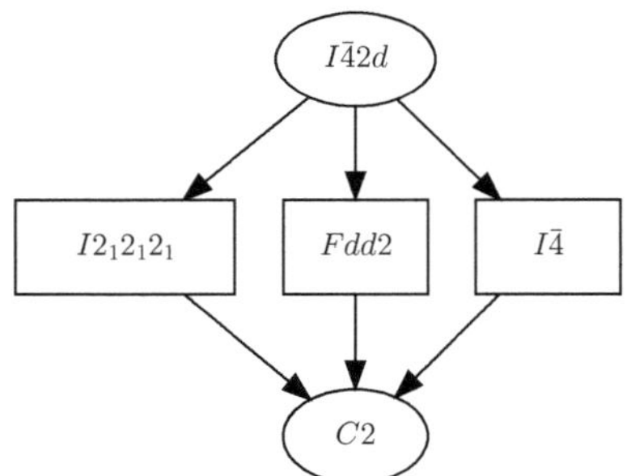

Fig. 24.2 Contracted graph for the pair $I\bar{4}2d$ (No. 122) > $C2$ (No. 5), index 4, as given by the program SUBGROUPGRAPH. The nodes of the graph correspond to space-group types. Each arrow of the graph corresponds to a maximal group–subgroup pair.

Group–subgroup pairs with specified indices

When the index of the subgroup $\mathcal{H}$ in the group $\mathcal{G}$ is specified, the program SUBGROUPGRAPH selects all possible chains of maximal subgroups relating $\mathcal{G}$ and $\mathcal{H}$ with this index. The number of different transformation matrices along with a link to a list of these matrices is given for each of the possible chains. In this case, the results are graphically represented by a contracted graph (Fig. 24.2), if the option Show contracted graph is selected.

SUBGROUPGRAPH can also calculate all different subgroups $\mathcal{H}_j$ of the same type as $\mathcal{H}$ and distribute them among classes of conjugate subgroups of $\mathcal{G}$ (option Classify (with a complete graph of all subgroups)). The subgroups of a conjugacy class are combined in a block where each subgroup is specified by the corresponding matrix–column pair $(\boldsymbol{P},\boldsymbol{p})_j$ for the transformation from the standard basis of $\mathcal{G}$ to that of $\mathcal{H}_j$ (Table 24.8).[12] The list of possible chains of maximal subgroups and the corresponding transformation matrices that results in the same identical subgroup can be obtained under a separate link (column 'Identical' in Table 24.8).

[12] The transformations involving the intermediate groups are not provided by the program.

Table 24.8 Group–subgroup relations for $I\bar{4}2d$ (No. 122) > $C2$ (No. 5), index 4. $I\bar{4}2d$ has three different subgroups of the type $C2$; here (not in the output of the program) they are distinguished by the orientational symbols [100], [010], and [001], stating the directions of the monoclinic axes with reference to the basis vectors of $I\bar{4}2d$. They are distributed into two classes of conjugate subgroups. The possible chains of maximal subgroups and the corresponding matrices $(\boldsymbol{P},\boldsymbol{p})_j$ as determined by SUBGROUPGRAPH are also shown. Upon a click on the link in the column 'Identical' another two different transformation matrices for the subgroup $C2_{[001]}$ emerge for one and the same subgroup $C2_{[001]}$; the program finds three different transformations for the different group–subgroup chains that relate the *very same* subgroup to $I\bar{4}2d$, cf. Fig. 24.3 and text.

	$\mathcal{H}_j$	Chains	$(\boldsymbol{P},\boldsymbol{p})_j$	Identical
Class 1	$C2_{[010]}$	$I\bar{4}2d > I2_12_12_1 > C2$	$\begin{pmatrix} -1 & 0 & 1 & \frac{1}{4} \\ 0 & 1 & 0 & \frac{1}{4} \\ -1 & 0 & 0 & \frac{3}{8} \end{pmatrix}$	-
	$C2_{[100]}$	$I\bar{4}2d > I2_12_12_1 > C2$	$\begin{pmatrix} 0 & 1 & 0 & 0 \\ -1 & 0 & 0 & \frac{1}{4} \\ -1 & 0 & 1 & \frac{5}{8} \end{pmatrix}$	-
Class 2	$C2_{[001]}$	$I\bar{4}2d > I2_12_12_1 > C2$	$\begin{pmatrix} -1 & 0 & 0 & 0 \\ -1 & 0 & 1 & \frac{1}{2} \\ 0 & 1 & 0 & \frac{3}{8} \end{pmatrix}$	to group

Chains	$(\boldsymbol{P},\boldsymbol{p})_j$
$I\bar{4}2d > I\bar{4} > C2$	$\begin{pmatrix} -1 & 0 & 0 & 0 \\ -1 & 0 & 1 & 0 \\ 0 & 1 & 0 & 0 \end{pmatrix}$
$I\bar{4}2d > Fdd2 > C2$	$\begin{pmatrix} 1 & 0 & 0 & 0 \\ 1 & 0 & -1 & \frac{1}{2} \\ 0 & 1 & 0 & 0 \end{pmatrix}$

The conjugacy class 2 in Table 24.8 consists of only *one* subgroup $C2_{[001]}$. This subgroup can be reached via the three different intermediate groups $I\bar{4}$, $I2_12_12_1$, and $Fdd2$, for which three different matrix–column pairs are displayed; they correspond to different orientations of the basis vectors and to different origin shifts for the very same (identical) subgroup.[13]

For each subgroup, the program provides a link to calculate the list of the symmetry operations of the subgroup transformed to the basis of the group $\mathcal{G}$, which allows the identification of those symmetry operations of $\mathcal{G}$ that are retained in the subgroup. By a click at Draw the graph the program exhibits the group–subgroup relations in a complete graph (Fig. 24.3). The graph contains the intermediate space groups $\mathcal{Z}_k$ for the pair $\mathcal{G} > \mathcal{H}$ with index i, but contrary to the graph of the previous step (Fig. 24.2), the different subgroups of the same type are represented by different nodes. Their labels consist of the standard symbol of the subgroup, followed by a number given in parentheses specifying the conjugacy class to which the subgroup $\mathcal{H}_j$ belongs.

[13] The additional different transformation matrices for one and the same (identical) subgroup listed by the program SUBGROUPGRAPH under the column 'Identical' in Table 24.8 may be confusing for the user. It shows that in general the transformation matrices are not unique. Even though the program does not search for it, it is always possible to find a common transformation matrix.

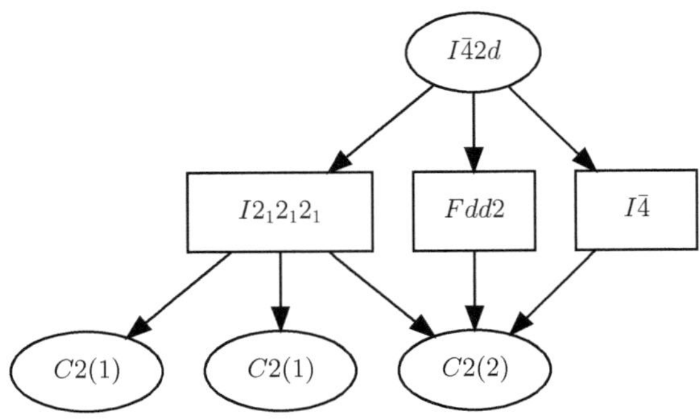

Fig. 24.3 Complete graph for the pair $I\bar{4}2d$ (No. 122) $>$ $C2$ (No. 5), index 4, as drawn by the program SUBGROUPGRAPH. The nodes represent space groups and not space-group types. There are three subgroups of the type $C2$ among two classes of conjugate subgroups, distinguished in parentheses after the space-group symbols. The two subgroups $C2(1)$ belong to the same conjugacy class and have their twofold axes along [100] and [010] of $I\bar{4}2d$. The subgroup $C2(2)$ of the second conjugacy class has twofold axes parallel to the tetragonal **c** axis. The program does not depict the basis transformations and the origin shifts in the graph.

For a group–subgroup pair with a high index the resulting complete graph with all subgroups $\mathcal{H}_j$ can be quite complicated and difficult to overview. In such cases, if the transformation matrix $\boldsymbol{P}$ is known, the use of the program SUBGROUPS can be of help.[14] For smaller indices (≤ 6), the program SUBGROUPGRAPH is helpful; for example, the tree shown in Fig. 8.9 can be reproduced with this program.

Alternatively, a more simple graph associated with a single specific subgroup $\mathcal{H}_j$ can be obtained by selecting Classify (with complete graphs for individual subgroups). The same graph corresponds to all subgroups in the same conjugacy class, that is, conjugate subgroups are not distinguished. Usually that is no disadvantage, and such graphs can be helpful to design Bärnighausen trees.

[14] The program SUBGROUPS is a more advanced and complex program that requires familiarity with the terminology in physics of phase transitions. It is available at the menu Group–subgroup relations of space groups; information and examples for its use can be found at https://www.cryst.ehu.es/html/cryst/tutorials/Tutorial_SUBGROUPS.pdf.

24.2.3 Relations of Wyckoff positions for a group–subgroup pair of space groups

In the menu Group–subgroup relations of space groups the program WYCKSPLIT [465] allows us to find out how the Wyckoff positions of the space group $\mathcal{G}$ are converted to Wyckoff positions of the subgroup $\mathcal{H}$ (called 'splitting of the Wyckoff positions', although not all positions are split; cf. Section 10.2, page 130). The program requires as input the space-group numbers of the group $\mathcal{G}$ and of its subgroup $\mathcal{H}$ and a matrix–column pair $(\boldsymbol{P},\boldsymbol{p})$ for the transformation that relates the standard basis of $\mathcal{G}$ to that of $\mathcal{H}$. In the output, after selecting one or more Wyckoff positions of $\mathcal{G}$, a click on 'Splitting' shows which Wyckoff positions of $\mathcal{H}$ result thereof. Optionally, the relations between the coordinate triplets of the Wyckoff positions of the group $\mathcal{G}$ and those of the subgroup $\mathcal{H}$ can be obtained by clicking on 'Relations'.

Example 24.1

The relation between the basis $(\mathbf{a}, \mathbf{b}, \mathbf{c})$ of $P31m$ (No. 157) and the basis $(\mathbf{a}', \mathbf{b}', \mathbf{c}')$ of its subgroup Cm (No. 8) can be retrieved by the program MAXSUB: $\mathbf{a}' = \mathbf{a}$, $\mathbf{b}' = \mathbf{a} + 2\mathbf{b}$, $\mathbf{c}' = \mathbf{c}$. The output for this group–subgroup pair, determined by WYCKSPLIT, is:

N	Group ($P31m$)	Subgroup (Cm)	More ...
1	$6d$	$4b\ 4b\ 4b$	Relations
2	$3c$	$4b\ 2a$	Relations
3	$2b$	$4b$	Relations
4	$1a$	$2a$	Relations

The general position $6d$ of $P31m$ splits into three orbits of the general position $4b$ of Cm. The Wyckoff position $3c$ splits into two independent positions $4b$ and $2a$. The special positions $1a$ and $2b$ do not split. Since the cell volume is increased by a factor of 2, the sums of the multiplicities of the positions of the subgroup Cm are twice of those of $P31m$ (cf. Section 10.2).

Note that subgroups of the same space group type that belong to different conjugacy classes show different splittings of their Wyckoff positions.

24.3 Crystal-structural data transformations

The Server contains programs that can perform calculations with crystal structure data. That includes transformations of atomic coordinates when the basis is changed and when the origin is shifted. As input, these programs accept both, the standard crystallographic file in CIF format [467], as well as the Server's default structure definition format called *BCS format*:

A comment line begins with a #	# Example
	# Valentinite Sb_2O_3, *Pccn*
Space-group number	56
$a\ b\ c\ \alpha\ \beta\ \gamma$	4.900 12.449 5.410 90 90 90
Number of independent atoms	3
Element Sequential Wyckoff $x\ y\ z$	Sb 1 8*e* 0.0406 0.1275 0.1778
atomic symbol	O 1 4*c* 0.2500 0.2500 0.0209
number	O 2 8*e* 0.1509 0.0587 −0.1454

The data in BCS format have to be entered manually in the input form, overwriting the values of the template. Unit-cell lengths and angles should be given in Ångström units and degrees, respectively.[15] If the Wyckoff symbol is unknown, use a hyphen '-' instead. The coordinates must be given as decimal numbers; fractional numbers are not allowed.

In general, the programs on the Server do not check the compatibility between the introduced cell parameters and the space group. No internal standardization of the unit cell and atomic coordinates of the input/output structural data (cf. Section 9.1, page 115) is performed by the programs of the Server.

[15] In crystallography lengths are usually specified in Ångström units, although this is not a unit of the international SI units system. Ångström is the only unit of length accepted by *Acta Crystallographica*. In Chemistry the use of picometres is widespread, and it is the unit we use in this book; physicists often use nanometres.
1 Å = 100 pm = 0.1 nm.
In the *Inorganic Crystal Structure Database* [3], the *Cambridge Structural Database* [6], and the *Open-Access Collection of Crystal Structures* [5] the lengths of the basis vectors are given in Å; in *Pearson's Crystal Data* [4] they are given in nm.

24.3.1 Coordinate transformations

The menu Structure Utilities gives access to two basic programs called `SETSTRU` and `TRANSTRU` for coordinate transformations (cf. Section 3.6). The program `SETSTRU` performs coordinate transformations between the so-called ITA settings, that is, to those conventional and non-conventional space-group settings listed in Table 1.5.4.4 of *International Tables A* 2016 [118],[16] including changes between origin choice 1 and 2. For monoclinic space group types, there are infinite ways to set up the unit cell (cf. Example 10.2, page 130, and the preceding paragraph). `SETSTRU` offers only transformations tabulated in *International Tables A*. In these cases, depending on the lattice parameters and the transformation, the transformed lattice parameters can result in rather oblique unit cells.

[16] Table 4.3.2.1 in previous editions [119].

Once the structural data have been introduced, a second input form appears with a list of settings of the corresponding space group. Here the initial setting, in which the input structure is described, and the final setting, into which the lattice parameters and the coordinates are to be transformed, have to be selected. As a result, the program provides the transformed structural data in BCS format, with an option to export them in CIF format. The Wyckoff letters of the transformed coordinates are assigned only if the final setting corresponds to the standard one.

Example 24.2

Potassium bromate, $KBrO_3$, crystallizes in the space group $R3m$ (No. 160). Its crystal structure was published using a rhombohedral cell [468], and also a hexagonal cell [469]. Do both refer to the same crystal structure? The crystal data in the BCS format are (ICSD = collection code in the *Inorganic Crystal Structure Database* [3]):

# (a) $KBrO_3$, $R3m$ rhombohedral setting						# (b) $KBrO_3$, $R3m$ hexagonal setting					
# ICSD: 74767						# ICSD: 47173					
160						160					
4.410 4.410 4.410 85.96 85.96 85.96						6.011 6.011 8.152 90 90 120					
3						3					
K	1	1*a*	0.0	0.0	0.0	K	1	3*a*	0	0	0.0
Br	1	1*a*	0.4828	0.4828	0.4828	Br	1	3*a*	0	0	0.4827
O	1	3*b*	0.5455	0.5455	0.1110	O	1	9*b*	0.1446	−0.1446	0.4002

To compare the two sets of coordinates, they have to be referred to the same setting. With the program `SETSTRU` the rhombohedral setting is transformed to the standard (hexagonal) setting:

# (c) $KBrO_3$, coordinate set (a) as transformed by `SETSTRU` to hexagonal setting					
160					
6.013 6.013 8.159 90 90 120					# excellent cell data match with data set (b)
3					
K	1	3*a*	0	0	0.0
Br	1	3*a*	0	0	0.4828
O	1	9*b*	0.1448	0.2897	0.4007

The output of the program includes a list of the coordinates of symmetry-equivalent atoms of the data set (c) within one unit cell:

	Wyckoff position	Site symmetry	x	y	z		Wyckoff position	Site symmetry	x	y	z
			0	0	0.0000				0.1448	0.2897	0.4007
K	3a	3	$\frac{2}{3}$	$\frac{1}{3}$	0.3333				0.7103	0.8552	0.4007
			$\frac{1}{3}$	$\frac{2}{3}$	0.6667				0.1448	0.8552	0.4007
			0	0	0.4828	O	9b	1	0.8115	0.6230	0.7340
Br	3a	3	$\frac{2}{3}$	$\frac{1}{3}$	0.8161				0.3770	0.1885	0.7340
			$\frac{1}{3}$	$\frac{2}{3}$	0.1495				0.8115	0.1885	0.7340
									0.4782	0.9563	0.0673
									0.0437	0.5218	0.0673
									0.4782	0.5218	0.0673

The origin of the space group $R3m$ is not fixed by symmetry and must be fixed by arbitrarily assigning a z value to one atom. In this particular case, in both structure determinations $z = 0$ was chosen for the potassium atom. There is an excellent match for every atom species of the data set (c) with the coordinates of the data set (b) (marked by white background). Therefore, both data sets refer to the very same crystal structure.

In the case of origin-shift transformations that may be necessary in the case of space groups with non-fixed origins, and for basis changes not available at SETSTRU, the program to use is TRANSTRU instead of SETSTRU. The program TRANSTRU can transform structural data to any space-group setting and allows for origin shifts. In addition to the structural data, the matrix–column pair $(\boldsymbol{P},\boldsymbol{p})$ for the transformation relating the initial and final setting is required.

TRANSTRU can also be used to transform the unit cell and the coordinates of a structure to those of a subgroup if the matrix–column pair $(\boldsymbol{P},\boldsymbol{p})$ for the basis transformation and the space-group number of the subgroup are provided. The program checks the validity of the transformation matrix and accounts for all possible splittings of the Wyckoff positions. The calculated coordinates refer to the subgroup basis, although they still fulfill the symmetry of the initial space group.

24.3.2 Equivalent crystal structure descriptions

As explained in Section 9.2, there are almost always several equivalent ways to describe the very same crystal structure, even if the setting of the space group remains unchanged. The equivalent sets of coordinates can be calculated with the help of the Euclidean normalizer of the space group, as listed in *International Tables A*[17] (cf. Table 9.1, page 117). The Euclidean normalizers are also retrievable with the program NORMALIZER,[18] available in the menu Space-group symmetry of the Server. Additionally, the program EQUIVSTRU in the menu Structure Utilities systematically calculates the sets of equivalent coordinates of a crystal structure, internally using the program NORMALIZER.

[17] Tables 3.5.2.3–3.5.2.5 in the sixth edition (2016); Tables 15.2.1.3 and 15.2.1.4 in the editions of 2002–2006; Table 15.3.2 in the editions of 1987–1998.

[18] The program NORMALIZER lists the affine, Euclidean, and chirality-preserving Euclidean normalizers of a given space group. Optionally, the program also calculates the enhanced Euclidean normalizers for any space group with specialized metric.

The program EQUIVSTRU requires the structural data referred to the standard setting of the space group as input. If that is not the case, the data first need to be transformed to the standard setting, which can be done with the programs SETSTRU or TRANSTRU. The program EQUIVSTRU returns all the equivalent sets of coordinates together with the corresponding transformations as output. For each set of coordinates, a downloadable CIF file and a link to visualize the structure with the program VISUALIZED are provided.[19] If the space group is chiral, the coordinate sets of the enantiomorphic structures are also calculated by the program.

[19] VISUALIZED is based on *JSmol* [470].

For example, the equivalent sets of coordinates for $WOBr_4$ were calculated in Example 9.2, page 118, and those of right quartz in Example 9.4, page 120. The same results can be reproduced using the program EQUIVSTRU. Since the space group $I4$ of $WOBr_4$ is a Sohncke space group, there is a warning in the output of the program for those sets of coordinates that refer to the inverted structure. In the case of right quartz the space group $P3_2 21$ itself is enantiomorphic (chiral), so that the calculation with the Euclidean normalizer yields only equivalent sets of coordinates with this handedness. The program further calculates all the equivalent coordinates of the inverted (enantiomorphic) structure i.e. those of left quartz in the space group $P3_1 21$.

Example 24.3

The similarity between the crystal structures of $KAsF_6$ [471] and $CsNbF_6$ [472] with space group $R\bar{3}$ (No. 148) is analysed with the help of the program EQUIVSTRU. The published structural data for $KAsF_6$ and $CsNbF_6$ are:

# $KAsF_6$, $R\bar{3}$, ICSD 59413					
148					
7.348 7.348 7.274 90 90 120					
# $c/a = 0.990$					
3					
K	1	3*b*	0.3333	0.6667	0.1667
As	1	3*a*	0	0	0
F	1	18*f*	0.1292	0.2165	0.1381

# $CsNbF_6$, $R\bar{3}$, ICSD 183851					
148					
7.866 7.866 8.162 90 90 120					
# $c/a = 1.038$					
3					
Cs	1	3*a*	0	0	0
Nb	1	3*b*	0	0	0.5
F	1	18*f*	0.1761	−0.2749	0.0319

We take $KAsF_6$ as the reference structure. After input of the data for $CsNbF_6$, the coordinates of all equivalent sets of coordinates of $CsNbF_6$ are calculated by EQUIVSTRU. There are four equivalent sets of coordinates calculated by *the additional generators of* $\mathcal{N}_{\mathcal{E}}(\mathcal{G})$ (cf. Section 9.2):

$$(1)\ x, y, z; \quad (2)\ x, y, z+\tfrac{1}{2}; \quad (3)\ y, x, \bar{z}; \quad (4)\ y, x, \bar{z}+\tfrac{1}{2}.$$

For each of them, a click on a button 'show' exhibits a list of all of its symmetry-equivalent coordinates. This list for the 'equivalent description 4' reads (abbreviated to three decimals):

	Wyckoff position	Site symmetry	Representative atom	Atomic orbit within one unit cell x y z	x y z
Cs	$3b$	$\bar{3}$	0.000 0.000 0.500	0.000 0.000 0.500	
				0.667 0.333 0.833	
				0.333 0.667 0.167	
Nb	$3a$	$\bar{3}$	0.000 0.000 0.000	0.000 0.000 0.000	
				0.667 0.333 0.333	
				0.333 0.667 0.667	
F	$18f$	1	0.725 0.176 0.468	0.725 0.176 0.468	0.942 0.157 0.865
				0.824 0.549 0.468	0.843 0.784 0.865
				0.451 0.275 0.468	0.216 0.058 0.865
				0.275 0.824 0.532	0.058 0.843 0.135
				0.176 0.451 0.532	0.157 0.216 0.135
				0.549 0.725 0.532	0.784 0.942 0.135
				0.392 0.509 0.801	0.608 0.491 0.199
				0.491 0.882 0.801	0.509 0.118 0.199
				0.118 0.608 0.801	0.882 0.392 0.199

The manual check shows that for every atom species there is one symmetry-equivalent position that closely matches the published coordinates of $KAsF_6$ (marked by white background). Apart from the different values of the lattice parameters (due to the different sizes of the K^+ and Cs^+ ions), the main differences between the two structures concern small deviations of the coordinates of the F atoms; these are due to the slightly different Si–F and Nb–F bond lengths (1.79 and 1.88 Å, respectively) in conjunction with the different unit cell lengths. The crystal structures of $KAsF_6$ and $CsNbF_6$ can be considered to be isotypic. This structure type is known as the $KOsF_6$ type.

If the metric of the unit cell is specialized, that is, if the lattice has a higher symmetry than that of the actual crystal class, the program EQUIVSTRU calculates the additional sets of coordinates that result from the enhanced symmetry of the lattice, using the enhanced Euclidean normalizer.

In order to check whether two differently documented crystal structures are isotypic or nearly equal, all symmetry-equivalent positions of all equivalent sets of coordinates of all atoms have to be compared. This task has to be carried out manually combining the results of SETSTRU, TRANSTRU, and EQUIVSTRU, or directly using the program COMPSTRU.

24.4 Comparison of crystal structures

24.4.1 Isotypic crystal structures

As pointed out in Section 9.5, crystal structures that are isotypic can be assigned to structure types. In Table 9.3, page 125, rules for such an assignment are summarized. One important rule is that the atoms must occupy the same sequence of Wyckoff positions, provided their structural data have been standardized in the same way. Another condition is that the geometrical and bonding interrelationships of the atoms must be similar. The meaning of the term 'similar' cannot be specified, since it varies from structure to structure. This condition is hence difficult to quantify and implement in algorithms for automated computer searches. However, the use of computer programs for the recognition of crystal structure types can be helpful when in conjunction with crystal-chemical expertise.

In the menu Structure Utilities, the program COMPSTRU can aid in deciding whether two crystal structures can be regarded as isotypic or nearly equal. As input, COMPSTRU requires the structural data of two crystal structures belonging to the same space-group type (or space groups that form an enantiomorphic pair), described in their standard settings, with the same or different chemical compositions, and having the same sequence of occupied Wyckoff positions. Preliminary standardization of the unit cells and atomic coordinates [72] is not required. The thresholds for the maximum allowed difference in the positions of the matching atoms, and the difference between the lattice parameters, have to be specified in the input of the program. By default, the program assigns some values (0.5 Å for a, b, and c; 5° for α, β, and γ; and 1 Å for the maximum difference in the position of matching atoms).

In the case of structures with different compositions, the program suggests what kind of atoms should be matched between the two structure descriptions. This can be manually modified by the user.

COMPSTRU keeps the data of the first crystal structure as the *reference description*. The program first checks if the lattice parameters of the second structure agree within the given threshold values; if not, using the affine normalizer, it tries to find a suitable transformation (i.e. an affine distortion). Then, using the Euclidean normalizer, it searches among the equivalent sets of atomic coordinates of the second structure for the one that best matches the reference description.

The program performs a calculated (conceptual) superposition of the two crystal structures and determines the distance of every atom of the reference description from its matching (superposed) atom in the description of the second structure. For each pair of superposed atoms their mutual distance d is calculated. The largest occurring of these values is d_{max} and their average is d_{av}. Among the sets of atomic coordinates of the second structure, the program chooses the one that results in the least d_{max} and terms it the most similar description.[20] The calculated atomic coordinates of this most similar description are not standardized; therefore, they cannot be used to calculate the deviation parameter Δ as defined in Section 9.5, eqn (9.3), page 124. The values d_{max} and d_{av} are calculated using the lattice parameters of the reference description. Therefore, depending on which of the two structure descriptions being compared is taken as the reference, the values may differ somewhat.

[20] The program COMPSTRU calculates two additional descriptors of the similarity, but these are not dealt with here.

Find more technical details about the procedure of the program COMPSTRU at [473].

The calculated parameters d_{max} and d_{av} can be considered as a preliminary step in the recognition of structure types. This purely geometric analysis carried out by COMPSTRU must be complemented by crystal chemical considerations, taking into account the coordination polyhedra and how they are linked with each other. Remember the definition of isotypism: the geometrical and bonding interrelationships must be similar.

Example 24.4

In Section 14.3.2, page 192, the α-$NaFeO_2$ type is presented as a substitution variant of the NaCl type. With the basis transformations mentioned in Fig. 14.12, using TRANSTRU, we calculate ideal cell parameters and ideal coordinates for the O atoms for α-$NaFeO_2$ assuming an undistorted NaCl arrangement with $a_{cub} = 4.17$ Å:

Ideal α-$NaFeO_2$ Space group $R\bar{3}2/m$
$a = 2.95$ Å, $c = 14.45$ Å; $c/a = 2\sqrt{6} = 4.90$
Na at 0, 0, 0; Fe at 0, 0, $\frac{1}{2}$; O at 0, 0, 0.25

Using COMPSTRU, we obtain d_{max} for the following compounds, as compared to the undistorted NaCl structure:

ABX_2	c/a	z(X)	d_{max} /Å	Δ	Bond lengths A–X/Å	Bond lengths B–X/Å	ICSD code
Ideal	4.90	0.25	0	0	6 × 2.09	6 × 2.09	
$LiCrO_2$	4.97	0.2416	0.12	0.09	6 × 2.14	6 × 1.99	97758
$LiCoO_2$	4.99	0.2397	0.15	0.03	6 × 2.09	6 × 1.96	156395
$NaErO_2$	4.90	0.236	0.20	0.01	6 × 2.54	6 × 2.67	97544
$SrCeN_2$	4.96	0.2361	0.20	0.02	6 × 2.72	6 × 2.44	95805
$NaFeO_2$	5.32	0.2334	0.24	0.10	6 × 2.37	6 × 2.05	420380
$NaYbS_2$	5.06	0.2572	0.10	0.04	6 × 2.88	6 × 2.71	73483
$TlNdSe_2$	5.38	0.267	0.25	0.11	6 × 3.38	6 × 2.90	106988
$CsNdTe_2$	5.75	0.2681	0.26	0.19	6 × 3.79	6 × 3.18	98660
Delafossite type							
$CuFeO_2$	5.66	0.1064	1.74	0.27	2 × 1.83	6 × 2.03	92184
$CuInO_2$	5.28	0.1061	1.73	0.19	2 × 1.85	6 × 2.17	91856
$AgCrO_2$	6.20	0.1119	1.75	0.39	2 × 2.07	6 × 2.00	4149
$HgCaO_2$	5.20	0.1037	1.73	0.17	2 × 1.93	6 × 2.38	80717

The bond lengths were calculated with the program available at the ICSD database [3]. The deviation parameters Δ according to eqn (9.3) were manually calculated. The first eight examples mentioned all belong to the α-$NaFeO_2$ type; all atoms have the same relative arrangement and the same kind of coordination. However, with the larger atoms, mutual deviations are more manifest, and the values d_{max} and Δ are less significant. That the α-$NaFeO_2$ type and the delafossite type (described on page 196) are not isotypic is indicated by the larger d_{max} and Δ, but more important is the different coordination of the atoms, especially that the Cu, Ag, and Hg atoms have only the coordination number 2.

An interactive visualization applet is also included in COMPSTRU. It allows visualization the reference structure and the best matching description of the second structure. A click on 'Compare descriptions' allows the simultaneous visualization of both structures and may help to decide whether two crystal structures are isotypic or not.

24.4.2 Group–subgroup related crystal structures

The menu Structure Utilities hosts the program STRUCTURE RELATIONS that is able to find the relation between two crystal structures having the same chemical composition, and whose space groups are group–subgroup related. This program can be applied for the analysis and characterization of crystalline-state phase transitions with no change of the chemical composition.[21] It is able to calculate the index of the symmetry reduction and the transformation matrix–column pair ($\boldsymbol{P}$,$\boldsymbol{p}$) relating the coordinate system of the group to that of the subgroup.

[21] A new version of the program STRUCTURE RELATIONS is under development (June 2023), which will allow the study of group–subgroup-related crystal structure descriptions for compounds with different chemical compositions.

The input of the program consists of the structural data for two group–subgroup-related crystal structures described in their standard settings, with the number of formula units per unit cell being optional. As in COMPSTRU, the program STRUCTURE RELATIONS assigns default values to the thresholds for the maximum allowed distance between the positions of matching atoms and for the difference between the lattice parameters. These values can be optionally configured by the user. By default, the program uses the data stored in the group–subgroup databases of space groups of the Server to determine the transformation matrix–column pair ($\boldsymbol{P}$,$\boldsymbol{p}$) that transforms the structure descriptions. The program allows the user to check the relationship for a predefined transformation matrix, if the option 'User-defined group–subgroup transformation matrix' is selected.

Example 24.5

Lead phosphate $Pb_3(PO_4)_2$ undergoes a phase transition from a paraelastic high-temperature phase with symmetry $R\bar{3}m$ (No. 166) to a ferroelastic low-temperature phase of symmetry $C2/c$ (No. 15) at 180 °C. The paraelastic phase is also observed at high pressures [476]. The crystal data at high and low temperatures are [474, 475]:

HT-structure (473 K) $R\bar{3}m$
data recalculated from [474]
166
5.530 5.530 20.277 90 90 120
5

Pb	1	$3a$	0	0	0
Pb	2	$6c$	0	0	0.214
P	1	$6c$	0	0	0.402
O	1	$6c$	0	0	0.327
O	2	$18h$	0.182	−0.182	0.094

LT-structure (298 K) $C2/c$ [475]
ICSD 66379 (in part, symmetry-equivalent
15 # atom sites are listed)
13.74 5.66 9.39 90 102.4 90
7

Pb	1	$4e$	0	0.2883	$\frac{1}{4}$
Pb	2	$8f$	0.6817	0.3102	0.1485
P	1	$8f$	0.3981	0.2501	0.0534
O	1	$8f$	0.8546	0.5302	0.1114
O	2	$8f$	0.3659	0.5304	0.6289
O	3	$8f$	0.8575	0.2212	0.3882
O	4	$8f$	0.5109	0.2290	0.0828

STRUCTURE RELATIONS determines the index of the symmetry reduction and the matrix–column pair ($\boldsymbol{P}$,$\boldsymbol{p}$) for the data transformation of the high-symmetry to the low-symmetry structure and it evaluates the similarity between the two structures internally using the module of the program COMPSTRU.

The transformation of the unit cell and the coordinates is rather intricate in this example. It involves not only the ever-difficult transformations from a rhombohedral cell, but also the selection of an appropriate setting, unit cell, orientation, and origin shift for the monoclinic cell. Trying to do this by hand with the aid of the formulas listed in *International Tables* A1 takes hours of pondering. The relative orientations and positions of the involved unit cells are shown in Fig. 24.4. The overall transformation relating the high- and the low-symmetry phases of $Pb_3(PO_4)_2$ as determined by `STRUCTURE RELATIONS` corresponds to the multiplication of the augmented matrices of the transformations given in the Bärnighausen tree in Fig. 24.5:

$$\mathbb{P} = \left(\begin{array}{ccc|c} -1 & 1 & \frac{1}{3} & 0 \\ 1 & 1 & -\frac{1}{3} & 0 \\ 0 & 0 & -\frac{1}{3} & 0 \\ \hline 0 & 0 & 0 & 1 \end{array}\right) \left(\begin{array}{ccc|c} 1 & 0 & -1 & \frac{1}{4} \\ 0 & 1 & 0 & -\frac{1}{4} \\ 2 & 0 & 0 & 0 \\ \hline 0 & 0 & 0 & 1 \end{array}\right) = \left(\begin{array}{ccc|c} -\frac{1}{3} & 1 & 1 & -\frac{1}{2} \\ \frac{1}{3} & 1 & -1 & 0 \\ -\frac{2}{3} & 0 & 0 & 0 \\ \hline 0 & 0 & 0 & 1 \end{array}\right)$$

The index of the symmetry reduction is 6. The corresponding transformation of the coordinates corresponds to $\mathbb{P}^{-1}$, that is, $(\boldsymbol{P}^{-1}, -\boldsymbol{P}^{-1}\boldsymbol{p})$:

$$-\tfrac{3}{2}z,\ \tfrac{1}{2}x+\tfrac{1}{2}y+\tfrac{1}{4},\ \tfrac{1}{2}x-\tfrac{1}{2}y-\tfrac{1}{2}z+\tfrac{1}{4};\ +(\tfrac{1}{2},0,\tfrac{1}{2})$$

During the phase transition, only the orbit of the O atoms occupying the Wyckoff position 18*h* of $R\bar{3}m$ splits into three orbits of the general position 8*f* of $C2/c$.

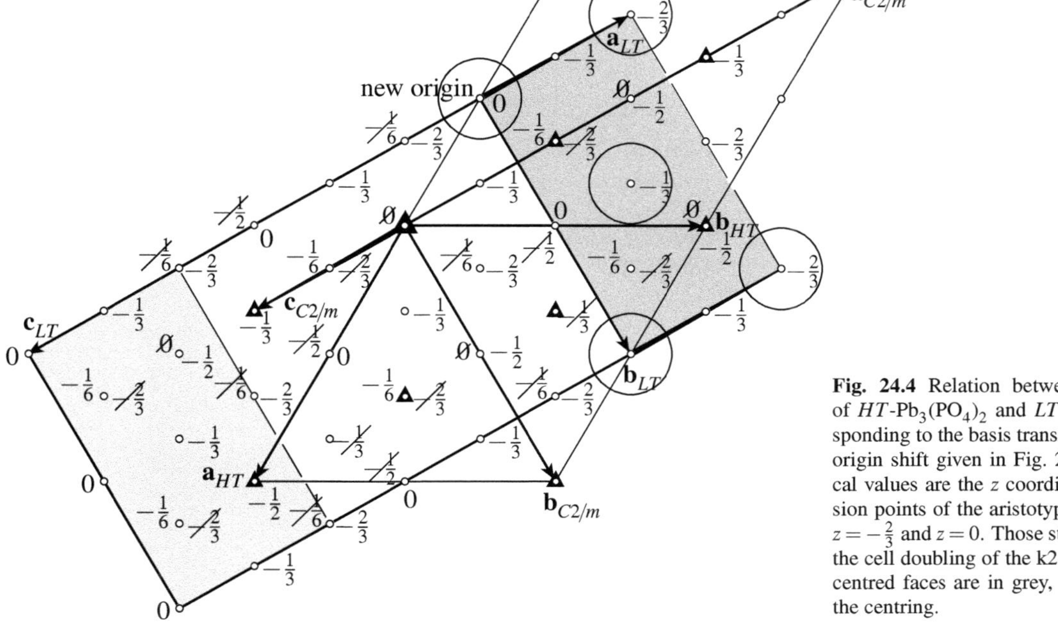

Fig. 24.4 Relation between the unit cells of HT-$Pb_3(PO_4)_2$ and LT-$Pb_3(PO_4)_2$ corresponding to the basis transformations and the origin shift given in Fig. 24.5. The numerical values are the z coordinates of the inversion points of the aristotype $R\bar{3}\,2/m$ between $z = -\frac{2}{3}$ and $z = 0$. Those struck out are lost at the cell doubling of the k2 step. The inclined centred faces are in grey, large circles mark the centring.

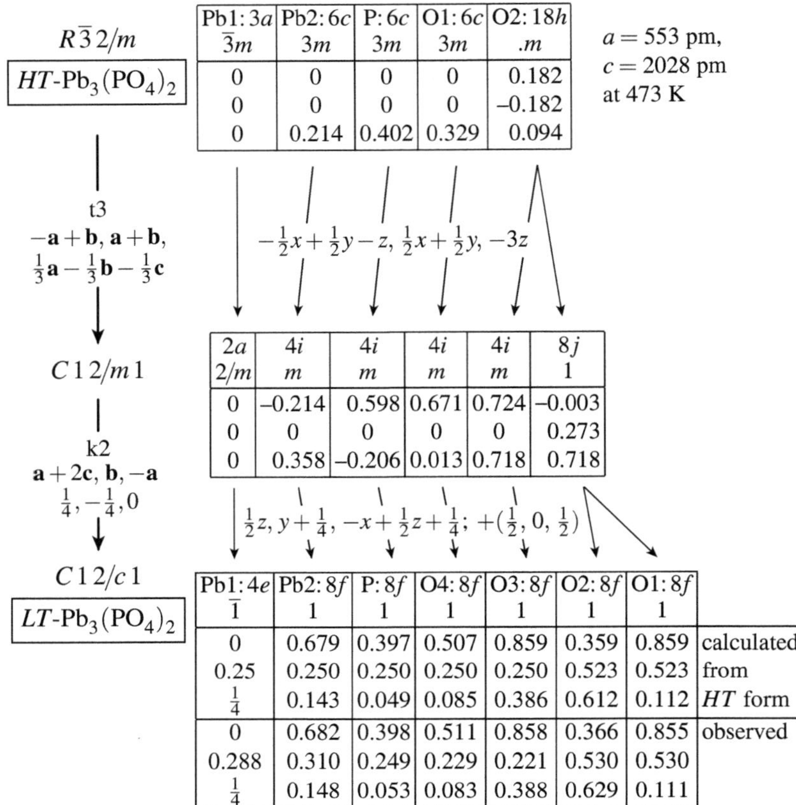

Calculated from *HT* form: $a = 1389$ pm, $b = 553$ pm, $c = 958$ pm, $\beta = 103.3°$
Observed ($\sim$298 K): $a = 1374$ pm, $b = 556$ pm, $c = 939$ pm, $\beta = 102.4°$ [475]

Fig. 24.5 Group–subgroup relation between *HT*-$Pb_3(PO_4)_2$ and *LT*-$Pb_3(PO_4)_2$. Due to the *translationengleiche* t3 step, the low-temperature modification forms twinned crystals with three kinds of domains.

Appendices

Solutions to the exercises
References
Glossary
Index

Solutions to the exercises

3.1 (a) For (8): $\tilde{x} = -y + \frac{1}{4},\ \tilde{y} = -x + \frac{1}{4},\ \tilde{z} = -z + \frac{3}{4}$

for (10): $\tilde{x} = x + \frac{1}{2},\ \tilde{y} = y,\ \tilde{z} = -z + \frac{1}{2}$

(b) For (8): $\boldsymbol{W}(8) = \begin{pmatrix} 0 & \bar{1} & 0 \\ \bar{1} & 0 & 0 \\ 0 & 0 & \bar{1} \end{pmatrix}$ and $\boldsymbol{w}(8) = \begin{pmatrix} \frac{1}{4} \\ \frac{1}{4} \\ \frac{3}{4} \end{pmatrix}$

for (10): $\boldsymbol{W}(10) = \begin{pmatrix} 1 & 0 & 0 \\ 0 & 1 & 0 \\ 0 & 0 & \bar{1} \end{pmatrix}$ and $\boldsymbol{w}(10) = \begin{pmatrix} \frac{1}{2} \\ 0 \\ \frac{1}{2} \end{pmatrix}$

(c) For (8): $\mathbb{W}(8) = \left(\begin{array}{ccc|c} 0 & \bar{1} & 0 & \frac{1}{4} \\ \bar{1} & 0 & 0 & \frac{1}{4} \\ 0 & 0 & \bar{1} & \frac{3}{4} \\ \hline 0 & 0 & 0 & 1 \end{array}\right)$

for (10): $\mathbb{W}(10) = \left(\begin{array}{ccc|c} 1 & 0 & 0 & \frac{1}{2} \\ 0 & 1 & 0 & 0 \\ 0 & 0 & \bar{1} & \frac{1}{2} \\ \hline 0 & 0 & 0 & 1 \end{array}\right)$

(d) $\mathbb{W}(8)\mathbb{W}(10) = \left(\begin{array}{ccc|c} 0 & \bar{1} & 0 & \frac{1}{4} \\ \bar{1} & 0 & 0 & -\frac{1}{4} \\ 0 & 0 & 1 & \frac{1}{4} \\ \hline 0 & 0 & 0 & 1 \end{array}\right)$

$\mathbb{W}(10)\mathbb{W}(8) = \left(\begin{array}{ccc|c} 0 & \bar{1} & 0 & \frac{3}{4} \\ \bar{1} & 0 & 0 & \frac{1}{4} \\ 0 & 0 & 1 & -\frac{1}{4} \\ \hline 0 & 0 & 0 & 1 \end{array}\right)$

The result depends on the sequence. The two results differ by the centring translation $\frac{1}{2}, \frac{1}{2}, -\frac{1}{2}$.

(e) The coordinate triplets are:

$\bar{y} + \frac{1}{4},\ \bar{x} - \frac{1}{4},\ z + \frac{1}{4}$ and $\bar{y} + \frac{3}{4},\ \bar{x} + \frac{1}{4},\ z - \frac{1}{4}$

After standardization $0 \le w_i < 1$ we have:

$\bar{y} + \frac{1}{4},\ \bar{x} + \frac{3}{4},\ z + \frac{1}{4}$ and $\bar{y} + \frac{3}{4},\ \bar{x} + \frac{1}{4},\ z + \frac{3}{4}$

That corresponds to the coordinate triplets $(15) + (\frac{1}{2}, \frac{1}{2}, \frac{1}{2})$ and (15).

3.2 (a)

$$\boldsymbol{P} = \begin{pmatrix} 1 & 0 & \frac{1}{2} \\ 0 & 1 & \frac{1}{2} \\ 0 & 0 & \frac{1}{2} \end{pmatrix} \quad \text{cf. eqn (3.28), page 31.}$$

Because of $\det(\boldsymbol{P}) = +\frac{1}{2}$ the volume of the primitive cell is half that of the conventional cell.

(b) $\boldsymbol{P}$ contains many zeros, so that inversion of $\boldsymbol{P}$ is simple. Temporarily, we designate the coefficients of $\boldsymbol{P}^{-1} = \boldsymbol{Q}$ by q_{ik}. Then, due to $\boldsymbol{QP} = \boldsymbol{I}$, we have:

$q_{11}\times 1 + q_{12}\times 0 + q_{13}\times 0 = 1;\quad q_{11}\times 0 + q_{12}\times 1 + q_{13}\times 0 = 0;$
$q_{11}\times \frac{1}{2} + q_{12}\times \frac{1}{2} + q_{13}\times \frac{1}{2} = 0;\quad q_{21}\times 1 + q_{22}\times 0 + q_{23}\times 0 = 0;$
$q_{21}\times 0 + q_{22}\times 1 + q_{23}\times 0 = 1;\quad q_{21}\times \frac{1}{2} + q_{22}\times \frac{1}{2} + q_{23}\times \frac{1}{2} = 0;$
$q_{31}\times 1 + q_{32}\times 0 + q_{33}\times 0 = 0;\quad q_{31}\times 0 + q_{32}\times 1 + q_{33}\times 0 = 0;$
$q_{31}\times \frac{1}{2} + q_{32}\times \frac{1}{2} + q_{33}\times \frac{1}{2} = 1$

From this we obtain:

$q_{11} = 1,\ q_{12} = 0,\ q_{13} = -1,\ q_{21} = 0,\ q_{22} = 1,\ q_{23} = -1,$
$q_{31} = 0,\ q_{32} = 0,\ q_{33} = 2$

$$\text{or} \quad \boldsymbol{P}^{-1} = \begin{pmatrix} 1 & 0 & \bar{1} \\ 0 & 1 & \bar{1} \\ 0 & 0 & 2 \end{pmatrix}$$

We make sure that $\boldsymbol{PP}^{-1} = \boldsymbol{P}^{-1}\boldsymbol{P} = \boldsymbol{I}$ holds. If the coordinates in the new basis are x_i', we have: $x' = x - z;\ y' = y - z;\ z' = 2z$.

(c) According to eqn (3.36), page 34, the symmetry operations are transformed by $\mathbb{W}' = \mathbb{P}^{-1}\mathbb{W}\mathbb{P}$. We obtain:

$$\mathbb{W}'(8) = \left(\begin{array}{ccc|c} 1 & 0 & \bar{1} & 0 \\ 0 & 1 & \bar{1} & 0 \\ 0 & 0 & 2 & 0 \\ \hline 0 & 0 & 0 & 1 \end{array}\right) \left(\begin{array}{ccc|c} 0 & \bar{1} & 0 & \frac{1}{4} \\ \bar{1} & 0 & 0 & \frac{1}{4} \\ 0 & 0 & \bar{1} & \frac{3}{4} \\ \hline 0 & 0 & 0 & 1 \end{array}\right) \left(\begin{array}{ccc|c} 1 & 0 & \frac{1}{2} & 0 \\ 0 & 1 & \frac{1}{2} & 0 \\ 0 & 0 & \frac{1}{2} & 0 \\ \hline 0 & 0 & 0 & 1 \end{array}\right)$$

$$= \left(\begin{array}{ccc|c} 0 & \bar{1} & 0 & -\frac{1}{2} \\ \bar{1} & 0 & 0 & -\frac{1}{2} \\ 0 & 0 & \bar{1} & \frac{3}{2} \\ \hline 0 & 0 & 0 & 1 \end{array}\right)$$

$$\mathbb{W}'(10) = \left(\begin{array}{ccc|c} 1 & 0 & \bar{1} & 0 \\ 0 & 1 & \bar{1} & 0 \\ 0 & 0 & 2 & 0 \\ \hline 0 & 0 & 0 & 1 \end{array}\right) \left(\begin{array}{ccc|c} 1 & 0 & 0 & \frac{1}{2} \\ 0 & 1 & 0 & 0 \\ 0 & 0 & \bar{1} & \frac{1}{2} \\ \hline 0 & 0 & 0 & 1 \end{array}\right) \left(\begin{array}{ccc|c} 1 & 0 & \frac{1}{2} & 0 \\ 0 & 1 & \frac{1}{2} & 0 \\ 0 & 0 & \frac{1}{2} & 0 \\ \hline 0 & 0 & 0 & 1 \end{array}\right)$$

$$= \left(\begin{array}{ccc|c} 1 & 0 & 1 & 0 \\ 0 & 1 & 1 & -\frac{1}{2} \\ 0 & 0 & \bar{1} & 1 \\ \hline 0 & 0 & 0 & 1 \end{array}\right)$$

$$\mathbb{W}'(15) = \left(\begin{array}{ccc|c} 1 & 0 & \bar{1} & 0 \\ 0 & 1 & \bar{1} & 0 \\ 0 & 0 & 2 & 0 \\ \hline 0 & 0 & 0 & 1 \end{array}\right) \left(\begin{array}{ccc|c} 0 & \bar{1} & 0 & \frac{3}{4} \\ \bar{1} & 0 & 0 & \frac{1}{4} \\ 0 & 0 & 1 & \frac{3}{4} \\ \hline 0 & 0 & 0 & 1 \end{array}\right) \left(\begin{array}{ccc|c} 1 & 0 & \frac{1}{2} & 0 \\ 0 & 1 & \frac{1}{2} & 0 \\ 0 & 0 & \frac{1}{2} & 0 \\ \hline 0 & 0 & 0 & 1 \end{array}\right)$$

$$= \left(\begin{array}{ccc|c} 0 & \bar{1} & \bar{1} & 0 \\ \bar{1} & 0 & \bar{1} & -\frac{1}{2} \\ 0 & 0 & 1 & \frac{3}{2} \\ \hline 0 & 0 & 0 & 1 \end{array}\right)$$

$$(\mathbb{W}(15) + (\tfrac{1}{2},\tfrac{1}{2},\tfrac{1}{2}))' =$$

$$\left(\begin{array}{ccc|c} 1 & 0 & \bar{1} & 0 \\ 0 & 1 & \bar{1} & 0 \\ 0 & 0 & 2 & 0 \\ \hline 0 & 0 & 0 & 1 \end{array}\right) \left(\begin{array}{ccc|c} 0 & \bar{1} & 0 & \frac{5}{4} \\ \bar{1} & 0 & 0 & \frac{3}{4} \\ 0 & 0 & 1 & \frac{5}{4} \\ \hline 0 & 0 & 0 & 1 \end{array}\right) \left(\begin{array}{ccc|c} 1 & 0 & \frac{1}{2} & 0 \\ 0 & 1 & \frac{1}{2} & 0 \\ 0 & 0 & \frac{1}{2} & 0 \\ \hline 0 & 0 & 0 & 1 \end{array}\right)$$

$$= \left(\begin{array}{ccc|c} 0 & \bar{1} & \bar{1} & 0 \\ \bar{1} & 0 & \bar{1} & -\frac{1}{2} \\ 0 & 0 & 1 & \frac{5}{2} \\ \hline 0 & 0 & 0 & 1 \end{array}\right)$$

(d) The standardized entries would be (the index n means standardized):

$(8)'_n \quad \bar{y}+\frac{1}{2},\ \bar{x}+\frac{1}{2},\ \bar{z}+\frac{1}{2};$

$(10)'_n \quad x+y,\ y+z+\frac{1}{2},\ \bar{z};$

$(15)'_n \quad \bar{y}-z,\ \bar{x}-z+\frac{1}{2},\ z+\frac{1}{2};$

$((15)+(\frac{1}{2},\frac{1}{2},\frac{1}{2}))'_n$ not different from $(15)'_n$

There is no difference between (15) and $(15)+(\frac{1}{2},\frac{1}{2},\frac{1}{2})$ because the centring translation consists of integral numbers when referred to the primitive basis and disappears by the standardization.

3.3 The transformation matrices are:

$$\boldsymbol{P} = \begin{pmatrix} 2 & \bar{1} & 0 \\ 1 & 1 & 0 \\ 0 & 0 & 1 \end{pmatrix} \quad \text{and} \quad \boldsymbol{P}^{-1} = \begin{pmatrix} \frac{1}{3} & \frac{1}{3} & 0 \\ -\frac{1}{3} & \frac{2}{3} & 0 \\ 0 & 0 & 1 \end{pmatrix}$$

Since $\det(\boldsymbol{P}) = 3$, the unit cell is tripled. With the coordinate transformation, the origin shift $\boldsymbol{p} = (\frac{2}{3}, \frac{1}{3}, 0)$ yields a column part of:

$$-\boldsymbol{P}^{-1}\boldsymbol{p} = -\begin{pmatrix} \frac{1}{3} & \frac{1}{3} & 0 \\ -\frac{1}{3} & \frac{2}{3} & 0 \\ 0 & 0 & 1 \end{pmatrix} \begin{pmatrix} \frac{2}{3} \\ \frac{1}{3} \\ 0 \end{pmatrix} = \begin{pmatrix} -\frac{1}{3} \\ 0 \\ 0 \end{pmatrix}$$

The coordinates are transformed according to $(\boldsymbol{P}^{-1}, -\boldsymbol{P}^{-1}\boldsymbol{p})$ or $\frac{1}{3}(x+y)-\frac{1}{3},\ \frac{1}{3}(-x+2y),\ z$.

3.4 The total transformation results from the product of the transformation matrices $\mathbb{P}_1$ and $\mathbb{P}_2$:

$$\mathbb{P} = \mathbb{P}_1\mathbb{P}_2 = \left(\begin{array}{ccc|c} 1 & -1 & 0 & 0 \\ 1 & 1 & 0 & \frac{1}{2} \\ 0 & 0 & 1 & 0 \\ \hline 0 & 0 & 0 & 1 \end{array}\right) \left(\begin{array}{ccc|c} 1 & 0 & -\frac{1}{2} & -\frac{1}{8} \\ 0 & -1 & 0 & \frac{1}{8} \\ 0 & 0 & -\frac{1}{2} & -\frac{1}{8} \\ \hline 0 & 0 & 0 & 1 \end{array}\right)$$

$$= \left(\begin{array}{ccc|c} 1 & 1 & -\frac{1}{2} & -\frac{1}{4} \\ 1 & -1 & -\frac{1}{2} & \frac{1}{2} \\ 0 & 0 & -\frac{1}{2} & -\frac{1}{8} \\ \hline 0 & 0 & 0 & 1 \end{array}\right)$$

Therefore, the total basis transformation and origin shift are: $\mathbf{a}' = \mathbf{a}+\mathbf{b}$, $\mathbf{b}' = \mathbf{a}-\mathbf{b}$, $\mathbf{c}' = -\frac{1}{2}(\mathbf{a}+\mathbf{b}+\mathbf{c})$ and $\boldsymbol{p} = (-\frac{1}{4}, \frac{1}{2}, -\frac{1}{8})$. The determinant of $\boldsymbol{P}$ (matrix part of $\mathbb{P}$) is 1, the volume of the unit cell does not change.

The easiest way to obtain $\boldsymbol{P}^{-1} = \boldsymbol{P}_2^{-1}\boldsymbol{P}_1^{-1}$ is to prepare a graph of the transformations (without origin shifts) and derive the reverse transformation. $\boldsymbol{P}^{-1}$ is correct if $\boldsymbol{P}\boldsymbol{P}^{-1} = \boldsymbol{I}$ holds. $\boldsymbol{P}^{-1}$ is mentioned in the following equations. The column part of $\mathbb{P}^{-1}$ results from:

$$-\boldsymbol{P}^{-1}\boldsymbol{p} = -\begin{pmatrix} \frac{1}{2} & \frac{1}{2} & -1 \\ \frac{1}{2} & -\frac{1}{2} & 0 \\ 0 & 0 & -2 \end{pmatrix} \begin{pmatrix} -\frac{1}{4} \\ \frac{1}{2} \\ -\frac{1}{8} \end{pmatrix} = \begin{pmatrix} -\frac{1}{4} \\ \frac{3}{8} \\ -\frac{1}{4} \end{pmatrix}$$

$$\mathbb{P}^{-1} = \mathbb{P}_2^{-1}\mathbb{P}_1^{-1} = \left(\begin{array}{ccc|c} \frac{1}{2} & \frac{1}{2} & -1 & -\frac{1}{4} \\ \frac{1}{2} & -\frac{1}{2} & 0 & \frac{3}{8} \\ 0 & 0 & -2 & -\frac{1}{4} \\ \hline 0 & 0 & 0 & 1 \end{array}\right)$$

The coordinates are transformed according to: $x' = \frac{1}{2}(x+y)-z-\frac{1}{4},\ y' = \frac{1}{2}(x-y)+\frac{3}{8},\ z' = -2z-\frac{1}{4}$.

3.5

$$\boldsymbol{p} = -\boldsymbol{P}\boldsymbol{p}' = -\begin{pmatrix} \frac{1}{2} & \frac{1}{2} & 0 \\ -\frac{1}{2} & \frac{1}{2} & 0 \\ 0 & 0 & 1 \end{pmatrix} \begin{pmatrix} \frac{1}{2} \\ 0 \\ 0 \end{pmatrix} = \begin{pmatrix} -\frac{1}{4} \\ \frac{1}{4} \\ 0 \end{pmatrix}$$

The origin shifts listed in Parts 2 and 3 of *International Tables A1* have been chosen differently.

4.1 (a) The results are:

		(8)	(10)	(15)	$(15)_2$	$(15)_{2n}$
(**a**)	determinant	$+1$	-1	-1	-1	-1
	trace	-1	$+1$	$+1$	$+1$	$+1$
(**b**)	type	2	m	m	m	m
(**c**)	$\boldsymbol{u}$	$[1\bar{1}0]$	[001]	[110]	[110]	[110]
(**d**)	screw, glide components	$0,0,0$	$\frac{1}{2},0,0$	$\frac{1}{4},-\frac{1}{4},\frac{3}{4}$	$\frac{1}{4},-\frac{1}{4},\frac{5}{4}$	$-\frac{1}{4},\frac{1}{4},\frac{1}{4}$
	$\boldsymbol{w}'$	$\frac{1}{4},\frac{1}{4},\frac{3}{4}$	$0,0,\frac{1}{2}$	$\frac{1}{2},\frac{1}{2},0$	$1,1,0$	$\frac{1}{2},\frac{1}{2},0$
(**e**)	Herm.–Maugin symbol	2	a	d	d	d
(**f**)	fixed points	$x,\bar{x}+\frac{1}{4},\frac{3}{8}$	$x,y,\frac{1}{4}$	$x,\bar{x}+\frac{1}{2},z$	$x,\bar{x},z$	$x,\bar{x}+\frac{1}{2},z$

Referring to (**c**), take eqn $\boldsymbol{Wu} = \pm\boldsymbol{u}$, $+$ for rotations, $-$ for rotoinversions and reflections. For example:

$$(8)\qquad \begin{pmatrix} 0 & \bar{1} & 0 \\ \bar{1} & 0 & 0 \\ 0 & 0 & \bar{1} \end{pmatrix} \begin{pmatrix} u \\ v \\ w \end{pmatrix} = \begin{pmatrix} u \\ v \\ w \end{pmatrix}$$

$$\text{yields}\qquad \begin{pmatrix} -v \\ -u \\ -w \end{pmatrix} = \begin{pmatrix} u \\ v \\ w \end{pmatrix}$$

From this follows $u = -v$; $w = -w$ yields $w = 0$. The solution is thus $u\bar{u}0$ or standardized $1\bar{1}0$.

Referring to (**d**), the screw and glide components are calculated with eqn (4.3), $\frac{1}{k}\boldsymbol{t} = \frac{1}{k}(\boldsymbol{W}+\boldsymbol{I})\boldsymbol{w}$ (page 48; for reflections and twofold rotations $k = 2$).
For example, for (15) we obtain:

$$\tfrac{1}{2}\boldsymbol{t} = \tfrac{1}{2}\begin{pmatrix} 1 & \bar{1} & 0 \\ \bar{1} & 1 & 0 \\ 0 & 0 & 2 \end{pmatrix} \begin{pmatrix} \frac{3}{4} \\ \frac{1}{4} \\ \frac{3}{4} \end{pmatrix} = \begin{pmatrix} \frac{1}{4} \\ -\frac{1}{4} \\ \frac{3}{4} \end{pmatrix}$$

The columns are calculated with the screw or glide components according to $\boldsymbol{w}' = \boldsymbol{w} - \frac{1}{2}\boldsymbol{t}$, for example, for (15):

$$\boldsymbol{w}' = \begin{pmatrix} \frac{3}{4} \\ \frac{1}{4} \\ \frac{3}{4} \end{pmatrix} - \begin{pmatrix} \frac{1}{4} \\ -\frac{1}{4} \\ \frac{3}{4} \end{pmatrix} = \begin{pmatrix} \frac{1}{2} \\ \frac{1}{2} \\ 0 \end{pmatrix}$$

Referring to (**f**), the columns $\boldsymbol{w}'$ are needed for the determination of the positions of the symmetry elements with the aid of eqn (4.5), page 48. For example, for (15) we have:

$$\begin{pmatrix} 0 & \bar{1} & 0 \\ \bar{1} & 0 & 0 \\ 0 & 0 & 1 \end{pmatrix} \begin{pmatrix} x \\ y \\ z \end{pmatrix} + \begin{pmatrix} \frac{1}{2} \\ \frac{1}{2} \\ 0 \end{pmatrix} = \begin{pmatrix} x \\ y \\ z \end{pmatrix}$$

From this follows the system of equations
$-y+\frac{1}{2} = x$, $-x+\frac{1}{2} = y$, $z = z$, and the result for the fixed points is $y = -x+\frac{1}{2}$ and z = arbitrary.
The screw and glide components $\frac{1}{2}\boldsymbol{t}$ show that the twofold rotation (8) is the only one having fixed points.

Question on page 55. Elements of order 2 have a *1* in the main diagonal.

5.1 If the elements ij and ji are the same in the group-multiplication table.

5.2

symm. oper.	order	symm. oper.	order
1	1	*m*	2
2	2	$4, 4^{-1}$	4

5.3 See Fig. 19.2, page 257. The order of the symmetry group of the trigonal prism is 12. Let m_1, m_2, and m_3 be the three vertical reflections, and 2_1, 2_2, and 2_3 the twofold rotations (in the sequence as in Fig. 19.2). In the following, two sequences of two consecutive permutations are mentioned:

$\bar{6}$ first, then m_z

	①	②	③	④	⑤	⑥
$\bar{6}$	↓	↓	↓	↓	↓	↓
	⑥	④	⑤	③	①	②
m_z	↓	↓	↓	↓	↓	↓
	③	①	②	⑥	④	⑤

that corresponds to the permutation (1 3 2)(4 6 5) which is 3^{-1}

m_1 first, then *3*

	①	②	③	④	⑤	⑥
m_1	↓	↓	↓	↓	↓	↓
	①	③	②	④	⑥	⑤
3	↓	↓	↓	↓	↓	↓
	②	①	③	⑤	④	⑥

that corresponds to the permutation (3)(6)(1 2)(4 5) which is m_3

The layout of the following group-multiplication table is such that the upper left quadrant corresponds to the subgroup $\bar{6}$:

	1	3	3^{-1}	$\bar{6}$	$\bar{6}^{-1}$	m_z	2_1	2_2	2_3	m_1	m_2	m_3
1	1	3	3^{-1}	$\bar{6}$	$\bar{6}^{-1}$	m_z	2_1	2_2	2_3	m_1	m_2	m_3
3	3	3^{-1}	1	m_z	$\bar{6}$	$\bar{6}^{-1}$	2_3	2_1	2_2	m_3	m_1	m_2
3^{-1}	3^{-1}	1	3	$\bar{6}^{-1}$	m_z	$\bar{6}$	2_2	2_3	2_1	m_2	m_3	m_1
$\bar{6}$	$\bar{6}$	m_z	$\bar{6}^{-1}$	3	1	3^{-1}	m_2	m_3	m_1	2_2	2_3	2_1
$\bar{6}^{-1}$	$\bar{6}^{-1}$	$\bar{6}$	m_z	1	3^{-1}	3	m_3	m_1	m_2	2_3	2_1	2_2
m_z	m_z	$\bar{6}^{-1}$	$\bar{6}$	3^{-1}	3	1	m_1	m_2	m_3	2_1	2_2	2_3
2_1	2_1	2_2	2_3	m_3	m_2	m_1	1	3	3^{-1}	m_z	$\bar{6}^{-1}$	$\bar{6}$
2_2	2_2	2_3	2_1	m_1	m_3	m_2	3^{-1}	1	3	$\bar{6}$	m_z	$\bar{6}^{-1}$
2_3	2_3	2_1	2_2	m_2	m_1	m_3	3	3^{-1}	1	$\bar{6}^{-1}$	$\bar{6}$	m_z
m_1	m_1	m_2	m_3	2_3	2_2	2_1	m_z	$\bar{6}^{-1}$	$\bar{6}$	1	3	3^{-1}
m_2	m_2	m_3	m_1	2_1	2_3	2_2	$\bar{6}$	m_z	$\bar{6}^{-1}$	3^{-1}	1	3
m_3	m_3	m_1	m_2	2_2	2_1	2_3	$\bar{6}^{-1}$	$\bar{6}$	m_z	3	3^{-1}	1

5.4 The group elements can be found in the first line of the preceding multiplication table.
Coset decomposition with respect to $\{1, 3, 3^{-1}\}$:
left cosets

$1 \circ \{1,3,3^{-1}\}$	$m_1 \circ \{1,3,3^{-1}\}$	$\bar{6} \circ \{1,3,3^{-1}\}$	$2_1 \circ \{1,3,3^{-1}\}$
$1, 3, 3^{-1}$	m_1, m_3, m_2	$\bar{6}, m_z, \bar{6}^{-1}$	$2_1, 2_3, 2_2$

right cosets

$\{1,3,3^{-1}\} \circ 1$	$\{1,3,3^{-1}\} \circ m_1$	$\{1,3,3^{-1}\} \circ \bar{6}$	$\{1,3,3^{-1}\} \circ 2_1$
$1, 3, 3^{-1}$	m_1, m_2, m_3	$\bar{6}, m_z, \bar{6}^{-1}$	$2_1, 2_2, 2_3$

The number of cosets being four, the index is 4. Left and right cosets are the same.

Coset decomposition with respect to $\{1, m_1\}$:
left cosets

$1\circ\{1,m_1\}$	$3\circ\{1,m_1\}$	$3^{-1}\circ\{1,m_1\}$	$\bar{6}\circ\{1,m_1\}$	$\bar{6}^{-1}\circ\{1,m_1\}$	$m_z\circ\{1,m_1\}$
$1, m_1$	$3, m_2$	$3^{-1}, m_3$	$\bar{6}, 2_3$	$\bar{6}^{-1}, 2_2$	$m_z, 2_1$

right cosets

$\{1,m_1\}\circ 1$	$\{1,m_1\}\circ 3$	$\{1,m_1\}\circ 3^{-1}$	$\{1,m_1\}\circ\bar{6}$	$\{1,m_1\}\circ\bar{6}^{-1}$	$\{1,m_1\}\circ m_z$
$1, m_1$	$3, m_3$	$3^{-1}, m_2$	$\bar{6}, 2_2$	$\bar{6}^{-1}, 2_3$	$m_z, 2_1$

There are six cosets each, the index is 6. Left and right cosets are different.

5.5 Only the first coset is a subgroup; it is the only one that contains the identity element.

5.6 A subgroup $\mathcal{H}$ of index 2 has only one more (second) coset in addition to itself; this second coset contains all elements of $\mathcal{G}$ that are not contained in $\mathcal{H}$. Therefore, right and left coset decompositions must yield the same cosets. That corresponds to the definition of a normal subgroup.

5.7 (a) Subgroups of order 4 are the cyclic group of elements of order 4, namely $\{1, 4, 2, 4^{-1}\}$, and the compositions from elements of order 2. The latter cannot contain the elements 4 and 4^{-1}. Therefore, they contain at least two reflections and, as a composition of these reflections, a rotation, which can only be 2. These subgroups must be of the kind $\{1, 2, m, m\}$; they are $\{1, 2, m_x, m_y\}$ and $\{1, 2, m_+, m_-\}$. The mentioned subgroups are maximal subgroups. They have further subgroups of order 2, consisting of the identity and one of the other elements: $\{1, 2\}$, $\{1, m_x\}$, $\{1, m_y\}$, $\{1, m_+\}$, $\{1, m_-\}$.

(b) m_x and m_+ cannot both appear in a subgroup of order 4 because any of their compositions always yields the elements 4 or 4^{-1} of order 4, so that the number of elements would be at least 6. According to LAGRANGE's theorem this generates the whole group $4mm$.

(c) $\{1, m_x\}$ and $\{1, m_y\}$ are conjugate, and so are $\{1, m_+\}$ and $\{1, m_-\}$. Conjugating elements are 4 and 4^{-1}: $4^{-1} \circ m_x \circ 4 = m_y$ as well as $4^{-1} \circ m_+ \circ 4 = m_-$. The mirror lines m_x and m_y are symmetry equivalent by a symmetry operation of the square (namely by 4 or 4^{-1}) and then the subgroups $\{1, m_x\}$ and $\{1, m_y\}$ are also symmetry equivalent. The mirror lines m_x and m_+, however, are not symmetry equivalent in the square (there is no symmetry operation of the square that maps m_x onto m_+).
Remark: m_x and m_y are symmetry equivalent only by a symmetry operation of the square, but not by a symmetry operation of $\{1, 2, m_x, m_y\}$. Therefore, $\{1, m_x\}$ and $\{1, m_y\}$ are conjugate in $4mm$, but not in $\{1, 2, m_x, m_y\}$.

(d) Subgroups of index 2 are always normal subgroups. The two trivial subgroups are also always normal subgroups. The subgroups $\{1, m_x\}$ and $\{1, m_+\}$ are not normal subgroups because their left and right cosets are different. The same applies to $\{1, m_y\}$ and $\{1, m_-\}$.

(e)

index	order					
1	8			$4mm$		
2	4	$2m_xm_y$		4		$2m_+m_-$
4	2	m_x —	m_y	2	m_+ —	m_-
8	1			1		

5.8 Left coset decomposition of $\mathbb{Z}$ with respect to the subgroup $\mathcal{H} = \{0, \pm 5, \pm 10, \ldots\}$:

first coset $\mathcal{H} =$	second coset $1+\mathcal{H} =$	third coset $2+\mathcal{H} =$	fourth coset $3+\mathcal{H} =$	fifth coset $4+\mathcal{H} =$
$\vdots$	$\vdots$	$\vdots$	$\vdots$	$\vdots$
-5	$1-5=-4$	$2-5=-3$	$3-5=-2$	$4-5=-1$
$e=0$	$1+0=1$	$2+0=2$	$3+0=3$	$4+0=4$
5	$1+5=6$	$2+5=7$	$3+5=8$	$4+5=9$
10	$1+10=11$	$2+10=12$	$3+10=13$	$4+10=14$
$\vdots$	$\vdots$	$\vdots$	$\vdots$	$\vdots$

There are five cosets, the index is 5. Right coset decomposition yields the second coset $0+1=1,\ 5+1=6,\ 10+1=11,\ \ldots$. The left cosets are identical with the right cosets. Therefore, the subgroup $\{0, \pm 5, \pm 10, \ldots\}$ is a normal subgroup.

5.9 $\mathcal{F}$ is an Abelian group. $g_i g_k = g_k g_i$ holds for all group elements.

6.1 $P4_132$: cubic; $P4_122$: tetragonal; $Fddd$: orthorhombic; $P12/c1$: monoclinic; $P\bar{4}n2$: tetragonal; $P\bar{4}3n$: cubic; $R\bar{3}m$: trigonal-rhombohedral; $Fm\bar{3}$: cubic.

6.2 Both space groups are hexagonal. $P6_3mc$ has reflection planes perpendicular to **a** and glide-reflection planes c perpendicular to **a** − **b**; for $P6_3cm$ it is the other way round.

6.3 $P2_12_12_1$: 222; $P6_3/mcm$: $6/mmm$; $P2_1/c$: $2/m$; $Pa\bar{3}$: $m\bar{3}$; $P4_2/m2_1/b2/c$: $4/m2/m2/m$, short $4/mmm$.

7.1 $^1_\infty P_6^{4-}$: $p\!\!\!/2/m11$ (or $p\!\!\!/12/m1$); $^1_\infty P_{12}$: $p\!\!\!/2/m2/c2_1/m$; $^1_\infty CrF_5^{2-}$: $p\!\!\!/2/m2/c2_1/m$; $^2_\infty Si_2O_5^{2-}$: $p31m$; black phosphorus: $p2/m2_1/a2/n$.

8.1 The four rhombohedral subgroups of a cubic space group are orientation-conjugate. Their threefold rotation axes run along the four body diagonals of the cube.

8.2

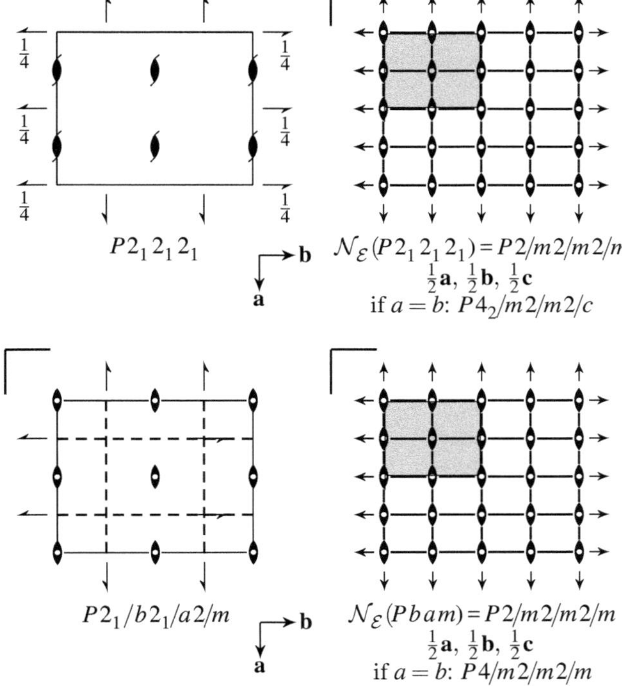

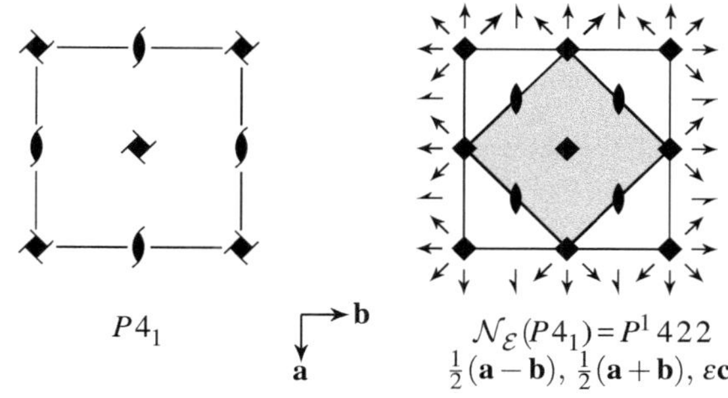

8.3 $\mathcal{N}_\mathcal{E}(\mathcal{H}) = P6/mmm$

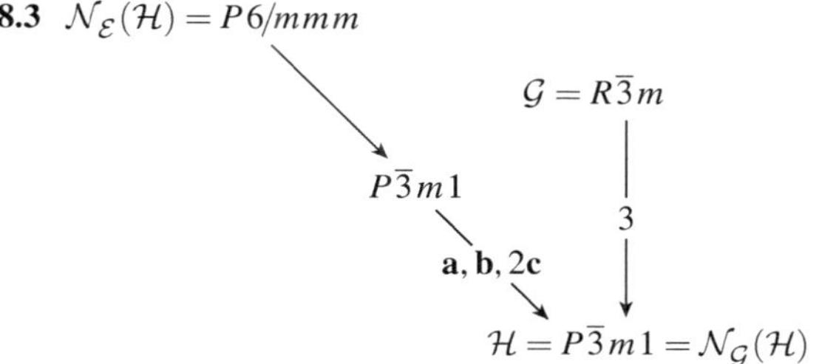

The index 3 of $\mathcal{N}_\mathcal{G}(\mathcal{H})$ in $\mathcal{G}$ indicates three conjugate subgroups of $\mathcal{H}$ in $\mathcal{G}$.

8.4

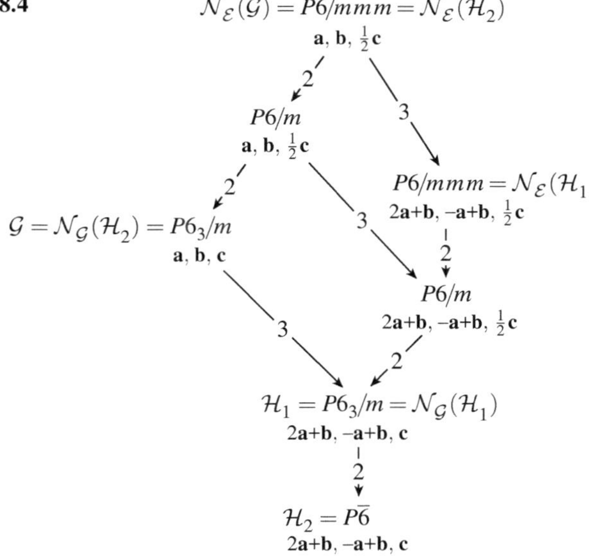

The index of $\mathcal{N}_\mathcal{G}(\mathcal{H}_1)$ in $\mathcal{G}$ is 3, there are three conjugates of $\mathcal{H}_1$. The index of $\mathcal{N}_\mathcal{G}(\mathcal{H}_2)$ in $\mathcal{G}$ is 1, $\mathcal{H}_2$ has no conjugates.

8.5 $\mathcal{N}_\mathcal{E}(\mathcal{G}) = P6/mmm$

$\mathbf{a}, \mathbf{b}, \frac{1}{2}\mathbf{c}$

$\downarrow$ 2

$\mathcal{G} = P6/mmm = \mathcal{N}_\mathcal{E}(\mathcal{H}) = \mathcal{N}_\mathcal{G}(\mathcal{H})$

$\mathbf{a}, \mathbf{b}, \mathbf{c}$

$\downarrow$ 2

$\mathcal{H} = P6_3/mmc$

$\mathbf{a}, \mathbf{b}, 2\mathbf{c}$

The index of $\mathcal{N}_\mathcal{G}(\mathcal{H})$ in $\mathcal{G}$ is 1, there are no conjugates. The index of $\mathcal{N}_\mathcal{E}(\mathcal{H})$ in $\mathcal{N}_\mathcal{E}(\mathcal{G})$ is 2, there are two subgroups on a par of type $\mathcal{H}$ with respect to $\mathcal{G}$.

9.1 The Euclidean normalizer of $P4/n$ is $P4/m2/m2/m$ with basis vectors $\frac{1}{2}(\mathbf{a}-\mathbf{b})$, $\frac{1}{2}(\mathbf{a}+\mathbf{b})$, $\frac{1}{2}\mathbf{c}$, index 8. There are eight sets of coordinates to describe the same structure. They result from the initial coordinates by addition of $\frac{1}{2}, \frac{1}{2}, 0$ and $0, 0, \frac{1}{2}$ and by the transformation y, x, z. The eight equivalent sets of coordinates are:

	x	y	z	x	y	z
P	$\frac{1}{4}$	$\frac{3}{4}$	0	$\frac{3}{4}$	$\frac{1}{4}$	0
C 1	0.362	0.760	0.141	0.862	0.260	0.141
C 2	0.437	0.836	0.117	0.937	0.336	0.117
Mo	$\frac{1}{4}$	$\frac{1}{4}$	0.121	$\frac{3}{4}$	$\frac{3}{4}$	0.121
N	$\frac{1}{4}$	$\frac{1}{4}$	−0.093	$\frac{3}{4}$	$\frac{3}{4}$	−0.093
Cl	0.400	0.347	0.191	0.900	0.847	0.191
P	$\frac{1}{4}$	$\frac{3}{4}$	$\frac{1}{2}$	$\frac{3}{4}$	$\frac{1}{4}$	$\frac{1}{2}$
C 1	0.362	0.760	0.641	0.862	0.260	0.641
C 2	0.437	0.836	0.617	0.937	0.336	0.617
Mo	$\frac{1}{4}$	$\frac{1}{4}$	0.621	$\frac{3}{4}$	$\frac{3}{4}$	0.621
N	$\frac{1}{4}$	$\frac{1}{4}$	0.407	$\frac{3}{4}$	$\frac{3}{4}$	0.407
Cl	0.400	0.347	0.691	0.900	0.847	0.691
P	$\frac{3}{4}$	$\frac{1}{4}$	0	$\frac{1}{4}$	$\frac{3}{4}$	0
C 1	0.760	0.362	0.141	0.260	0.862	0.141
C 2	0.836	0.437	0.117	0.336	0.937	0.117
Mo	$\frac{1}{4}$	$\frac{1}{4}$	0.121	$\frac{3}{4}$	$\frac{3}{4}$	0.121
N	$\frac{1}{4}$	$\frac{1}{4}$	−0.093	$\frac{3}{4}$	$\frac{3}{4}$	−0.093
Cl	0.347	0.400	0.191	0.847	0.900	0.191
P	$\frac{3}{4}$	$\frac{1}{4}$	$\frac{1}{2}$	$\frac{1}{4}$	$\frac{3}{4}$	$\frac{1}{2}$
C 1	0.760	0.362	0.641	0.260	0.862	0.641
C 2	0.836	0.437	0.617	0.336	0.937	0.617
Mo	$\frac{1}{4}$	$\frac{1}{4}$	0.621	$\frac{3}{4}$	$\frac{3}{4}$	0.621
N	$\frac{1}{4}$	$\frac{1}{4}$	0.407	$\frac{3}{4}$	$\frac{3}{4}$	0.407
Cl	0.347	0.400	0.691	0.847	0.900	0.691

9.2 Considering the similarity of the lattice parameters, the same space group and the approximately coincident coordinates of the vanadium atoms, β'-$Cu_{0.26}V_2O_5$ and β-$Ag_{0.33}V_2O_5$ seem to be isotypic. In that case there should exist an equivalent set of coordinates for both compounds, such that the coordinates of the Cu and Ag atoms also coincide. The Euclidean normalizer is $P2/m$. With the coordinate transformation $+(\frac{1}{2}, 0, 0)$ taken from the normalizer we obtain possible alternative coordinates for the Cu atom: 0.030 0 0.361; they agree acceptably with the Ag coordinates. However, this transformation would have to be applied also to the V atoms, and their coordinates would then show no agreement. The compounds are neither isotypic nor homeotypic.

9.3 The basis vectors **a** and **b** and the coordinates x and y of Pr_2NCl_3 have to be interchanged first. This causes no change of the space-group symbol; the simultaneous inversion of one of the signs, for example z by $-z$, is not necessary in this case because of the reflection plane; $Ibam$ always has pairs x, y, z and $x, y, -z$. After the interchange, the lattice parameters and all coordinates of Na_3AlP_2 and Pr_2NCl_3 show approximate agreement; however, for Cl 2 a symmetry-equivalent position of $Ibam$ has to be taken: $\frac{1}{2}-x, \frac{1}{2}+y, z$ yields 0.320, 0.299, 0 for Cl 2 (after interchange of the axes), in agreement with the coordinates of Na2 of Na_3AlP_2. Na_3AlP_2 and Pr_2NCl_3 are isotypic. However, $NaAg_3O_2$ is not isotypic; there is no transformation that yields an equivalent set of coordinates. This can also be deduced from the different kinds of Wyckoff positions: Position $4c$ has site symmetry $2/m$, whereas $4a$ and $4b$ have 222; $8e$ has site symmetry $\bar{1}$, in contrast to $..m$ of $8j$.

9.4 In space group $P6_3mc$, the positions $x, \bar{x}, z$ and $\bar{x}, x, z+\frac{1}{2}$ are symmetry equivalent. An equivalent position for Cl 1 therefore is 0.536, 0.464, 0.208. The Euclidean normalizer of $P6_3mc$ ($P^1 6/mmm$) has the arbitrary translation $0, 0, t$ as an additional generator. After addition of 0.155 to all z coordinates of Ca_4OCl_6 its coordinates nearly agree with those of Na_6FeS_4. The Na positions correspond to the Cl positions, the S positions to the Ca positions. The structures are isotypic.

9.5 The Euclidean normalizer of $P6_3mc$ ($P^1 6/mmm$) has the inversion at the origin as an additional generator. The coordinates of wurtzite and rambergite are equivalent; it is the same structure type. Space group $P6_3mc$ has mirror planes, it is not a Sohncke space group; the structure is not chiral, and therefore there exists no absolute configuration.

9.6 GeS_2: The Euclidean normalizer of $I\bar{4}2d$ is $P4_2/nmm$, index 4. There are four equivalent sets of coordinates that are obtained by addition of $0, 0, \frac{1}{2}$ and by inversion at $\frac{1}{4}, 0, \frac{1}{8}$ (i.e. $\frac{1}{2}-x, -y, \frac{1}{4}-z$). Equivalent descriptions for GeS_2 are thus:

	x	y	z	x	y	$\frac{1}{2}+z$	$\frac{1}{2}-x$	$-y$	$\frac{1}{4}-z$	$\frac{1}{2}-x$	$-y$	$\frac{3}{4}-z$
Ge	0	0	0	0	0	$\frac{1}{2}$	$\frac{1}{2}$	0	$\frac{1}{4}$	$\frac{1}{2}$	0	$\frac{3}{4}$
S	0.239	$\frac{1}{4}$	$\frac{1}{8}$	0.239	$\frac{1}{4}$	$\frac{5}{8}$	0.261	$\frac{3}{4}$	$\frac{1}{8}$	0.261	$\frac{3}{4}$	$\frac{5}{8}$

None of these coordinate sets can be transformed to one of the others by a symmetry operation of $I\bar{4}2d$. The number of equivalent sets of coordinates is actually four; the space group $I\bar{4}2d$ is correct.

Na_2HgO_2: The Euclidean normalizer of $I422$ is $P4/mmm$ with axes $\frac{1}{2}(\mathbf{a}-\mathbf{b})$, $\frac{1}{2}(\mathbf{a}+\mathbf{b})$, $\frac{1}{2}\mathbf{c}$, index 4. There are four equivalent sets of coordinates that are obtained by inversion at 0, 0, 0 and by addition of 0, 0, $\frac{1}{2}$. For Na_2HgO_2, this results in:

	x	y	z	$-x$	$-y$	$-z$	x	y	$\frac{1}{2}+z$	$-x$	$-y$	$\frac{1}{2}-z$
Na	0	0	0.325	0	0	−0.325	0	0	0.825	0	0	0.175
Hg	0	0	0	0	0	0	0	0	$\frac{1}{2}$	0	0	$\frac{1}{2}$
O	0	0	0.147	0	0	−0.147	0	0	0.647	0	0	0.353

However, the first and the second as well as the third and the fourth sets of coordinates are symmetry equivalent in $I422$ (x, y, z and $-x, y, -z$) and thus are no new descriptions; there are only two coordinate sets. The space group $I422$ is wrong ($I4/mmm$ is correct).

9.7 The Euclidean normalizer of $P12_1/m1$ with specialized metric $a=c$ of the unit cell is $Bmmm$ with $\frac{1}{2}(\mathbf{a}+\mathbf{c})$, $\frac{1}{2}\mathbf{b}$, $\frac{1}{2}(-\mathbf{a}+\mathbf{c})$, index 16. An equivalent coordinate set is obtained by the transformation z, y, x (cf. Table 9.1):

	x	y	z	z	y	x
Au	0	0	0	0	0	0
Na1	0.332	0	0.669	0.669	0	0.332
Na2	0.634	$\frac{3}{4}$	0.005	0.005	$\frac{3}{4}$	0.634
Na3	0.993	$\frac{1}{4}$	0.364	0.364	$\frac{1}{4}$	0.993
Co	0.266	$\frac{3}{4}$	0.266	0.266	$\frac{3}{4}$	0.266
O1	0.713	0.383	0.989	0.989	0.383	0.713
O2	0.989	0.383	0.711	0.711	0.383	0.989
O3	0.433	$\frac{1}{4}$	0.430	0.430	$\frac{1}{4}$	0.433

The second coordinate set is not a new coordinate set. $P12_1/m1$ being centrosymmetric, x, y, z and $\bar{x}, \bar{y}, \bar{z}$ are symmetry equivalent in the first place. The coordinates of O1 and O2 are the same, and this also applies to Na2 and Na3 after inversion of the signs; therefore, these atoms are actually symmetry equivalent. The number of equivalent sets of coordinates is less than the index in the Euclidean normalizer; the space group $P2_1/m$ is wrong.

The basis vectors **a** and **c**, having the same length, span a rhombus whose diagonals are mutually perpendicular and thus allow for a B-centred, orthorhombic cell.

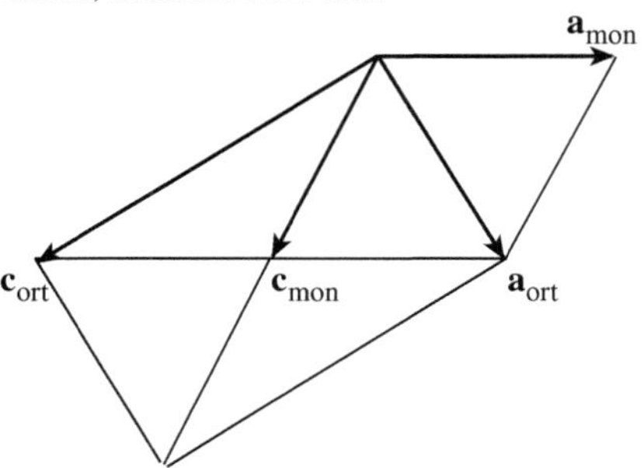

The monoclinic → orthorhombic transformation matrices are:

$$\boldsymbol{P}=\begin{pmatrix}1&0&-1\\0&1&0\\1&0&1\end{pmatrix}\quad\text{and}\quad\boldsymbol{P}^{-1}=\begin{pmatrix}\frac{1}{2}&0&\frac{1}{2}\\0&1&0\\-\frac{1}{2}&0&\frac{1}{2}\end{pmatrix}$$

The coordinate transformation $x'=\frac{1}{2}x+\frac{1}{2}z$, $y'=y$, $z'=-\frac{1}{2}x+\frac{1}{2}z$ then yields the coordinates in the orthorhombic cell:

	x'	y'	z'		x'	y'	z'
Au	0	0	0	Co	0.266	$\frac{3}{4}$	0
Na1	$\frac{1}{2}$	0	0.168	O1	0.851	0.383	0.138
Na2	0.320	$\frac{3}{4}$	−0.314	O3	0.432	$\frac{1}{4}$	0

The actual space group must be a B-centred, orthorhombic supergroup of $P12_1/m1$. From *International Tables A1* we see that there is only one centred orthorhombic supergroup of $P12_1/m1$: $Cmcm$ is mentioned. However, in *International Tables* supergroups are only listed with their standard settings (see Example 7.2, page 97). In the table of space group $Cmcm$, $P112_1/m$ is mentioned as a subgroup (i.e. with monoclinic c axis), and not $P12_1/m1$. By cyclic exchange of axes $\mathbf{a}\leftarrow\mathbf{b}\leftarrow\mathbf{c}\leftarrow\mathbf{a}$ $P112_1/m$ is transformed to $P12_1/m1$ and $Cmcm$ to $Bbmm$. Therefore, the actual supergroup of $P12_1/m1$ is $Bbmm$, a non-conventional setting of $Cmcm$; this is the correct space group of Na_4AuCoO_5.

10.1 The inverse of the transformation matrix $\boldsymbol{P}$ is $\boldsymbol{P}^{-1}$:

$$\boldsymbol{P}=\begin{pmatrix}1&0&1\\0&1&1\\0&0&1\end{pmatrix}\qquad\boldsymbol{P}^{-1}=\begin{pmatrix}1&0&-1\\0&1&-1\\0&0&1\end{pmatrix}$$

The coordinates are transformed according to $\boldsymbol{P}^{-1}$ (i.e. $x-z, y-z, z$). This causes the coordinates of Wyckoff position $1b$ $(0,0,\frac{1}{2})$ to be converted to those of $1h$ $(\frac{1}{2},\frac{1}{2},\frac{1}{2})$ and vice versa, and those of $1f$ $(\frac{1}{2},0,\frac{1}{2})$ to $1g$ $(0,\frac{1}{2},\frac{1}{2})$ and vice versa. The positions $1b$ and $1h$ as well as $1f$ and $1g$ are interchanged.

10.2 $\mathbf{abc}$: $P2_1/n2_1/m2_1/a$ $\quad 0.24, \frac{1}{4}, 0.61$

$\mathbf{cab}$: $P2_1/b2_1/n2_1/m$ $\quad 0.61, 0.24, \frac{1}{4}$

$\mathbf{abc}$: $P2_1/n2_1/m2_1/a$ $\quad 0.24, \frac{1}{4}, 0.61$

$\mathbf{b\bar{a}c}$: $P2_1/m2_1/n2_1/b$ $\quad \frac{1}{4}, -0.24, 0.61$

10.3 The space-group symbol becomes $C2/c11$, the monoclinic axis is **a**. To ensure a right-handed coordinate system, the direction of one basis vector has to be inverted, preferably $\mathbf{a}'=-\mathbf{b}$ because then the monoclinic angle keeps its value, $\alpha'=\beta$; inverting $\mathbf{b}'=-\mathbf{a}$ or $\mathbf{c}'=-\mathbf{c}$ would entail $\alpha'=180^\circ-\beta$. The new glide-reflection plane c is at $x = 0$. The coordinates have to be exchanged correspondingly, $-y, x, z$ (if $\mathbf{a}' = -\mathbf{b}$). If the transformation to the monoclinic **a** setting were made by cyclic exchange, the space group symbol would be $B2/b11$ and the Wyckoff symbols would be retained. Doubts can arise with the setting $C2/c11$ as to which Wyckoff symbols belong to which positions; they should be named explicitly.

10.4 $C4_2/e2/m2_1/c$.

10.5 The three cell choices for $P12_1/c1$ are $P12_1/c1$, $P12_1/n1$, and $P12_1/a1$. With coinciding axes directions, the supergroup has to have a 2_1 axis parallel to **b** and a glide-reflection plane perpendicular to **b**. With the setting to be chosen, the lattice parameters of the group and its supergoup have to be (nearly) the same and the monoclinic angle has to be $\beta \approx 90°$.
Supergroup $Pnna$ ($P2/n2_1/n2/a$): $P2/n2_1/n2/a$ is a supergroup if the initial setting $P12_1/n1$ instead of $P12_1/c1$ is chosen; with the setting $P12_1/n1$ the monoclinic angle must be (nearly) $\beta = 90°$.
Supergroup $Pcca$ ($P2_1/c2/c2/a$): $2_1/c$ takes the first position which is the direction of **a**; the directions of **a** and **b** have to be interchanged, **c** must be retained; the symbol of the supergroup is then $P2/c2_1/c2/b$.
Supergroup $Pccn$ ($P2_1/c2_1/c2/n$): $2_1/c$ takes the first and second positions; there are two supergroups of type $Pccn$; one of them is $P2_1/c2_1/c2/n$ with unchanged axes setting; the other one requires an interchange of **a** and **b**, but this does not change the Hermann–Mauguin symbol (the interchanged axes have to be mentioned explicitly); which one of the two possibilities is chosen depends on the lattice parameters.
Supergroup $Cmce$ ($C2/m2/c2_1/e$): $2_1/e$ means $2_1/a$ and $2_1/b$ referred to the direction **c**; interchange of **b** and **c** converts the Hermann–Mauguin symbol to $B2/m2_1/e2/b$; by cyclic exchange $\mathbf{c} \leftarrow \mathbf{a} \leftarrow \mathbf{b} \leftarrow \mathbf{c}$ it becomes $B2/b2_1/e2/m$; in both cases the e contains the necessary c glide reflection. The conventional space-group symbol of the supergroup $Cmce$ therefore refers to two B-centred orthorhombic supergroups. The B-centred cells are twice as large as that of $P12_1/c1$, so that the cell of the latter cannot be retained; it must be transformed as shown in the graph, and this must result in a right angle β_{ort}. $\mathbf{a}_{\text{ort}}$ and $\mathbf{c}_{\text{ort}}$ are the basis vectors of the B-centred cells.

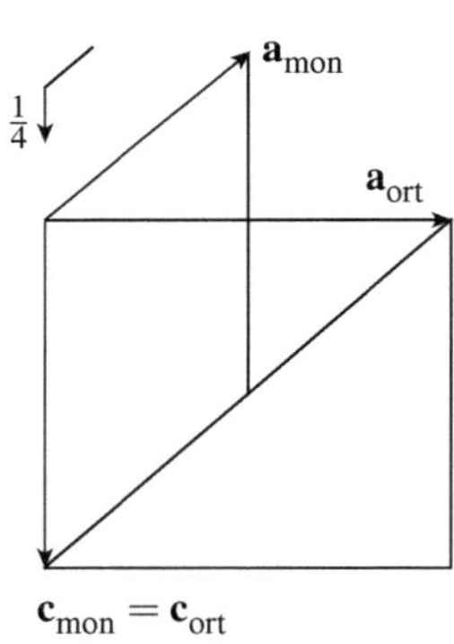

12.1 The symmetry reduction from $P2_1/a\bar{3}$ to $P2_1/b2_1/c2_1/a$ involves the loss of the threefold axes. From $P2_1/b2_1/c2_1/a$ to $Pbc2_1$ (non-conventional for $Pca2_1$), in addition, the centres of inversion, the 2_1 axes parallel to **a** and **b**, and the glide planes perpendicular to **c** are lost. From $P2_1/a\bar{3}$ to $P2_13$ the centres of inversion and the glide planes are lost.

12.2 and **12.3**

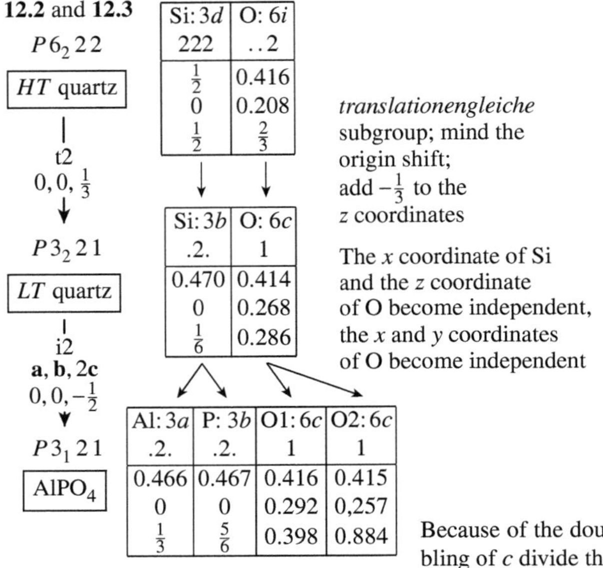

Because of the doubling of c divide the z coordinates by 2, then, due to the origin shift, add $\frac{1}{4}$ and $\frac{3}{4}$ to the z coordinates.

12.4 Indium: The relation $Fm\bar{3}m \to I4/mmm$ with the basis transformation $\mathbf{a}' = \frac{1}{2}(\mathbf{a} - \mathbf{b})$ yields a ratio of $c/a = \sqrt{2}$. For indium it is $c/a = 1.52$ which is slightly greater than $\sqrt{2}$. It can be described as a cubic-closest packing of spheres that has been slightly expanded in the direction of c.
Protactinium: With $c/a = 0.83$ it is to be described as a compressed body-centred packing of spheres.

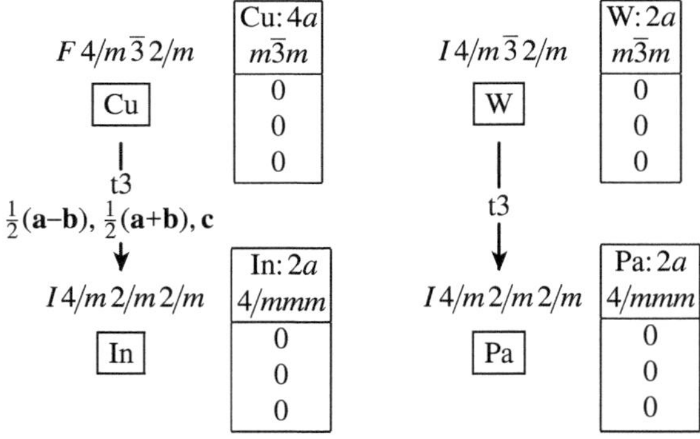

Mercury: The relation $Fm\bar{3}m \to R\bar{3}m^{(\text{hex})}$ requires the basis transformation $\frac{1}{2}(-\mathbf{a}+\mathbf{b})$, $\frac{1}{2}(-\mathbf{b}+\mathbf{c})$, $\mathbf{a}+\mathbf{b}+\mathbf{c}$; from this follows a ratio of $c/a = \sqrt{3}/(\frac{1}{2}\sqrt{2}) = 2.45$. That is slightly greater than $c/a = 1.93$ of mercury; it can be described as a trigonally distorted cubic-closest packing of spheres. For $Im\bar{3}m \to$

$R\bar{3}m^{(\text{hex})}$ the transformation would be $-\mathbf{a}+\mathbf{b}$, $-\mathbf{b}+\mathbf{c}$, $\frac{1}{2}(\mathbf{a}+\mathbf{b}+\mathbf{c})$ and thus $c/a = \frac{1}{2}\sqrt{3}/\sqrt{2} = 0.61$.
Uranium: We have $b/a = 2.06$ instead of $\sqrt{3}$ and $c/a = 1.74$ instead of 1.633; it is a hexagonal-closest packing of spheres that has been expanded in the direction of b of the orthorhombic cell.

$F\,4/m\,\bar{3}\,2/m$ — Cu

Cu: $4a$
$m\bar{3}m$
0
0
0

t4
$\frac{1}{2}(-\mathbf{a}+\mathbf{b})$, $\frac{1}{2}(-\mathbf{b}+\mathbf{c})$, $\mathbf{a}+\mathbf{b}+\mathbf{c}$

$R\,\bar{3}\,2/m$ — α-Hg

Hg: $3a$
$\bar{3}m$
0
0
0

$P\,6_3/m\,2/m\,2/c$ — Mg

Mg: $2c$
$\bar{6}m2$
$\frac{1}{3}$
$\frac{2}{3}$
$\frac{1}{4}$

t3
$\mathbf{a}$, $\mathbf{a}+2\mathbf{b}$, $\mathbf{c}$ $\quad x-\frac{1}{2}y,\ \frac{1}{2}y,\ z$

$C\,2/m\,2/c\,2_1/m$ — α-U

U: $4c$
$m2m$
0
0.398
$\frac{1}{4}$

12.5 The lattice parameters and the coordinates of $\sim\frac{1}{3}$ show that the cell has to be enlarged by $3\times3\times3$. The index 27 is the smallest possible index for an isomorphic subgroup of $I\,m\bar{3}m$.

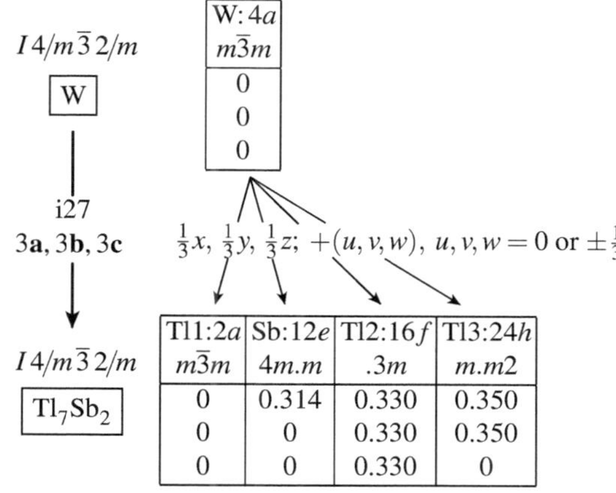

12.6 The first step of the symmetry reduction rutile $\rightarrow$ $CaCl_2$ type is as in Fig. 1.2.

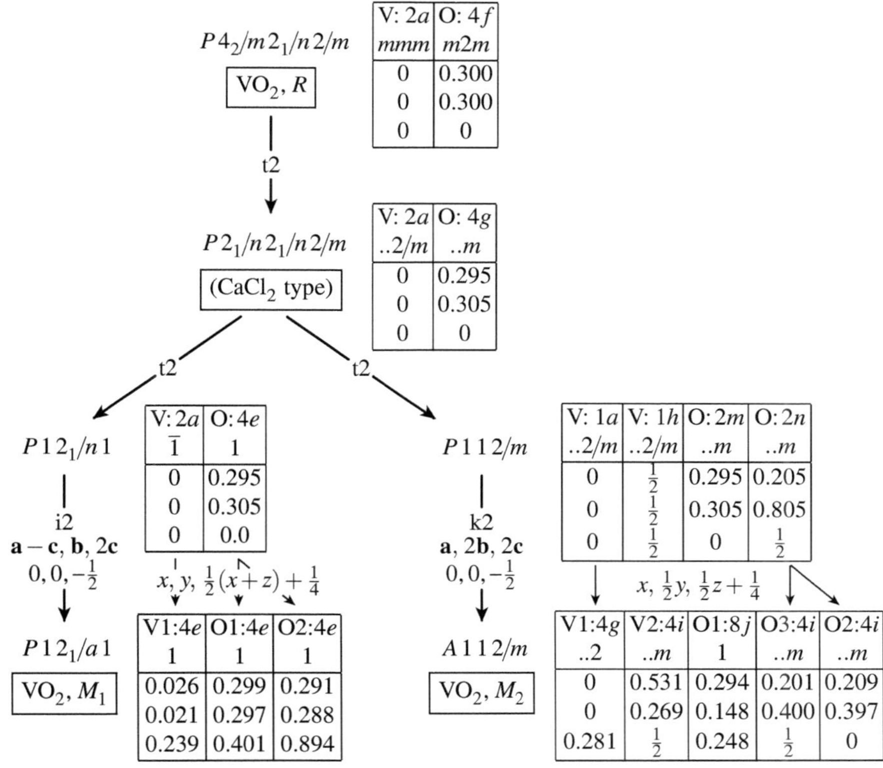

12.7 The lattice parameters of $TlAlF_4$-*tP*6 are transformed to those of $TlAlF_4$-*tI*24 according to $a\sqrt{2} = 364.9\sqrt{2}$ pm $= 516.0$ pm ≈ 514.2 pm and $2c = 2\times 641.4$ pm $= 1282.8$ pm ≈ 1280.7 pm. Since the cell of $TlAlF_4$-*tI*24 is body centred, it is doubly primitive; the primitive cell has only a doubled volume, in accordance with the index 2. $TlAlF_4$-*tI*24 no longer has the F2 atoms on mirror planes. The coordination octahedra about the Al atoms are mutually turned about the direction of *c*.

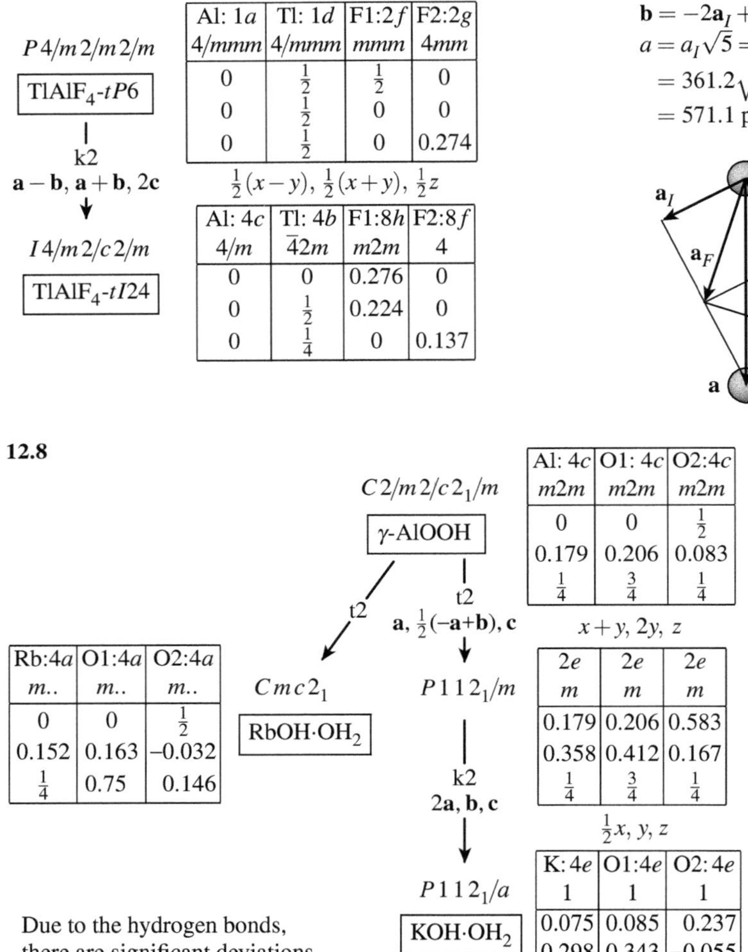

12.9 Halfway between the reflection planes that run diagonally through the cell of $P4/nmm$, there are glide planes with the glide component $\frac{1}{2}(\mathbf{a}+\mathbf{b})$; this automatically means also a glide component of $\frac{1}{2}(\mathbf{a}+\mathbf{b})+\mathbf{c}$. Upon doubling of *c*, the glide component $\frac{1}{2}(\mathbf{a}+\mathbf{b})$ is lost, but the glide component $\frac{1}{2}(\mathbf{a}+\mathbf{b})+\frac{1}{2}\mathbf{c}$ is retained ($\frac{1}{2}\mathbf{c}$ instead of $\mathbf{c}$ due to the doubling of *c*). That is an *n* glide plane.

12.10 The face-centred cell of the Cu type has an oblique orientation relative to the cell of $MoNi_4$. First, we transform the face-centred cell to a body-centred cell and then to the cell of $MoNi_4$:

$$\mathbf{a}_I = \tfrac{1}{2}(\mathbf{a}_F - \mathbf{b}_F),$$
$$\mathbf{b}_I = \tfrac{1}{2}(\mathbf{a}_F + \mathbf{b}_F),\ \mathbf{c}_I = \mathbf{c}_F$$
$$a_I = a_F/\sqrt{2}$$
$$\mathbf{a} = \mathbf{a}_I + 2\mathbf{b}_I,$$
$$\mathbf{b} = -2\mathbf{a}_I + \mathbf{b}_I,\quad \mathbf{c} = \mathbf{c}_I$$
$$a = a_I\sqrt{5} = (a_F/\sqrt{2})\sqrt{5} = 361.2\sqrt{\tfrac{5}{2}}\ \text{pm} = 571.1\ \text{pm}$$

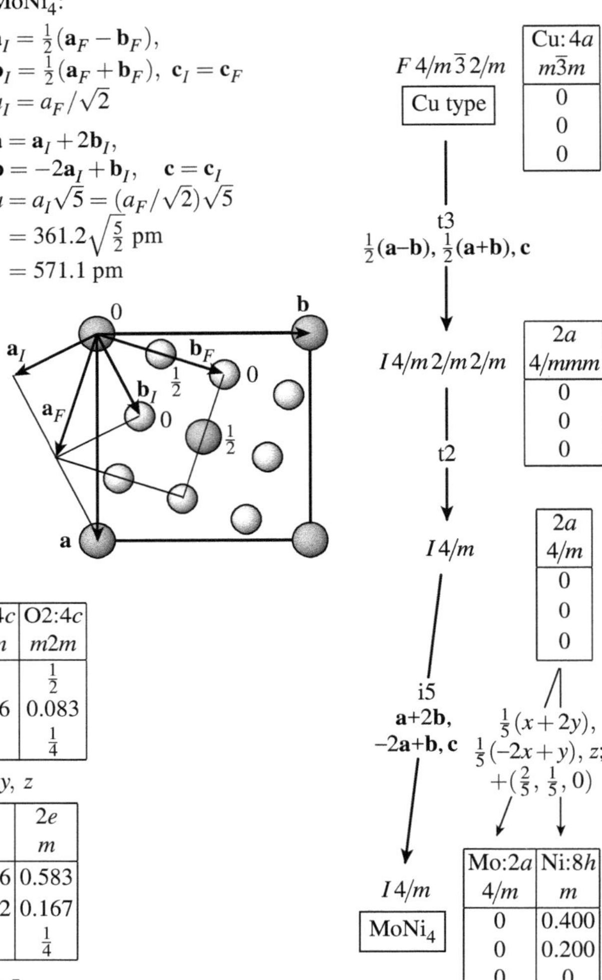

The face-centred cell of the Cu type is fourfold-primitive and contains $Z = 4$ Cu atoms; the body-centred cell is half as large ($Z = 2$). Compared to the latter, the cell of $MoNi_4$ is enlarged by a factor of 5; it contains 10 atoms. This agrees with the experimental value of $a = 572.0$ pm. Therefore, a step of symmetry reduction of index 5 is needed. An index of 5 is possible only as an isomorphic subgroup of $I4/m$; before, first a symmetry reduction down to $I4/m$ is required; this step involves only site symmetry reductions.

12.11

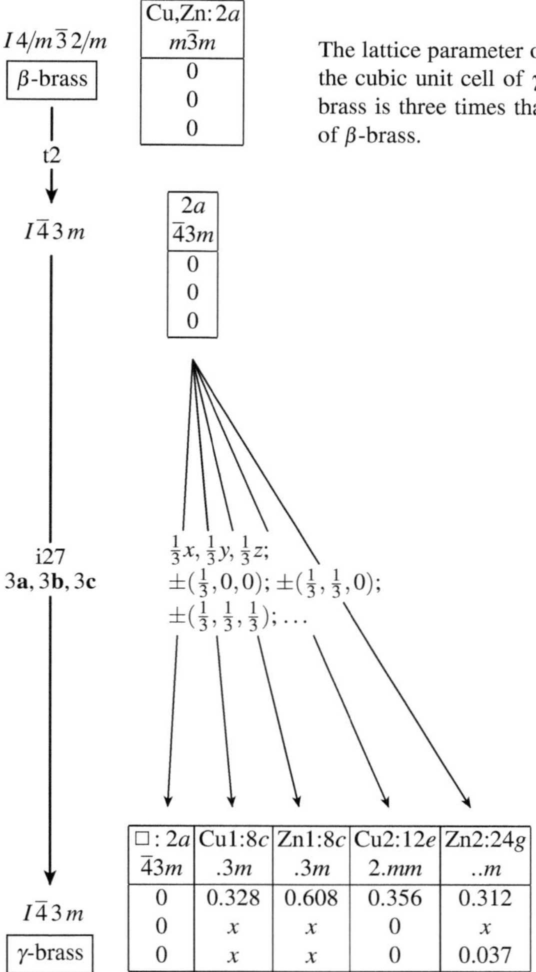

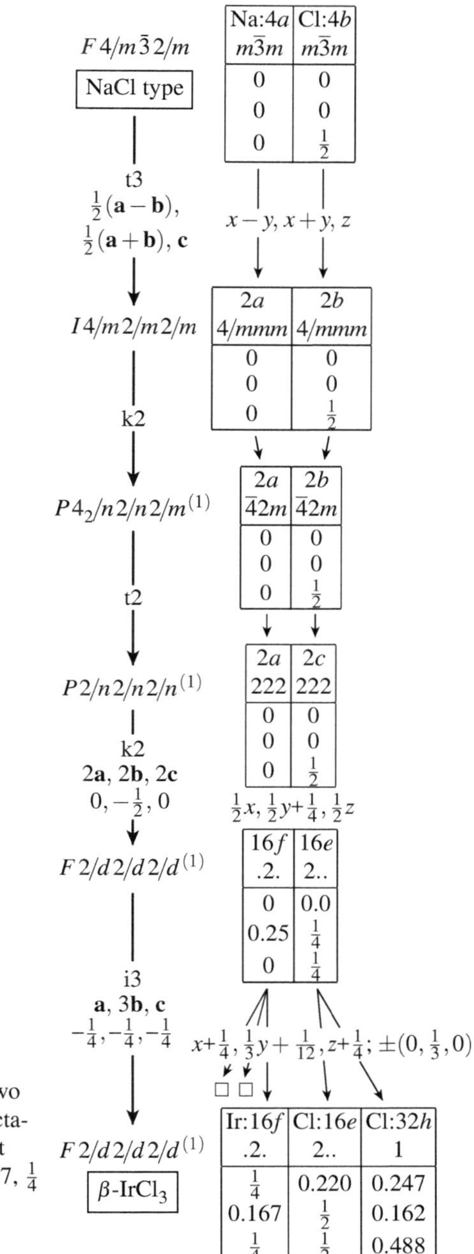

□ designate two unoccupied octahedral voids at $16f$, $\frac{1}{4}, -0.167, \frac{1}{4}$ and $\frac{1}{4}, 0.5, \frac{1}{4}$

13.1 The common supergroup has to have 2_1 axes in all three axis directions, and it must have *a* glide planes perpendicular to **c**. Only two supergroups fulfil these conditions: $P2_1/b2_1/c2_1/a$ and $P2_1/n2_1/m2_1/a$. The relation $P2_1/n2_1/m2_1/a$ —t2→ $P2_12_12_1$ requires an origin shift $(0, 0, -\frac{1}{4})$ and is thus excluded. $P2_1/b2_1/c2_1/a$ is the common supergroup.

13.2 In the right branch of Fig. 13.3 the cell decrease $\frac{1}{2}(\mathbf{a}-\mathbf{b})$, $\frac{1}{2}(\mathbf{a}+\mathbf{b})$, $\mathbf{c}$ is missing. In addition, there is no step that permits the necessary origin shift.
Pnnn has eight subgroups on a par of type *F ddd* with doubled lattice parameters. Only the one with the mentioned origin shift yields the correct coordinates for the Ir and Cl atoms.

13.3 $Pm\bar{3}m$ has two subgroups on a par of the same type $Pm\bar{3}m$ that are not maximal subgroups (two conjugacy classes of four conjugates each). Either $Fm\bar{3}m$ or $Im\bar{3}m$ has to be considered as an intermediate group with an eightfold enlarged unit cell.

13.4 The right branch of the tree is correct. The left branch leads to the same cell; however, by the triplication of the cell in the first step, translational symmetry is removed that is reinserted in the second step; this is not permitted. The intermediate group $C2/m2/c2_1/m$ is wrong. This can easily be recognized by drawing the unit cells (only the inversion centres have been drawn):

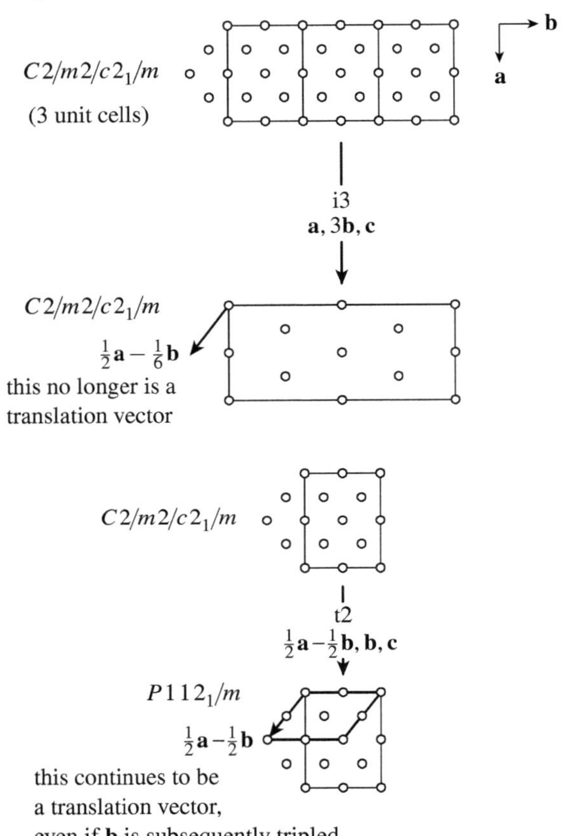

In addition, it is recommended to draw the vertical distances between the space group symbols proportional to the logarithms of the indices; the distance at the i3 step should be $\lg 3/\lg 2 = 1.58$ times greater than at the t2 step.

13.5 $P2/m2/c2_1/m$ is not centred, there is no translation vector $\frac{1}{2}(\mathbf{a}-\mathbf{b})$. $P112_1/m$ is only a subgroup of $P2/m2/c2_1/m$ without cell transformation.

14.1 Compared to a sphere packing of iodine atoms with $a = 2r(\text{I})$, a is larger by a factor of approximately $\sqrt{3}$: $a = 699$ pm $\approx 2\sqrt{3} \times 198$ pm $=$ 686 pm. The packing of spheres has $c/a = 1.633$; the expected value is thus $c \approx 2 \times 1.633 \times 198$ pm $= 647$ pm. Actually, $c = 736$ pm is somewhat larger because there are no chemical bonds in this direction.

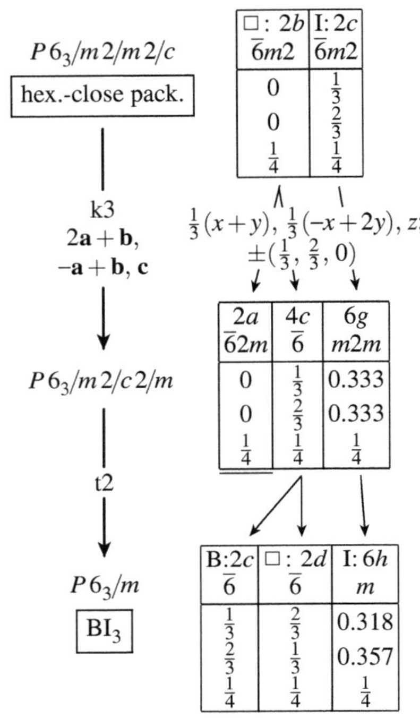

14.2 First, symmetry must be reduced in two *translationengleiche* steps from $F4/m\bar{3}2/m$ to $R\bar{3}$, then follows an isomorphic subgroup of index 13 or 19, respectively, where $\frac{12}{13}$ and $\frac{18}{19}$ of the symmetry operations are lost. 13 and 19 are prime numbers of the kind $6n+1$ that are permitted as indices of isomorphic subgroups of $R\bar{3}$. By comparison with *International Tables* A1, the number of symmetry-independent Cl atoms (2 and 3, respectively) indicates that 13 and 19 are the sought prime numbers: for isomorphic subgroups of $R\bar{3}$ of index p, the position $3a$ splits into $1 \times 3a$ and $\frac{p-1}{6} \times 18f$, that is, $\frac{13-1}{6} = 2 \times 18f$ and $\frac{19-1}{6} = 3 \times 18f$.

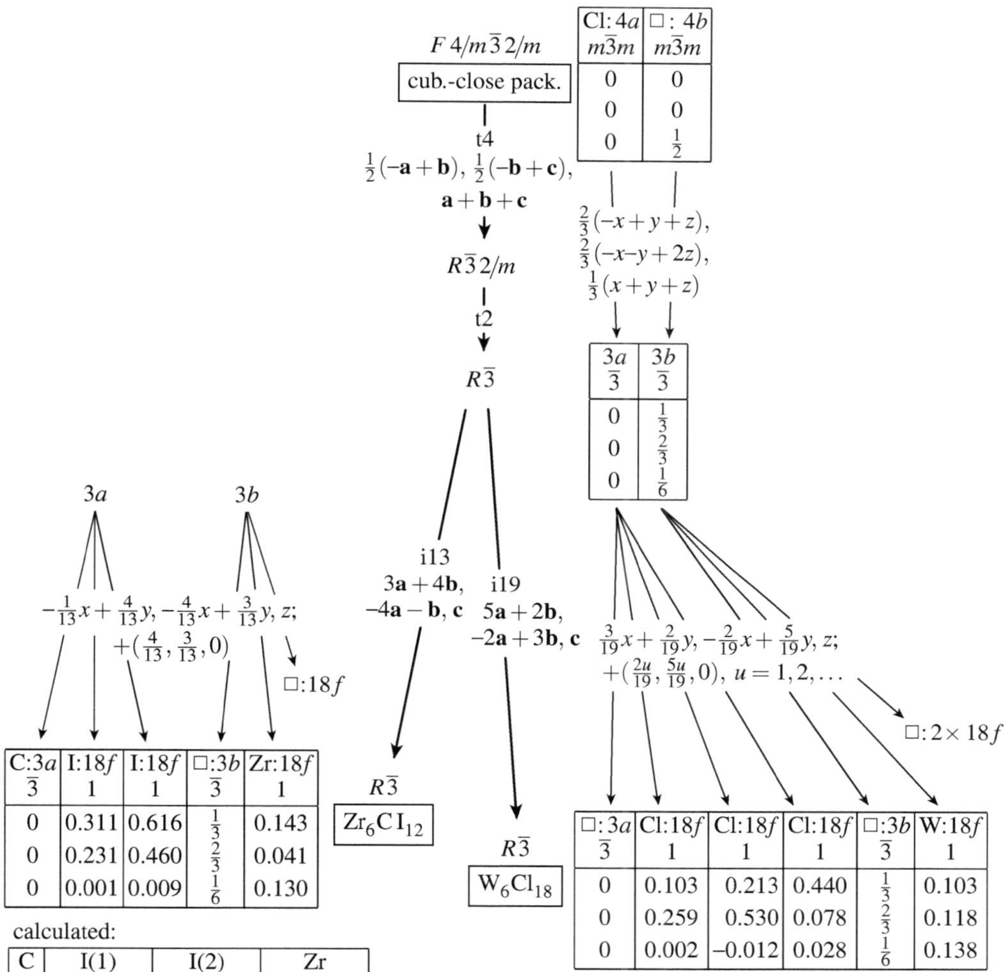

calculated:

C	I(1)	I(2)	Zr
0	$\frac{4}{13}=0.308$	$\frac{8}{13}=0.615$	$\frac{7}{39}=0.180$
0	$\frac{3}{13}=0.231$	$\frac{6}{13}=0.462$	$\frac{2}{39}=0.051$
0	0	0	$\frac{1}{6}=0.167$

calculated:

Cl(2)	Cl(1)	Cl(3)	W
$\frac{2}{19}=0.105$	$\frac{4}{19}=0.211$	$\frac{8}{19}=0.421$	$\frac{7}{57}=0.123$
$\frac{5}{19}=0.263$	$\frac{10}{19}=0.526$	$\frac{20}{19}=1.053$	$\frac{8}{57}=0.140$
0	0	0	$\frac{1}{6}=0.167$

14.3

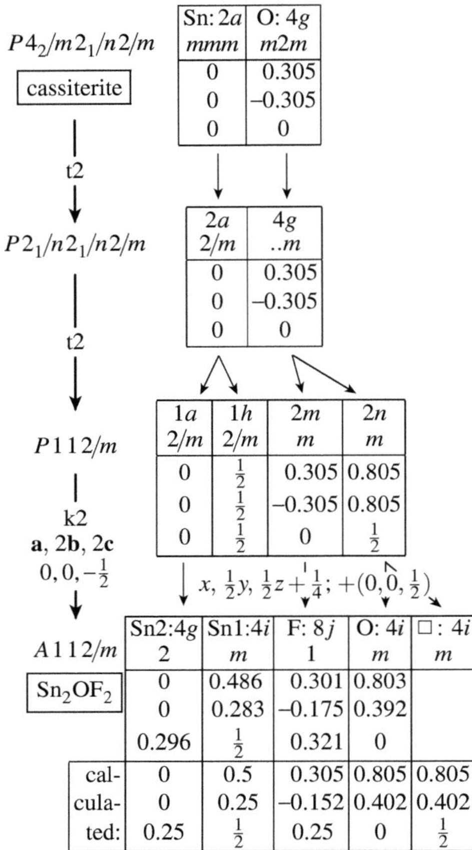

calculated:
$a = 474$ pm, $b = 947$ pm, $c = 637$ pm, $\gamma = 90°$
observed:
$a = 507$ pm, $b = 930$ pm, $c = 808$ pm, $\gamma = 97.9°$

14.4

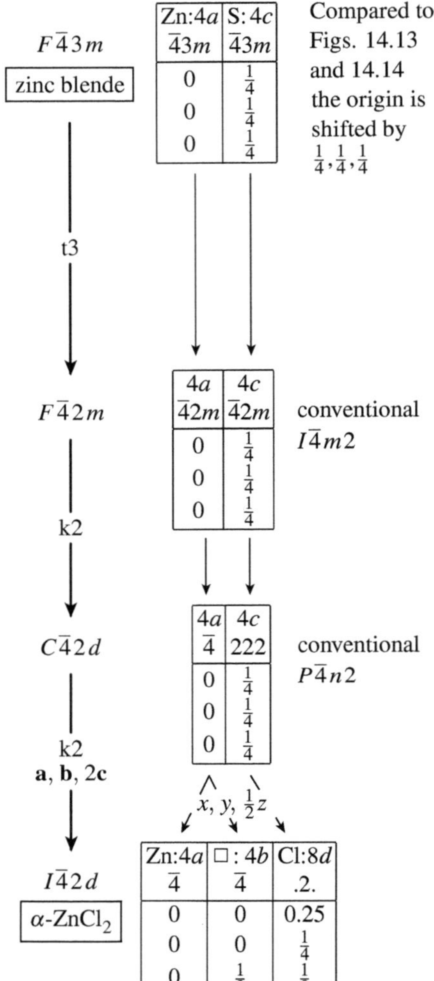

14.5

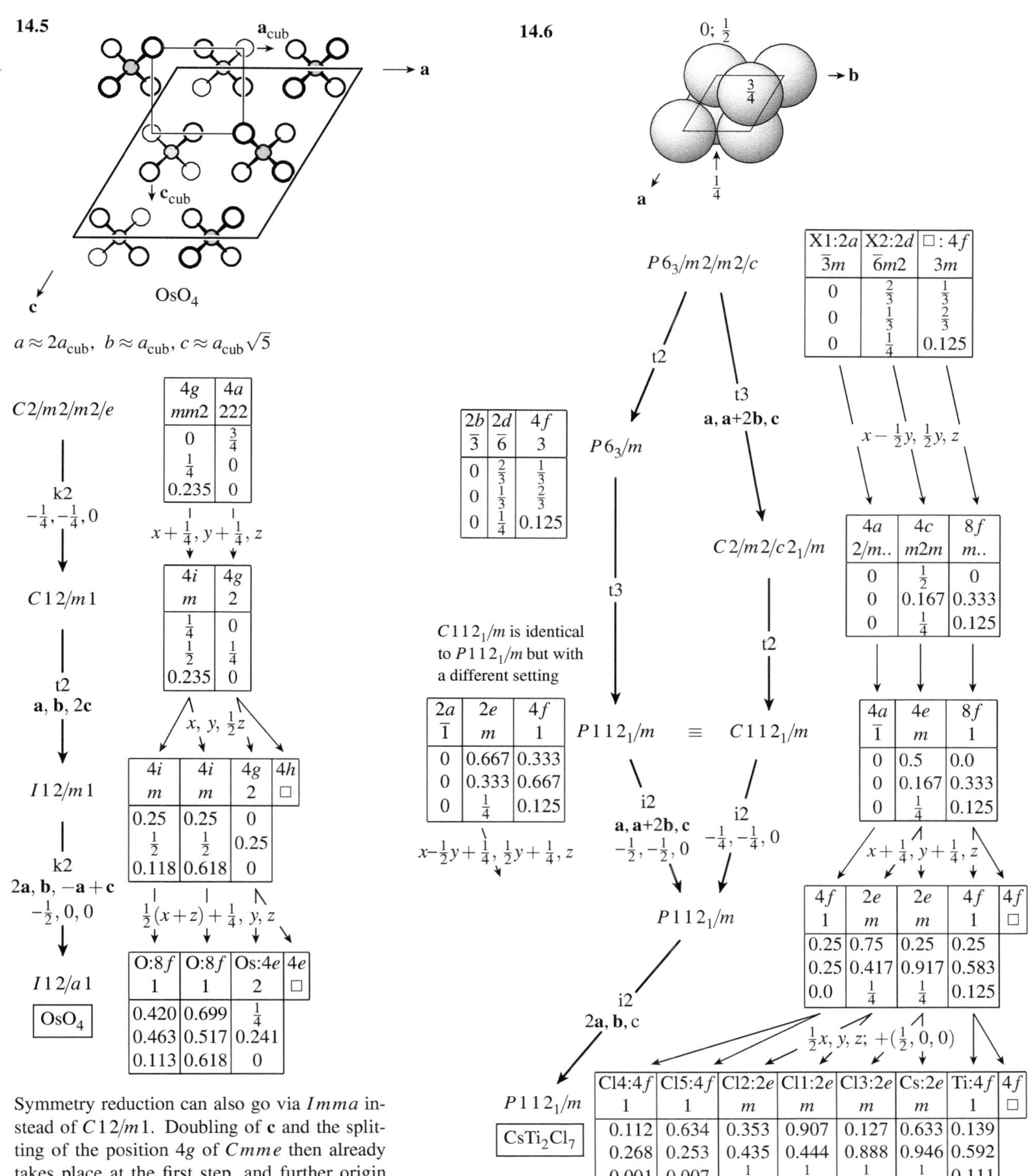

Symmetry reduction can also go via $Imma$ instead of $C12/m1$. Doubling of **c** and the splitting of the position 4g of $Cmme$ then already takes place at the first step, and further origin shifts have to be considered.

15.1 Let the basis vectors of $C_{16}H_{34}\cdot[OC(NH_2)_2]_{12}$-I be **a**, **b**, and **c**. Then the lattice parameters of the other compounds are:

$C_{16}H_{34}\cdot[OC(NH_2)_2]_{12}$-II	$a, b, 2c$
$C_{16}H_{34}\cdot[OC(NH_2)_2]_{12}$-III	$a, a\sqrt{3}, c$
$C_{10}H_{18}O_2\cdot[OC(NH_2)_2]_8$	$a, b, 4c$
$C_8H_{16}O_2\cdot[OC(NH_2)_2]_7$	$a, b, 7c$

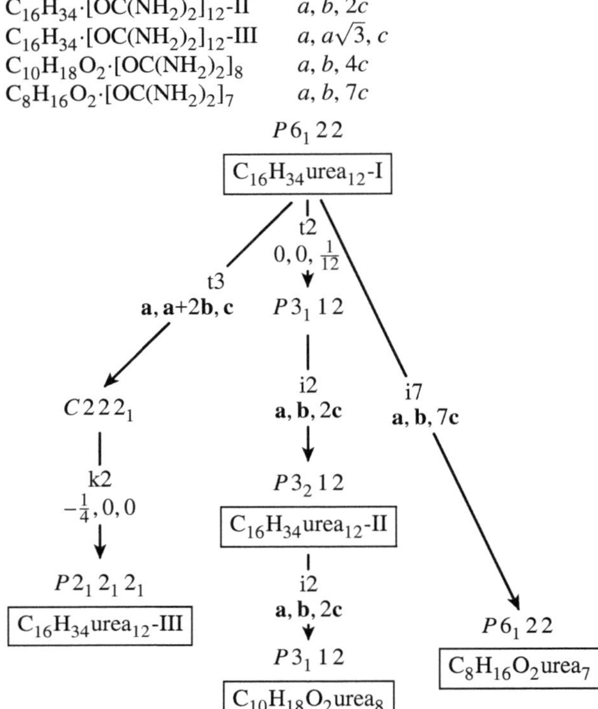

15.2 The Bärnighausen tree is like that of Exercise 14.2, page 333, but the position $4b$ of $Fm\bar{3}m$ can be deleted. Instead of the i13 and i19 steps there is an i7 step with basis transformation $3\mathbf{a}+\mathbf{b}, -\mathbf{a}+2\mathbf{b}, \mathbf{c}$ and coordinate transformation $\frac{2}{7}x+\frac{1}{7}y, -\frac{1}{7}x+\frac{3}{7}y, z$; $\pm(\frac{1}{7},\frac{3}{7},0)$. Position $3a$ of $R\bar{3}$ splits into $3a$ and $18f$. If hexachlorbenzene, C_6Cl_6, were to crystallize with this packing, the space group would be $R\bar{3}$ (with the enlarged unit cell). Because $(BN)_3Cl_6$ does not have inversion centres, a step $R\bar{3}$ —t2→ $R3$ follows, with splitting of $18f$ to $2\times 9b$.

16.1 After EHRENFEST: At the point of transition there may be no discontinuous change of density and volume and no latent heat. Continuous: There may be no latent heat and no hysteresis; There must exist some variable that can be used as an order parameter (e.g. the position of a key atom) and that adopts the value of zero in the high-symmetry form and follows a power law in the low-symmetry phase (eqn 22.5, page 290); the frequency of a vibrational mode of appropriate symmetry must also follow the power law and tend toward zero when approaching the point of transition (soft mode).

16.2 At an isosymmetric phase transition the space group does not change. There is no group–subgroup relation. Therefore, a continuous transition is excluded.

16.3 Volume is a first derivative of the Gibbs free energy, $V = \partial G/\partial p$. It may not exhibit a discontinuity at a second-order transition. The second derivative $\partial^2 G/(\partial p \partial T) = \partial V/\partial T$ is the slope of the curve V over T in Fig. 16.4. A kink of the curve corresponds to a discontinuity of the second derivative. At least one second derivative of G must have a discontinuity.

16.4 All modifications belong to different branches of the Bärnighausen tree. If there are phase transitions, they can only be discontinuous (first order).

16.5 The symmetry reduction $Pm\bar{3}m$ —t3→ $P4/mmm$ —t2→ $P4mm$ involves a t3 and a t2 step. The formation of twins of triplets is to be expected. The +**c** axis of $P4mm$ can be oriented parallel to $\pm\mathbf{a}$, $\pm\mathbf{b}$, or $\pm\mathbf{c}$ of $Pm\bar{3}m$.

16.6 The symmetry reduction $Pm\bar{3}m$ —t3→ $P4/mmm$ —k2→ $I4/mcm$ involves a t3 and k2 step. The formation of triplets is to be expected, each of which has two kinds of antiphase domains.

16.7 $I\bar{4}2m$ is a *translationengleiche* subgroup of index 2 of $I4/mcm$; twins with two orientations are expected, +**c** and −**c**. The X-ray pattern will have reflections from both domains that are exactly coincident, including the extinct reflections ($h+k+l=$ odd); the high-symmetry space group $I4/mcm$ will not be feigned because it has c glide reflections that cause different extinctions. Twinning will not be noticed before structure refinement, where it will show up by seemingly split atomic positions or abnormal ellipsoids of 'thermal motion'.

16.8 The symmetry reduction $Fd\bar{3}m \rightarrow F\bar{3}m$ ($R\bar{3}m$) is *translationengleiche* of index 4. Twins with four orientations may appear, with their $\bar{3}$ axes in the directions of the four cube diagonals of the initial cubic unit cell. A reflection *hkl* of the cubic phase splits into four different reflections hkl, $\bar{h}kl$, $h\bar{k}l$, and $hk\bar{l}$.

16.9 The curves show hysteresis, the transition is discontinuous. The same follows from the missing group–subgroup relation. Since the atoms are only slightly displaced, the transformation could be termed displacive. Because of the increase of the coordination number of the Si atom from 4 to 5 it should be reconstructive. The example shows that a clear distinction between reconstructive and displacive is not always possible.

17.1 Let $\boldsymbol{P}$ be the transformation matrix A → B for the basis vectors; then we have:

$$(\mathbf{a}_B, \mathbf{b}_B, \mathbf{c}_B) = (\mathbf{a}_A, \mathbf{b}_A, \mathbf{c}_A)\boldsymbol{P} \qquad (h_B, k_B, l_B) = (h_A, k_A, l_A)\boldsymbol{P}$$

$$(\mathbf{a}_A, \mathbf{b}_A, \mathbf{c}_A) = (\mathbf{a}_B, \mathbf{b}_B, \mathbf{c}_B)\boldsymbol{P}^{-1} \qquad (h_A, k_A, l_A) = (h_B, k_B, l_B)\boldsymbol{P}^{-1}$$

$$\begin{pmatrix} \mathbf{a}_B^* \\ \mathbf{b}_B^* \\ \mathbf{c}_B^* \end{pmatrix} = \boldsymbol{P}^{-1} \begin{pmatrix} \mathbf{a}_A^* \\ \mathbf{b}_A^* \\ \mathbf{c}_A^* \end{pmatrix} \qquad \begin{pmatrix} x_B \\ y_B \\ z_B \end{pmatrix} = \boldsymbol{P}^{-1} \begin{pmatrix} x_A \\ y_A \\ z_A \end{pmatrix}$$

$$\begin{pmatrix} \mathbf{a}_A^* \\ \mathbf{b}_A^* \\ \mathbf{c}_A^* \end{pmatrix} = \boldsymbol{P} \begin{pmatrix} \mathbf{a}_B^* \\ \mathbf{b}_B^* \\ \mathbf{c}_B^* \end{pmatrix} \qquad \begin{pmatrix} x_A \\ y_A \\ z_A \end{pmatrix} = \boldsymbol{P} \begin{pmatrix} x_B \\ y_B \\ z_B \end{pmatrix}$$

$$\begin{pmatrix} u_A \\ v_A \\ w_A \end{pmatrix} = \boldsymbol{P} \begin{pmatrix} u_B \\ v_B \\ w_B \end{pmatrix}$$

(a) From the image of the reciprocal lattice shown with the exercise we deduce (note that only even h_B indices have been included):

$$\mathbf{a}_A^* \approx 2\mathbf{a}_B^* + 2\mathbf{c}_B^*;\ \mathbf{c}_A^* \approx -2\mathbf{a}_B^* + \mathbf{c}_B^*$$

$\mathbf{b}_A^*$ has the same value as $\mathbf{a}_A^*$, it is perpendicular to $\mathbf{c}_A^*$ and inclined by 60° against the plane of projection. In the projection, $\mathbf{b}_A^*$ is half as long as $\mathbf{a}_A^*$. We obtain:

$\mathbf{b}_A^* \approx \mathbf{a}_B^* + \mathbf{b}_B^* + \mathbf{c}_B^*$;

In matrix notation this is:

$$\begin{pmatrix} \mathbf{a}_A^* \\ \mathbf{b}_A^* \\ \mathbf{c}_A^* \end{pmatrix} = \boldsymbol{P} \begin{pmatrix} \mathbf{a}_B^* \\ \mathbf{b}_B^* \\ \mathbf{c}_B^* \end{pmatrix} = \begin{pmatrix} 2 & 0 & 2 \\ 1 & 1 & 1 \\ -2 & 0 & 1 \end{pmatrix} \begin{pmatrix} \mathbf{a}_B^* \\ \mathbf{b}_B^* \\ \mathbf{c}_B^* \end{pmatrix}$$

(b) The transformation of the basis vectors A → B requires the same matrix $\boldsymbol{P}$ as the transformation of the reciprocal basis vectors in the opposite direction B → A:

$$(\mathbf{a}_B, \mathbf{b}_B, \mathbf{c}_B) = (\mathbf{a}_A, \mathbf{b}_A, \mathbf{c}_A) \begin{pmatrix} 2 & 0 & 2 \\ 1 & 1 & 1 \\ -2 & 0 & 1 \end{pmatrix}$$

The metric relations can be taken from the figure in the next column. With $\mathbf{a}' = 2\mathbf{a}_A + \mathbf{b}_A$ we obtain $a' = a_A\sqrt{3} = 654.4$ pm. The diagonal in the $a'c'$ plane has a length of $\sqrt{654.4^2 + 594.0^2}$ pm = 883.8 pm and corresponds to $c_B = 885.6$ pm. For a_B we expect $a_B = \sqrt{a'^2 + 4c_A^2} = \sqrt{654.4^2 + 4 \times 594.0^2}$ pm = 1356.3 pm (observed: 1420 pm). The monoclinic angle follows from β_B = arc tan$(2c_A/a')$ + arc tan(c_A/a') = 103.4° (observed: 100.0°). The enlargement of the cell corresponds to the fourth image for $\Xi = 3$ in Fig. 19.1, page 255.

(c) The *translationengleiche* symmetry reduction of index 3 yields an orthohexagonal cell (grey in the figure) which, however, is monoclinic for reasons of symmetry ($C12/m1$). Then follows an isomorphic subgroup with triplication of the unit cell. The figure shows that the enlarged cell is C centred.

(d) The transformation matrices $\boldsymbol{P}_1$ and $\boldsymbol{P}_2$ for both steps of symmetry reduction can be taken from the figure. Their product $\boldsymbol{P} = \boldsymbol{P}_1\boldsymbol{P}_2$ is the above-mentioned transformation matrix for the total transformation A-Ln_2O_3 → B-Ln_2O_3. As there are no origin shifts, we only need 3 × 3 matrices:

$$\boldsymbol{P}_1\boldsymbol{P}_2 = \begin{pmatrix} 2 & 0 & 0 \\ 1 & 1 & 0 \\ 0 & 0 & 1 \end{pmatrix} \begin{pmatrix} 1 & 0 & 1 \\ 0 & 1 & 0 \\ -2 & 0 & 1 \end{pmatrix} = \begin{pmatrix} 2 & 0 & 2 \\ 1 & 1 & 1 \\ -2 & 0 & 1 \end{pmatrix}$$

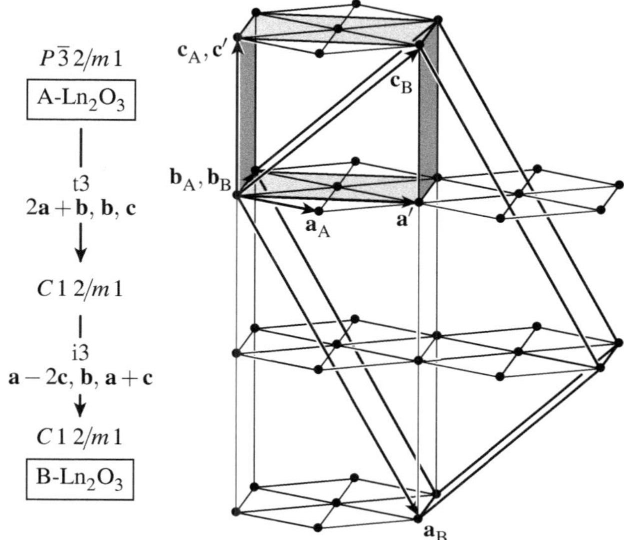

(e) The transformation of the atomic coordinates of the B form to the coordinate system of the A form requires the same matrix $\boldsymbol{P}$:

$$\begin{pmatrix} x_A \\ y_A \\ z_A \end{pmatrix} = \begin{pmatrix} 2 & 0 & 2 \\ 1 & 1 & 1 \\ -2 & 0 & 1 \end{pmatrix} \begin{pmatrix} x_B \\ y_B \\ z_B \end{pmatrix}$$

The results are:

	Sm(1)	Sm(2)	Sm(3)	O(1)	O(2)	O(3)	O(4)	O(5)
x	1.250	0.656	1.310	0.830	0.296	0.344	1.366	0
y	0.625	0.328	0.655	0.915	0.648	0.672	0.683	0
z	0.220	−0.242	−0.746	0.028	−0.377	0.778	0.605	0

The values agree acceptably with the coordinates of the A form, cf. the table given with the exercise; note that atoms at x, y, z are symmetry equivalent to $\bar{x}, \bar{y}, \bar{z}$ and to $x \pm m, y \pm n, z \pm q$ with $m, n, q =$ integral.

(f) The lattice direction (vector) [1 3 2] refers to the B form. Transformation to a lattice direction of the A form, B → A, requires the same matrix $\boldsymbol{P}$:

$$\begin{pmatrix} u_A \\ v_A \\ w_A \end{pmatrix} = \boldsymbol{P} \begin{pmatrix} u_B \\ v_B \\ w_B \end{pmatrix} = \begin{pmatrix} 2 & 0 & 2 \\ 1 & 1 & 1 \\ -2 & 0 & 1 \end{pmatrix} \begin{pmatrix} 1 \\ 3 \\ 2 \end{pmatrix} = \begin{pmatrix} 6 \\ 6 \\ 0 \end{pmatrix}$$

The axis of twinning [1 3 2] of the B form corresponds to the lattice direction [1 1 0] of the A form. This is the direction of a twofold rotation axis of the space group $P\bar{3}2/m1$ that is being lost by the symmetry reduction to $C12/m1$, but which is conserved indirectly by the twinning.

(g) The transformation of the Miller indices B → A requires the inverse matrix $\boldsymbol{P}^{-1}$. This can be deduced from the figure of the basis vectors on the preceding page; it corresponds to the transformation B → A of the basis vectors:

$$(h,k,l)_\mathrm{A} = (h,k,l)_\mathrm{B} \begin{pmatrix} \frac{1}{6} & 0 & -\frac{1}{3} \\ -\frac{1}{2} & 1 & 0 \\ \frac{1}{3} & 0 & \frac{1}{3} \end{pmatrix}$$

That it is correct can be checked by multiplication $\boldsymbol{P} \times \boldsymbol{P}^{-1} = \boldsymbol{I}$. The crystal faces of the B form are $\{\bar{2}01\}_\mathrm{B}$, $\{101\}_\mathrm{B}$, $\{111\}_\mathrm{B}$, and $\{1\bar{1}1\}_\mathrm{B}$; transformation yields $\{001\}_\mathrm{A}$, $\{100\}_\mathrm{A}$, $\{010\}_\mathrm{A}$, and $\{1\bar{1}0\}_\mathrm{A}$. This corresponds to a hexagonal prism in the coordinate system of the A form.

18.1 In space group $P\bar{6}2m$ the Gd atom would occupy the position $3g$ with the site symmetry $m2m$. The large parameter $U_{11} = U_{22}$ is an indication that the Gd atom is placed on a plane perpendicular to **c** close to this position, with reduced site symmetry. In that case, the actual space group must be a subgroup of $P\bar{6}2m$ that has a reduced site symmetry at this position. U_{33} not being suspicious, the plane of reflection perpendicular to **c** can be retained, but not the one parallel **c** nor the twofold rotation axis (cf. the figure of the symmetry elements of $P\bar{6}2m$ in *International Tables A*). As there are no superstructure reflections, we have to look for a *translationengleiche* subgroup. $P\bar{6}2m$ has only one *translationengleiche* subgroup that meets these conditions: $P\bar{6}$. $P\bar{6}$ is also the only subgroup that results in a splitting of the position $2c$ of $P\bar{6}2m$ into two independent positions ($1c$, $1e$), such that an ordered occupation by Rh and In becomes possible. However, $P\bar{6}$ does not belong to the Laue class $6/mmm$. Evidently, this is feigned by twinning. Refinement as a twin in space group $P\bar{6}$ yields the correct structure [406].

18.2. Doubling of the a and b axes permits an ordered distribution of the palladium and silicon atoms within a plane of the hexagons, as shown in the following figure. The subgroup is isomorphic of index 4.

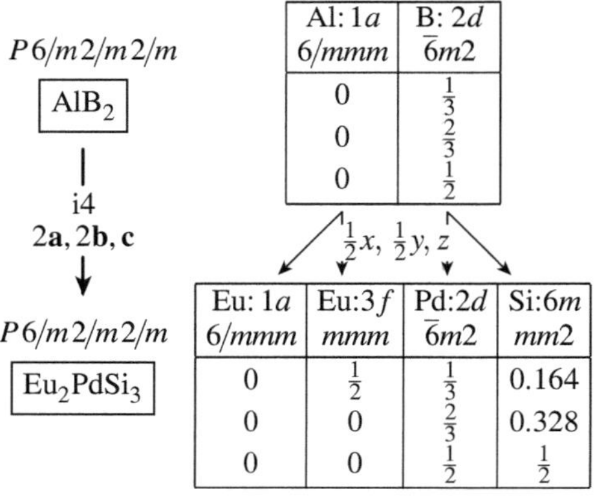

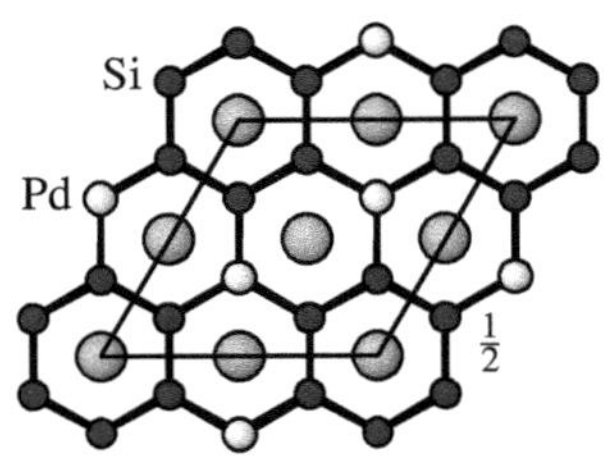

18.3. In space group $P4/nmm$, atom F1 is forced to occupy a position on a mirror plane at $x = \frac{1}{4}$. By removal of the mirror plane, F1 can shift to $x \approx 0.23$, which is closer to one and farther away from the other bonded Mn atom. This way, four short and two long Mn–F bonds result for each Mn atom. The correct space group is $P4/n$, a maximal *translationengleiche* subgroup of index 2 of $P4/nmm$ [410]. The wrong space group is feigned by merohedral twins. The wrong position of F1 yields the large vibrational ellipsoids.

19.1 If the prism has the point group $2mm$, the permutation group has order 4:

$(1)(2)(3)(4)(5)(6)$	identity	$= s_1^6$
$(1)(4)(23)(56)$	vertical reflection	$= s_1^2 s_2^2$
$(14)(26)(35)$	twofold rotation	$= 2s_2^3$
$(14)(25)(36)$	horizontal reflection	

$$Z = \tfrac{1}{4}(s_1^6 + s_1 s_2^2 + 2s_2^3)$$

With one colour x_1 and colourless vertices $x_2 = x^0 = 1$ the generating function is:

$$\begin{aligned} C &= \tfrac{1}{4}[(x_1+1)^6 + (x_1+1)^2(x_1^2+1^2)^2 + 2(x_1^2+1^2)^3] \\ &= x_1^6 + 2x_1^5 + 6x_1^4 + 6x_1^3 + 6x_1^2 + 2x_1 + 1 \end{aligned}$$

The coefficients show two possible ways to mark one vertex, six ways to mark two, and six ways to mark three vertices.

19.2 Permutation group of the square pyramid (point group $4mm$):

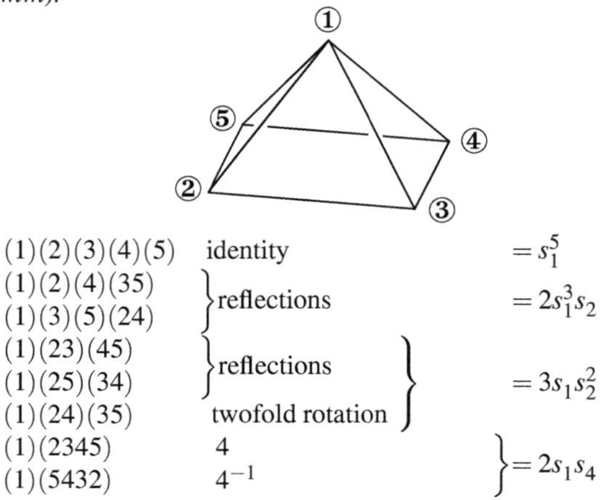

$(1)(2)(3)(4)(5)$	identity	$= s_1^5$
$(1)(2)(4)(35)$	reflections	$= 2s_1^3 s_2$
$(1)(3)(5)(24)$		
$(1)(23)(45)$	reflections	$= 3s_1 s_2^2$
$(1)(25)(34)$		
$(1)(24)(35)$	twofold rotation	
$(1)(2345)$	4	$= 2s_1 s_4$
$(1)(5432)$	4^{-1}	

$$Z = \tfrac{1}{8}(s_1^5 + 2s_1^3 s_2 + 3s_1 s_2^2 + 2s_1 s_4)$$

Marking the vertices with three colours (or occupation with three kinds of atoms) can be treated the same way as with two colours in addition to colourless (unmarked) vertices; we can calculate with $s_1 = x_1 + x_2 + 1$ instead of $s_1 = x_1 + x_2 + x_3$. The generating function is then:

$$\begin{aligned}
C &= \tfrac{1}{8}[(x_1+x_2+1)^5 + 2(x_1+x_2+1)^3(x_1^2+x_2^2+1^2) \\
&\quad +3(x_1+x_2+1)(x_1^2+x_2^2+1^2)^2 \\
&\quad +2(x_1+x_2+1)(x_1^4+x_2^4+1^4)] \\
&= x_1^5 + 2x_1^4x_2 + 3x_1^3x_2^2 + 3x_1^2x_2^3 + 2x_1x_2^4 + x_2^5 \\
&\quad +2x_1^4 + 4x_1^3x_2 + 6x_1^2x_2^2 + 4x_1x_2^3 + 2x_2^4 \\
&\quad +3x_1^3 + 6x_1^2x_2 + 6x_1x_2^2 + 3x_2^3 \\
&\quad +3x_1^2 + 4x_1x_2 + 3x_2^2 \\
&\quad +2x_1 + 2x_2 \\
&\quad +1
\end{aligned}$$

The coefficients $4x_1^3x_2$ and $6x_1^2x_2^2$ show four and six possible distributions of atoms (isomers) for the compositions MX_3YZ and MX_2Y_2Z, respectively. Pairs of enantiomers are counted as one isomer each.

The number of chiral pairs of enantiomers is determined by calculating the cycle index Z' and the generating function C' considering only rotations:

$$\begin{aligned}
Z' &= \tfrac{1}{4}(s_1^5 + s_1 s_2^2 + 2s_1 s_4) \\
C' &= \tfrac{1}{4}[(x_1+x_2+1)^5 + (x_1+x_2+1)(x_1^2+x_2^2+1^2)^2 \\
&\quad +2(x_1+x_2+1)(x_1^4+x_2^4+1^4)] \\
&= x_1^5 + 2x_1^4x_2 + 3x_1^3x_2^2 + 3x_1^2x_2^3 + 2x_1x_2^4 + x_2^5 \\
&\quad +2x_1^4 + 5x_1^3x_2 + 8x_1^2x_2^2 + 5x_1x_2^3 + 2x_2^4 \\
&\quad +3x_1^3 + 8x_1^2x_2 + 8x_1x_2^2 + 3x_2^3 \\
&\quad +3x_1^2 + 5x_1x_2 + 3x_2^2 \\
&\quad +2x_1 + 2x_2 \\
&\quad +1
\end{aligned}$$

The number of pairs of enantiomers follows from:

$$C' - C = x_1^3x_2 + 2x_1^2x_2^2 + x_1x_2^3 + 2x_1^2x_2 + 2x_1x_2^2 + x_1x_2$$

There is one pair of enantiomers for MX_3YZ and two for MX_2Y_2Z.

19.3 We designate the space groups by $\mathcal{G}_1$ to $\mathcal{G}_4$. We set up the following tree of group–subgroup relations that includes the Euclidean normalizers (cf. Section 8.3).

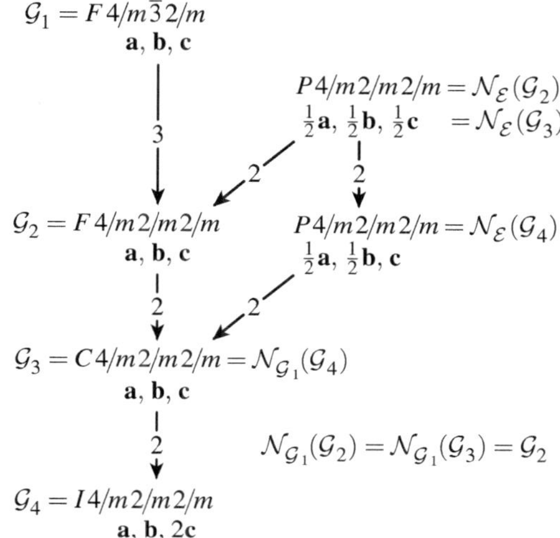

The index 3 of $\mathcal{G}_2$ in $\mathcal{G}_1$ shows three conjugate subgroups $\mathcal{G}_2$ in $\mathcal{G}_1$, $[\mathcal{G}_2] = 3$ (with **c** along **a**, **b**, and **c** of $\mathcal{G}_1$, respectively). The same applies to $\mathcal{G}_3$, $[\mathcal{G}_3] = 3$. There are six conjugate subgroups $\mathcal{G}_4$ in $\mathcal{G}_1$, $[\mathcal{G}_4] = 6$ (index of $\mathcal{N}_{\mathcal{G}_1}(\mathcal{G}_4)$ in $\mathcal{G}_1$); in addition to the three orientations, their origins can be at $0,0,0$ and $0,\tfrac{1}{2},\tfrac{1}{2}$ of $\mathcal{G}_1$. Two of them are also subgroups of $\mathcal{G}_2$ and $\mathcal{G}_3$. With eqn (19.5), page 260, we have:

$$\mathbf{M} = \begin{pmatrix} 1 & & & \\ 1 & 1 & & \\ 1 & 1 & 2 & \\ 1 & 1 & 2 & 2 \end{pmatrix}$$

$$\mathbf{B} = \mathbf{M}^{-1} = \begin{pmatrix} 1 & & & \\ -1 & 1 & & \\ 0 & -\tfrac{1}{2} & \tfrac{1}{2} & \\ 0 & 0 & -\tfrac{1}{2} & \tfrac{1}{2} \end{pmatrix}$$

In the space groups $\mathcal{G}_1$ and $\mathcal{G}_2$, there are no positions that permit an occupation resulting in the composition ABX_2. There are two combinatorial distributions for $\mathcal{G}_3$ ($C4/mmm$): A at $2a$, B at $2d$, and vice versa ($2b$ and $2c$ are occupied by X). There are also two combinatorial distributions for $\mathcal{G}_4$ ($I4/mmm$): A at $2a$ and $2b$, B at $4d$, and vice versa. We thus have $\mathbf{v} = (0,\ 0,\ 2,\ 2)^{\mathrm{T}}$ and $\mathbf{z} = \mathbf{B}\mathbf{v} = (0,\ 0,\ 1,\ 0)$. There exists only one possible structure in space group $\mathcal{G}_3$ (where **c** is not yet doubled) and none in space group $\mathcal{G}_4$.

For the composition $AB\square_2X_4$ there are only two combinatorial distributions in space group $\mathcal{G}_4$: A at $2a$, B at $2b$, and vice versa. $\mathbf{v} = (0,\ 0,\ 0,\ 2)^{\mathrm{T}}$; $\mathbf{z} = \mathbf{B}\mathbf{v} = (0,\ 0,\ 0,\ 1)$. There exists only one possible structure in space group $\mathcal{G}_4$.

21.1 $P\,1\,2/c\,1$ has a glide-reflection plane with the glide direction **c**. Therefore, **c** cannot be doubled. Isomorphic subgroups of index 2 are possible by doubling of **a** and **b**.
$P2_1\,2_1\,2$ has 2_1 axes parallel to **a** and **b** and, therefore, cannot be doubled in these directions. Isomorphic subgroups of index 2 are possible by doubling of **c**.
The n glide-reflections of $P2_1/n\,2_1/n\,2/m$ have glide components in the directions of **a**, **b**, and **c**; there are no subgroups of index 2.
$P4/m\,2_1/b\,2/m$ has 2_1 axes parallel to **a** (and **b**); a doubling of the cell in the *a–b* plane is excluded. A doubling of **c** is possible.
The **c** vector of $P6_1\,22$ can be enlarged only by factors of $6n+1$ or $6n-1$; in the *a–b* plane, the only possible transformations are $2\mathbf{a}+\mathbf{b}, -\mathbf{a}+\mathbf{b}$ (index 3) and $p\mathbf{a}, p\mathbf{b}$ (Index p^2). There is no subgroup of index 2.

21.2

$I\,4/m\,2/m\,2/m$	$I\bar{4}m2$
↓ k2	↓ k2
$P\,4/m\,2/m\,2/m$	$P\bar{4}m2$
↓ k2 $\mathbf{a}-\mathbf{b}, \mathbf{a}+\mathbf{b}, 2\mathbf{c}$	↓ k2 $\mathbf{a}-\mathbf{b}, \mathbf{a}+\mathbf{b}, \mathbf{c}$
$I\,4/m\,2/m\,2/m$	$P\bar{4}2m$
	↓ k2 $\mathbf{a}+\mathbf{b}, -\mathbf{a}+\mathbf{b}, 2\mathbf{c}$
	$I\bar{4}m2$

21.3 A subgroup with $i=4$ is possible, but it is not maximal.
$i=9$, $3\mathbf{a}, 3\mathbf{b}, \mathbf{c}$ is one of the permitted isomorphic subgroups $p\mathbf{a}, p\mathbf{b}, \mathbf{c}$, $p=4n-1$.
$i=17$ corresponds to the basis transformation $\mathbf{a}'=q\mathbf{a}+r\mathbf{b}$, $\mathbf{b}'=-r\mathbf{a}+q\mathbf{b}$, $p=q^2+r^2=4n+1$, with $q,r=4,1$ and with $q,r=1,4$.
$11\neq 4n+1$ is not a possible index.

21.4 The index of 65 has the divisors 1, 5, 13, and 65, all of which are of the kind $4n+1$. There are four conjugacy classes with 65 conjugate subgroups each. The basis transformations are $\mathbf{a}'=q\mathbf{a}+r\mathbf{b}$, $\mathbf{b}'=-r\mathbf{a}+q\mathbf{b}$, with $q,r=8,1$, $q,r=1,8$, $q,r=7,4$ and $q,r=4,7$. $8^2+1^2=1^2+8^2=7^2+4^2=4^2+7^2=65$. A subgroup of index 65 is reached by two steps of maximal subgroups, one of index 5, the other of index 13 (or vice versa).

21.5 From the basis transformations from $Fm\bar{3}m$ to the intermediate group $R\bar{3}^{(\text{hex})}$ we have:
$a_{\text{hex}}=\frac{1}{2}\sqrt{2}\,a_{\text{cub}}\approx\frac{1}{2}\sqrt{2}\times 494\text{ pm}=349.3\text{ pm};$
$c_{\text{hex}}=\sqrt{3}\,a_{\text{cub}}\approx\sqrt{3}\times 494\text{ pm}=855.6\text{ pm}.$
c_{hex} agrees with $c=860$ pm of $PtCl_3$. Since $\frac{1}{37}$ of the atom positions of the packing of spheres are vacant, the unit cell has to be enlarged by a factor of 37 (or a multiple thereof). There must be an isomorphic group–subgroup relation of index 37. $37=6n+1$ is one of the permitted prime numbers for an isomorphic subgroup with hexagonal axes and cell enlargement in the *a–b* plane. $a_{\text{hex}}\sqrt{37}=349.3\sqrt{37}$ pm $=2125$ pm agrees with $a=2121$ pm of $PtCl_3$. $q^2-qr+r^2=37$ is fulfilled for $q,r=7,3$ and $q,r=7,4$. Therefore, the basis transformation is $\mathbf{a}_{PtCl_3}=7\mathbf{a}_{\text{hex}}+3\mathbf{b}_{\text{hex}}$, $\mathbf{b}_{PtCl_3}=-3\mathbf{a}_{\text{hex}}+4\mathbf{b}_{\text{hex}}$ or $\mathbf{a}_{PtCl_3}=7\mathbf{a}_{\text{hex}}+4\mathbf{b}_{\text{hex}}$, $\mathbf{b}_{PtCl_3}=-4\mathbf{a}_{\text{hex}}+3\mathbf{b}_{\text{hex}}$.

21.6

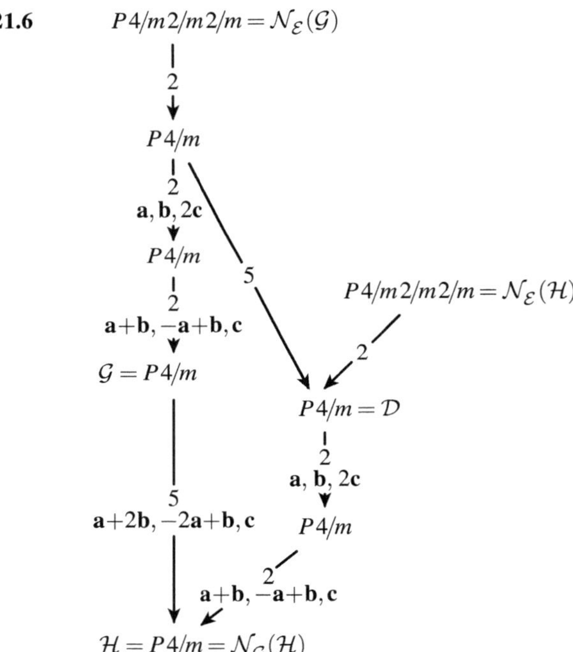

Index 5 of $\mathcal{N}_{\mathcal{G}}(\mathcal{H})$ in $\mathcal{G}$ shows five conjugates of $\mathcal{H}$ in $\mathcal{G}$. According to the index 2×5 of $\mathcal{D}$ in $\mathcal{N}_{\mathcal{E}}(\mathcal{G})$ there are five conjugates in two conjugacy classes each.

Question in Table 24.5

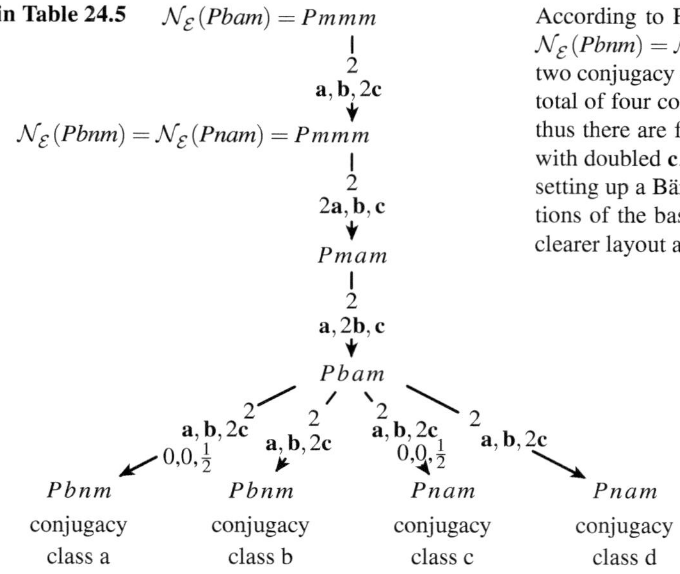

$\mathcal{N}_{\mathcal{E}}(Pbam) = Pmmm$

2
$\mathbf{a}, \mathbf{b}, 2\mathbf{c}$

$\mathcal{N}_{\mathcal{E}}(Pbnm) = \mathcal{N}_{\mathcal{E}}(Pnam) = Pmmm$

2
$2\mathbf{a}, \mathbf{b}, \mathbf{c}$

$Pmam$

2
$\mathbf{a}, 2\mathbf{b}, \mathbf{c}$

$Pbam$

2 | 2 | 2 | 2
$\mathbf{a}, \mathbf{b}, 2\mathbf{c}$; $0, 0, \frac{1}{2}$ | $\mathbf{a}, \mathbf{b}, 2\mathbf{c}$ | $\mathbf{a}, \mathbf{b}, 2\mathbf{c}$; $0, 0, \frac{1}{2}$ | $\mathbf{a}, \mathbf{b}, 2\mathbf{c}$

$Pbnm$	$Pbnm$	$Pnam$	$Pnam$
conjugacy class a	conjugacy class b	conjugacy class c	conjugacy class d

According to Fig. 8.7 (left) the index of 2 of the normalizers $\mathcal{N}_{\mathcal{E}}(Pbnm) = \mathcal{N}_{\mathcal{E}}(Pnam)$ in $\mathcal{N}_{\mathcal{E}}(Pbam)$ shows the existence of two conjugacy classes of *Pbnm* and of *Pnam* in *Pbam*. That is a total of four conjugacy classes for the subgroup type *Pnma* and thus there are four different *klassengleiche* subgroups of *Pbam* with doubled **c**. An eye has to be kept on all four of them when setting up a Bärnighausen tree. In the tree on the left, the directions of the basis vectors have not been changed; that yields a clearer layout as compared to the use of standard settings.

References

[1] *Strukturbericht* **1–7** (1931–1943) (supplementary volumes to *Z. Kristallogr.*). Leipzig: Akademische Verlagsgesellschaft.

[2] *Structure Reports* **8–43** (1956–1979), Utrecht: Oosthoek, Scheltema & Holkema; **44–48** (1980–1984), Dordrecht: D. Reidel; **49–58** (1989–1993), Dordrecht: Kluwer Academic Publishers.

[3] *Inorganic Crystal Structure Database* (ICSD). Fachinformationszentrum Karlsruhe, `https://icsd.fiz-karlsruhe.de`

[4] *Pearson's Crystal Data* (PCD). Bonn: Crystal Impact, `www.crystalimpact.de/pcd`

[5] *Open-Access Collection of Crystal Structures* (COD). `www.crystallography.net`

[6] *Cambridge Structural Database* (CSD). Cambridge Crystallographic Data Centre, `www.ccdc.cam.ac.uk`

[7] Groom, C. R., Bruno, I. J., Lightfoot, M. P., and Ward S. C. (2016). The Cambridge Structural Database. *Acta Crystallogr.* **B 72**, 640. `https://doi.org/10.1107/S2052520616003954`

[8] *Protein Data Bank*. Brookhaven National Laboratory, `https://rcsb.org`

[9] *Nucleic Acid Knowledgebase*. Rutgers University, `https://www.nakb.org`

[10] Goldschmidt, V. M. (1926, 1928). Untersuchungen über den Bau und Eigenschaften von Krystallen. *Skrifter Norsk Vidensk. Akad. Oslo, I Mat. Naturv. Kl.* 1926, Nr. 2, and 1928 Nr. 8.

[11] Pauling, L. (1960). *The Nature of the Chemical Bond* (3rd edn). Cornell: Cornell University Press.

[12] Landau, L. D. and Lifshitz, E. M. (1980). *Statitstical Physics* (3rd edn), Part 1, pp. 459–471. London: Pergamon.
[Translation from Russian: *Statitsticheskaya Fizika*, chast 1. Moskva: Nauka (1976). German: *Lehrbuch der theoretischen Physik*, 6. Aufl., Band 5, Teil 1, pp. 436–447. Berlin: Akademie-Verlag (1984).]

[13] Bärnighausen, H. (1980). Group–subgroup relations between space groups: a useful tool in crystal chemistry. *MATCH, Commun. Math. Chem.* **9**, 139. `https://match.pmf.kg.ac.rs/electronic_versions/Match09/match9_139-175.pdf`

[14] *International Tables for Crystallography*, Vol. *A*, *Space-group symmetry* (1st–5th edns, 1983, 1987, 1992, 1995, 2002, corrected reprint 2005; electronic version 2006; ed. Th. Hahn). Dordrecht: Kluwer Academic Publishers. `https://it.iucr.org/Ab`

[15] *International Tables for Crystallography*, Vol. *A*: *Space-group symmetry*, 6th edn (2016; ed. M. I. Aroyo). Chichester: Wiley. `https://it.iucr.org/A`

[16] *International Tables for Crystallography*, Vol. *A*1, *Symmetry relations between space groups* (1st edn, 2004; eds. Wondratschek, H. and Müller, U.). Part 2:

Subgroups of the Space Groups. Part 3: Relations of the Wyckoff Positions. Dordrecht: Kluwer Academic Publishers. 2nd edn (2010), Chichester: Wiley. https://it.iucr.org/A1

[17] Cotton, F. A. (1990). *Chemical Applications of Group Theory* (3rd edn). New York: Wiley.

[18] Kettle, S. F. A. (2007). *Symmetry and Structure* (3rd edn). Chichester: Wiley.

[19] Willock, D. J. (2009). *Molecular Symmetry*. Chichester: Wiley.

[20] Jacobs, P. (2005). *Group Theory with Applications to Chemical Physics*. Cambridge: Cambridge University Press.

[21] Ludwig, W. and Falter, C. (1996). *Symmetries in Physics: Group Theory Applied to Physical Problems* (2nd edn). Berlin: Springer.

[22] Cornwell, J. F. (1997). *Group Theory in Physics—An Introduction*. San Diego: Academic Press.

[23] Sternberg, S. (1994). *Group Theory and Physics*. Cambridge: Cambridge University Press.

[24] Dresselhaus, M., Dresselhaus, G., and Jorio, A. (2008). *Group Theory*. Berlin: Springer.

[25] Powell, R. C. (2010). *Symmetry, Group Theory, and the Physical Properties of Crystals*. New York: Springer.

[26] *Internationale Tabellen zur Bestimmung von Kristallstrukturen* (1935; ed. Hermann, C.; trilingual, German, English, French). Berlin: Geb. Borntraeger.

[27] *International Tables for X-Ray Crystallography*, Vol. I (1952; eds. Henry, N. F. M. and Lonsdale, K.). Birmingham: Kynoch.

[28] Neubüser, J. and Wondratschek, H. (1966). Untergruppen der Raumgruppen. *Kristall u. Technik* **1**, 529. Supplementary tables: *Maximal subgroups of the space groups*, 2nd typing (1969, corrected 1977); *Minimal supergroups of the space groups* (1970). Copied material distributed by Institut für Kristallographie, Universität Karlsruhe, Germany.

[29] Bertaut, E. F. and Billiet, Y. (1979). On equivalent subgroups and supergroups of the space groups. *Acta Crystallogr.* **A 35**, 733.

[30] Brunner, G. O. (1971). An unconventional view of the closest sphere packings. *Acta Crystallogr.* **A 27**, 388. https://doi.org//10.1107/S0567739471000858

[31] Bernal, J. D. (1938). Conduction in solids and diffusion and chemical change in solids. Geometrical factors in reactions involving solids. *Trans. Faraday Soc.* **34**, 834.

[32] Lotgering, F. K. (1959). Topotactical reactions with ferrimagnetic oxides having hexagonal crystal structures. *J. Inorg. Nucl. Chem.* **9**, 113.

[33] Giovanoli, D. and Leuenberger, U. (1969). Über die Oxidation von Manganoxidhydroxid. *Helv. Chim. Acta* **52**, 2333.

[34] Bernal, J. D. and Mackay, A. L. (1965). Topotaxy. *Tschermaks mineralog. petrogr. Mitt., Reihe* 3, (*Mineralogy and Petrology*) **10**, 331.

[35] Megaw, H. D. (1973). *Crystal Structures. A Working Approach.* Philadelphia: Saunders Co.

[36] Buerger, M. J. (1947). Derivative crystal structures. *J. Chem. Phys.* **15**, 1.

[37] Buerger, M. J. (1951). *Phase Transformations in Solids*, Ch. 6. New York: Wiley. *Fortschr. Mineral.* **39** (1961), 9.

[38] Anselment, B. (1985). *Die Dynamik der Phasenumwandlung vom Rutil- in den $CaCl_2$-Typ am Beispiel des $CaBr_2$ und zur Polymorphie des $CaCl_2$*. Dissertation, Universität Karlsruhe. ISBN 3-923161-13-1.

[39] Purgahn, J., Pillep, B., and Bärnighausen, H. (1998). Röntgenographische Untersuchung der Phasenumwandlung von $CaCl_2$ in den Rutil-Typ. *Z. Kristallogr. Suppl.* **15**, 112.

[40] Purgahn, J. (1999). *Röntgenographische Untersuchungen zur Dynamik struktureller Phasenumwandlungen für zwei Fallbeispiele: Rb_2ZnI_4 und $CaCl_2$*. Dissertation, Universität Karlsruhe. ISBN 3-932136-37-3.

[41] Brown, H., Bülow, R., Neubüser, J., Wondratschek, H., and Zassenhaus, H. (1978). *Crystallographic groups of four-dimensional space*. New York: Wiley.

[42] Chuprunov, E. V. and Kuntsevich, T. S. (1988). n-Dimensional space groups and regular point systems. *Comput. Math. Applic.* **16**, 537.

[43] Janssen, T., Janner, A., Looijenga-Vos, A., and de Wolff, P. M. (2006). Incommensurate and commensurate modulated structures. In *International Tables for Crystallography*, Vol. *C*, *Mathematical, physical and chemical tables* (3rd edn; ed. Prince, E.), Chap. 9.8. Chichester: Wiley. `https://it.iucr.org/Cb/ch9o8v0001/ch9o8.pdf`

[44] Souvignier, B. (2006). The four-dimensional magnetic point and space groups. *Z. Kristallogr.* **221**, 77. `https://doi.org/10.1524/zkri.2006.221.1.77`

[45] Van Smaalen, S. (2007). *Incommensurate Crystallography*. Oxford: Oxford University Press.

[46] Janssen, T. (2014). Development of symmetry concepts for aperiodic crystals. *Symmetry* **6**, 171.

[47] Janssen, T. and Janner, A. (2014). Aperiodic crystals and superspace concepts. *Acta Crystallogr.* **B 70**, 617. `https://doi.org/10.1107/S2052520614014917`

[48] Massa, W. (2016). *Crystal Structure Determination* (3rd edn). Norderstedt: BoD—Books on Demand. `https://www.bod.de/buchshop/crystal-structure-determination-werner-massa-9783741241246`

[49] Giacovazzo, C., Monaco, H. L., Artioli, G., Viterbo, D., Milanesio, M., Gilli, G., Gilli, P., Zanotti, G., Ferraris, G., and Catti, M. (2010). *Fundamentals of Crystallography* (3rd edn). Oxford: Oxford University Press.

[50] Blake, A. J., Clegg, W., Cole, J. M., Evans, J. S. O., Main, P., Parsons, S., and Watkin, D. J. (2009). *Crystal Structure Analysis: Principles and Practice* (2nd edn). Oxford: Oxford University Press.

[51] Glusker, J. P., Trueblood, K. N. (2010). *Crystal Structure Analysis. A Primer* (3rd ed). Oxford: Oxford University Press.

[52] Tilley, R. (2006). *Crystals and Crystal Structures*. Chichester: Wiley.

[53] Ledermann, W. and Weir, A. J. (1996). *Introduction to Group Theory*. Harlow: Addison-Wesley Longman.

[54] Hall, S. R. (1981). Space group notation with an explicit origin. *Acta Crystallogr.* **A 37**, 517.

[55] Fedorov, E. S. (1971). *Symmetry of Crystals*. American Crystallographic Association monograph No. 7. Pittsburgh: Polycrystal Book Service. [Translation from Russian, *Simmetriya i struktura kristallov*. Akademiya Nauk SSSR (1949)].

[56] Hermann, C. (1928). Zur systematischen Strukturtheorie I. Eine neue Raumgruppensymbolik. *Z. Kristallogr.* **68**, 257.

[57] Mauguin, C. (1931). Sur le symbolisme des groupes de répétition ou de symétrie des assemblages cristallins. *Z. Kristallogr.* **76**, 542.

[58] Hermann, C. (1929). Zur systematischen Strukturtheorie II. Ableitung der 230 Raumgruppen aus ihren Kennvektoren. *Z. Kristallogr.* **69**, 226.

[59] Litvin, D. B. (2016): Magnetic subperiodic groups and magnetic space groups. In: *International Tables for Crystallography*, Vol. *A*, *Space-group symmetry*, (6th edn; ed. M. I. Aroyo), Ch. 3.6. Chichester: Wiley. `https://it.iucr.org/Ac/ch3o6v0001/ch3o6.pdf`

[60] Hermann, C. (1929). Zur systematischen Strukturtheorie III. Ketten- und Netzgruppen. *Z. Kristallogr.* **69**, 250.

[61] *International Tables for Crystallography*, Vol. *E*, *Subperiodic Groups* (2nd edn, 2010; eds. Kopský, V. and Litvin, D. B.). Chichester: Wiley. `https://it.iucr.org` `https://it.iucr.org/E`

[62] Spruiell, J. E. and Clark, E. S. (1980). In: *Methods in Experimetal Physics*, Vol. 16, *Polymers* (ed. R. A. Fava), Part B, Ch. 6, pp. 19–22. New York: Academic Press.

[63] Müller, U. (2017). Die Symmetrie von Spiralketten. Acta Crystallogr. **B 73**, 443. In English: `https://dx.doi.org/10.17192/es2017.0002`

[64] Natta, G. and Corradini, P. (1960). General cosideration on the structure of crystalline polyhydrocarbons. *Nuovo Cimento*, Suppl. **15**, 9. `https://doi.org//10.1007/BF02731858`

[65] Pfister, D., Schäfer, K., Ott, C., Gerke, B., Pöttgen, R., Janka, O., Baumgartner, M., Efimova, A., Hohmann, A., Schmidt, P., Venkachalam, S., von Wüllen, L., Schüermann, U., Kienle, L., Duppel, V., Parzinger, E., Miller, B., Becker, J., Holleitner, A., Weihrich, R., and Nilges, T. (2016). Inorganic double helices in semiconducting SnIP. *Advan. Materials* **28**, 9783. `https://doi.org/10.1002/adma.201603135`

[66] Meille, S. V., Allegra, G., Heil, P. H., He, J., Hess, M., Jin, J., Kratochvíl, P., Mormann, M., and Stepto, R. (2011). Definitions and terms relating to crystalline polymers (IUPAC recommendations 2011), *Pure Appl. Chem.* **83**, 1831. `https://doi.org/10.1351/PAC-REC-10-11-13`

[67] Bohm, J. and Dornberger-Schiff, K. (1966). The nomenclature of crystallographic symmetry groups. *Acta Crystallogr.* **21**, 1004.

[68] Bohm, J. and Dornbeger-Schiff, K. (1967). Geometrical symbols for all crystallographic symmetry groups up to three dimension. *Acta Crystallogr.* **23**, 913. `https://doi.org/10.1107/S0365110X67004050`

[69] Hirshfeld, F. L. (1968). Symmetry in the generation of trial structures. *Acta Crystallogr.* **A 24**, 301.

[70] Koch, E. and Müller, U. (1990). Euklidische Normalisatoren für trikline und monokline Raumgruppen bei spezieller Metrik des Translationengitters. *Acta Crystallogr.* **A 46**, 826.

[71] Koch, E. (1984). The implications of normalizers on group–subgroup relations between space groups. *Acta Crystallogr.* **A 40**, 593.

[72] Parthé, E. and Gelato, L. (1984). The standardization of inorganic crystal structures. *Acta Crystallogr.* **A 40**, 169. `https://doi.org/10.1107/S0108767384000416`

[73] Parthé, E., Cenzual, K., and Gladyshevskii, R. E. (1993). Standardization of crystal structure data as an aid to the classification of crystal structure types. *J. Alloys Comp.* **197**, 291.

[74] Gelato, L. M. and Parthé, E. (1987). STRUCTURE TIDY—a computer program to standardize crystal structure data. *J. Appl. Crystallogr.* **20**, 139. The Program has been incorporated to A. L. Spek's program PLATON, `www.platonsoft.nl/platon/pl000613.html`

[75] Fischer, W. and Koch, E. (1983). On the equivalence of point configurations due to Euclidean normalizers (Cheshire groups) of space groups. *Acta Crystallogr.* **A 39**, 907.

[76] Koch, E. and Fischer, W. (2006). Normalizers of space groups, a useful tool in crystal description, comparison and determination. *Z. Kristallogr.* **221**, 1.

[77] Müller, U. (1984). Wolframtetrabromidoxid, $WOBr_4$. *Acta Crystallogr.* **C 40**, 915.

[78] Moss, G. P. (1996). Basic terminology of stereochemistry. *Pure Appl. Chem.* **68**, 2193. (IUPAC Recommendations for the terminology in stereo chemistry; `https://www.degruyter.com/document/doi/10.1351/pac199668122193/html`)

[79] Flack, H. D. (2003). Chiral and achiral crystal structures. *Helv. Chim. Acta* **86**, 905. `https://doi.org/10.1002/hlca.200390109`

[80] Von Schnering, H. G. and Hönle, W. (1979). Zur Chemie und Strukturchemie der Phosphide und Polyphosphide. 20. Darstellung, Struktur und Eigenschaften der Alkalimetallmonophosphide NaP und KP. *Z. Anorg. Allg. Chem.* **456**, 194.

[81] Baur, W. H. and Kassner, D. (1992). The perils of *Cc*: comparing the frequencies of falsely assigned space groups with their general population. *Acta. Crystallogr.* **B 48**, 356.

[82] Marsh, R. E. (1999, 2002, 2004, 2005). $P1$ or $P\bar{1}$? Or something else? *Acta Crystallogr.* **B 55**, 931; The space groups of point symmetry C_3: some corrections, some comments. *Acta Crystallogr.* **B 58**, 893; Space group *Cc*: an update. *Acta Crystallogr.* **B 60**, 252; Space group $P1$: an update. *Acta Crystallogr.* **B 61**, 359.

[83] Herbstein, F. H. and Marsh, R. E. (1998). More space group corrections: from triclinic to centred monoclinic and to rhombohedral; also from $P1$ to $P\bar{1}$ and from *Cc* to *C*2/*c*. *Acta Crystallogr.* **B 54**, 677.

[84] Herbstein, F. H., Hu, S., and Kapon, M. (2002). Some errors from the crystallographic literature, some amplifications and a questionable result. *Acta Crystallogr.* **B 58**, 884.

[85] Marsh, R. E. and Spek, A. L. (2001). Use of software to search for higher symmetry: space group *C*2. *Acta Crystallogr.* **B 57**, 800.

[86] Marsh, R. E., Kapon, M., Hu, S., and Herbstein, F. H. (2002). Some 60 new space-group corrections. *Acta Crystallogr.* **B 58**, 62.

[87] Clemente, D. A. and Marzotto, A. (2003, 2004). 22 space group corrections. *Acta Crystallogr.* **B 59**, 43; 30 space-group corrections: two examples of false polymorphism and one of incorrect interpretation of the fine details of an IR spectrum. *Acta Crystallogr.* **B 60**, 287.

[88] Clemente, D. A. (2005, 2003). A survey of the 8466 structures reported in *Inorganica Chimica Acta*: 52 space group changes and their consequences. *Inorg. Chim. Acta* **358**, 1725; 26 space group changes and 6 crystallographic puzzles found in Tetrahedron journals. *Tetrahedron* **59**, 8445.

[89] Marsh, R. E. and Clemente, D. A. (2007). A survey of crystal structures published in the Journal of the American Chemical Society. *Inorg. Chim. Acta* **360**, 4017.

[90] Cenzual, K., Gelato, L. M., Penzo, M., and Parthé, E. (1991). Inorganic structure types with revised space groups. *Acta Crystallogr.* **B 47**, 433.

[91] Bauer, J. and Bars, O. (1980). The ordering of boron and carbon atoms in the LaB_2C_2 structure. *Acta. Crystallogr.* **B 36**, 1540.

[92] Lima-de-Faria, J., Hellner, E., Liebau, F., Makovicky, E., and Parthé, E. (1990). Nomenclature of inorganic structure types. Report of the International Union of Crystallography Commission on Crystallographic Nomenclature Subcommittee on the nomenclature of inorganic structure types. *Acta Crystallogr.* **A 46**, 1. https://doi.org/10.1107/S0108767389008834

[93] Allmann, R. and Hinek, R. (2007). The introduction of structure types into the Inorganic Crystal Structure Database ICSD. *Acta Crystallogr.* **A 63**, 412.

[94] Brostigen, G. and Kjekshus, A. (1969). Redetermined crystal structure of FeS_2 (pyrite). *Acta Chem. Scand.* **23**, 2186.

[95] Simon, A. and Peters, K. (1980). Single-crystal refinement of the structure of carbon dioxide. *Acta Crystallogr.* **B 36**, 2750.

[96] Burzlaff, H. and Rothammel, W. (1992). On quantitative relations among crystal structures. *Acta Crystallogr.* **A 48**, 483. https://doi.org/10.1107/S0108767392000072

[97] Burzlaff, H. and Malinovsky, Y. (1997). A Procedure for the Clasification of Non-Organic Crystal Structures. I. Theoretical Background *Acta Crystallogr.* **A 53**, 217.

[98] Bergerhoff, G., Berndt, M., Brandenburg, K., and Degen, T. (1999). Concerning inorganic crystal structure types. *Acta Crystallogr.* **B 55**, 147. https://doi.org/10.1107/S0108768198010969

[99] Parthé, E. and Rieger, W. (1968). Nowotny phases and apatites. *J. Dental Res.* **47**, 829.

[100] Müller, U., Schweda, E., and Strähle, J. (1983). Die Kristallstruktur vom Tetraphenylphosphoniumnitridotetrachloromolybdat $PPh_4[MoNCl_4]$. *Z. Naturforsch.* **38 b**, 1299.

[101] Kato, K., Takayama-Muromachi, E., and Kanke, Y. (1989). Die Struktur der Kupfer-Vanadiumbronze $Cu_{0,261}V_2O_5$. *Acta Crystallogr.* **C 45**, 1845.

[102] Ha-Eierdanz, M.-L. and Müller, U. (1993). Ein neues Syntheseverfahren für Vanadiumbronzen. Die Kristallstruktur von β-$Ag_{0,33}V_2O_5$. Verfeinerung der Kristallstruktur von ε-$Cu_{0,76}V_2O_5$. Z. *Anorg. Allg. Chem.* **619**, 287.

[103] Jansen, M. (1976). Über $NaAg_3O_2$. *Z. Naturforsch.* **31 b**, 1544.

[104] Somer, M. Carrillo-Cabrera, W., Peters, E. M., Peters, K., and von Schnering, H. G. (1995). Crystal structure of trisodium di-μ-phosphidoaluminate, Na_3AlP_2. *Z. Kristallogr.* **210**, 777.

[105] Uhrlandt, S. and Meyer, G. (1995). Nitride chlorides of the early lanthanides $[M_2N]Cl_3$. *J. Alloys Comp.* **225**, 171.

[106] Bronger, W. Balk-Hardtdegen, H., and Ruschewitz, U. (1992). Darstellung, Struktur und magnetische Eigenschaften der Natriumeisenchalkogenide Na_6FeS_4 und Na_6FeSe_4. *Z. Anorg. Allg. Chem.* **616**, 14.

[107] Meyer, H. J., Meyer, G., and Simon, M. (1991). Über ein Oxidchlorid des Calciums: Ca_4OCl_6. *Z. Anorg. Allg. Chem.* **596**, 89.

[108] Reckeweg, O. and DiSalvo, F. J. (2008). Alkaline earth metal oxyhalides revisited—syntheses and crystal structures of Sr_4OBr_6, Ba_4OBr_6 and Ba_2OI_2. *Z. Naturforsch.* **63 b**, 519.

[109] Eriksson, L. and Kalinowski, M. P. (2001). $Mn_{1-x}Fe_xS$, $x \approx 0.05$, an example of an anti-wurtzite structure. *Acta Crystallogr.* **E 57**, i92.

[110] Prewitt, C. T. and Young, H. S. (1965). Germanium and silicon disulfides: structures and synthesis. *Science* **149**, 535.

[111] Hoppe, R. and Roehrborn, H. S. (1964). Oxomercurate(II) der Alkalimetalle M_2HgO_2. *Z. Anorg. Allg. Chem.* **329**, 110.

[112] Mader, K. and Hoppe, R. (1992). Neuartige Mäander mit Co^{3+} und Au^{3+}: $Na_4[AuCoO_5]$. *Z. Anorg. Allg. Chem.* **612**, 89.

[113] Fischer, W. and Koch, E. (1975). Automorphismengruppen von Raumgruppen und die Zuordnung von Punktlagen zu Konfigurationslagen. *Acta Crystallogr.* **A 31**, 88.

[114] Fischer, W. and Koch, E. (1985). Lattice complexes and limiting complexes versus orbit types and non-characteristic orbits: a comparartive discussion. *Acta Crystallogr.* **A 41**, 421.

[115] Wondratschek, H. (1980). Crystallographic orbits, lattice complexes, and orbit types. *MATCH, Commun. Math. Chem.* **9**, 121. https://match.pmf.kg.ac.rs/electronic_versions/Match09/match9_121-125.pdf

[116] Wondratschek, H. (1983). Special topics on space groups. In *International Tables for Crystallography*, Vol. *A* (1st edn), p. 723. Dordrecht: Kluwer Academic Publishers.

[117] Wondratschek, H. (1993). Splitting of Wyckoff positions. *Mineral. Petrol.* **48**, 87.

[118] Souvigner, B., Chapuis, G., and Wondratschek, H. (2016). Synoptic tables of plane and space groups. In *International Tables for Crystallography*, Vol. *A* (6th edn; ed. M. I. Aroyo), Sect. 1.5.4. Chichester: Wiley. https://it.iucr.org/Ac/ch1o5v0001/sec1o5o4.pdf

[119] Bertaut, E. F. (2002, 2005). Symbols for space groups. In *International Tables for Crystallography*, Vol. *A* (5th edn; ed. Th. Hahn), Ch. 4.3. Dordrecht: Kluwer Academic Publishers. https://it.iucr.org/Ab/ch4o3v0001/ch4o3.pdf

[120] Armbruster, T., Lager, G. A., Ihringer, J., Rotella, F. J., and Jorgenson, J. D. (1983). Neutron and X-ray powder study of phase transitions in the spinel $NiCr_2O_4$. *Z. Kristallogr.* **162**, 8.

[121] Janovec, V. and Přívratská, J. (2013). Domain structures. In: *International Tables for Crystallography*, Vol. *D, Physical properties of crystals* (ed. Authier, A.), Sect. 3.4.2. Dordrecht: Kluwer Academic Publishers. https://it.iucr.org/Db/ch3o4v0001/sec3o4o2.pdf

[122] Foecker, A. J. and Jeitschko, W. (2001). The atomic order of the pnictogen and chalcogen atoms in equiatomic ternary compounds TPnCh (T = Ni, Pd; Pn = P, As, Sb; Ch = S, Se, Te). *J. Solid State Chem.* **162**, 69.

[123] Grønvold, F. and Rost, E. (1957). The crystal structure of PdS_2 and $PdSe_2$. *Acta Crystallogr.* **10**, 329.

[124] Fleet, M. E. and Burns, P. C. (1990). Structure and twinning of cobaltite. *Canad. Miner.* **28**, 719.

[125] Hofmann, W. and Jäniche, W. (1935, 1936). Der Strukturtyp von Aluminiumborid (AlB_2). *Naturwiss.* **23**, 851; *Z. Phys. Chem.* **31 B**, 214.

[126] Nielsen, J. W. and Baenziger, N. C. (1954). The crystal structure of ZrBeSi and $ZrBe_2$. *Acta Crystallogr.* **7**, 132.

[127] Iandelli, A. (1964). MX_2-Verbindungen der Erdalkali- und seltenen Erdmetalle mit Gallium, Indium und Thallium. *Z. Anorg. Allg. Chem.* **330**, 221.

[128] Nuspl, G., Polborn, K., Evers, J., Landrum, G. A., and Hoffmann, R. (1996). The four-connected net in the $CeCu_2$ structure and its ternary derivatives. *Inorg. Chem.* **35**, 6922.

[129] Christensen, A. N., Broch, N. C., von Heidenstam, O., and Nilsson, A. (1967). Hydrothermal investigation of the systems In_2O_3–H_2O–Na_2O. The crystal structure of rhombohedral In_2O_3 and $In(OH)_3$. *Acta Chem. Scand.* **21**, 1046.

[130] Mullica, D. F., Beall, G. W., Milligan, W. O., Korp, J. D., and Bernal I. (1979). The crystal structure of cubic $In(OH)_3$ by X-ray and neutron diffraction methods. *J. Inorg. Nucl. Chem.* **41**, 277.

[131] Cohen-Addad, C. (1968). Etude structurale des hydroxystannates $CaSn(OH)_6$ et $ZnSn(OH)_6$ par diffraction neutronique, absorption infrarouge et résonance magnetique nucléaire. *Bull. Soc. Franç. Minér. Cristallogr.* **91**, 315.

[132] Bastiano, L. C., Peterson, R. T., Roeder, P. L., and Swainson, I. (1998). Description of schoenfliesite, $MgSn(OH)_6$, and roxbyite, $Cu_{1.22}S$, from a 1375 BC ship wreck, and rietveld neutron diffraction refinement of synthetic schoenfliesite, wickmanite, $MnSn(OH)_6$, and burtite, $CaSn(OH)_6$. *Canad. Mineral.* **36**, 1203.

[133] Mandel, N. and Donohue, J. (1971). The refinement of the crystal structure of skutterudite, $CoAs_3$. *Acta Crystallogr.* **B 27**, 2288.

[134] Kjekshus, A. and Rakke, T. (1974). Compounds with the skutterudite type crystal structure. III. Structural data for arsenides and antimonides. *Acta Chem. Scand.* **A 28**, 99.

[135] Billiet, Y. and Bertaut, E. F. (2005). Isomorphic subgroups of space groups. In *International Tables for Crystallography*, Vol. *A* (ed. Hahn, T.), Ch. 13. Dordrecht: Kluwer Academic Publishers. https://it.iucr.org/Ab/ch13o1v0001/ch13o1.pdf

[136] Müller, U. and Brelle, A. (1994). Über isomorphe Untergruppen von Raumgruppen der Kristallklassen 4, $\bar{4}$, $4/m$, 3, $\bar{3}$, 6, $\bar{6}$ und $6/m$. *Acta Crystallogr.* **A 51**, 300. https://doi.org/10.1107/S0108767394012614

[137] Fischer, P., Hälg, W., Schwarzenbach, D., and Gamsjäger, H. (1974). Magnetic and crystal structure of copper(II) fluoride. *J. Phys. Chem. Solids* **35**, 1683.

[138] Longo, J. M. and Kirkegaard, P. (1970). A refinement of the structure of VO_2. *Acta Chem. Scand.* **24**, 420.

[139] Baur, W. H. (1956). Über die Verfeinerung der Kristallstrukturbestimmung einiger Vertreter des Rutiltyps: TiO_2, SnO_2, GeO_2 und MgF_2. *Acta Crystallogr.* **9**, 515.

[140] Reimers, J. N., Greedan, J. E., Stager, C. V., and Kremer, R. (1989). Crystal structure and magnetism in $CoSb_2O_6$ and $CoTa_2O_6$. *J. Solid State Chem.* **83**, 20.

[141] Billiet, Y. (1973). Les sous-groupes isosymboliques des groupes spatiaux. *Bull. Soc. Franç. Minér. Cristallogr.* **96**, 327.

[142] Meyer, A. (1981). *Symmetriebeziehungen zwischen Kristallstrukturen des Formeltyps AX_2, ABX_4 und AB_2X_6 sowie deren Ordnungs- und Leerstellenvarianten.* Dissertation, Universität Karlsruhe. ISBN 3-923161-02-6.

[143] Baur, W. H. (1994). Rutile type derivatives. *Z. Kristallogr.* **209**, 143.

[144] Tremel, W., Hoffmann, R., and Silvestre, J. (1986). Transitions between NiAs and MnP type phases: an electronically driven distortion of triangular 3^6 nets. *J. Am. Chem. Soc.* **108**, 5174.

[145] Motizuki, K., Ido, H., Itoh, T., and Morifuji, M. (2010). Overview of magnetic properties of NiAs-type (MnP-Type) and Cu_2Sb-type compounds. *Springer Series in Materials Science* **131**, 11.

[146] Fjellvag, H. and Kjekshus, A. (1984). Magnetic and structural properties of transition metal substituted MnP. *Acta Chem. Scand.* **A 38**, 563.

[147] Strähle, J. and Bärnighausen, H. (1970). Die Kristallstruktur von Rubidiumtetrachloroaurat $RbAuCl_4$. *Z. Naturforsch.* **25 b**, 1186.

[148] Strähle, J. and Bärnighausen, H. (1971). Kristallchemischer Vergleich der Strukturen von $RbAuCl_4$ und $RbAuBr_4$. *Z. Kristallogr.* **134**, 471.

[149] Bulou, A. and Nouet, J. (1987). Structural phase transitions in ferroelastic $TlAlF_4$. *J. Phys. C* **20**, 2885.

[150] Bock, O. and Müller, U. (2002). Symmetrieverwandtschaften bei Varianten des ReO_3-Typs. *Z. Anorg. Allg. Chem.* **628**, 987. https://doi.org/10.1002/1521-3749(200206)628:5<987::AID-ZAAC987>3.0.CO;2-P

[151] Glazer, A. M. (1972). The classification of tilted octahedra in perovskites. *Acta Crystallogr.* **B 28**, 3384. https://doi.org/10.1107/S0567740872007976

[152] Burns, G. and Glazer, A. M. (1990). Space groups for solid state scientists, 2nd edn. San Diego: Acadamic Press.

[153] Locherer, K. R., Swainson, I. P., and Salje, E. K. H. (1999). Transition to a new tetragonal phase of WO_3: crystal structure and distortion parameters. *J. Phys. Cond. Mat.* **11**, 4143.

[154] Vogt, T., Woodward, P. M., and Hunter, B. A. (1999). The high-temperature phases of WO_3. *J. Solid State Chem.* **144**, 209.

[155] Xu, Y., Carlson, B., and Norrestam, R. (1997). Single crystal diffraction studies of WO_3 at high pressures and the structure of a high-pressure WO_3 phase. *J. Solid State Chem.* **132**, 123.

[156] Salje, E. K. H., Rehmann, S., Pobell, F., Morris, D., Knight, K. S., Herrmannsdoerfer, T., and Dove, M. T. (1997). Crystal structure and paramagnetic behaviour of ε-WO_{3-x}. *J. Phys. Cond. Matt.* **9**, 6563.

[157] Woodward, P. M., Sleight, A. W., and Vogt, T. (1995). Structure refinement of triclinic tungsten trioxide. *J. Phys. Chem. Solids* **56**, 1305.

[158] Aird, A., Domeneghetti, M. C., Nazzi, F., Tazzoli, V., and Salje, E. K. H. (1998). Sheet conductivity in WO_{3-x}: crystal structure of the tetragonal matrix. *J. Phys. Cond. Matt.* **10**, L 569.

[159] Howard, C. J., Luca, V., and Knight, K. S. (2002). High-temperature phase transitions in tungsten trioxide—the last word? *J. Phys. Cond. Mat.* **14**, 377.

[160] Bärnighausen, H. (1975). Group–subgroup relations between space groups as an ordering principle in crystal chemistry: the 'family tree' of perovskite-like structures. *Acta Crystallogr.* **A 31**, part S3, 01.1–9.

[161] Bock, O. and Müller, U. (2002). Symmetrieverwandtschaften bei Varianten des Perowskit-Typs. *Acta Crystallogr.* **B 58**, 594. https://doi.org/10.1107/S0108768102001490

[162] Howard, C. J. and Stokes, H. T. (2005). Structures and phase transitions in perovskites — a group-theoretical approach. *Acta Crystallogr.* **A 61**, 93.

[163] Talanov, M. V. (2019). Group-theoretical analysis of 1:3 A-site-ordered perovskite formation. *Acta Crystallogr.* **A 75**, 379.

[164] Baur, W. H. (2007). The rutile type and its derivatives. *Crystallogr. Rev.* **13**, 65. `https://doi.org/10.1080/08893110701433435`

[165] Pöttgen, R. and Hoffmann, R.-D. (2001). AlB_2-related intermetallic compounds—a comprehensive review based on a group–subgroup scheme. *Z. Kristallogr.* **216**, 127. `https://doi.org/10.1524/zkri.216.3.127.20327`

[166] Kussmann, S., Pöttgen, R., Rodewald, U.-C., Rosenhahn, C., Mosel, B. D., Kotzyba, G., and Künnen, B. (1999). Structure and properties of the stannide $Eu_2Au_2Sn_5$ and its relationship with the family of $BaAl_4$-related structures. *Z. Naturforsch.* **54 b**, 1155.

[167] Pöttgen, R. (2014). Coloring, distortions, and puckering in selected intermetallic structures from the perspective of group–subgroup relations. *Z. Anorg. Allg. Chem.* **640**, 869. `https://doi.org/10.1002/zaac.201400023`

[168] Hübner, J.-M., Carrillo-Cabrera, W., Cardoso-Gil, R., Koželj, P., Burkhardt, U., Etter, M., Akselrud, L., Grin, Y. and Schwarz, U. (2020). High-pressure synthesis of $SmGe_3$. *Z. Kristallogr.* **235**, 333.

[169] Schmidt, T. and Jeitschko, W. (2002). Preparation and crystal structure of the ternary lanthanoid platinum antimonides $Ln_3Pt_7Sb_4$ with $Er_3Pd_7P_4$ type structure. *Z. Anorg. Allg. Chem.* **628**, 927.

[170] Landolt-Börnstein, *Numerical Data and Functional Relationships in Science and Technology*, New Series, Group IV (eds. Baur, W. H. and Fischer, R. X.), Vol. **14** (2000, 2002, 2006).

[171] Baur, W. H. (2011). Rigid frameworks of zeolite-like compounds of the pharmacosiderite structure-type. *Mesoporous Materials* **151**, 13.

[172] Fischer, R. X. and Baur, W. H. (2009). Symmetry relationships of sodalite type crystal structures. *Z. Kristallogr.* **224**, 185.

[173] Graf, T., Casper, F., Winterlik, J., Balke, B., Fecher, G. H., and Felser, C. (2009). Crystal structure of new Heusler compounds. *Z. Anorg. Allg. Chem.* **635**, 976.

[174] Seidlmayer, S., Bachhuber, F., Anusca, I., Rothballer, J., Bräu, M., Peter, P., Weihrich, R. (2010). Half antiperovskites: V. Systematics in ordering and group–subgroup relations for $Pb_2Pd_3S_2$, $Bi_2Pd_3Se_2$, and $Bi_2Pd_3S_2$. *Z. Kristallogr.* **225**, 371.

[175] Kohlmann, H. (2009). Structural relationships in complex hydrides of the late transition metals. *Z. Kristallogr.* **224**, 454.

[176] Kohlmann, H. (2020). Hydrogen order in hydrides of Laves phases metals. *Z. Kristallogr.* **235**, 319. `https://doi.org//10.1515/zkri-2020-0043`

[177] Falkowski, V., Zeigner, A., Seidel, S., Pöttgen, R., Wurst, K., Ruck, M., and Huppertz, H. (2020). Syntheses and crystal structures of the manganese hydroxide halides $Mn_5(OH)_6Cl_4$, $Mn_5(OH)_7I_3$ and $Mn_5(OH)_{10}I_4$. *Z. Kristallogr.* **235**, 375.

[178] Müller, U. (1980). Strukturverwandtschaften unter den EPh_4^+-Salzen. *Acta Crystallogr.* **B 36**, 1075. `https://doi.org/10.1107/S0567740880005328`

[179] Müller, U. (2004). Kristallographische Gruppe-Untergruppe-Beziehungen und ihre Anwendung in der Kristallchemie. *Z. Anorg. Allg. Chem.* **630**, 1519. https://doi.org/10.1002/zaac.200400250

[180] Block, T., Seidel, S., and Pöttgen, R. (2022). Bärnighausen trees—A group–subgroup reference database. *Z. Kristallogr.* **237** 215. https://doi.org/10.1515/zkri-2022-0021

[181] Kihara, K. (1990). An X-ray study of the temperature dependence of the quartz structure. *Eur. J. Miner.* **2**, 63.

[182] Sowa, H., Macavei, J., and Schulz, H. (1990). Crystal structure of berlinite $AlPO_4$ at high pressure. *Z. Kristallogr.* **192**, 119.

[183] Stokhuyzen, R., Chieh, C., and Pearson, W. B. (1977). Crystal structure of Sb_2Tl_7. *Canad. J. Chem.* **55**, 1120.

[184] Marezio, M., McWhan, D. B., Remeika, J. P., and Dernier, P. D. (1972). Structural aspects of some metal–insulator transitions in Cr-doped VO_2. *Phys. Rev.* **B 5**, 2541.

[185] Farkas, L., Gadó, P., and Werner, P. E. (1977). The structure refinement of boehmite (γ-AlOOH) and the study of its structural variability. *Mat. Res. Bull.* **12**, 1213.

[186] Christoph, G. G., Corbato, C. E., Hofmann, D. A., and Tettenhorst, R. T. (1979). The crystal structure of boehmite. *Clays Clay Miner.* **27**, 81.

[187] Jacobs, H., Tacke, T., and Kockelkorn, J. (1984). Hydroxidmonohydrate des Kaliums und Rubidiums; Verbindungen, deren Atomanordnungen die Schreibweise $K(H_2O)OH$ bzw. $Rb(H_2O)OH$ nahelegen. *Z. Anorg. Allg. Chem.* **516**, 67.

[188] Harker, D. (1944). The crystal structure of Ni_4Mo. *J. Chem. Phys.* **12**, 315.

[189] Gourdon, O., Gouti, D., Williams, D. J., Proffen, T., Hobbs, S., and Miller, G. J. (2007). Atomic distributions in the γ-brass structure of the Cu–Zn system. *Inorg. Chem.* **46**, 251.

[190] Jansen, M. and Feldmann, C. (2000). Strukturverwandtschaften zwischen *cis*-Natriumhyponitrit und den Alkalimetallcarbonaten M_2CO_3 (M = Na, K, Rb, Cs), dargestellt durch Gruppe-Untergruppe-Beziehungen. *Z. Kristallogr.* **215**, 343.

[191] Swainson, I. P., Dove, M. T., and Harris, M. J. (1995). Neutron powder-diffraction study of the ferroelastic phase transition and lattice melting in sodium carbonate. *J. Phys. Cond. Matt.* **7**, 4395.

[192] Dinnebier, R. E., Vensky, S., Jansen, M., and Hanson, J. C. (2005). Crystal structures and topochemical aspects the high-temperature phases and decomposition products of the alkali-metal oxalates $M_2C_2O_4$ (M = K, Rb, Cs). *Chem. Europ. J.* **11**, 1119.

[193] Becht, H. Y. and Struikmans, R. (1976). A monoclinic high-temperature modification of potassium carbonate. *Acta Crystallogr.* **B 32**, 3344.

[194] Idemoto, Y., Ricardson, J. M., Koura, N., Kohara, S., and Loong, C.-K. (1998). Crystal structure of $(Li_xK_{1-x})_2CO_3$ ($x = 0, 0.43, 0.5, 0.62, 1$) by neutron powder diffraction analysis. *J. Phys. Chem. Solids* **59**, 363.

[195] Dubbledam, G. C. and de Wolff, P. M. (1969). The average crystal structure of γ-Na_2CO_3. *Acta Crystallogr.* **B 25**, 2665.

[196] Aalst, W. V., Hollander, J. D., Peterse, W. J. A., and de Wolff, P. M. (1976). The modulated structure of γ-Na_2CO_3 in a harmonic approximation. *Acta Crystallogr.* **B 32**, 47.

[197] Dušek, M., Chapuis, G., Meyer, M., and Petříček, V. (2003). Sodium carbonate revisited. *Acta Crystallogr.* **B 59**, 337.

[198] Arakcheeva, A. V., Bindi, L., Pattison, O., Meisser, N., Chapuis, G., and Pekov, I. V. (2010). The incommensurably modulated structures of natural natrite at 120 and 293 K. *Amer. Mineralogist* **B 95**, 574.

[199] Babel, D. and Deigner, P. (1965). Die Kristallstruktur von β-Iridium(III)-chlorid. *Z. Anorg. Allg. Chem.* **339**, 57.

[200] Coppens, P., Eibschütz, M. (1965). Determination of the crystal structure of yttrium othoferrite and refinement of gadolinium orthoferrite. *Acta Crystallogr.* **19**, 524.

[201] Ross, N. L., Zhao, J., Angel, R. J. (2004). High-pressure structural behavior of $GdAlO_3$ and $GdFeO_3$ perovskites. *J. Solid State Chem.* **177**, 3768.

[202] Brese, N. E., O'Keeffe, M., Ramakrishna, B. L., and van Dreele, R. B. (1990). Low-temperature structures of CuO and AgO and their relationships to those of MgO and PdO. *J. Solid State Chem.* **89**, 184.

[203] Jansen, M. and Fischer, P. (1988). Eine neue Darstellungsmethode für monoklines Silber(I,III)-oxid, Einkristallzüchtung und Röntgenstrukturanalyse. *J. Less Common Metals* **137**, 123.

[204] Grzelak, A., Gawraczinski, J., Jaron, T., Somayazulu, M., Derzsi, M., Struhkin, V., and Grochala, W. (2017). Persistence of mixed and non-intermediate valence in the high-pressure structure of silver(I,III)-oxide, AgO. *Inorg. Chem.* **56**, 5804.

[205] Hauck, J. and Mika, K. (1995). Close-packed structures. In: *Intermetallic Compounds*, Vol. 1—Principles (eds. Westbrook, J. H. and Fleischer, R. L.). Chichester: Wiley.

[206] Müller, U. (1998). Strukturverwandtschaften zwischen trigonalen Verbindungen mit hexagonal-dichtester Anionen-Teilstruktur und besetzten Oktaederlücken. Berechnung der Anzahl möglicher Strukturtypen II. *Z. Anorg. Allg. Chem.* **624**, 529.

[207] Michel, C., Moreau, J., and James, W. J. (1971). Structural relationships in compounds with $R\bar{3}c$ symmetry. *Acta Crystallogr.* **B 27**, 501.

[208] Meisel, K. (1932). Rheniumtrioxid. III. Über die Kristallstruktur des Rheniumtrioxids. *Z. Anorg. Allg. Chem.* **207**, 121.

[209] Daniel, P., Bulou, A., Rosseau, M., Nouet, J., Fourquet, J., Leblanc, M., and Burriel, R, (1992). A study of the structural phase transitions in AlF_3. *J. Physics Cond. Matt.* **2**, 5663.

[210] Leblanc, M., Pannetier, J., Ferey, G., and De Pape, R. (1985). Single-crystal refinement of the structure of rhombohedral FeF_3. *Rev. Chim. Minér.* **22**, 107.

[211] Sowa, H. and Ahsbas, H. (1998). Pressure-induced octahedron strain in VF_3 type compounds. *Acta Crystallogr.* **B 55**, 578.https://doi.org//10.1107/S0108768198001207

[212] Jørgenson, J. E. and Smith, R. I. (2006). On the compression mechanism of FeF_3. *Acta Crystallogr.* **B62**, 987.

[213] Daniel, P., Bulou, A., Leblanc, M., Rousseau, M., and Nouet, J. (1990). Structural and vibrational study of VF_3. *Mat. Res. Bull.* **25**, 413.

[214] Hector, A. L., Hope, E. G., Levason, W., and Weller, M.T. (1998). The mixed-valence structure of R-NiF_3 *Z. Anorg. Allg. Chem.* **624**, 1982.

[215] Roos, M. and Meyer, G. (2001). Refinement of the crystal structure of GaF_3. *Z. Kristallogr. NCS* **216**, 18. NCS 409507.

[216] Hoppe, H. and Kissel, R. (1984). Zur Kenntnis von AlF_3 und InF_3. *J. Fluorine Chem.* **24**, 327.

[217] Jørgenson, J. E., Marshall, W. G., and Smith, R. I. (2004). The compression mechanism of CrF_3. *Acta Crystallogr.* **B 60**, 669. `https://doi.org//10.1107/S010876810402316X`

[218] Goncharenko, I. N., Glazkov, V. P., Irodova, A. V., and Somenkov, V. A. (1991). Neutron diffraction study of crystal structure and equation of state AlD_3 up to pressure of 7.2 GPa. *Physica B, Cond. Matt.* **174**, 117.

[219] Averdunk, F. and Hoppe, R. (1990). Zur Kristallstruktur von MoF_3. *J. Less Common Metals* **161**, 135.

[220] Dušek, M. and Loub, J. (1988). X-ray powder diffraction data and structure refinement of TeO_3. *Powder Diffract.* **3**, 175.

[221] Grosse, L. and Hoppe, R. (1987). Zur Kenntnis von Sr_2RhF_7. Mit einer Bemerkung zur Kristallstruktur von RhF_3. *Z. Anorg. Allg. Chem.* **552**, 123.

[222] Beck, H. P. and Gladrow, E. (1983). Neue Hochdruckmodifikationen im RhF_3-Typ bei Seltenerd-Trichloriden. *Z. Anorg. Allg. Chem.* **498**, 75.

[223] Sass, R. L., Vidale, R., and Donohue, I. (1957). Interatomic distances and thermal anisotropy in sodium nitrate and calcite. *Acta Crystallogr.* **10**, 567.

[224] Ruck, M. (1995). Darstellung und Kristallstruktur von fehlordnungsfreiem Bismuttriiodid. *Z. Kristallogr.* **210**, 650.

[225] Wechsler, B. A. and Prewitt, C. T. (1984). Crystal structure of ilmenite ($FeTiO_3$) at high temperature and high pressure. *Amer. Mineral.* **69**, 176.

[226] Ketelaar, J. A. and van Oosterhout, G. W. (1943). Die Kristallstruktur des Wolframhexachlorids. *Rec. Trav. Chim. Pays-Bas* **62**, 197.

[227] Burns, J. H. (1962). The Crystal structure of lithium fluoroantimonate(V). *Acta Crystallogr.* **15**, 1098.

[228] Abrahams, S. C., Hamilton, W. C., and Reddy, J. M. (1966). Ferroelectric lithium niobate 3, 4, 5. *J. Phys. Chem. Solids* **27**, 997, 1013, 1019.

[229] Hsu, R., Maslen, E. N., du Boulay, D., and Ishizawa, N. (1997). Synchrotron X-ray studies of $LiNbO_3$ and $LiTaO_3$. *Acta Crystallogr.* **B 53**, 420.

[230] Steinbrenner, U. and Simon, A. (1998). Ba_3N—a new binary nitride of an alkaline earth metal. *Z. Anorg. Allg. Chem.* **624**, 228.

[231] Angelkort, J., Schönleber, A., and van Smaalen, S. (2009). Low- and high-temperature crystal structures of TiI_3. *J. Solid State Chem.* **182**, 525.

[232] Beesk, W., Jones, P. G., Rumpel, H., Schwarzmann, E., and Sheldrick G. M. (1981). X-ray crystal structure of Ag_6O_2. *J. Chem. Soc., Chem. Comm.*, 664.

[233] Brunton, G. (1973). Li_2ZrF_6. *Acta Crystallogr.* **B 29**, 2294.

[234] Lachgar, A., Dudis, D. S., Dorhout, P. K., and Corbett, J. D. (1991). Synthesis and properties of two novel line phases that contain linear scandium chains, lithium scandium iodide ($LiScI_3$) and sodium scandium iodide ($Na_{0.5}ScI_3$). *Inorg. Chem.* **30**, 3321.

[235] Leinenweber, A., Jacobs, H., and Hull, S. (2001). Ordering of nitrogen in nickel nitride Ni_3N determined by neutron diffraction. *Inorg. Chem.* **40**, 5818.

[236] Troyanov, S. I., Kharisov, B. I., and Berdonosov, S. S. (1992). Crystal structure of $FeZrCl_6$— a new structural type for ABX_6. *Russian J. Inorg. Chem.* **37**, 1250; *Zh. Neorgan. Khim.* **37**, 2424.

[237] Kuze, S., du Boulay, D., Ishizawa, N., Kodama, N., Yamagu, M., and Henderson, B. (2004). Structures of $LiCaAlF_6$ and $LiSrAlF_6$ at 120 and 300 K by synchrotron X-ray single crystal diffraction. *J. Solid State Chem.* **177**, 3505.

[238] Ouili, Z., Leblanc, A., and Colombet, P. (1987). Crystal structure of a new lamellar compound: $Ag_{1/2}In_{1/2}PS_3$. *J. Solid State Chem.* **66**, 86.

[239] Hillebrecht, H., Ludwig, T., and Thiele, G. (2004). About trihalides with TiI_3 chain structure: proof of pair forming cations in β-$RuCl_3$ and $RuBr_3$ by temperature-dependent single-crystal X-ray analyses. *Z. Anorg. Allg. Chem.* **630**, 2199. https://doi.org/10.1002/zaac.200400106

[240] Lance, E. T., Haschke, J. M., and Peacor, D. R. (1976). Crystal and molecular structure of phosphorus triiodide. *Inorg. Chem.* **15**, 780.

[241] Svensson, C., Albertsson, J., Liminga, R., Kvick, A., and Abrahams, S. C. (1983). Structural temperature dependence in alpha lithium iodate. *K. Chem. Phys.* **78**, 7343.

[242] Blake, A. J., Ebsworth, E. A. V., and Welch, A. J. (1984). Structure of trimethylamine, C_3H_9N, at 118 K. *Acta Crystallogr.* **C 40**, 413.

[243] Horn, M., Schwerdtfeger, C. F., and Meagher, E. P. (1972). Refinement of the structure of anatase at several temperatures. *Z. Kristallogr.* **136**, 273.

[244] Cox, D. E., Shirane, G., Flinn, P. A., Ruby, S. L., and Takei, W. (1963). Neutron diffraction and Mössbauer study of ordered and disordered $LiFeO_2$. *Phys. Rev.* **132**, 1547.

[245] Bork, M. and Hoppe, R. (1996). Zum Aufbau von PbF_4 mit Strukturverfeinerung an SnF_4. *Z. Anorg. Allg. Chem.* **622**, 1557.

[246] Leciejewicz, J., Siek, S., and Szytula, A. (1988). Structural properties of ThT_2Si_2 compounds. *J. Less Common Metals* **144**, 9.

[247] Sens, I. and U. Müller, U. (2003). Die Zahl der Substitutions- und Leerstellenvarianten des NaCl-Typs bei verdoppelter Elementarzelle. *Z. Anorg. Allg. Chem.* **629**, 487. https://doi.org//10.1002/zaac.200390080

[248] Sumin, V. V. (1989). Study of NbO by neutron diffraction of inelastic scattering of neutrons. *Soviet Phys. Crystallogr.* **34**, 391; *Kristallografiya* **34**, 655.

[249] Rodic, D., Spasojeviv, V., Kusigerski, R., Tellgren, R., and Rundlof, H. (2000). Magnetic ordering in polycrystalline $Ni_xZn_{1-x}O$ solid solutions. *Phys. Stat. Sol.* (b) **218**, 527.

[250] Krüger, E. (2020). Structural distortion stabilizing the antiferromgnetic and insulating ground state of NiO. Symmetry **12**, 56.

[251] Partin, D. E., and O'Keefe, M. (1991). The structures and crystal chemistry of magnesium chloride and cadmium chloride. *J. Solid State Chem.* **95**, 176.

[252] McQueen, T. M., Huang, Q., Lynn, P. W., Berger, R. F., Klimczuk, T., Ueland, B. G., Schiffer, P., and Cava, R. J. (2007). Magentic structures and properties of the antiferromagnet α-$NaFeO_2$. *Phys. Rev. Serie 3, B—Condensed Matter.* **76**, 024420.

[253] Sofin, M. and Jansen, M. (2005). Synthesis and properties of $NaNiO_2$. *Z. Naturforsch.* **60 b**, 701.

[254] Restori, R. and Schwarzenbach, D. (1986). Charge density in cuprite, Cu_2O. *Acta Crystallogr.* **B 42**, 201.

[255] Gohle, R. and Schubert, K. (1964). Das System Platin–Silicium. *Z. Metallkunde* **55**, 503.

[256] Boher, P., Garnier, P., Gavarri, J. R., and Hewat, A. W. (1985). Monoxyde quadratique PbO-α: description de la transition structurale ferroélastique. *J. Solid State. Chem.* **57**, 343.

[257] Grøvold, F., Haraldsen, H., and Kjekshus, A. (1960). On the sulfides, selenides and tellurides of platinum. *Acta Chem. Scand.* **14**, 1879.

[258] Delgado, J. M., McMullan, R. K., and Wuensch, B. J. (1987). Anharmonic refinement of the crystal structure of HgI_2 with neutron diffraction data. *Trans. Amer. Crystallogr. Assoc.* **23**, 93.

[259] Schwarzenbach, D., Birkedal, H., Hostettler, M., and Fischer, P. (2007). Neutron diffraction investigation of the temperature dependence of crystal structure and thermal motions of red HgI_2. *Acta Crystallogr.* **B 63**, 828.

[260] Peters, J. and Krebs, B. (1982). Silicon disulfide and silicon diselenide: a reinvestigation. *Acta Crystallogr.* **B 38**, 1270.

[261] Evers, J., Mayer, P., Möckl, L., Oehlinger, G., Köppe, R., and Schnöckel, H. (2015). Two high-pressure phases of SiS_2 as missing links between the extremes of only edge-sharing and only corner-sharing tetrahedra. *Inorg. Chem.* **54**, 1240.

[262] Prewitt, C. T., Shannon, R. D., and Rogers, D. B. (1971). Chemistry of noble metal oxides. Crystal structures of platinum cobalt oxide, palladium cobal oxide, copper iron oxide, and silver iron oxide. *Inorg. Chem.* **10**, 719.

[263] Shan, Y., Li, G., Tian, G., Han, J., Wang, C., Liu, S., Du, H., and Yang, Y. (2009). Description of the phase transitions of cuprous iodide. *J. Alloys Comp.* **477**, 403.

[264] Albert, B. and Schmitt, K. (2001). Die Kristallstruktur von Bortriiodid, BI_3. *Z. Anorg. Allg. Chem.* **627**, 809.

[265] Smith, J. D. and Corbett, J. (1985). Stabilization of clusters by interstitial atoms. Three carbon-centered zirconium iodide clusters, $Zr_6I_{12}C$, $Zr_6I_{14}C$, and $MZr_6I_{14}C$ (M = K, Rb, Cs). *J. Amer. Chem. Soc.* **107**, 5704.

[266] Dill, S., Glaser, J., Ströbele, M., Tragl, S., and Meyer, H.-J. (2004). Überschreitung der konventionellen Zahl von Clusterelektronen in Metallhalogeniden des M_6X_{12}-Typs: W_6Cl_{18}, $(Me_4N)_2[W_6Cl_{18}]$ und $Cs_2[W_6Cl_{18}]$. *Z. Anorg. Allg. Chem.* **630**, 987.

[267] Santamaría-Pérez, D., Vegas, A., and Müller, U. (2005). A new description of the crystal structures of tin oxide fluorides. *Solid State Sci.* **7**, 479.

[268] Oswald, H. R. and Jaggi, H. (1960). Die Struktur der wasserfreien Zinkhalogenide. *Helv. Chim. Acta* **43**, 72.

[269] Ueki, T., Zalkin, A., and Templeton, D. H. (1965). The crystal structure of osmium tetroxide. *Acta Crystallogr.* **19**, 157.

[270] Krebs, B. and Hasse, K.-D. (1976). Refinements of the crystal structures of $KTcO_4$, $KReO_4$ and OsO_4. *Acta Crystallogr.* **B 32**, 1334.

[271] Hinz, D., Gloger, T., Möller, A., and Meyer, G. (2000). $CsTi_2Cl_7$-II, Synthese, Kristallstruktur und magnetische Eigenschaften. *Z. Anorg. Allg. Chem.* **626**, 23.

[272] Rekis, T. (2020). Crystallization of chiral molecular compounds: what can be learned from the Cambridge Structural Database? *Acta Crystallogr.* **B 76**, 307.

[273] Kitaigorodskii, A. I. (1961). *Organic Chemical Crystallography.* New York: Consultants Bureau.
[Translation from Russian: *Organicheskaya Kristallokhimiya.* Moskva: Akademiya Nauk SSSR (1955).]

[274] Brock, C. P. and Dunitz, J. D. (1994). Towards a grammar of molecular packing. *Chem. Mater.* **6**, 1118.

[275] Wilson, A. J. C., Karen, V. L., and Mighell, A. (2006). The space group distribution of molecular organic structures. In *International Tables for Crystallography*, Vol. *C, Mathematical, physical and chemical tables* (3rd edn; ed. Prince, E.), Ch. 9.7. Dordrecht: Kluwer Academic Publishers. `https://it.iucr.org/Cb/ch9o7v0001/ch9o7.pdf`

[276] Gavezzotti, A. (1998). The crystal packing of organic molecules: a challenge and fascination below 1000 Da. *Crystallogr. Rev.* **7**, 5.

[277] Hargittai, M. and Hargittai, I. (2009). *Symmetry through the Eyes of a Chemist* (3rd edn). Berlin: Springer.

[278] Woodley, S. M. and Catlow, R. (2008). Crystal structure prediction from first principles. *Nature Mat.* **7**, 937.

[279] Oganov, A. R (ed.) (2010). Modern methods of crystal structure prediction. Berlin: Wiley-VC.

[280] Zhu, Q. and Hattori, S. (2022). Organic crystal structure prediction and ist application to to materials design. *J. Materials Res.* `https://doi.org//10.1557/s43578-022-00698-9`

[281] Day, G. M. and 30 other authors (2009). Significant progress in predicting the crystal structures of small organic molecules—the fourth blind test. *Acta Crystallogr.* **B 65**, 107.

[282] Bardwell, D. A. and 41 other authors (2011). Toward crystal structure prediction of complex organic compounds—a report on the fifth blind test. *Acta Crystallogr.* **B 67**, 535. `https://doi.org/10.1107/S0108768111042868`

[283] Reilly, A. M. and 100 other authors (2016). Report on the sixth blind test of organic structure predicion methods. *Acta Crystallogr.* **B 72**, 439. `https://doi.org/10.1107/S2052520616007447`

[284] Kitaigorodskii, A. I. (1973). *Molecular Crystals and Molecules*. New York: Academic Press.

[285] Pidcock, E., Motherwell, W. D. S., and Cole, J. C. (2003). A database survey of molecular and crystallographic symmetry. *Acta Crystallogr.* **B 59**, 634.

[286] Yao, J. W., Cole, J. C., Pidcock, E., Allen, F. H., Howard, J. A. K., and Motherwell, W. D. S. (2002). CSDSymmetry: The definitive database of point group and space group relationships in small-molecule crystal structures. *Acta Crystallogr.* **B 58**, 640. `https://doi.org/10.1107/S0108768102006675`

[287] Maloney, A. and Pidcock, E. (2022). Unpublished.

[288] Müller, U. (1978). Kristallisieren zentrosymmetrische Moleküle immer in zentrosymmetrischen Raumgruppen? – Eine statistische übersicht. *Acta Crystallogr.* **B 34**, 1044.

[289] Urusov, V. S. and Nadezhina, T. N. (2009). Frequency distribution and selection of space groups in inorganic crystal chemistry. *J. Structural Chem.* **50** *Suppl.*, S22. *Zhurnal Strukt. Khimii* **50** *Suppl.*, S26.

[290] Pascal, R. A., Wang, C. M., Wang, G. C., and Koplitz, L. V. (2012). Ideal molecular conformation versus crystal site symmetry. *Crystal Growth & Design* **12**, 4367.

[291] Müller, U. (1972). Verfeinerung der Kristallstrukturen von KN_3, RbN_3, CsN_3 und TlN_3. *Z. Anorg. Allg. Chem.* **392**, 159.

[292] Levy, J. H., Sanger, P. L., Taylor, J. C., and Wilson, P. W. (1975). The structures of fluorides. XI. Kubic harmonic analysis of the neutron diffraction pattern of the body-centred cubic phase of MoF_6 at 266 K. *Acta Crystallogr.* **B 31**, 1065.

[293] Levy, J. H., Taylor, J. C., and Waugh, A B. (1983). Neutron powder structural studies of UF_6, MoF_6 and WF_6 at 77 K. *J. Fluorine Chem.* **23**, 29.

[294] Levy, J. H., Taylor, J. C., and Wilson, P. W. (1975). The structures of fluorides. IX. The orthorhombic form of molybdenum hexafluoride. *Acta Crystallogr.* **B 31**, 398.

[295] Bürgi, H.-B., Restori, R., and Schwarzenbach, D. (1993). Structure of C_{60}: partial orientational order in the room-temperature modification of C_{60}. *Acta Crystallogr.* **B 49**, 832.

[296] Dorset, D. L. and McCourt, M. P. (1994). Disorder and molecular packing of C_{60} buckminsterfullerene: a direct electron-crystallographic analysis. *Acta Crystallogr.* **A 50**, 344.

[297] David, W. I. F., Ibberson, R. M., Matthewman, J. C., Prassides, K., Dennis, T. J. S., Hare, J. P., Kroto, H. W., Taylor, R., and Walton, D. R. (1991). Crystal structure of ordered C_{60}. *Nature* **353**, 147.

[298] Panthöfer, M., Shopova, D., and Jansen. M. (2005). Crystal structure and stability of the fullerene–chalcogene co-crystal $C_{60}Se_8CS_2$. *Z. Anorg. Allg. Chem.* **631**, 1387.

[299] Gruber, H. and U. Müller, U. (1997). γ-P_4S_3, eine neue Modifikation von Tetraphosphortrisulfid. *Z. Kristallogr.* **212**, 662.

[300] Müller, U. and Noll, A. (2003). (Na-15-Krone-5)$_2$$ReCl_6 \cdot 4\,CH_2Cl_2$, eine Struktur mit CH_2Cl_2-Molekülen in pseudohexagonalen Kanälen. *Z. Kristallogr.* **218**, 699.

[301] Phillips, F. L. and Skapski, A. C. (1975). The crystal structure of tetraphenylarsonium nitridotetrachlororuthenate(IV): a square-pyramidal ruthenium complex. *Acta Crystallogr.* **B 31**, 2667.

[302] Müller, U., Sieckmann, J., and Frenzen, G. (1996). Tetraphenylphosphonium-pentachlorostannat, $PPh_4[SnCl_5]$, und Tetraphenylphosphonium-pentachlorostannat Monohydrat, $PPh_4[SnCl_5 \cdot H_2O]$. *Acta Crystallogr.* **C 52**, 330.

[303] Müller, U., Mronga, N., Schumacher, C., and Dehnicke, K. (1982). Die Kristallstrukturen von $PPh_4[SnCl_3]$ und $PPh_4[SnBr_3]$. *Z. Naturforsch.* **37 b**, 1122.

[304] Rabe, S. and Müller, U. (2001). Die Kristallstrukturen von $PPh_4[MCl_5(NCMe)] \cdot MeCN$ (M = Ti, Zr), zwei Modifikationen von $PPh_4[TiCl_5(NCMe)]$ und von *cis*-$PPh_4[TiCl_4(NCMe)_2] \cdot MeCN$. *Z. Anorg. Allg. Chem.* **627**, 201.

[305] Weller, F., Müller, U., Weiher, U., and Dehnicke, K. (1980). N-chloralkylierte Nitridochlorokomplexe des Molybdäns $[Cl_5Mo{\equiv}N{-}R]^-$. Die Kristallstruktur von $(AsPh_4)_2[(MoOCl_4)_2CH_3CN]$. *Z. Anorg. Allg. Chem.* **460**, 191.

[306] Rabe, S. and Müller, U. (2001). Die Kristallpackung von $(PPh_4)_2[NiCl_4] \cdot 2\,MeCN$ und $PPh_4[CoCl_{0,6}Br_{2,4}(NCMe)]$. *Z. Anorg. Allg. Chem.* **627**, 742.

[307] Rutherford, J. S. (2001). A combined structure and symmetry classification of the urea series channel inclusion compounds. *Crystal Eng.* **4**, 269.

[308] Yeo, L., Kariuki, B. K., Serrano-González, H., and Harris, K. D. M. (1997). Structural properties of the low-temperature phase of hexadecane/urea inclusion compound. *J. Phys. Chem.* **B 101**, 9926.

[309] Forst, R., Boysen, H., Frey, F., Jagodzinski, H., and Zeyen, C. (1986). Phase transitions and ordering in urea inclusion compounds with n-paraffins. *J. Phys. Chem. Solids* **47**, 1089.

[310] Hollingsworth, M. D., Brown, M. E., Hillier, A. C., Santarsiero, B. D., and Chaney, J. D. (1996). Superstructure control in the crystal growth and ordering of urea inclusion compounds. *Science* **273**, 1355.

[311] Brown, M. E., Chaney, J. D., Santarsiero, B. D., and Hollingsworth, M. D. (1996). Superstructure topologies and host–guest interactions in commensurate inclusion compounds of urea with bis(methyl)ketones. *Chem. Mater.* **8**, 1588.

[312] Müller, U. (1971). Die Kristallstruktur von Hexachlorborazol. *Acta Crystallogr.* **B 27**, 1997.

[313] Deen, P. P., Braithwaite, D., Kernavanois, N., Poalasini, L., Raymond, S., Barla, A., Lapertot, G., and Sánchez, J. P. (2005). Structural and electronic transitions of the low-temperature, high-pressure phase of SmS. *Phys. Rev.* **B 71**, 245118.

[314] Sousanis, A., Smet, P. F., and Poelman, D. (2017). Samarium monosulfide (SmS): Reviewing properties and application. *Materials* **10**, 953.

[315] Boldyreva, E. V. (2003). High-pressure induced structural changes in molecular crystals preserving the space group symmetry. *Crystal Eng.* **6**, 235.

[316] Dove, M. T. (1997). Theory of displacive phase transitions in minerals. *Amer. Mineral.* **82**, 213.

[317] Pirc, R. and Blinc, R. (2004). Off-center Ti model of barium titanate. *Phys. Rev.* **B 70**, 134107.

[318] Völkel, G. and Müller, K. A. (2007). Order–disorder phenomena in the low-temperature phase of $BaTiO_3$. *Phys. Rev.* **B 76**, 094105.

[319] Decius, J. C. and Hexter, R. M. (1977). *Molecular Vibrations in Crystals*. New York: McGraw Hill.

[320] Sherwood, P. M. A. (1972). *Vibrational Spectroscopy of Solids*. Cambridge: Cambridge University Press.

[321] Weidlein, J., Müller, U., and Dehnicke, K. (1988). *Schwingungsspektroskopie* (2. Aufl.). Stuttgart: Thieme.

[322] Sathnayanarayana, D. N. (2004). *Vibrational Spectroscopy: Theory and Applications*. New Delhi: New Age International.

[323] Franzen, H. F. (1982). *Second-order Phase Transitions and the Irreducible Representations of Space Groups*. Berlin: Springer.

[324] Placzek, G. (1959). *The Rayleigh and Raman Scattering*. Translation series, US Atomic Energy Commission, UCRL Tansl. 256. Berkeley: Lawrence Radiation Laboratory. https://babel.hathitrust.org/cgi/pt?id=uc1.31210003079991&view=1up&seq=126
[Translation from German: Rayleigh-Streuung und Raman-Effekt. In *Handbuch der Radiologie*, Vol. VI, pp. 205–374. Leipzig: Akademische Verlagsgesellschaft (1934).]

[325] G. Herzberg (1945). *Molecular Spectra and Molecular Structure*, Vol. II: *Infrared and Raman Spectra of Polyatomic Molecules*, pp. 104–131. New York: Van Norstrand Reinhold.

[326] Mulliken, R. S. (1933). Electronic structures of polyatomic molecules and valence. IV. Electronic states, quantum theory and the double bond. *Phys. Rev.* **43**, 279.

[327] Izyumov, Yu. A. and Syromyatnikov, V. N. (1990). *Phase Transitions and Crystal Symmetry.* Dordrecht: Kluwer Academic Publishers.
[Translation from Russian: *Fazovie perekhodi i simmetriya kristallov.* Moskva: Nauka (1984).]

[328] Lyubarskii, G. Y. (1960). *Application of Group Theory in Physics.* Oxford: Pergamon.
[Translation from Russian: *Teoriya grupp i e'e primenenie v fizike.* Moskva: Gostekhizdat (1957). German: *Anwendungen der Gruppentheorie in der Physik*, Berlin: Deutscher Verlag der Wissenschaften (1962).]

[329] Chaikin, P. M. and Lubensky, T. C. (1995; reprint 2000). *Principles of Condensed Matter in Physics.* Cambridge: Cambridge University Press.

[330] Tolédano, J.-C. and P. Tolédano, P. (1987). *The Landau Theory of Phase Transitions.* Singapore: World Scientific.

[331] Tolédano, J.-C., Janovec, V., Kopský, V., Scott, J.-F., and Boček, P. (2013). Structural phase transitions. In *International Tables for Crystallography*, Vol. *D, Physical properties of crystals* (2nd edn; ed. Authier, A.). Ch. 3.1. Dordrecht: Kluwer Academic Publishers. https://it.iucr.org/Db/ch3o1v0001/ch3o1.pdf

[332] Papon, P., Leblond, L. and Meijer, P. H. E. (2006). *The Physics of Phase Transitions. Concepts and Applications.* 2nd ed. Berlin–Heidelberg–New York: Springer.

[333] Unruh, H. G. (1993). Ferroelastic phase transitions in calcium chloride and calcium bromide. *Phase Transitions* **45**, 77.

[334] Salje, E. K. H. (1990). *Phase Transitions in Ferroelastic and Co-elastic Crystals.* Cambridge: Cambridge University Press.

[335] Kovalev, O. V. (1993). *Representations of the Crystallographic Space Groups: Irreducible Representaions, Induced Representations and Corepresentations* (2nd edn). New York: Gordon & Breach.

[336] Stokes, H. T. and Hatch, D. M. (1988). *Isotropy Subgroups of the 230 Crystallographic Space Groups.* Singapore: World Scientific. https://doi.org/10.1142/0751

[337] Stokes, H. T., Hatch, D. M., and Campbell, B. J. (2007). ISOTROPY. https://stokes.byu.edu/isotropy.html

[338] Stokes, H. T., Hatch, D. M., and Wells, J. D. (1991). Group-theoretical methods for obtaining distortions in crystals. Applications to vibrational modes and phase transitions. *Phys. Rev.* **B 43**, 11010.

[339] Pérez-Mato, J. M., Orobengoa, C., and Aroyo, M. I. (2010). Mode crystallography of distorted structures. *Acta Crystallogr.* **A 66**, 558.

[340] Kleber, W. (1967). Über topotaktische Gefüge. *Kristall u. Technik* **2**, 5. https://doi.org/10.1002/crat.19670020102

[341] Wadhawan, V. K. (2006). Towards a rigorous definition of ferroic phase transitions. *Phase Transitions* **64**, 165.

[342] Van Tendeloo, G. and Amelinckx, S. (1974). Group-theoretical considerations concerning domain formation in ordered alloys. *Acta Crystallogr.* **A 30**, 431.

[343] Wondratschek, H. and Jeitschko, W. (1976). Twin domains and antiphase domains. *Acta Crystallogr.* **A 32**, 664.

[344] Chen, Ch.-Ch., Herhold, A. B., Johnson, C. S., and Alivisatos, A. P. (1997). Size dependence and structural metastability in semiconductor nanocrystals. *Science* **276**, 398.

[345] McMahon, M. I., Nelmes, R. J., Wright, N. G., and Allan, D. R. (1994). Pressure dependence of the *Imma* phase of silicon. *Phys. Rev. B* **50**, 739.

[346] Capillas, C., Aroyo, M. I., and Pérez-Mato, J. M. (2007). Transition paths in reconstructive phase transitions. In: *Pressure-induced Phase Transitions.* (ed. Grzechnik, A.). Trivandrum, Kerala: Transworld Research Network.

[347] Capillas, C., Pérez-Mato, J. M., and Aroyo, M. I. (2007). Maximal symmetry transition paths for reconstructive phase transitions. *J. Phys. Cond. Matt.* **19**, 275203.

[348] Sowa, H. (2010). Phase transitions in AB systems—symmetry aspects. In: *High-Pressure Crystallography* (eds. Boldyreva, E. and Dera, P.), p. 183. NATO Science for Peace and Security Series B—Physics. Dordrecht: Springer.

[349] Leoni, S., Boulfelfel, S. E., and Baburin, I. A. (2011). A walk on the chemical landscape: the role of B33 along the B1–B2 phase transition in RbF and NaBr. *Z. Anorg. Allg. Chem.* **637**, 864.

[350] Hahn, Ch. and Unruh, H.-G. (1991). Comment on temperature-induced structural phase transitions in $CaBr_2$ studied by Raman spectroscopy. *Phys. Rev. B* **43**, 12665. `https://dx.doi.org/10.1103/PhysRevB.43.12665`

[351] Hahn, T. and Klapper, H. (2013). Twinning of crystals. In: *International Tables for Crystallography*, Vol. *D, Physical properties of crystals* (ed. Authier, A.), Ch. 3.3. Chichester: Wiley. `https://it.iucr.org/Db/ch3o3v0001/`

[352] McGinnety, J. A. (1972). Redetermination of the structures of potassium sulphate and potassium chromate. *Acta Crystallogr.* **B 28**, 2845.

[353] Ojima, K., Nishihata, Y., and Sawada, A. (1995). Structure of potassium sulfate at temperatures from 296 K down to 15 K. *Acta Crystallogr.* **B 51**, 287.

[354] Arnold, H., Kurtz, W., Richter-Zinnius, A., Bethke, J., and Heger, G. (1981). The phase transition of K_2SO_4 at about 850 K. *Acta Crystallogr.* **B 37**, 1643.

[355] González-Silgo, C., Solans, X., Ruiz-Pérez, C., Martínez-Sarrión, M. L., Mestres, L., and Bocanegra, E. H. (1997). Study on mixed crystals $(NH_4)_{2-x}K_xSO_4$. *J. Phys. Cond. Matt.* **9**, 2657.

[356] Hasebe, K. (1981). Studies of the crystal structure of ammonium sulfate in connection with its ferroelectric phase transition. *J. Phys. Soc. Jap.* **50**, 1266.

[357] Makita, Y., Sawada, A., and Takagi, Y. (1976). Twin plane motion in $(NH_4)_2SO_4$. *J. Phys. Soc. Jap.* **41**, 167.

[358] Sawada, A., Makita, Y., and Takagi, Y. (1976). The origin of mechanical twins in $(NH_4)_2SO_4$. *J. Phys. Soc. Jap.* **41**, 174.

[359] Perrotta, A. J. and Smith, J. V. (1968). The crystal structure of $BaAl_2O_4$. *Bull. Soc. Chim. Franç. Minér. Cristallogr.* **91**, 85.

[360] Hörkner, W. and Müller-Buschbaum, H. (1979). Zur Kristallsrtuktur von $BaAl_2O_4$. *Z. Anorg. Allg. Chem.* **451**, 40.

[361] Abakumov, A. M, Lebedev, O. I., Nistor, L., van Tendeloo, G., and Amelinckx, S. (2000). The ferroelectric phase transition in tridymite type $BaAl_2O_4$ studied by electron microscopy. *Phase Transitions* **71**, 143. `https://doi.org/10.1080/01411590008224545`

[362] Ðuriš, K., Müller, U., and Jansen, M. (2012). K_3NiO_2 revisited, phase transition, and crystal structure refinement. *Z. Anorg. Allg. Chem.* **638**, 737.

[363] Sparta, K., Redhammer, G. J., Roussel, P., Heger, G., Roth, G., Lemmens, P., Ionescu, A., Grove, M., Güntherodt, G., Hüning, F., Lueken, H., Kageyama, H., Onizuka, K., and Ueda, Y. (2001). Structural phase transitions in the 2D spin dimer compound $SrCu(BO_3)_2$. *Europ. Phys. J.* **B 19**, 507.

[364] Daniš, S., Javorský, P., Rafaja, D., and Sechovský, V. (2002). Low-temperature transport and crystallographic studies of $Er(Co_{1-x}Si_x)_2$ and $Er(Co_{1-x}Ge_x)_2$. *J. Alloys Comp.* **345**, 54.

[365] Oganov, A. R., Price, G. D., and Brodholt, J. P. (2001). Theoretical investigation of metastable Al_2SiO_5 polymorphs. *Acta Crystallogr.* **A 57**, 548.

[366] Muller, O., Wilson, R., Colijn, H., and Krakow, W. (1980). δ-FeO(OH) and its solutions. *J. Mater. Sci.* **15**, 959. `https://doi.org/10.1007/BF00552109`

[367] Green, J. (1983). Calcination of precipitated $Mg(OH)_2$ to active MgO in the production of refractory and chemical grade MgO. *J. Mater. Sci.* **18**, 637.

[368] Kim, M. G., Dahmen, U., and Searcy, A. W. (1987). Structural transformations in the decomposition of $Mg(OH)_2$ and $MgCO_3$. *J. Am. Ceram. Soc.* **70**, 146.

[369] Verheijen, M. A., Meekes, H., Meijer, G., Bennema, P., de Boer, J. L., van Smaalen, S., van Tendeloo, G., Amelinckx, S., Muto, S., and van Landuyt, J. (1992). The structure of different phases of pure C_{70} crystals. *Chem. Phys.* **166**, 287.

[370] Van Smaalen, S., Petříček, V., de Boer, J. L., Dušek, M., Verheijen, M. A., and G. Meijer, G. (1994). Low-temperature structure of solid C_{70}. *Chem. Phys. Lett.* **223**, 323.

[371] Soldatov, A. V., Roth, G., Dzyabchenko, A., Johnels, D., Lebedkin, S., Meingast, C., Sundqvist, B., Haluska, M., and Kuzmany, H. (2001). Topochemical polymerization of C_{70} controlled by monomer crystal packing. *Science* **293**, 680.

[372] Bärnighausen, H. (1976). Mixed-valence rare-earth halides and their unusual crystal structures. In *Rare Earths in Modern Science and Technology* **12** (eds. McCarthy, G. J. and Rhyne, J. J.), pp. 404–413. New York: Plenum Press.

[373] Wenz, O. E. (1992). *Die Interpretation von Vernier-Strukturen des Formeltyps M_mX_{m+1} als eindimensional modulierte Überstrukturen des Fluorit-Typs (M = Ln^{3+}, Ln^{2+} oder Sr^{2+}; X = Cl oder Br; m = 4–7).* Dissertation, Universität Karlsruhe.

[374] Bärnighausen, H. and Haschke, J. M. (1978). Compositions and crystal structures of the intermediate phases in the samarium bromine system. *Inorg. Chem.* **17**, 18.

[375] Rinck, C. (1982). *Röntgenographische Strukturuntersuchungen an Seltenerdmetall(II,III)-chloriden des Formeltyps Ln_mCl_{2m+1} (Ln = Dy oder Tm; m = 5, 6 und 7).* Dissertation, Universität Karlsruhe. ISBN 3-923161-08-5.

[376] Bachmann, R. (1987). *Die Kristallstruktur des neuen Strukturtyps Ln_4X_9 (Ln = Eu, Nd; X = Cl, Br).* Dissertation, Universität Karlsruhe. ISBN 3-923161-15-8.

[377] Lange, F. T. (1992). *Darstellung von Europium(II,III)-chloriden durch chemischen Transport und Untersuchungen zum Valenzzustand des Europiums in diesen Verbindungen.* Dissertation, Universität Karlsruhe. ISBN 3-923161-34-4.

[378] Haselhorst, R. W. (1994). *Die Darstellung der Phasen Yb_mCl_{m+1} mit $m = 6$–9 und Untersuchungen zur Strukturdynamik dieser Verbindungen.* Dissertation, Universität Karlsruhe. ISBN 3-923161-38-7.

[379] Stegmüller, P. (1997). *Strukturelle Untersuchungen an Verbindungen Yb_mCl_{m+1} ($m = 6, 8$) und Chloroaluminaten der Erdalkali-Elemente Sr und Ba sowie der Lanthanoide Yb, Sm und Eu.* Dissertation, Universität Karlsruhe.

[380] Lumpp, A. and Bärnighausen, H. (1988). Die Kristallstruktur von Nd_3Cl_7 und ihre Beziehung zum Fluorit-Typ. *Z. Kristallogr.* **182**, 174.

[381] Bärnighausen, H. (1979). Unpublished; partly cited by Haschke, J. M. in: *Handbook of the Physics and Chemistry of Rare Earths* (eds. Gschneidner, K. A. and Eyring, L. R.), Vol. 4, pp. 117–130. Amsterdam: North-Holland; and in *Gmelin Handbuch der anorganischen Chemie, Seltenerdelemente*, Vol. C 4a (1982), pp. 47–50.

[382] Sass, R. L., Brackett, T. E., and Brackett, E. B. (1963). The crystal structure of strontium bromide. *J. Phys. Chem.* **67**, 2862.

[383] Eick, H. A. and Smeggil, J. G. (1971). The crystal structure of strontium dibromide. *Inorg. Chem.* **10**, 1458.

[384] Astakhova, I. S., Goryushkin, V. F., and Poshevneva, A. I. (1991). An X-ray diffraction study of dysprosium dichloride. *Russ. J. Inorg. Chem.* **36**, 568; *Zh. Neorgan. Khim.* **36**, 1000.

[385] Rietschel, E. T. and Bärnighausen, H. (1969). Die Kristallstruktur von Strontiumjodid SrJ_2. *Z. Anorg. Allg. Chem.* **368**, 62.

[386] Naterstad, T. and Corbett, J. (1976). The crystal structure of thulium(II) chloride. *J. Less Common Metals* **46**, 291.

[387] Beyer, A. (1978). *Präparative und röntgenographische Untersuchungen im Thulium-Chlor-System.* Diplomarbeit, Universität Karlsruhe.

[388] Voronin, B. M. and Volkov, S. V. (2001). Ionic conductivity of fluorite type crystals CaF_2, SrF_2, BaF_2, and $SrCl_2$ at high temperatures. *J. Phys. Chem. Solids* **62**, 1349.

[389] Dworkin, A. S. and Bredig, M. A. (1963). The heats of fusion and transition of alkaline earth and rarer earth metals halides. *J. Phys. Chem.* **67**, 697.

[390] Gouteron, J., Michel, D., Lejus, A. M., and Zarembovitch, J. (1981). Raman spectra of lanthanide sesquioxide single crystals: correlation between A and B-type structures. *J. Solid State Chem.* **38**, 288.

[391] Bärnighausen, H. and Schiller. G. (1985). The crystal structure of A-Ce_2O_3. *J. Less Common Metals* **110**, 385.

[392] Müller-Buschbaum, H. (1967). Untersuchung der Phasenumwandlungen der monoklinen B- in die hexagonale A-Form an Einkristallen der Oxide der Seltenen Erdmetalle. *Z. Anorg. Allg. Chem.* **355**, 41.

[393] Schleid, T. and Meyer, G. (1989). Single crystals of rare earth oxides from reducing halide melts. *J. Less Common Metals* **149**, 73.

[394] Bärnighausen, H., unpublished.

[395] Müller, U., Dübgen, R., and Dehnicke, K. (1980). Diazidoiodat(I): Darstellung, IR-Spektrum und Kristallstruktur von $PPh_4[I(N_3)_2]$. *Z. Anorg. Allg. Chem.* **463**, 7.

[396] De Costanzo, L., Forneris, F., Geremia, S., and Randaccio, L. (2003). Phasing protein structures using the group–subgroup relation. *Acta Crystallogr.* **D 59**, 1435.

[397] Zumdick, M. F., Pöttgen, R. Müllmann, R., Mosel, B. D., Kotzyba, G., and Kühnen, B. (1998). Syntheses, crystal structure, and properties of Hf_2Ni_2In, Hf_2Ni_2Sn, Hf_2Cu_2In, and Hf_2Pd_2In with ordered Zr_3Al_2 type structure. *Z. Anorg. Allg. Chem.* **624**, 251.

[398] Buzek, P., Klapötke, T. M., Schleyer, P. v. R., Thornieport-Oetting, P. S., and White, P. S. (1993). Iodine azide. *Angew. Chem. Int. Ed.* **32**, 275; *Angew. Chem.* **105**, 299.

[399] Lyhs, B., Bläser, D., Wölper, C., Schulz, S., and Jansen, G. (2012). A comparison of the solid-state structures of halogen azides XN_3 (X = Cl, Br, I). *Angew. Chem. Int. Ed.* **51**, 12859; *Angew. Chem.* **124**, 13031;

[400] Müller, U., Ivlev, S., Schulz, S., and Wölper, C. (2021). Automated crystal structure determination has its pitfalls: Correction to the crystal structures of iodine azide. *Angew. Chem. Int. Ed.* **60**, 17592; *Angew. Chem.* **133**, 17452. `https://doi.org//10.1002/anie.202105666`

[401] Rodewald, U. C., Lukachuk, M., Heying, B., and Pöttgen, R. (2006). The indide $Er_{2.30}Ni_{1.84}In_{0.70}$—a new superstructure of the U_3Si_2 family. *Monatsh. Chem.* **137**, 7.

[402] Wu, K. K. and Brown, I. D. (1973). Refinement of the crystal structure of $CaCrF_5$. *Mater. Res. Bull.* **8**, 593.

[403] Wandner, K.-H. and Hoppe, R. (1986). Zum Jahn-Teller-Effekt bei Mn(III)-Fluoriden: $CaMnF_5$. *Rev. Chim. Minér.* **23**, 520.

[404] Wandner, K.-H. and Hoppe, R. (1988). Die Kristallstruktur von $CdMnF_5$. *Z. Anorg. Allg. Chem.* **557**, 153.

[405] Müller, U. and Hoppe, R. (1990). Korrektur zu den Kristallstrukturen von $CaMnF_5$ und $CdMnF_5$. *Z. Anorg. Allg. Chem.* **583**, 205.

[406] Rodewald, U. C., Lukachuk, M., Hoffmann, R. D., and Pöttgen, R. (2005). Syntheses and structure of $Gd_3Rh_{1.94}In_4$. *Monatsh. Chem.* **136**, 1985.

[407] Rodewald, U. C., Hoffmann, R. D., Pöttgen, R., and Sampathkumaran, E. V. (2003). Crystal structure of Eu_2PdSi_3. *Z. Naturforsch.* **58 b**, 971.

[408] Massa, W. and Steiner, M. (1980). Crystal and magnetic structure of the planar ferromagnet $CsMnF_4$. *J. Solid State Chem.* **32**, 137.

[409] Köhler, P., Massa, W., Reinen, D., Hofmann, B., and Hoppe, R. (1978). Der Jahn-Teller-Effekt des Mn^{3+}-Ions in oktaederischer Fluorkoordination. Ligandenfeldspektroskopische und magnetische Untersuchungen. *Z. Anorg. Allg. Chem.* **446**, 131.

[410] Molinier, M. and Massa, W. (1992). Die Kristallstrukturen der Tetrafluoromanganate $AMnF_4$ (A = Rb, Cs). *Z. Naturforsch.* **47 b**, 783.

[411] Müller, U. (1992). Berechnung der Anzahl möglicher Strukturtypen für Verbindungen mit dichtest gepackter Anionenteilstruktur. I. Das Rechenverfahren. *Acta Crystallogr.* **B 48**, 172. `https://doi.org/10.1107/S010876819101340X`

[412] Schön, J. C. and Jansen, M. (2001). Determination, prediction and understanding of structures, using the energy landscapes of chemical systems. *Z. Kristallogr.* **216**, 307 and 361.

[413] Schön, J. C. and Jansen, M. (2007). Global exploration of the energy landscapes of chemical systems. *Chem. Phys. Phys. Chem.* **9**, 6128.

[414] Jansen, M. and Schön, J. C. (2006). 'Design' in chemical synthesis—an illusion? *Angew. Chem. Int. Ed.* **45**, 3406. *Angew. Chem.* **118**, 3484.

[415] Schön, J. C., Doll, K., and Jansen, M. (2010). Predicting solid compounds via global exploration of the energy landscape of solids on the ab initio level without recourse to experimental information. *Phys. Status Solidi* B **247**, 23.

[416] Conradi, E. (1987). *Herleitung möglicher Kristallstrukturtypen.* Dissertation, Universität Marburg.

[417] El-Kholi, A. (1989). *Berechnung der Anzahl von Polytypen für Verbindungen* $A_aB_bC_cX_n$. Dissertation, Universität Marburg.

[418] Sens, I. (1993). *Polytypenberechnung bei Verbindungen des Typs MX_n*. Dissertation, Universität Marburg.

[419] Strenger, I. (1999). *Berechnung der Anzahl möglicher Kristallstrukturtypen für anorganische Verbindungen.* Dissertation, Universität Kassel.

[420] Pólya, G. (1937). Kombinatorische Anzahlbestimmung für Gruppen, Graphen und chemische Verbindungen. *Acta Math.* **68**, 145.

[421] Pólya, G. (1936). Algebraische Berechnung der Anzahl der Isomeren einiger organischer Verbindungen. *Z. Kristallogr.* **93**, 414.

[422] Harary, F. and Palmer, F. M. (1973). *Graphical Enumeration.* New York: Academic Press.

[423] Harary, F., Palmer, E. M., and Robinson, R. W. (1976). Pólya's contributions to chemical enumeration. In *Chemical Applications of Graph Theory* (ed. Balaban, A. T.), Ch. 3. London: Academic Press.

[424] Knop, O., Barker, W. W., and White, P. S. (1975). Univalent (monodentate) substitution on convex polyhedra. *Acta Crystallogr.* **A 31**, 461.

[425] White, D. E. (1975). Counting patterns with a given automorphism group. *Proc. Amer. Math. Soc.* **47**, 41; Classifying patterns by automorphism group: an operator theoretic approach. *Discr. Math.* **13** (1975) 277.

[426] McLarnan, T. J. (1981). Mathematical tools for counting polytypes. *Z. Kristallogr.* **155**, 227. https://doi.org/10.1524/zkri.1981.155.14.227

[427] McLarnan, T. J. (1981). The numbers of polytypes in close packings and related structures. *Z. Kristallogr.* **155**, 269.

[428] McLarnan, T. J. and Baur, W. H. (1982). Enumeration of wurtzite derivatives and related dipolar tetrahedral structures. *J. Solid State Chem.* **42**, 283.

[429] McLarnan, T. J. (1981). The numbers of polytypes in sheet silicates. *Z. Kristallogr.* **155**, 247.

[430] Iglesias, J. E. (2006). Enumeration of polytypes MX and MX_2 through the use of symmetry of the Zhdanov symbol. *Acta Crystallogr.* **A 62**, 178.

[431] Müller, U. (1988). Berechnung der Anzahl von Kombinationen, um eine gegebene Menge von unterschiedlichen Atomen auf gegebene kristallographische Positionen zu verteilen. *Z. Kristallogr.* **182**, 189.

[432] Müller, U. (1978). Strukturmöglichkeiten für Pentahalogenide mit Doppeloktaeder-Molekülen $(MX_5)_2$ bei dichtester Packung der Halogenatome. *Acta Crystallogr.* **A 34**, 256. https://doi.org/10.1107/S0567739478000492

[433] Müller, U. (1979). Mögliche Kristallstrukturen für oktaedrische Moleküle MX_6 bei dichtester Packung der X-Atome. *Acta Crystallogr.* **A 35**, 188.

[434] Müller, U. (1981). MX_4-Ketten aus kantenverknüpften Oktaedern: mögliche Kettenkonfigurationen und Kristallstrukturen. *Acta Crystallogr.* **B 37**, 532.

[435] Müller, U. (1986). MX_5-Ketten aus eckenverknüpften Oktaedern. Mögliche Kettenkonfigurationen und mögliche Kristallstrukturen bei dichtester Packung der X-Atome. *Acta Crystallogr.* **B 42**, 557. https://doi.org/10.1107/S0108768186097707

[436] Beck, J. and Wolf, F. (1997). Three new polymorphic forms of molybdenum pentachloride. *Acta Crystallogr.* **B 53**, 895.

[437] Müller, U. and Klingelhöfer. P. (1983). β-$NbBr_5$, eine Modifikation von Niobpentabromid mit eindimensionaler Lagefehlordnung. *Z. Naturforsch.* **38 b**, 559.

[438] Müller, U. (1979). Die Kristallstrukturen von Tantalpentaiodid und ihre Fehlordnung. *Acta Crystallogr.* **B 35**, 2502.

[439] Müller, U. (1981). Hexameric molybdenum tetrachloride. *Angew. Chem. Int. Ed.* **20**, 692; *Angew. Chem.* **93**, 697. https://doi.org/10.1002/anie.198106921

[440] Wiener, C. (1863). *Grundzüge der Weltordnung*, pp. 82–89. Leipzig: Winter.

[441] Sohncke, L. (1879). *Entwickelung einer Theorie der Krystallstruktur*. Leipzig: Teubner.

[442] Niggli, P. (1919). *Geometrische Kristallographie des Diskontinuums*. Leipzig: Gebr. Borntraeger.

[443] Wyckoff, R. G. W. (1922, 1930). *An Analytical Expression of the Results of the Theory of Space Groups*. Washington: Carnegie Institution.

[444] Astbury, W. T. and Yardley, K. (1924). Tabulated data for the examination of the 230 space groups. *Phil. Trans. Roy. Soc. London A* **224**, 221.

[445] Fedorov, E. S. and Artemiev, D. (1920). *Das Krystallreich. Tabellen zur krystallochemischen Analyse*. Petrograd: Zapiski Rossiiskoi Akademii Nauk.

[446] Fedorov, E. S. (1903). Allgemeine Krystallisationsgesetze und die darauf fussende eindeutige Aufstellung der Krystalle. *Z. Kristallogr. Miner.* **38**, 321.

[447] Niggli, P. (1926). *Lehrbuch der Mineralogie II. Spezielle Mineralogie* (2nd edn), pp. 645–646 and 662–663. Berlin: Gebr. Borntraeger.

[448] Laves, F. (1939). Kristallographie der Legierungen. *Naturwiss.* **27**, 65.

[449] Laves, F. (1956). *Theory of Alloy Phases*, pp. 124–198. Cleveland: Amer. Soc. for Metals.

[450] Laves, F. (1967). *Phase Stability in Metals and Alloy Phases*, pp. 85–99. New York: McGraw–Hill.

[451] Von Schnering, H. G., J. Chang, J., Freiberg, M., Peters, K., Peters, E. M., Ormeci, A., Schröder, L., Thiele, G., and Röhr, C. (2004). Structure and bonding of the mixed-valent platinum trihalides $PtCl_3$ and $PtBr_3$. *Z. Anorg. Allg. Chem.* **630**, 109.

[452] Zhao, Y. and Holzapfel, W. B. (1997). Structural studies on the phase diagram of cerium. *J. Alloys Comp.* **246**, 216.

[453] Schiwek, A., Porsch, F., and Holzapfel, W. B. (2002). High temperature–high pressure structural studies of cerium. *High Pressure Res.* **22**, 407.

[454] Johansson, B., Abrikosov, I. A., Aldén, M., Ruban, A. V., and Skriver, H. L. (1995). Calculated phase diagram for the $\gamma \rightleftharpoons \alpha$ transition in Ce. *Phys. Rev. Lett.* **74**, 2335.

[455] Samara, G. A. (1987). Pressure dependence of the static and dynamic properties of KH_2PO_4 and related ferroelectric and antiferroelectric crystals. *Ferrolectrics* **71**, 161.

[456] Western, A. B., Baker, A. G., Bacon, C. R., and Schmidt, V. H. (1978). Pressure-induced tricritical point in the ferroelectric phase transition of KH_2PO_4. *Phys. Rev.* **B 17**, 4461.

[457] Troussaut, J. and Vallade. M. (1985). Birefringence study of the tricritical point of KDP. *J. Physique* **46**, 1173.

[458] Wilson, K. G. (1983). The renormalization group and critical phenomena. *Rev. Mod. Phys.* **55**, 583.

[459] Wilson, K. G. (1979). Problems in physics with many scales of length. *Scientific American* **241**, August, 140.

[460] Griffiths, R. B. (1970). Dependence of critical indices on a parameter. *Phys. Rev. Lett.* **24**, 1479.

[461] Aroyo, M. I., Pérez-Mato, J. M., Capillas, C., Kroumova, E., Ivantchev, S., Madariaga, G., Kirov, A., and Wondratschek, H. (2006). Bilbao Crystallographic Server. I. Databases and crystallographic computer programs. *Z. Kristallogr.* **221**, 15.

[462] Aroyo, M. I., Kirov, A., Capillas, A., Pérez-Mato, J. M., and Wondratschek, H. (2006). Bilbao Crystallographic Server. II. Representations of crystallographic point groups and space groups. *Acta Crystallogr.* **A 62**, 115.

[463] Aroyo, M. I., Pérez-Mato, J. M., Capillas, C., and Wondratschek, H. (2010). The Bilbao Crystallographic Server. In *International Tables for Crystallography*, Vol. *A*1 (2nd edn; eds. Wondratschek, H. and Müller, U.), Ch. 1.7. Chichester: Wiley. `https://it.iucr.org/A1b/ch1o7v0001/ch1o7.pdf`

[464] Aroyo, M. I., Pérez-Mato, J. M., Orobengoa, D., Tasci, E. S., de la Flor, G., and Kirov, A. (2011). Crystallography online: Bilbao Crystallographic Server. *Bulg. Chem. Commun.* **43**, 183.

[465] Kroumova, E., Pérez-Mato, J. M., and Aroyo, M. I. (1998). WYCKSPLIT, computer program for the determination of the relations of Wyckoff positions for a group–subgroup pair. *J. Appl. Crystallogr.* **31**, 646.

[466] Ivantchev, S., Kroumova, E., Madariaga, G., Pérez-Mato, J. M., and Aroyo, M. I. (2000). SUBGROUPGRAPH, a computer program for analysis of group–subgroup relations between space groups. *J. Appl. Crystallogr.* **33**, 1190.

[467] Hall, S. R., Allen, F. H., and Brown I. D. (1991). The Crystallographic Information File (CIF): A New Standard Archive File for Crystallography. *Acta Crystallogr.* **A 47**, 655.

[468] Szafranski, M., and Stahl, K. (1994). Refinements of the crystal structures and UV-absorption properties $KBrO_3$, $RbBrO_3$ and $CsBrO_3$. *Z. Kristallogr.* **209**, 491.

[469] Templeton, D. H., and Templeton, L. K. (1985). Crystal behavior of potassium bromate under compression. *Acta Crystallogr.* **A 41**, 133.

[470] Hanson, R. M., Prilusky, J., Renjian, Z., Nakane, T., and Sussman, J. L. (2013). JSmol and the next generation web-based representation of 3D molecular structure as applied to proropedia. *Israel J. Chem.* **53**, 207.

[471] Peterková, J., Dušek, M., Petříček, V., and Loub, J. (1998). Structures of fluoroarsenates $KAsF_{6-n}(OH)_n$, $n = 0, 1, 2$. *Acta Crystallogr.* **B 54**, 809.

[472] Berg, R. W., von Barner, J. H., Christensen, E., and Nielsen K. (2000). Crystal structure and IR and Raman spectra of α-$CsNbF_6$. *Progr. Molten Salts Chem.* **1** (2000) 75.

[473] de la Flor, G., Orobengoa, D., Tasci, E., Pérez-Mato, J. M., and Aroyo, M. I. (2016). Comparison of structures applying the tools available at the Bilbao Crystallographic Server. *J. Appl. Crystallogr.* **49**, 653.

[474] Guimaraes, D. M. Z. (1979). Ferroelectric transformations in lead orthophosphate and its structure as a function of temperature. *Acta Crystallogr.* **A 35**, 108.

[475] Kiat, J. M., Yamada, Y., Chevrier, G., Uesu, E., Y., Boutrouille, P., and Calvarin, G. (1992). Reinvestigation of the ferroelectric phase transition of lead phosphate $Pb_3(PO_4)_2$. *J. Phys. Cond. Matter* **4**, 4915.

[476] Angel, R. J., Bismayer, U., and Marshall, W. G. (2001). Renormalization of the phase transition in lead phosphate, $Pb_3(PO_4)_2$, by high pressure: structure. *J. Phys. Cond. Matter* **13**, 5353.

Glossary

Abelian group (commutative group; Chap. 5). Group whose elements fulfil the condition $g_i g_k = g_k g_i$.

Affine mapping (Sects. 3.2–3.3). Mapping that allows an arbitrary distortion, provided that parallel straight lines are mapped onto parallel straight lines. The ratios of distances between points on a straight line are preserved.

Antiphase domains (Sect. 16.6). Domains in a crystal having mutually shifted origins.

Aristotype (basic structure; Sect. 1.2, Chap. 11). High-symmetry crystal structure from which the *hettotypes* can be derived, i.e. structures with reduced symmetry.

Augmented matrix (Sect. 3.2). 4×4 matrix used to convert the coordinates related to a mapping.

Bärnighausen tree (Sect. 1.2, Chaps. 11, 12). Tree of group–subgroup relations between space groups which discloses relationships between crystal structures.

Bilbao Crystallographic Server (Chap. 24). Server accesible via internet with databases and computer programs for crystallographic applications.

Bravais type (Bravais lattice; Sect. 6.2). Space-group type of a point lattice.

Brillouin zone (Sect. 16.2.1). Polyhedron about the origin of '*k* space', delimited by faces normal to and across the midpoints of the connecting lines to adjacent lattice points in reciprocal space. Vibrational modes are characterized by points in the Brillouin zone.

Centred lattice (Sects. 2.3, 6.2). Lattice whose basis is not primitive. Lattice vectors may be linear combinations from certain fractions of the basis vectors. The conventional centred lattices are base centred, face centred, body centred, and trigonal-rhombohedral.

Chirality (Sect. 9.3). Property of an object that cannot be superposed by pure rotation and translation on its image formed by inversion through a point.

Column part (Sect. 3.2). Column of numbers that contains the translational components of a mapping.

Conjugacy class (Sects. 5.4, 8.1, 8.3). The set of subgroups of a group $\mathcal{G}$ that are conjugate in $\mathcal{G}$.

Conjugate subgroups (Sects. 5.4, 8.1, 8.3). Subgroups of a space group $\mathcal{G}$ that are equivalent under symmetry operations of $\mathcal{G}$.

Conventional setting (Sects. 2.3, 10.3). Setting of a space group that is completely listed in *International Tables A*.

Coset (Sect. 5.3). The elements of a group $\mathcal{G}$ can be decomposed into cosets with respect to a subgroup $\mathcal{H}$. The subgroup itself is the first coset; the other cosets consist of elements $g\mathcal{H}$ (left cosets) or $\mathcal{H}g$ (right cosets), $g \in \mathcal{G}$, $g \notin \mathcal{H}$, and contain each the same number of different group elements $g \notin \mathcal{H}$. The number of cosets is the *index* of the subgroup.

Critical exponent (Sect. 16.2.2, 22.2). Exponent in the power law that describes the dependence of the order parameter on a variable of state during a continuous phase transition.

Crystal class (Sect. 6.1). Point-group type of a crystal.

Crystal pattern (infinite ideal crystal; Sect. 2.2). Flawless, infinitely extended three-dimensional periodic array of atoms.

Crystal structure (Sect. 2.2). The spatial distribution of the atoms in a crystal.

Crystal system (Sect. 6.1). Point group type of the lattice of a crystal structure (holohedry).

Crystallographic basis (Sect. 2.3). Three non-coplanar lattice vectors that serve to determine the lattice and the coordinate system of a crystal structure.

Daughter phase (distorted structure; Sect. 1.2). Hettotype at a phase transition.

Domain (Sects. 16.3, 16.5, 16.6). Region within a crystal separated from other regions by domain boundaries.

Enantiomorphic space group (chiral space group; Sects. 6.1.3, 9.3). Space group whose symmetry elements cannot be superposed by pure rotation and translation with those of the space group formed by inversion through a point. There exists 11 pairs of enantiomorphic space group types.

Euclidean normalizer (Sect. 8.2). $\mathcal{N}_{\mathcal{E}}(\mathcal{G})$, the normalizer of the group $\mathcal{G}$ with respect to the Euclidean group $\mathcal{E}$ (group of all isometries of three-dimensional space). It is used to determine the normalizers between space groups and to determine equivalent sets of coordinates of a crystal structure.

Family of structures (Sect. 1.2, Chap. 11). A family of structures consists of an aristotype and its hettotypes.

General position (Sects. 6.1, 6.4, 6.5). Point with the site symmetry 1.

Generating function (Sect. 19.2.1). Power series that discloses how many inequivalent configurations exist for a structure.

Generators (Sect. 5.2). Set of group elements from which the complete group (all group elements) can be generated by repeated composition.

Glide reflection (Sect. 4.2). Symmetry operation where a reflection is coupled with a translation parallel to a glide-reflection plane.

Group (Sects. 5.1, 5.2). A set (of numbers, symmetry operations, or any other elements) which satisfies the *group axioms*.

Group element (Sects. 5.1, 5.2). One representative out of the set of elements of a group.

Hermann–Mauguin symbol (international symbol; Sects. 4.2, 6.3.1). Symbol common in crystallography to designate symmetry, specifying the present symmetry operations according to certain rules.

Hettotype. See aristotype.

Index (Sect. 5.3). Ratio $|\mathcal{G}|/|\mathcal{H}|$, $|\mathcal{G}|$ being the order of a group and $|\mathcal{H}|$ being the order of its subgroup. See also coset.

Inversion (Sect. 4.2). Symmetry operation that generates an inverted image at the opposite side and at the same distance from a centre of inversion.

Irreducible representation (symmetry species; Sect. 16.2.1, Chap. 23). Effect of a symmetry operation referred to the local conditions at a point (e.g. for a vibrational displacement).

Isometry (Sects. 3.4, 3.5). Mapping that leaves all distances and angles unchanged.

Isomorphic group (Sect. 5.2). Two groups are isomorphic if they have the same multiplication table.

Isomorphic subgroup (Sect. 7.2, Chap. 11, Sect. 12.3, Chap. 21). *Klassengleiche* subgroup belonging to the same or the enantiomorphic space-group type as the supergroup.

Isotropy subgroup (Sect. 16.2). Subgroup of a space group that results when a structure is distorted according to a certain irreducible representation.

Isotypism (Sect. 9.5). Two crystal structures are isotypic if their atoms are distributed in a like manner and if they have the same space group.

***Klassengleiche* subgroup** (Sect. 7.2, Chap. 11, Sect. 12.2). Subgroup of a space group belonging to the same crystal class. It has fewer translations (enlarged primitive unit cell).

Landau theory (Sects. 16.2.2, 22.2). Theory that describes the processes taking place during a continuous phase transition.

Lattice (vector lattice; Sects. 2.2, 6.2). The set of all translation vectors of a crystal structure.

Lattice parameters (lattice constants; Sects. 2.3, 2.5). The lengths a, b, c of the basis vectors and the angles α, β, γ between them.

Layer group (Sect. 7.4). Symmetry group of an object in three-dimensional space that has translational symmetry in only two dimensions.

Mapping (Sects. 3.1–3.2). Instruction that assigns exactly one image point to any point in space.

Matrix–column pair (Sect. 3.2). Matrix and column that serve to convert the coordinates at a mapping, including translational components.

Matrix part (Sect. 3.2). Matrix that serves to convert the coordinates at a mapping, not considering translational components.

Miller indices (Sect. 2.4). Three integral numbers hkl used to designate net planes. From a set of parallel planes, that one which is closest to the origin without running itself through the origin intersects the coordinate axes at distances of a_1/h, a_2/k, a_3/l from the origin.

Mode (vibrational mode, Sect. 16.2.1). Collective and correlated vibrational motion of atoms having certain symmetry properties.

Multiplicity. For point groups (Sect. 6.1): The number of points that are symmetry equivalent to one point.
For space groups (Sect. 6.4): The number of points that are symmetry equivalent to one point within one unit cell.

Normal subgroup (Sect. 5.4). Subgroup $\mathcal{H}$ of a group $\mathcal{G}$ that satisfies the condition $\mathcal{H}g_m = g_m\mathcal{H}$ for all $g_m \in \mathcal{G}$.

Normalizer (Sect. 8.2). $\mathcal{N}_{\mathcal{G}}(\mathcal{H})$, the normalizer of $\mathcal{H}$ in $\mathcal{G}$, consists of all those elements of $\mathcal{G}$ that map $\mathcal{H}$ onto itself. $\mathcal{H} \trianglelefteq \mathcal{N}_{\mathcal{G}}(\mathcal{H}) \leq \mathcal{G}$ holds.

Orbit. See point orbit.

Order of an axis (Sect. 3.5). For axes of rotation and even-fold rotoinversion: The least number of rotations to be performed until a point at a general position returns to its initial position.
For odd-fold axes of rotoinversion: One half of the least number of rotations to be performed until a point at a general position returns to its initial position.
For screw-rotation axes: The least number of rotations to be performed until a point reaches a translation-equivalent position.

Order of a group (Sect. 5.2). The number of group elements in a group.

Order parameter (Sects. 16.2.2, 22.2, 22.4). Appropriate measurable quantity apt to disclose the course of a phase transition in a quantitative manner.

Parent clamping approximation (Sect. 12.1). The basis vectors of a crystal structure are treated as being the same as or integer multiples of the basis vectors of another crystal structure even though they are not exactly the same. The lattices of two crystal structures are taken as *translationengleiche* even if their lattices coincide only approximately.

Penetration rod group (rod site symmetry; Sect. 7.4) Group of symmetry operations of a space group that leaves a traversing straight line invariant.

Phase transition (Sect. 16.1, Chap. 22). Event which entails a discontinuous (sudden) change of at least one property of a material.

Point group (Sect. 6.1). For molecules: The group of the symmetry operations of a molecule.
For crystals: The group of the symmetry operations relating the normals on the faces of a crystal.

Point lattice (Sect. 2.2). The set of all points that are translation equivalent to a point.

Point orbit (crystallographic point orbit; Sects. 5.6, 6.5, 10.1). The set of all points that are symmetry equivalent to a point in a point group or a space group.

Prototype (Sects. 1.2; 9.5). Aristotype at a phase transition (parent phase); or substance that gives the name to a series of isotypic crystal structures.

Pseudosymmetry (Sect. 1.1). Symmetry that is fulfilled approximately.

Reciprocal lattice (Sect. 2.4). Vector lattice having the basis vectors $\mathbf{a}_1^*, \mathbf{a}_2^*, \mathbf{a}_3^*$, which are normal to the net planes (100), (010), and (001) and have the lengths $1/d_{100}$, $1/d_{010}$, $1/d_{001}$, respectively, d_{100} being the distance between neighbouring net planes (100).

Reduced cell (Sects. 2.3, 9.1). Unique unit cell settled according to certain rules.

Reflection (Sect. 4.2). Symmetry operation associated with a reflection plane.

Rod group (Sect. 7.4). Symmetry group of an object in three-dimensional space which has translational symmetry in only one dimension.

Rotation (Sect. 4.2). Symmetry operation associated with a turn of $360°/N$ around a rotation axis, N = integer (order of the rotation).

Rotoinversion (Sect. 4.2). Symmetry operation where a rotation is coupled with an inversion through a point on the rotation axis.

Schoenflies symbol (Sect. 6.3.2). Symbol to designate symmetry. It is of common usage for molecules, but hardly ever used in crystallography.

Screw rotation (Sect. 4.2). Symmetry operation where a rotation is coupled with a translation parallel to a screw axis.

Sectional layer group (layer site symmetry; Sect. 7.4) Group of symmetry operations of a space group that leaves a transecting plane invariant.

Site symmetry (Sects. 6.1, 6.4, 6.5). Symmetry operations of the point group or space group that leave a point of a molecule or a crystal unchanged.

Soft mode (freezing lattice mode; Sects. 16.2, 22.2). Vibrational lattice mode whose frequency tends toward zero when approaching the point of transition of a phase transition, and which determines the atomic motions during a continuous phase transition.

Sohncke space group (Sect. 9.3). Space group that has no points of inversion, no rotoinversion axes, no reflection planes, and no glide planes. Chiral crystal structures are compatible only with Sohncke space groups.

Space group (Sect. 3.1, Chap. 6). The set of all symmetry operations of a crystal structure, including the translations with given lattice parameters.

Space-group type (Sects. 6.1.3, 6.6). One out of 230 possibilities to combine crystallographic symmetry operations to a space group (with arbitrary lattice parameters).

Special position (Sects. 6.1.1, 6.4, 6.5). Point whose site symmetry is higher than 1.

Standard setting (Sect. 10.3, Chap. 24). Setting of a space group or of crystal structure data according to certain rules.

Standardization (Sect. 2.3). Conversion of coordinates to values $0 \le x, y, z < 1$.

Subgroup (Sects. 5.2, 7.1, 7.2). Subset of the elements of a group which itself satisfies the group axioms. The subgroup is *maximal* if there does not exist an intermediate group between the group and the subgroup.
Subgroups on a par (Sect. 8.3). Subgroups $\mathcal{H}_1$, $\mathcal{H}_2, \dots$ of a space group $\mathcal{G}$, $\mathcal{H}_1, \mathcal{H}_2, \dots < \mathcal{Z} \leq \mathcal{G}$, that are *not* conjugate in $\mathcal{G}$, but are conjugate in one of the Euclidean normalizers $\mathcal{N}_{\mathcal{E}}(\mathcal{Z})$ or $\mathcal{N}_{\mathcal{E}}(\mathcal{G})$. They have the same lattice dimensions and the same space-group type.
Supergroup (Sects. 5.2, 7.3). A supergroup contains additional group elements.
Superspace (Sect. 2.2. Space having more than three dimensions.
Superstructure (Sects. 12.2, 18.3). Crystal structure whose unit cell is enlarged as compared to an ideal structure. Its space group is a *klassengleiche* subgroup.
Symmetry-adapted cell (Sect. 6.3.1). Unit cell having basis vectors parallel to symmetry axes or to normals on symmetry planes.
Symmetry Database (Chap. 24. Database accessible by internet with computer programs concerning space and point groups.
Symmetry direction (Sect. 6.3.1). Special direction parallel to symmetry axes or to normals on symmetry planes.
Symmetry element (Sects. 4.3, 6.4.1). Point, line, or plane that remains unchanged when a symmetry operation is performed.
Symmetry operation (Sect. 3.1.1). Mapping that brings an object to superposition with itself or with its mirror image.
Symmetry principle (Sect. 1.1). In crystal structures the arrangement of atoms reveals a pronounced tendency towards the highest possible symmetry.
Symmetry species. See irreducible representation.
Topotactic reaction (Chap. 17). Chemical reaction in the solid state such that the orientations of the domains of the product crystal are predetermined by the orientation of the initial crystal.
Topotactic texture (Sect. 16.3). Texture of domains of crystallites that are intergrown in a regular manner.
Translation (symmetry translation; Sect. 2.2). Shift by a *translation vector* that results in a superposition of a crystal structure.
***Translationengleiche* subgroup** (Sect. 7.2, Chap. 11, Sect. 12.1). Subgroup of a space group whose lattice is unchanged (unchanged primitive unit cell). It belongs to a lower-symmetry crystal class.
Twinned crystal (Sects. 16.3, 16.5, 18.4). Intergrowth of two or more congruent or enantiomorphic individuals of the same crystal species having a crystal-symmetric relative orientation between the individuals.
Unit cell (Sect. 2.3). The parallelepiped in which the coordinates of all points are $0 \leq x, y, z < 1$.
Wyckoff position (Sects. 6.1, 6.4, 10.1). The set of all point orbits whose site symmetries are conjugate.
Wyckoff symbol (Sects. 1.2, 6.4.2, 10.1). Symbol to designate a Wyckoff position. It consists of the multiplicity of one of its points and the Wyckoff letter (an alphabetical label).

Index